ALGEBRAIC STRUCTURES OF SYMMETRIC DOMAINS

PUBLICATIONS OF THE MATHEMATICAL SOCIETY OF JAPAN

1. The Construction and Study of Certain Important Algebras. By Claude Chevalley.
2. Lie Groups and Differential Geometry. By Katsumi Nomizu.
3. Lectures on Ergodic Theory. By Paul R. Halmos.
4. Introduction to the Problem of Minimal Models in the Theory of Algebraic Surfaces. By Oscar Zariski.
5. Zur Reduktionstheorie Quadratischer Formen. Von Carl Ludwig Siegel.
6. Complex Multiplication of Abelian Varieties and its Applications to Number Theory. By Goro Shimura and Yutaka Taniyama.
7. Équations Différentielles Ordinaires du Premier Ordre dans le Champ Complexe. Par Masuo Hukuhara, Tosihusa Kimura et Mme Tizuko Matuda.
8. Theory of Q-varieties. By Teruhisa Matsusaka.
9. Stability Theory by Liapunov's Second Method. By Taro Yoshizawa.
10. Fonctions Entières et Transformées de Fourier. Application. Par Szolem Mandelbrojt.
11. Introduction to the Arithmetic Theory of Automorphic Functions. By Goro Shimura. (Kanô Memorial Lectures 1)
12. Introductory Lectures on Automorphic Forms. By Walter L. Baily, Jr. (Kanô Memorial Lectures 2)
13. Two Applications of Logic to Mathematics. By Gaisi Takeuti. (Kanô Memorial Lectures 3)
14. Algebraic Structures of Symmetric Domains. By Ichiro Satake. (Kanô Memorial Lectures 4)

PUBLICATIONS OF THE MATHEMATICAL SOCIETY OF JAPAN
14

ALGEBRAIC STRUCTURES OF SYMMETRIC DOMAINS

by
Ichiro Satake

KANÔ MEMORIAL LECTURES 4

Iwanami Shoten, Publishers
and
Princeton University Press

1980

Kanô Memorial Lectures

In 1969, the Mathematical Society of Japan received an anonymous donation to encourage the publication of lectures in mathematics of distinguished quality in commemoration of the late Kôkichi Kanô (1865–1942).

K. Kanô was a remarkable scholar who lived through an era when Western mathematics and philosophy were first introduced to Japan. He began his career as a scholar by studying mathematics and remained a rationalist for his entire life, but enormously enlarged the domain of his interest to include philosophy and history.

In appreciating the sincere intentions of the donor, our Society has decided to publish a series of "Kanô Memorial Lectures" as a part of our Publications. This is the fourth volume in the series.

Publications of the Mathematical Society of Japan, volumes 1 through 10, should be ordered directly from the Mathematical Society of Japan. Volume 11 and subsequent volumes should be ordered from Princeton University Press, except in Japan, where they should be ordered from Iwanami Shoten, Publishers.

Printed in the United States of America

Dedicated to
S. Iyanaga and K. Iwasawa

Preface

數回細寫愁仍破
萬顆匀圓訝許同

The symmetric domain is rich in algebraic structures. It provides an important common ground for various branches of mathematics, not only for function theory and geometry, but also for number theory and algebraic geometry. However, until recently, the structures associated with symmetric domains (Lie groups, Jordan algebras, Siegel domains, etc.) have been studied independently from separate view-points with little mutual recognition. It is one of our aims to give a unified treatment, clarifying the relationship between these structures.

Our main theme in this book is a study of "morphisms", i. e., equivariant holomorphic maps, of symmetric domains. This enables us to establish, on the one hand, the equivalences between various categories related to symmetric domains (I, § 9; II, § 8; V, § 7). On the other hand, we study in Chapters III and IV two important special cases of the morphisms, i. e., (1) the case of (H_1)-homomorphisms of $\mathfrak{sl}_2(\boldsymbol{R})$ into $\mathfrak{g}=$ Lie Hol$(\mathcal{D})$, which leads to the theory of Wolf and Korányi on the Siegel domain realizations of symmetric domains; (2) the case of symplectic representations $\mathfrak{g}\to\mathfrak{sp}_{2n}(\boldsymbol{R})$ satisfying (H_1), which leads to the analytic construction of Kuga's fiber varieties. In addition to these, we will, in Chapter V, study infinitesimal automorphisms of Siegel domains in general, give a characterization of symmetric Siegel domains, and classify an intermediate class of Siegel domains, called "quasi-symmetric". This portion contains detailed proofs of the results announced in earlier papers of the author ([17], [18]).

This book is basically designed for graduate students or non-specialists who have a sound background in the theory of Lie groups. In the hope of making the book more self-contained, we included two introductory chapters I and II, in which a summary of basic results on related topics: algebraic groups, Jordan triple systems and symmetric spaces, is given. The Appendix contains a brief account on classical groups. For more detailed instructions on the reading of this book, the reader should consult the *Instructions to the reader*.

Some materials of this book have been lectured in various occasions at various places: at Chicago (1966), Tokyo (1968), Berkeley (1972), and Nancy (1977–78). Especially, a set of lecture notes at University of California, Berkeley, prepared by H. Yamaguchi in 1972, served as a base of Chapter III. I wish to express my thanks to Yamaguchi here. I would also like to acknowledge gratefully the helps given by many colleagues and friends, especially by M. Takeuchi, T. Ochiai, and Zelow (Lundquist), who read a large portion of the manuscript and gave me useful suggestions. In preparing Chapter IV, occasional conversations with M. Kuga have been most helpful. I am also indebted to several institutions, especially to the University of California at Berkeley for secretarial assistance (and for comfortable climate there), to the University of Nancy for rendering facilities while the final version of the manuscript was being prepared, and to the National Science Foundation for continuous financial support. Finally, I am happy to express my thanks to Professor S. Iyanaga who invited me to write a volume in this series of the Publications of Mathematical Society of Japan, and to Mr. H. Arai of Iwanami whose professional cooperation in the process of printing has been invaluable. This book is respectfully dedicated to my two eminent teachers, Professors S. Iyanaga and K. Iwasawa.

Berkeley, December 1979

I. Satake

Contents

Instructions to the reader

1. The logical interdependence of the chapters is as follows :

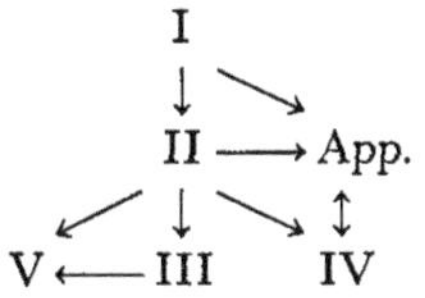

After reading Chapters I and II (together with Appendix), the reader may proceed in either one of the three directions, depending on his (or her) interest.

2. As mentioned in the *Preface*, some knowledge on the theory of Lie groups (say, up to the structure theory of semi-simple Lie algebras) is an indispensable prerequisite for the reading of this book. For convenience of the reader, we give here a general guidance on the literature, classified by subjects. Further bibliographical comments will be given occasionally in the text. In many cases, the reader would be better off to accept the statements in the text without spending too much time for these references (because there are so many). Some of the missing proofs may be recovered by solving the *Exercises*.

(A) **Algebraic groups and Lie groups.** The first half of Chapter I is a summary of what we need from these theories throughout the book. Chapter I, §§ 1∼3 contain basic definitions and results (mostly without proofs) on linear algebraic groups. The general reference is Borel [13] or Humphreys [2], but for the understanding of the elementary part (except for the relation to Lie algebras) the original paper of Borel [4] will be sufficient (and perhaps more accessible). For the theory of algebraic tori (I, § 2), see Ono [1], and for the topics in I, § 5, see Chevalley [5], Satake [10], and Borel-Tits [1]. For basic concepts on Lie groups, we refer the reader to Chevalley [1], Varadarajan [1], and Helgason [1], [1a] ; especially, for the structure and representations of (complex) semi-simple Lie algebras, to Matsushima [1] (in Japanese), Serre [3] and Humphreys [1]. For lack of immediate reference, we give in I, § 3 a sketch of proof for the existence and the uniqueness of algebraic group structure on a certain class of reductive Lie groups, essentially due to Chevalley [1], VI (and also to Harish-Chandra, Cartier and Hochschild-Mostow [1]). The properties of (infinitesimal and global) Cartan involutions of reductive $\boldsymbol{R}$-groups given in I, § 4 are mainly due to Mostow [1], [2] ; cf. also Borel-Harish-Chandra [1]. The proof of Theorem 4. 2 is an adaptation of the one given in Helgason [1], III, § 7 (for the semi-simple case).

(B) **Jordan algebras and Jordan triple systems.** The second half of Chapter I

(§§ 6∼8) together with V, § 6 is a short introduction to Jordan triple systems (abbreviated as JTS), which generalize the notion of Jordan algebras. (For simplicity, we always assume that the ground field F is of characteristic zero, though most of the results remain valid over any field of characteristic different from 2 and 3.) For Jordan algebras in general, cf. Albert [6], [7], or Braun-Koecher [1], Jacobson [7]; for a quick review, Koszul [1], Jacobson [9] and McCrimmon [3] will be useful. For JTS's, cf. Koecher [9], Loos [3], [8], and Meyberg [5]. In I, §§ 6, 7, we define the "structure group" (or algebra) and the symmetric Lie algebra (or the "superstructure algebra") attached to a JTS. These constructions of Lie algebras from a JTS were first given by Koecher [9]; cf. also Tits [2], Koecher [7], Meyberg [1] and Kantor [2]. In I, § 8 we explain the relation between formally real Jordan algebras and self-dual homogeneous cones, which was given (in a more general context of "ω-domains") by Koecher [5] ([1], [2], [3]) and (in the context of homogeneous cones) by Vinberg [1], [5] (cf. also Rothaus [2] and Dorfmeister [2], [6], [7]). One can also find a comprehensible account on this topic in Ash et al. [1].

(C) **Symmetric spaces and symmetric domains.** In II, §§ 1∼3 we give a summary of basic facts on symmetric spaces (domains), essentially due to É. Cartan (especially [5]). Here the standard reference is Kobayashi-Nomizu [1] or Helgason [1], [1a]. The relations between symmetric domains and positive definite hermitian JTS's given in II, §§ 3, 5, 8 were first established by Koecher [9].

(D) **Siegel domains.** The notion of Siegel domains, due to Piatetskii-Shapiro [2], is introduced in III, § 6. A detailed study of the "infinitesimal automorphisms" (i. e., the Lie algebra of the holomorphic automorphism group) of a Siegel domain is given in Chapter V. In this part, we assume the results of Kaup-Matsushima-Ochiai [1] and Murakami [5], which are reviewed in V, § 1. In V, §§ 2∼4, we give proofs of the results announced in Satake [17], [18]; closely related results (by different approaches) have been obtained by W. Kaup, Dorfmeister and Rothaus.

(E) **Classical groups and classical domains.** The Appendix contains the definitions (constructions) of classical groups, especially those corresponding to classical domains, in terms of algebras with involutions, along with brief accounts on quaternion algebras, Clifford algebras and spin representations. For these topics, we refer the reader to Bourbaki [1], Weil [3], and Siegel [8]. The most important case, the "Siegel space", is discussed in II, § 7.

3. Next we give some comments on the notations and conventions which are adopted throughout the book.

The cardinality of a set X is denoted by $|X|$. Thus, when G is a group, $|G|$ is the "order" of G and, when H is a subgroup of G, $(G:H)=|G/H|$ is the "index" of H in G. The normalizer (resp. centralizer) of H in G is denoted by $N_G(H)$ or $N(H)$ (resp. $C_G(H)$ or $C(H)$):

$$N_G(H)=\{g\in G \mid gHg^{-1}=H\},$$
$$C_G(H)=\{g\in G \mid gh=hg \text{ for all } h\in H\}.$$

The center of G is denoted by Cent G.

For a topological group G, G° denotes always the connected component of G containing the identity element. For an algebraic group $\mathcal{G}$ (or an F-group G), $\mathcal{G}^z$ (or G^z) denotes the Zariski connected component of $\mathcal{G}$ (or G) containing the identity element (I, § 1).

The identity map of a set X onto itself is denoted by id_X or 1_X, or simply by id or 1. Also, the unit element of a group or a ring is usually denoted by e or 1. The restriction of a map φ to a subset Y is denoted by $\varphi|Y$. The composite of two maps φ and ψ is denoted by $\varphi\circ\psi$ or (when there is no fear of confusion) simply by $\varphi\psi$, i. e., $(\varphi\psi)(x)=(\varphi\circ\psi)(x)=\varphi(\psi(x))$. When φ is a homomorphism, the "kernel" of φ is denoted by Ker φ. The module or ring consisting of only the zero element is often denoted by 0 (instead of $\{0\}$).

As usual, $\boldsymbol{Q}$, $\boldsymbol{R}$, and $\boldsymbol{C}$ denote the fields of rational numbers, real numbers, and complex numbers, respectively. $\boldsymbol{Z}$ is the ring of rational integers, and $\boldsymbol{H}$ is the (non-commutative) field of Hamilton quaternions. The real and imaginary parts of a complex number α are denoted by $\mathrm{Re}\,\alpha$ and $\mathrm{Im}\,\alpha$. $\bar{\alpha}$ is the complex conjugate of α. (Bar is also used to denote the closure in topological sense.) For $\alpha\in\boldsymbol{C}$, we put $\mathbf{e}(\alpha)=\exp(2\pi i\alpha)$. We also use the following notation :

$$\boldsymbol{R}_+ = \{\lambda\in\boldsymbol{R}\,|\,\lambda\geq 0\},$$
$$\boldsymbol{C}^{(1)} = \{\alpha\in\boldsymbol{C}\,|\,|\alpha|=1\}.$$

For any associative ring R, $R^\times$ denotes the group of "units" (i.e., invertible elements) in R. We put $\boldsymbol{R}_+^\times=\boldsymbol{R}^\times\cap\boldsymbol{R}_+$.

When F' is an extension field of a field F, the "degree" of the extension is denoted by $[F':F]$, and when F'/F is a Galois extension, the Galois group is denoted by $\mathrm{Gal}(F'/F)$.

Let V be a vector space over a field F. For a subset $A=\{\cdots\}$ of V, $A_F=\{\cdots\}_F$ denotes the linear subspace of V spanned by A. For a linear transformation φ of V and $\alpha\in F$, we denote by $V(\varphi\,;\alpha)$ the eigenspace of φ corresponding to the eigenvalue α, i. e.,

$$V(\varphi\,;\alpha) = \{v\in V\,|\,\varphi v=\alpha v\}.$$

More generally, for $\alpha_1,\cdots,\alpha_m\in F$, we set

$$V(\varphi\,;\alpha_1,\cdots,\alpha_m) = \sum_{i=1}^{m} V(\varphi\,;\alpha_i).$$

The trace of φ is denoted by $\mathrm{tr}_V\,\varphi$ (or $\mathrm{tr}\,\varphi$). By definition, the "dual space" V^* of V is the space of all linear functionals on V. For a linear map $\varphi: V\to W$, the "dual map" ${}^t\varphi: W^*\to V^*$ is defined by ${}^t\varphi(f)=f\circ\varphi$ $(f\in W^*)$. If F' is an extension field of F, the tensor product $V\otimes_F F'$ (viewed as a vector space over F') is called the *vector space obtained from V by scalar extension F'/F* and is denoted by $V_{F'}$. When an algebra R is acting on a vector space V, we sometimes denote the action of $\alpha\in R$ on V by $(\alpha)_V$.

The following notations are introduced in I, §§ 1, 3 : $\mathrm{End}(V)$, $GL(V)$, $\mathfrak{gl}(V)$, $\mathcal{M}_n(F)$, $GL_n(F)$, $SL_n(F)$, $\mathfrak{gl}_n(F)$, $\mathfrak{sl}_n(F)$, $\mathrm{Sym}_n(F)$, $O_n(F)$. For further notations concerning classical groups, see App., §§ 1, 3. $\mathcal{M}_{m,n}(F)$ denotes the module of all $m\times n$ matrices with entries in F. The identity matrix of degree n is denoted by 1_n or simply by 1, and the diagonal matrix of degree n with the i-th diagonal entry α_i is denoted by $\mathrm{diag}(\alpha_1,\cdots,\alpha_n)$. We write $1_{p,q}=\begin{pmatrix}1_p & 0\\ 0 & -1_q\end{pmatrix}$. For a real symmetric bilinear form S (resp. a hermitian form h) or the corresponding matrix, $S \gg 0$ (resp. $h \gg 0$) means that S (resp. h) is positive definite.

For any algebra R (associative or not), Aut(R) and Der(R) denote the group of automorphisms of R and the Lie algebra of derivations of R, respectively. In general, the Lie algebra of a Lie group (or algebraic group) G is denoted by Lie G. For instance, one has Der(R)=Lie Aut(R) (I, § 1, Exerc. 3).

Let G and G' be Lie groups with Lie algebras $\mathfrak{g}$ and $\mathfrak{g}'$. Then an analytic homomorphism $\varphi : G \rightarrow G'$ gives rise to a Lie algebra homomorphism $d\varphi : \mathfrak{g} \rightarrow \mathfrak{g}'$ (I, § 1). In this book, whenever there is no fear of confusion, the "differential" $d\varphi$ is also denoted by the same letter φ. In particular, *ad* denotes the "adjoint representation" of the Lie algebra $\mathfrak{g}$ (I, § 3) as well as that of the Lie group G (i. e., for $g_1 \in G$, we set $ad(g_1) = d(\nu_{g_1})$, ν_{g_1} denoting the "inner automorphism" of G defined by $\nu_{g_1}(g) = g_1 g g_1^{-1}$). We also define two kinds of adjoint groups of $\mathfrak{g}$: the "algebraic adjoint group" Ad $\mathfrak{g}$ and the "analytic adjoint group" Inn $\mathfrak{g}$ (I, § 3).

More special symbols are listed in the *List of symbols* in the order of their first appearance. (For some symbols, introduced more than once, the places to be consulted are indicated in parenthesis.)

In Chapters III and V, different notations are introduced for the same objects, depending on their interpretations. The relations between them are explained in V, §§ 1, 3, and 4 (pp. 213, 224–225, 232–233).

4. Lemmas, Propositions and Theorems are numbered together in each section. The cross reference to results in another chapter is given in the form, such as Prop. II. 3. 4 or (II. 3. 4), which mean, respectively, Proposition 3. 4 or the formula (3. 4) in Chapter II, § 3. The name of author followed by a number in bracket, such as Albert [5], refers to the *References* at the end of the book.

List of symbols

I

§ 1 : $\boldsymbol{F}$, $\mathscr{G}^z$, G^z, $\mathscr{G}(F)$, G°, $F[G]$, λ_g, Lie G, $T_e(G)$, exp X, $d\varphi$, $G_{F'}$, $G'(F)$, $R_{F'/F}(G')$ § 2 : g_s, g_u, $\boldsymbol{G}_a$, $G^\wedge$, $H^\perp$, G^c, $\boldsymbol{C}^{(1)}$ § 3 : Cent $\mathfrak{g}$, $ad^{\mathfrak{g}}\, X$, $B^{\mathfrak{g}}(X, Y)$, $\mathfrak{g}_1^\perp$, Inn $\mathfrak{g}$, Ad $\mathfrak{g}$, $\mathfrak{g}^a$, $\mathfrak{g}^s$, G^a, G^s, $\mathrm{Sym}_n(F)$, $O_n(F)$, $\mathscr{H}_n(F)$, $\mathscr{P}_n(F)$, $U_n(F)$ $(F=\boldsymbol{R}, \boldsymbol{C})$ (App., § 3) § 4 : θ, $\mathfrak{k}$, $\mathfrak{p}$ (II, § 1), σ_0, ν_g, K, P § 5 : F-rank, $\mathfrak{g}_\gamma$, $\mathfrak{r}$, s_α, W, (BC_r) (II, § 4), Δ, A_Γ, $\mathfrak{a}_\Gamma$, $\mathfrak{r}_\Gamma$, $\mathfrak{b}_\Gamma$, B_Γ, $\tilde{\mathfrak{r}}$, $\tilde{W}$, $\tilde{\Delta}$ (II, § 4), Δ_0, w_σ, $\mathfrak{a}^+$ § 6 : $\{x, y, z\}$, $x \square y$, T_x, $\tau(x, y)$, T^*, $\Gamma(V, \{\ \})$, $\mathrm{Der}(V, \{\ \})$, exp x, $P(x)$, $x \underset{a}{\cdot} y$, $x \underset{a}{\square} y$, $\{x, y, z\}_a$ § 7 : $\mathfrak{P}_\nu$, $p\dfrac{\partial}{\partial x}$, $[p, q]$, p_b, $\mathfrak{G}(V, \{\ \})$, $\mathfrak{G}_\nu$, $\mathfrak{G}_0^+$ § 8 : $G(\Omega)$, Ω^*, $\mathfrak{g}(\Omega)$, $\mathfrak{gl}_n^0(\mathscr{K})$, $J(U, S, e)$, $\mathscr{P}(1, m-1)$, $\mathscr{P}_3(\boldsymbol{O})$, $\phi(x)$.

II

§ 1 : $I(M)$, $T_x(M)$, $q_x(X, Y)$, $d_x\varphi$, s_x, $ad_{\mathfrak{p}}$, $B_{\mathfrak{p}}$ § 2 : θ_x, ad^G § 3 : J_x, $h_x(X, Y)$, H_0, $\mathfrak{p}_\pm$ § 4 : $\tilde{\mathfrak{r}}_c$, $\tilde{\mathfrak{r}}_\pm$, $P_\pm$, $\mathscr{D}$, M^*, h, h', h'', $e_\pm$ (III, § 1), $\mathscr{D}^1$, o_k, H_k, X_k, Y_k, $\tilde{\kappa}_k$, $\|\ \|$ § 5 : $(g)_0$, $(g)_\pm$, $J(g, z)$, $K(z, w)$, Jac, j_z, k_z, χ_1, g_z § 6 : $\mathscr{H}^2(\mathscr{D})$, $k_{\mathscr{D}}$, jac, ∂_z, $q_{\mathscr{D}}$, ω, $\mathrm{Hol}(\mathscr{D})$ § 7 : $Sp(V, A)$, $Sp_{2n}(F)$ (App., § 1), $V_\pm$, h_A, $J_+(g, z)$, $GSp(V, A)$, $K_+(z, w)$, v_z § 8 : $\rho_{\mathscr{D}}$, ρ_+, ρ_G.

III

§ 1 : $\mathfrak{g}^1$, H_0^1, ρ_ν, $V^{[\nu]}$ (IV, § 1), $\mathfrak{g}^{[\nu]}$, $\mathfrak{k}^{[\nu]}$, $\mathfrak{p}^{[\nu]}$, $\mathfrak{g}^{[\mathrm{even}]}$, X_κ, Y_κ, H_κ, c_κ, γ, $\mathfrak{g}_\kappa^{(1)}$, $\mathfrak{l}_2$, $\mathfrak{g}_\kappa$, $\mathfrak{g}_\kappa^*$, $\mathfrak{l}_1$, $\mathfrak{c}(\rho)$, $\mathfrak{c}^*(\rho)$ § 2 : $\mathfrak{b}_\kappa$, V_κ, U_κ, $\mathfrak{k}_\kappa^*$, $\mathfrak{g}_\kappa^{(2)}$, $G_\kappa^{(i)}$, $\mathfrak{p}_+^{[\nu]}$ (§ 3), e_κ, Ω_κ § 3 : $H_0^{(1)}$, $A_u(v, v')$, $A(v, v')$ (§ 6), S_e, $\mathscr{H}(V, I_0, h)$, $\mathscr{P}(V, I_0, h)$, $R_u(=R(u))$ (§ 6; V, § 2), X_u (V, § 4) § 4 : c_1, $\tilde{\kappa}$, o_κ, κ_j, $m(\kappa)$, $GL^0(n, \boldsymbol{C})$ § 5 : $\mathscr{J}_U$, $\mathscr{K}_U$, $\Phi(u, w, z_1)$ § 6 : $\mathscr{S}(U, V, \Omega, H)$, $\mathrm{Aff}(\mathscr{S})$, $H_u(v, v')$, $\mathfrak{S}(V, A, \Omega)$, I_z, $H_z(v, v')$, H_z^+, $\mathscr{L}_z(w, w')$, $\mathscr{S}_\varphi$ § 7 : $\tilde{V}_\kappa$, $\mathscr{S}_\kappa$ § 8 : $\mathscr{F}_\kappa$, $X^{(b)}$, $o^{(b)}$, $\mathfrak{g}_b$, $\mathfrak{a}_b$, $\boldsymbol{m}(\kappa)$, $\kappa^{(b)}$, $o^{(b)}$, $B(\mathscr{D})$ § 9 : $V^{[\lambda,\mu]}$, $\mathfrak{g}^{[\lambda,\mu]}$, $\mathfrak{g}_\nu^{[1,1]}$, $\mathfrak{g}^{(\lambda,\mu)}$, $V_{\kappa',\kappa''}$, $U_{\kappa',\kappa''}$, $\mathscr{S}_{\kappa',\kappa''}$, $\mathscr{D}_\kappa^*$, $\mathscr{S}_{\kappa,\kappa'}^*$ § 10 : $M(\rho)$.

IV

§ 1 : $\mathrm{End}_G(V_1)$, F_1, $\mathrm{Hom}_G(V_1, V)$, $\bar{D}_1$, $U_1 \otimes_{D_1} V_1$, $V^{[i]}$, $(\ )_W$ § 2 : F_{10}, $\mathrm{tr}_{D_1/F}$, $M(\ )$ § 3 : $F_1^{(i)}$, $V_1^{(i)}$, $\rho_1^{(i)}$, $D_1^{(i)}$, $h^{(i)}$, $A^{(i)}$ § 4 : C_ρ, $\mathscr{D}_\rho$, $\rho^{[i]}$, $A^{[i]}$, $I^{[i]}$ § 5 : $\tilde{V}_1$, $\tilde{\rho}_1$, λ_1, $\wedge^k$ § 6 : G_1', G_1'' § 7 : $\sim$, G_L, $G_L(N)$, $I_{\varphi(z)}$, $w(z, v)$, $\tilde{M}$, $G(\mathfrak{A})$, $\mathscr{D}(\mathfrak{A})$ § 8 : $K(M)$, $c_{\boldsymbol{R}}(E)$, $E(j)$, $\eta(\gamma, z)$, $\kappa(z, z')$, $\eta_\phi(\gamma, z)$, $d^{z_1}w$.

V

§ 1 : $\mathfrak{g}(\mathscr{S})$, $\mathfrak{P}_{\mu,\nu}$, $\mathfrak{Q}_{\mu,\nu}$, $p\dfrac{\partial}{\partial z}$, $q\dfrac{\partial}{\partial w}$, ∂_a, $\tilde{\partial}_b$, $X_{A,B}(=X(A, B))$, ∂, ∂', $\mathfrak{g}_\nu$, $\mathfrak{r}_\nu$, φ_e, ψ_e, $\mathfrak{g}_0^0$ § 2 :

Φ_v, c_Φ, $R_u(=R(u))$, Φ^*, Y_Φ, Ψ_v, $Z_{a,b}$, A_u, B_u § 3 : π_U, π_V, θ_Ω, ∂_u^*, $\tilde{\partial}_v^*$, a^u, b^u, $A_{u',u}$, $B_{w',u}$, Φ^v, $\Phi_{v',v}$, $\Psi_{v',v}$ § 4 : X_u, $\Phi_{v,v'}^0$, $\Psi_{v,v'}^0$ § 5 : $\mathcal{S}(\Omega, e, \beta)$, β_r, $\beta_{r,s}$ § 6 : τ_1, $\tau_{1/2}$, τ'
§ 7 : $\mathfrak{g}_-$, ρ_-, $X(A)$, ρ_Ω, $\rho_{\mathcal{S}}$.

Appendix

§ 1 : $\mathfrak{A}^\times$, $\mathfrak{A}^{(1)}$, $GL_n(D)$, $SL_n(D)$, $\mathfrak{sl}_n(D)$, $D^\pm$, $U(V, h)$, $SU(V, h)$, $SU_n(D, h)$, $\mathfrak{su}_n(D, h)$
§ 2 : (α_1, α_2), $\mathcal{B}(F_1)$ § 3 : $U(p, q, D)$, $SU(p, q, D)$, $O(p, q)$, $SO(p, q)$, $U(p, q)$, $SU(p, q)$, $\mathcal{H}_n(D)$, $\mathcal{P}_n(D)$, $\mathcal{P}_n^{(1)}(D)$, $\mathcal{D}(V, h)$, (III_n), $(I_{p,q})$, (II_n) (III, § 4), $SU_n^-(\boldsymbol{H})$, $SU^-(n, \boldsymbol{H})$, ${}_{\boldsymbol{R}}V$ § 4 : $\Delta(S)$, $\mathcal{T}(V)$, $C(V, S)$, $C^\pm$, $\check{e}$, $Spin(V, S)$ § 5 : e_+, e_-, C_+^+, C_-^+, $\mathcal{P}(C^+)$ § 6 : h_S, (IV_p) (III, § 4).

Chapter I
Algebraic Preliminaries

§ 1. Linear algebraic groups.

Let F be a field of characteristic zero and let V be a vector space of dimension n over F. We denote by $\mathrm{End}(V)$ the algebra of all F-linear transformations of V and by $GL(V)$ the group of units in $\mathrm{End}(V)$, i. e., the group of all non-singular F-linear transformations of V. When a basis of V is fixed, V is identified with the space of n-tuples F^n, and so $\mathrm{End}(V)$ and $GL(V)$ are also identified with $\mathcal{M}_n(F)$, the full matrix algebra of degree n over F, and $GL_n(F)$, the general linear group of degree n over F, respectively.

In the following, we fix an algebraically closed extension $\boldsymbol{F}$ of F once and for all. A subgroup $\mathcal{G}$ of $GL_n(\boldsymbol{F})$ is called a (linear) *algebraic group* defined over F, if $\mathcal{G}$ is the set of common zeros of a (finite) system of polynomial equations in n^2 matrix entries with coefficients in F. If we embed $\mathcal{G}$ in $\boldsymbol{F}^{n^2+1}$ by the map

$$\mathcal{G} \ni g = (g_{ij}) \longmapsto (g_{11}, g_{12}, \cdots, g_{nn}, \det(g)^{-1}) \in \boldsymbol{F}^{n^2+1},$$

$\mathcal{G}$ may be viewed as an affine variety in $\boldsymbol{F}^{n^2+1}$. The dimension of an algebraic group $\mathcal{G}$ is the dimension of the underlying affine variety. For instance, $GL_n(\boldsymbol{F})$ and $SL_n(\boldsymbol{F})\ (=\{g \in GL_n(\boldsymbol{F}) \mid \det(g)=1\})$ are algebraic groups of dimension n^2 and n^2-1, respectively.

Let $\mathcal{G}$ and $\mathcal{G}'$ be algebraic groups defined over F in $GL_n(\boldsymbol{F})$ and $GL_{n'}(\boldsymbol{F})$, respectively. A group homomorphism $\varphi: \mathcal{G} \rightarrow \mathcal{G}'$ is called a (rational) homomorphism defined over F, or for short, an *F-homomorphism,* if φ is a restriction to $\mathcal{G}$ of a polynomial map with coefficients in F of $\boldsymbol{F}^{n^2+1}$ into $\boldsymbol{F}^{n'^2+1}$. φ is called a (rational) isomorphism defined over F, or an *F-isomorphism,* if φ is a group isomorphism and both φ and φ^{-1} are F-homomorphisms. A bijective F-homomorphism is necessarily an F-isomorphism. (This is not true in general when the characteristic is positive.) Linear algebraic groups which are F-isomorphic to each other may be viewed as matrix expressions of one and the same (affine) algebraic group defined over F.

A subgroup $\mathcal{H}$ of an algebraic group $\mathcal{G}$ defined over F is called *F-closed* if $\mathcal{H}$ itself is an algebraic group defined over F, i. e., defined by polynomial equations with coefficients in F (for one and hence all matrix expression of $\mathcal{G}$). When $\mathcal{H}$ is an F-closed normal subgroup of $\mathcal{G}$, it is known (e. g., Borel [13]) that the factor group $\mathcal{G}/\mathcal{H}$ has a natural structure of (linear) algebraic group defined over F (determined up to an F-isomorphism) such that the canonical homomorphism $\mathcal{G} \rightarrow \mathcal{G}/\mathcal{H}$ is an F-homomorphism; one then has the relation

$$\dim \mathcal{G}/\mathcal{H} = \dim \mathcal{G} - \dim \mathcal{H}.$$

It is also known (Borel, loc. cit.) that, if $\varphi:\mathcal{G}\to\mathcal{G}'$ is an F-homomorphism, the image $\varphi(\mathcal{G})$ is an F-closed subgroup of $\mathcal{G}'$, the kernel $\mathcal{N}=\varphi^{-1}(e')$ is an F-closed normal subgroup of $\mathcal{G}$, and one has an F-isomorphism $\mathcal{G}/\mathcal{N}\cong\varphi(\mathcal{G})$. The notions of a direct product $\mathcal{G}_1\times\mathcal{G}_2$ and a semi-direct product $\mathcal{G}_1\cdot\mathcal{G}_2$ of two algebraic groups $\mathcal{G}_1$ and $\mathcal{G}_2$ are defined in the natural manner.

For an algebraic group $\mathcal{G}$ defined over F, we denote by $\mathcal{G}^z$ the Zariski connected component of the underlying affine variety of $\mathcal{G}$ containing the unit element e. Then $\mathcal{G}^z$ is an F-closed normal subgroup of $\mathcal{G}$ of finite index, and the coset decomposition of $\mathcal{G}$ with respect to $\mathcal{G}^z$ coincides with the decomposition of the underlying affine variety of $\mathcal{G}$ into the union of irreducible components. Thus $\mathcal{G}^z$ is the unique irreducible component of $\mathcal{G}$ containing e; an algebraic group $\mathcal{G}$ is Zariski connected if and only if it is irreducible as affine variety.

For an algebraic group $\mathcal{G}$ defined over F in $GL_n(\boldsymbol{F})$, the subgroup $\mathcal{G}(F)=\mathcal{G}\cap GL_n(F)$ is called the group of "F-rational points" in $\mathcal{G}$. Clearly the group $\mathcal{G}(F)$ is well-determined, independently of the matrix expression of $\mathcal{G}$. More generally, if $\varphi:\mathcal{G}\to\mathcal{G}'$ is an F-homomorphism, φ induces an (abstract) group homomorphism $\varphi_F:\mathcal{G}(F)\to\mathcal{G}'(F)$. It is known that, when $\mathcal{G}$ is Zariski connected, or when F is algebraically closed, $\mathcal{G}(F)$ is Zariski dense in $\mathcal{G}$.

In general, let G be an (abstract) subgroup of $GL_n(F)$ and let $\mathcal{G}$ be the Zariski closure of G in $GL_n(\boldsymbol{F})$. Then $\mathcal{G}$ is an algebraic group defined over F and one has $G\subset\mathcal{G}(F)$. When we have the equality $G=\mathcal{G}(F)$, we call G an *F-group* and $\mathcal{G}$ the *associated algebraic group*. Let G and G' be F-groups with the associated algebraic groups $\mathcal{G}$ and $\mathcal{G}'$. By definition, an "F-homomorphism" of G into G' is the restriction φ_F to G of an F-homomorphism $\varphi:\mathcal{G}\to\mathcal{G}'$ (which is uniquely determined by φ_F). The notions of "F-isomorphisms", "F-subgroups", etc. of F-groups are defined in a similar manner. For an F-group G with the associated algebraic group $\mathcal{G}$, $G^z=\mathcal{G}^z(F)$ will be called the "Zariski connected component" of G; G^z is an F-group with the associated algebraic group $\mathcal{G}^z$. It should be noted that some properties of F-homomorphisms of algebraic groups mentioned above break down in general for F-homomorphisms of F-groups. To see this, let $\varphi_F: G\to G'$ be an F-homomorphism of F-groups coming from an F-homomorphism $\varphi: \mathcal{G}\to\mathcal{G}'$ of the associated algebraic groups, and let $N=\mathrm{Ker}\ \varphi_F$, $\mathcal{N}=\mathrm{Ker}\ \varphi$. Then it is clear that $N=\mathcal{N}\cap G$, so that N is a normal F-subgroup of G. However, if $\mathcal{N}_1$ denotes the algebraic group associated with N, one has only $\mathcal{N}^z\subset\mathcal{N}_1\subset\mathcal{N}$. On the other hand, $\varphi_F(G)$ is Zariski dense in $\varphi(\mathcal{G})$, but may not be equal to $\varphi(\mathcal{G})(F)$. Thus we have only an injective homomorphism $G/N\to\varphi(\mathcal{G})(F)$ and, going to the Zariski closures, the induced F-homomorphism $\mathcal{G}/\mathcal{N}_1\to\varphi(\mathcal{G})$, which may not be injective. For instance, for $G=G'=GL_1(\boldsymbol{R})=\boldsymbol{R}^\times$ and the $\boldsymbol{R}$-homomorphism $\varphi: x\mapsto x^4$, one has $|\mathcal{N}|=4$, $|N|=|\mathcal{N}_1|=2$, and $\varphi(\mathcal{G})(\boldsymbol{R})=\boldsymbol{R}^\times$, $\varphi_{\boldsymbol{R}}(\boldsymbol{R}^\times)=\boldsymbol{R}_+^\times$ (the multiplicative group of positive real numbers).

When $F=\boldsymbol{R}$ or $\boldsymbol{C}$, an F-group G in $GL_n(F)$ is a closed subgroup of $GL_n(F)$ in the usual topology, so that G has a unique structure of real (or complex) Lie group of the same dimension. We denote the identity connected component of

G in the usual topology by G°. Then clearly $G^\circ \subset G^z$, and by a theorem of Whitney [1] the index $(G^z : G^\circ)$ is finite. In particular, in the case $F=\boldsymbol{C}$, one has $G^\circ = G^z$, and so G is Zariski connected (i. e., the associated algebraic group $\mathcal{G}$ is Zariski connected) if and only if G is connected in the usual topology. This is false for $F=\boldsymbol{R}$; e. g., $G=\boldsymbol{R}^\times$ and $SO(p, q)$ $(p>0, q>0)$ are Zariski connected, but one has $(G : G^\circ)=2$.

For a Zariski connected F-group G we denote by $F[G]$ the "affine ring" of the associated algebraic group $\mathcal{G}$ over F, i. e., the ring of polynomial functions on $\mathcal{G}$ (viewed as an affine variety in $\boldsymbol{F}^{n^2+1}$) with coefficients in F. $F[G]$ is then an integral domain, finitely generated over F, and G acts on $F[G]$ by the left translation $\lambda_{g_1} (g_1 \in G)$:

$$(1.1) \qquad (\lambda_{g_1} f)(g) = f(g_1^{-1} g) \qquad (f \in F[G]).$$

By definition, a *derivation* of $F[G]$ is an F-linear transformation X of $F[G]$ satisfying the condition

$$(1.2) \qquad X(f_1 \cdot f_2) = (Xf_1) \cdot f_2 + f_1 \cdot (Xf_2) \qquad (f_1, f_2 \in F[G]).$$

We denote by Der $F[G]$ the Lie algebra of all derivations of $F[G]$ with the bracket product $[X_1, X_2]=X_1X_2-X_2X_1$. A derivation X is *left invariant* if it commutes with all left translations λ_{g_1}, i. e.,

$$(1.3) \qquad X \circ \lambda_{g_1} = \lambda_{g_1} \circ X \qquad \text{for all } g_1 \in G.$$

The set of all left invariant derivations of $F[G]$ forms a Lie subalgebra $\mathfrak{g}$, called the *Lie algebra* of the F-group G and denoted by Lie G. The dimension of $\mathfrak{g}=$ Lie G (as vector space over F) is equal to the dimension of the associated algebraic group $\mathcal{G}$. When an F-group G is not Zariski connected, we define Lie G to be Lie G^z.

Let $T_e(G)$ denote the "tangent space" over F to $\mathcal{G}$ at the unit element e, i. e., the vector space over F formed of all F-valued derivations of $F[G]$ with respect to the homomorphism $F[G] \ni f \mapsto f(e) \in F$. If we set

$$(1.4) \qquad X_e(f) = (Xf)(e) \qquad \text{for} \quad X \in \mathfrak{g},$$

then the map $X \mapsto X_e$ gives an F-linear isomorphism of $\mathfrak{g}$ onto $T_e(G)$, by which the Lie algebra $\mathfrak{g}$ of G is often identified with the tangent space $T_e(G)$. For instance, the Lie algebra $\mathfrak{gl}_n(F)$ of $GL_n(F)$ is identified with the tangent space $T_{1_n}(GL_n(F))$, which in turn can naturally be identified with $\mathcal{M}_n(F)$. The bracket product $[X_1, X_2]$ in $\mathfrak{gl}_n(F)$ (as derivations) coincides with the usual bracket product in $\mathcal{M}_n(F)$ (as matrices). Thus, for an F-group G in $GL_n(F)$, the Lie algebra $\mathfrak{g}$ may be viewed as a subalgebra of $\mathfrak{gl}_n(F)=\mathcal{M}_n(F)$. It should be noted that not all Lie subalgebras of $\mathfrak{gl}_n(F)$ correspond to F-subgroups of $GL_n(F)$; one corresponding to an F-subgroup of $GL_n(F)$ is called an "algebraic" Lie algebra (cf. Chevalley [2], [4]).

When $F=\boldsymbol{R}$ or $\boldsymbol{C}$, the Lie algebra $\mathfrak{g}$ and the tangent space $T_e(G)$ can also be identified with the ones defined in the theory of Lie groups. In these cases, one has the "exponential map" $\exp : \mathfrak{g} \to G$. For $X \in \mathfrak{g}$, $\exp tX$ $(t \in F)$ is the unique one-parameter subgroup of G such that one has in matrix expression

$$\lim_{t\to 0}\frac{1}{t}((\exp tX)-1_n) = X, \tag{1.5}$$

or equivalently,

$$\frac{d}{dt}f(g(\exp tX))\Big|_{t=0} = (Xf)(g) \qquad (f\in F[G],\ g\in G), \tag{1.6}$$

where X on the right-hand side is interpreted as a derivation of $F[G]$.

Let G and G' be Zariski connected F-groups and let $\mathfrak{g}=\text{Lie } G$ and $\mathfrak{g}'=\text{Lie } G'$. Given an F-homomorphism $\varphi_F: G\to G'$ (coming from an F-homomorphism of the associated algebraic groups, $\varphi: \mathcal{G}\to\mathcal{G}'$), one has an F-algebra homomorphism φ^*: $F[G']\to F[G]$ given by $\varphi^*(f)=f\circ\varphi$ $(f\in F[G'])$. One defines the *differential* $d\varphi$: $\mathfrak{g}\to\mathfrak{g}'$ by setting

$$(d\varphi)(X) = X\circ\varphi^* \qquad \text{for}\quad X\in\mathfrak{g} = T_e(G),$$

(where X is viewed as an F-valued derivation of $F[G]$). Then $d\varphi$ is a Lie algebra homomorphism, and the correspondence $\varphi\mapsto d\varphi$ is clearly functorial. In the case $F=\boldsymbol{R}$ or $\boldsymbol{C}$, this (algebraic) definition of differentials coincides with the analytic one. In particular, for $X\in\mathfrak{g}$, the "one-parameter subgroup" $\varphi(t)=\exp tX$ $(t\in F)$ is defined by an F-homomorphism φ of the additive group of F into the F-group G and the relation (1.5) or (1.6) is equivalent to saying that $(d\varphi)(1)=X$. In this book, whenever there is no fear of confusion, the differential $d\varphi$ will also be denoted by φ.

Let F' be an extension field of F, contained in $\boldsymbol{F}$. Then from an F-group G in $GL_n(F)$ one obtains an F'-group G' in $GL_n(F')$ by putting $G'=\mathcal{G}(F')$ where $\mathcal{G}$ is the algebraic group associated with G (and hence also with G'). The F'-isomorphism class of G' is uniquely determined by the F-isomorphism class of G. We call G' the F'-group obtained from G by *scalar extension* F'/F and write $G'=G_{F'}$. The Lie algebra of $G_{F'}$ can naturally be identified with the Lie algebra $\mathfrak{g}_{F'}=\mathfrak{g}\otimes_F F'$ obtained from $\mathfrak{g}=\text{Lie } G$ by scalar extension (in the sense of linear algebra). Conversely, suppose there is given an F'-group G' in $GL_n(F')$. Then $G=G'\cap GL_n(F)$ is an F-group and one has $G_{F'}\subset G'$. G is called an *F-form* of G' if $G_{F'}=G'$, or equivalently, if G and G' have the same associated algebraic group. When this is the case, we say that G' has an F-form in $GL_n(F)$ (with respect to the given matrix expression) and write $G=G'(F)$ (or G'_F). Let $\mathcal{G}$ be the algebraic group in $GL_n(\boldsymbol{F})$ associated with G' and let $I(\mathcal{G})$ be the ideal corresponding to $\mathcal{G}\subset\boldsymbol{F}^{n^2+1}$, i. e., the ideal in the polynomial ring $\boldsymbol{F}[x_{ij}(1\le i,j\le n),y]$ consisting of all polynomials annihilating $\mathcal{G}$. Then in order that G' has an F-form in $GL_n(F)$ the following two conditions are necessary and sufficient:

(i) $\mathcal{G}$ is defined over F, i. e., $I(\mathcal{G})$ has a basis consisting of polynomials with coefficients in F.

(ii) Every coset in $\mathcal{G}/\mathcal{G}^z$ contains an F-rational point.

When these conditions are satisfied, $G=\mathcal{G}(F)$ is an F-form of G' (relative to the given matrix expression).

Similarly, for a Lie algebra $\mathfrak{g}'$ over F', a Lie algebra $\mathfrak{g}$ over F such that $\mathfrak{g}'=\mathfrak{g}\otimes_F F'$ is called an F-form of $\mathfrak{g}'$. If G is an F-form of an F'-group G', then $\mathfrak{g}=\mathrm{Lie}\ G$ is an F-form of $\mathfrak{g}'=\mathrm{Lie}\ G'$, and if G' is Zariski connected the correspondence $G\mapsto \mathrm{Lie}\ G$ gives a one-to-one correspondence between F-forms G of G' and F-forms $\mathfrak{g}$ of $\mathfrak{g}'$ which are algebraic.

When $[F' : F]=d<\infty$, let $\rho : F'\to\mathscr{M}_d(F)$ be a regular representation of F', viewed as an algebra over F. For any positive integer n, let ρ_n denote an F-algebra homomorphism $\mathscr{M}_n(F')\to\mathscr{M}_{nd}(F)$ defined by $\rho_n((\xi_{ij}))=(\rho(\xi_{ij}))$. Then, for an F'-group G' in $GL_n(F')$, $\rho_n(G')$ is an F-group in $GL_{nd}(F)$, whose F-isomorphism class is uniquely determined by the F'-isomorphism class of G'. The F-group $\rho_n(G')$ is called the F-group obtained from G' by *scalar restriction* and is denoted by $R_{F'/F}(G')$ (Weil [5]). One has the relation

(1.7) $$\dim R_{F'/F}(G') = d\cdot\dim G'.$$

When F'/F is a finite Galois extension with Galois group $\mathrm{Gal}(F'/F)=\{\sigma_i\,(1\leq i\leq d)\}$, one has a natural F'-isomorphism

(1.8) $$(R_{F'/F}(G'))_{F'}\cong\prod_{i=1}^{d} G'^{\sigma_i},$$

where $G'^{\sigma_i}=\{g'^{\sigma_i}\,|\,g'\in G'\}\ (1\leq i\leq d)$ are the "conjugates" of G' in $GL_n(F')$.

Exercises

1. Let G be an F-group, N a normal F-subgroup of G and let $\mathscr{G}$ and $\mathscr{N}$ be the associated algebraic groups. Assume that $\mathscr{N}$ has the following property: Let (g_σ) be a (continuous) "1-cocycle" of $\mathrm{Gal}(\bar{F}/F)$ in $N_{\bar{F}}$ (where $\bar{F}$ is the algebraic closure of F in $\boldsymbol{F}$), i. e., $\sigma\mapsto g_\sigma$ is a map from $\mathrm{Gal}(\bar{F}/F)$ into $N_{\bar{F}}$ such that 1) $g_\tau g_\sigma^\tau=g_{\sigma\tau}$ for all $\sigma,\tau\in\mathrm{Gal}(\bar{F}/F)$ and 2) there exists a finite extension F' of F (contained in $\boldsymbol{F}$) such that g_σ depends only on $\sigma|F'$. Then there exists $h\in N_{\bar{F}}$ such that $g_\sigma=hh^{-\sigma}$. [We express this property by saying that the first Galois cohomology of $\mathscr{N}$ over F is trivial. For instance, this property holds for $\mathscr{N}=G_a$ (cf. § 2) and GL_m over any F.] Show that under this assumption one has $(\mathscr{G}/\mathscr{N})_F\cong G/N$, i. e., every F-rational coset in $\mathscr{G}/\mathscr{N}$ has an F-rational representative in G.

2. Let G be an F-group in $GL_n(F)$. We identify $\mathfrak{g}=\mathrm{Lie}\ G$ with a subalgebra of $\mathfrak{gl}_n(F)$ as explained in the text [i. e., $X\in\mathfrak{g}$ is identified with a matrix (ξ_{ij}) defined by $\xi_{ij}=(Xx_{ij})(1_n)$].

2. 1) Show that the left invariant derivation corresponding to $X=(\xi_{ij})$ is given by a (formal) vector field $\sum_{i,j,k} x_{ik}\xi_{kj}\frac{\partial}{\partial x_{ij}}$. Check that the Poisson bracket corresponds to the bracket of matrices.

2. 2) Show that, if G is defined by polynomial equations $f_\nu(x)=0$ $(1\leq\nu\leq N)$, then $\mathfrak{g}$ is defined by linear equations

$$f_\nu(1_n+\delta X)\equiv 0 \pmod{\delta^2} \qquad (1\leq\nu\leq N),$$

where δ is a new variable.

3. Let $A=(V,\cdot)$ be an algebra over F (distributive, but not necessarily associative) and let $G=\mathrm{Aut}(A)=\{g\in GL(V)\,|\,g(a\cdot b)=(ga)\cdot(gb)\ (a,b\in A)\}$. Show that

$$\mathrm{Lie}\ G=\mathrm{Der}(A)=\{X\in\mathrm{End}(V)\,|\,X(a\cdot b)=(Xa)\cdot b+a\cdot(Xb)\ (a,b\in A)\}.$$

§ 2. Tori and unipotent algebraic groups.

Let G be an F-group in $GL_n(F)$. An element $g \in G$ is called *semi-simple*, if the matrix g is diagonalizable in $GL_n(\boldsymbol{F})$, i. e., if the minimal polynomial of g has no multiple roots, g is called *unipotent*, if $g-1_n$ is nilpotent, i. e., if $(g-1_n)^n=0$. In general, every element $g \in G$ can be decomposed uniquely in the form $g=g_s g_u$ where g_s (resp. g_u) is a semi-simple (resp. unipotent) element in G such that $g_s g_u = g_u g_s$. This decomposition (called "multiplicative Jordan decomposition") and the notion of semi-simple and unipotent elements do not depend on the matrix realization of G. More generally, it is known that, if $\varphi: G \rightarrow G'$ is an F-homomorphism, then for $g \in G$ one has $\varphi(g_s)=\varphi(g)_s$, $\varphi(g_u)=\varphi(g)_u$; i. e., $\varphi(g)=\varphi(g_s)\cdot\varphi(g_u)$ is the multiplicative Jordan decomposition of $\varphi(g)$ in G'.

An algebraic group $\mathcal{G}$ is called an (algebraic) *torus* if $\mathcal{G}$ is Zariski connected and if the following mutually equivalent conditions are satisfied:

(a) All elements in $\mathcal{G}$ are semi-simple.

(b) $\mathcal{G}$ has a matrix expression such that all elements in $\mathcal{G}$ are of diagonal form.

(c) $\mathcal{G}$ is $\boldsymbol{F}$-isomorphic to $GL_1(\boldsymbol{F})^m$ for some m.

In particular, a torus $\mathcal{G}$ is abelian. An F-group G is called an *F-torus* if the associated algebraic group $\mathcal{G}$ is a torus. The one-dimensional F-torus $GL_1(F)$ is nothing but the multiplicative group $F^\times$ of the field F.

A (Zariski connected) algebraic group or F-group consisting of only unipotent elements is called *unipotent*. An F-group G is unipotent if and only if the associated algebraic group $\mathcal{G}$ is unipotent. A unipotent algebraic group $\mathcal{G}$ defined over F has a composition series $\{\mathcal{G}_i\}$ consisting of F-closed normal subgroups such that every factor group $\mathcal{G}_{i-1}/\mathcal{G}_i$ is F-isomorphic to the one-dimensional unipotent algebraic group $\left\{\begin{pmatrix}1 & \xi \\ 0 & 1\end{pmatrix} \middle| \xi \in \boldsymbol{F}\right\}$, which is denoted by $\boldsymbol{G}_a$. In particular, a unipotent group $\mathcal{G}$ is nilpotent. If we set $G_i=\mathcal{G}_i(F)$, then $\{G_i\}$ is a composition series of $G=\mathcal{G}(F)$ consisting of normal F-subgroups and, in this case, the natural homomorphism $G_{i-1}/G_i \rightarrow (\mathcal{G}_{i-1}/\mathcal{G}_i)(F)$ is bijective (§ 1, Exerc. 1); thus G_{i-1}/G_i is isomorphic to the F-group $\boldsymbol{G}_a(F)$, which is nothing but the additive group of the field F.

Let G be an F-group. An F-homomorphism $\chi: G \rightarrow GL_1(F)=F^\times$ is called an *F-character*. We define the sum of two F-characters χ and χ' of G by

$$(2.1) \qquad (\chi+\chi')(g) = \chi(g)\cdot\chi'(g) \qquad (g \in G).$$

Then the set $G^\wedge$ of all F-characters of G becomes a ($\boldsymbol{Z}$-)module, called the *F-character module* of G. For any extension F', $F \subset F' \subset \boldsymbol{F}$, one has a natural injection $G^\wedge \rightarrow (G_{F'})^\wedge$ by which $G^\wedge$ is identified with a submodule of $(G_{F'})^\wedge$. Let $\bar{F}$ be the algebraic closure of F in $\boldsymbol{F}$. Then one has $(G_{\bar{F}})^\wedge=(G_{\boldsymbol{F}})^\wedge$; $\boldsymbol{X}=(G_{\bar{F}})^\wedge$ is called the (absolute) character module of $\mathcal{G}=G_{\boldsymbol{F}}$. The Galois group $\mathrm{Gal}(\bar{F}/F)$ acts (continuously in Krull topology) on $\boldsymbol{X}$ by

$$(2.2) \qquad (\sigma\chi)(g) = (\chi(g^\sigma))^{\sigma^{-1}} \qquad (\sigma \in \mathrm{Gal}(\bar{F}/F),\ \chi \in \boldsymbol{X},\ g \in G_{\bar{F}})$$

and $G^\wedge (\subset \boldsymbol{X})$ coincides with the submodule of $\boldsymbol{X}$ consisting of all $\mathrm{Gal}(\bar{F}/F)$-invariant elements.

Let G be an F-torus of dimension m. Then, since $G_{\bar{F}} \cong (\bar{F}^\times)^m$, the character module $\boldsymbol{X}$ is a free ($\boldsymbol{Z}$-)module of rank m. If H is an F-subtorus of G (i. e., an F-subgroup of G, which is an F-torus), then the "annihilator" of H in $\boldsymbol{X}$:

$$H^\perp = \{\chi \in \boldsymbol{X} \mid \chi(h) = 1 \text{ for all } h \in H\}$$

is a submodule of $\boldsymbol{X}$, which is "cotorsion-free", i. e., the factor module $\boldsymbol{X}/H^\perp$ is torsion-free, and stable under $\mathrm{Gal}(\bar{F}/F)$. $\boldsymbol{X}/H^\perp$ is then naturally identified with the character module of $H_{\boldsymbol{F}}$. The correspondence $H \mapsto H^\perp$ gives a one-to-one correspondence between the set of F-subtori of G and that of $\mathrm{Gal}(\bar{F}/F)$-stable, cotorsion-free submodules of $\boldsymbol{X}$. An F-torus G is called *F-split*, if G is F-isomorphic to $(F^\times)^m$, or what amounts to the same thing, if $\mathrm{Gal}(\bar{F}/F)$ acts trivially on $\boldsymbol{X}$, i. e., $\boldsymbol{X} = G^\wedge$. An F-torus G is called *F-compact* (or F-anisotropic), if $G^\wedge$ consists of only the trivial character $\chi_0 : g \mapsto 1$. In general, for an F-torus G, let G^c and A denote the F-subtori of G corresponding to $G^\wedge$ and the submodule consisting of all characters "of trace zero", i. e., characters χ such that one has

$$\sum_{\sigma \in \mathrm{Gal}(F'/F)} \sigma\chi = 0, \tag{2.3}$$

where F' is any finite Galois extension of F such that $\chi \in (G_{F'})^\wedge$. Then A (resp. G^c) is the largest F-split (resp. F-compact) subtorus of G, and G is an "almost" direct product of A and G^c, i. e., $A \cdot G^c$ is Zariski dense in G and the intersection $A \cap G^c$ is finite; or, in other words, the natural map $A \times G^c \to G$ is an "F-isogeny" (i. e., an F-homomorphism with finite kernel and Zariski dense image). For instance, in the case $F = \boldsymbol{R}$ and $\boldsymbol{F} = \boldsymbol{C}$, a complex character $\chi \in \boldsymbol{X} = (G_{\boldsymbol{C}})^\wedge$ is of trace zero if and only if $\bar{\chi} = -\chi$, i. e., $|\chi(g)| = 1$ for all $g \in G$, and in the above notation one has

$$A \cong (\boldsymbol{R}^\times)^r, \qquad G^c \cong (\boldsymbol{C}^{(1)})^{m-r}, \tag{2.4}$$

where $\boldsymbol{C}^{(1)}$ is the compact part of $R_{\boldsymbol{C}/\boldsymbol{R}}(\boldsymbol{C}^\times)$, i. e., the group of complex numbers of absolute value one. Thus, for a real torus G, viewed as a Lie group, the identity connected component G° is the direct product of the "vector part" $A^\circ \cong (\boldsymbol{R}_+^\times)^r$ and the "compact part" G^c.

Now let $\{\chi_1, \cdots, \chi_m\}$ be a basis of the free module $\boldsymbol{X}$. Then the affine ring $\boldsymbol{F}[\mathcal{G}]$ is the polynomial ring $\boldsymbol{F}[\chi_1^{\pm 1}, \cdots, \chi_m^{\pm 1}]$ (allowing negative exponents), and the Lie algebra $\mathfrak{g}_{\boldsymbol{F}}$ of $\mathcal{G} = G_{\boldsymbol{F}}$ is an abelian Lie algebra spanned by the invariant derivations $\chi_i \dfrac{\partial}{\partial \chi_i}$ $(1 \leq i \leq m)$. Therefore the dual space $\mathfrak{g}_{\boldsymbol{F}}^*$ of $\mathfrak{g}_{\boldsymbol{F}}$ is spanned by the invariant differentials $\chi_i^{-1} d\chi_i$ $(1 \leq i \leq m)$. The natural map

$$\chi \longmapsto \chi^{-1} d\chi$$

gives an injective homomorphism of the character module $\boldsymbol{X}$ into $\mathfrak{g}_{\boldsymbol{F}}^*$, so that one has $\mathfrak{g}_{\boldsymbol{F}}^* = \boldsymbol{X} \otimes_{\boldsymbol{Z}} \boldsymbol{F}$. In the case $F = \boldsymbol{R}$ and $\boldsymbol{F} = \boldsymbol{C}$, it is easy to see that Lie A (resp. Lie G^c) consists of all elements X in $\mathfrak{g} = \mathrm{Lie}\, G$ such that $d\chi(X)$ (evaluated at the unit element) is real (resp. purely imaginary) for all $\chi \in \boldsymbol{X}$, or equivalently, all eigenvalues of X (in any matrix expression of G) are real (resp. purely imaginary).

§ 3. Semi-simple and reductive algebraic groups.

Let $\mathfrak{g}$ be a Lie algebra over F. For each $X\in\mathfrak{g}$, we denote by $ad\,X$ a linear transformation of $\mathfrak{g}$ defined by $Y\mapsto[X, Y]$ $(Y\in\mathfrak{g})$. Then by Jacobi identity one has

$$(3.1) \qquad ad([X, Y]) = [ad\,X, ad\,Y] \qquad (X, Y\in\mathfrak{g}).$$

This implies that $ad\,X$ is a derivation of $\mathfrak{g}$ and that the map $X\mapsto ad\,X$ is a Lie algebra homomorphism of $\mathfrak{g}$ into $\mathrm{Der}(\mathfrak{g})\subset\mathfrak{gl}(\mathfrak{g})$, called the *adjoint representation* of $\mathfrak{g}$. Clearly the kernel of ad is Cent $\mathfrak{g}$ (the center of $\mathfrak{g}$).

The *Killing form* of $\mathfrak{g}$ is a symmetric bilinear form on $\mathfrak{g}\times\mathfrak{g}$ defined by

$$(3.2) \qquad B(X, Y) = \mathrm{tr}(ad\,X\cdot ad\,Y) \qquad (X, Y\in\mathfrak{g}).$$

From (3.1) one has

$$(3.3) \qquad B([X, Y_1], Y_2)+B(Y_1, [X, Y_2]) = 0,$$

i.e., $ad\,X$ is skew-symmetric with respect to B. We will write $ad^{\mathfrak{g}}$ and $B^{\mathfrak{g}}$ for ad and B when $\mathfrak{g}$ has to be indicated.

A Lie algebra $\mathfrak{g}$ is called *semi-simple* if there exist no non-trivial abelian ideals, or equivalently, if the "radical" (i.e., the largest solvable ideal) of $\mathfrak{g}$ reduces to $\{0\}$. It is known that $\mathfrak{g}$ is semi-simple if and only if the Killing form $B^{\mathfrak{g}}$ is non-degenerate (Cartan's criterion).

Let $\mathfrak{g}$ be a semi-simple Lie algebra over F. From (3.3) one sees that, if $\mathfrak{g}_1$ is an ideal of $\mathfrak{g}$, then the orthogonal complement $\mathfrak{g}_1^{\perp}$ of $\mathfrak{g}_1$ in $\mathfrak{g}$ with respect to B is also an ideal of $\mathfrak{g}$, and one has a direct sum decomposition

$$\mathfrak{g} = \mathfrak{g}_1 \oplus \mathfrak{g}_1^{\perp}.$$

Thus the adjoint representation of $\mathfrak{g}$ is fully reducible. It follows that $\mathfrak{g}$ is decomposed uniquely into the direct sum of simple (non-abelian) ideals:

$$(3.4) \qquad \mathfrak{g} = \mathfrak{g}_1 \oplus \cdots \oplus \mathfrak{g}_r \qquad (\mathfrak{g}_i \text{ simple}).$$

Conversely, any Lie algebra which is the direct sum of simple, non-abelian ideals is semi-simple. It is also known that any representation of a semi-simple Lie algebra (over F) is fully reducible (Weyl's theorem).

For any Lie algebra $\mathfrak{g}$ over F, the automorphism group $\mathrm{Aut}(\mathfrak{g})$ of $\mathfrak{g}$ is an F-group in $GL(\mathfrak{g})$, and the Lie algebra of $\mathrm{Aut}(\mathfrak{g})$ is equal to the derivation algebra $\mathrm{Der}(\mathfrak{g})$ (§ 1, Exerc. 3). When $F=\boldsymbol{R}$ or $\boldsymbol{C}$, the analytic subgroup of $\mathrm{Aut}(\mathfrak{g})$ corresponding to $ad(\mathfrak{g})(\subset\mathrm{Der}(\mathfrak{g}))$ is called the (analytic) *adjoint group* or the group of "inner automorphisms" of $\mathfrak{g}$ and is denoted by $\mathrm{Inn}(\mathfrak{g})$. It is known that, if $\mathfrak{g}$ is semi-simple, then the adjoint representation gives an isomorphism $\mathfrak{g}\cong\mathrm{Der}(\mathfrak{g})$, so that one has $\mathrm{Inn}(\mathfrak{g})=\mathrm{Aut}(\mathfrak{g})^{\circ}$. In this case, $\mathrm{Aut}(\mathfrak{g})^{z}$ is called the "algebraic adjoint group" and will be denoted, in this book, by $\mathrm{Ad}(\mathfrak{g})$. Thus one has $\mathrm{Inn}(\mathfrak{g})\subset \mathrm{Ad}(\mathfrak{g})\subset\mathrm{Aut}(\mathfrak{g})$.

It is known (Weyl) that for any semi-simple Lie algebra $\tilde{\mathfrak{g}}$ over $\boldsymbol{C}$, there exists a "compact" real form, i.e., a real form $\mathfrak{u}$ for which the Killing form $B^{\mathfrak{u}}$ is negative definite. This condition implies that $\mathrm{Inn}(\mathfrak{u})$ is compact. Moreover, any analytic group U with Lie algebra $\mathfrak{u}$ is compact (Weyl).

A Lie algebra $\mathfrak{g}$ over F (of characteristic zero) is called *reductive* if the following mutually equivalent conditions are satisfied:

(a) $\mathfrak{g}=\mathfrak{g}^a\oplus\mathfrak{g}^s$ where $\mathfrak{g}^a$ is abelian and $\mathfrak{g}^s$ is semi-simple.

(b) The adjoint representation of $\mathfrak{g}$ is fully reducible.

(c) $\mathfrak{g}$ has a faithful, fully reducible representation.

(For the equivalence of these conditions, see, e. g., Bourbaki [2], Ch. I, § 6, n° 4.) $\mathfrak{g}^a$ (resp. $\mathfrak{g}^s$) in the decomposition in (a) is called the *abelian* (resp. *semi-simple*) *part* of $\mathfrak{g}$; clearly $\mathfrak{g}^a=\mathrm{Cent}\ \mathfrak{g}$ and $\mathfrak{g}^s=[\mathfrak{g},\mathfrak{g}]$. It is easy to see (by Weyl's theorem) that, for a reductive Lie algebra $\mathfrak{g}$ over F, a representation ρ is fully reducible if and only if $\rho(X)$ is semi-simple for all $X\in\mathfrak{g}^a$.

A Zariski connected F-group G, or the associated algebraic group $\mathcal{G}$, is called (algebraically) *reductive* if the following mutually equivalent conditions are satisfied:

(a) G is an almost direct product of two (Zariski connected) normal F-subgroups G^a and G^s, where G^a is an F-torus and G^s is semi-simple (i. e., Lie G^s is semi-simple).

(b) All F-representations of G are fully reducible.

(c) G has a matrix realization, which is fully reducible.

We sketch a proof of the equivalence of these conditions. First, assume the condition (a). Let $\mathfrak{g}$ be the Lie algebra of G and let $\mathfrak{g}^a=\mathrm{Lie}\ G^a$, $\mathfrak{g}^s=\mathrm{Lie}\ G^s$. Then, clearly, one has a direct sum decomposition $\mathfrak{g}=\mathfrak{g}^a\oplus\mathfrak{g}^s$, so that $\mathfrak{g}$ is a reductive Lie algebra. If ρ is an F-representation of G, $\rho(g)$ is semi-simple for all $g\in G^a$. Hence, for the corresponding representation of $\mathfrak{g}$, denoted also by the same letter ρ, $\rho(X)$ is semi-simple for all $X\in\mathfrak{g}^a$. Therefore ρ is fully reducible. (b)$\Rightarrow$(c) is obvious. Assume the condition (c). Then the Lie algebra $\mathfrak{g}$ is reductive. Since $\mathfrak{g}^a$ and $\mathfrak{g}^s$ are "algebraic", there exist F-subgroups of G corresponding to $\mathfrak{g}^a$ and $\mathfrak{g}^s$, which we denote by G^a and G^s. Then G^a is an F-torus, because all $g\in G^a$ are simultaneously diagonalizable, and we see that the condition (a) is satisfied.

If F' is an extension field of F, an F-group G is reductive if and only if $G_{F'}$ is. For an F-torus G, all F-representations are fully reducible, and all (absolutely) irreducible $\boldsymbol{F}$-representations of $\mathcal{G}=G_{\boldsymbol{F}}$ are given by the characters $\chi\in\boldsymbol{X}=\mathcal{G}^\wedge$. Note that a reductive $\boldsymbol{R}$-group may have a transcendental representation which is *not* fully reducible. $\left(\text{E. g., } \boldsymbol{R}^\times\ni x\mapsto\begin{pmatrix}1 & \log|x|\\ 0 & 1\end{pmatrix}.\right)$

In the case $F=\boldsymbol{R}$ or $\boldsymbol{C}$, we use the following notation:

$$\begin{aligned}\mathcal{H}_n(F) &= \{x\in\mathcal{M}_n(F)\,|\,{}^t\bar{x}=x\},\\ \mathcal{P}_n(F) &= \{x\in\mathcal{H}_n(F)\,|\,x\gg 0\},\\ U_n(F) &= \{g\in GL_n(F)\,|\,{}^t\bar{g}g=1_n\}\ \text{(the unitary group)}.\end{aligned}$$

Especially, we have $\mathcal{H}_n(\boldsymbol{R})=\mathrm{Sym}_n(\boldsymbol{R})$ and $U_n(\boldsymbol{R})=O_n(\boldsymbol{R})$ (the orthogonal group). The exponential map gives a homeomorphism $\mathcal{H}_n(F)\to\mathcal{P}_n(F)$. For $h\in\mathcal{P}_n(F)$, we denote by $\log h$ the unique element x in $\mathcal{H}_n(F)$ such that $\exp x=h$.

Lemma 3.1. *Let G be an $\boldsymbol{R}$-group in $GL_n(\boldsymbol{C})$ and let $h \in G \cap \mathcal{P}_n(\boldsymbol{C})$. Then one has* $\log h \in \operatorname{Lie} G$.

Proof. One can suppose that h is of diagonal form: $h = \operatorname{diag}(\alpha_1, \cdots, \alpha_n)$, $\alpha_i > 0$ $(1 \leq i \leq n)$. Let T be the smallest $\boldsymbol{R}$-subgroup of G containing h. Then, by the theory of tori (§ 2), T consists of all diagonal matrices $\operatorname{diag}(\xi_1, \cdots, \xi_n)$ $(\xi_i \in \boldsymbol{R}^\times)$ satisfying the monomial equations of the form

$$\xi_1^{m_1} \cdots \xi_n^{m_n} = 1,$$

where $(m_1, \cdots, m_n)$ runs over all n-tuples of integers such that $\alpha_1^{m_1} \cdots \alpha_n^{m_n} = 1$. Hence, $\operatorname{Lie} T$ is defined by the corresponding linear equations $\sum_{i=1}^{n} m_i \xi_i = 0$. Therefore, one has $\log h = \operatorname{diag}(\log \alpha_1, \cdots, \log \alpha_n) \in \operatorname{Lie} T \subset \operatorname{Lie} G$, q. e. d.

Corollary 3.2. *If G is an $\boldsymbol{R}$-group in $GL_n(\boldsymbol{C})$ and if $g \in G \cap {}^t\bar{G}$, then g can be written uniquely in the form*

$$g = u \cdot \exp x \tag{3.5}$$

with $u \in G \cap U_n(\boldsymbol{C})$, $x \in \operatorname{Lie} G \cap \mathcal{H}_n(\boldsymbol{C})$.

Clearly, it is enough to put $x = \frac{1}{2}\log({}^t\bar{g}g)$ and $u = g(\exp x)^{-1}$.

Proposition 3.3. *Let G be a compact Lie group and let $\rho_0 : G \to GL_n(\boldsymbol{R})$ be a faithful representation of G. Then $\rho_0(G)$ is a reductive $\boldsymbol{R}$-group. Any (continuous) representation ρ of G is an $\boldsymbol{R}$-homomorphism with respect to this structure of $\boldsymbol{R}$-group. In particular, the structure of $\boldsymbol{R}$-group of G is uniquely determined, independently of the choice of the faithful representation ρ_0.*

Proof. We sketch a proof, following Chevalley [1], Ch. VI. Let $\mathcal{R}$ denote the "representative ring" of G over $\boldsymbol{C}$, i. e., the algebra over $\boldsymbol{C}$ generated by all $\boldsymbol{C}$-valued functions on G that occur as matrix entries of some (continuous) representations of G over $\boldsymbol{C}$. Then $\mathcal{R}$ is an affine ring over $\boldsymbol{C}$, i. e., a finitely generated commutative algebra over $\boldsymbol{C}$. Let $\tilde{G} = \operatorname{Spec}_{\boldsymbol{C}}(\mathcal{R})$ be the corresponding (not necessarily irreducible) complex affine variety. Using the interpretation of $\tilde{g} \in \tilde{G}$ as a "representation of representations" of G, one can define a group structure on $\tilde{G}$, so that $\tilde{G}$ becomes a (not necessarily connected) affine algebraic group over $\boldsymbol{C}$. Then, to any (continuous) representation $\rho : G \to GL_n(\boldsymbol{C})$, there corresponds a rational representation $\tilde{\rho} : \tilde{G} \to GL_n(\boldsymbol{C})$ defined by $\tilde{\rho}(\tilde{g}) = (\tilde{g}(\rho_{ij}))$, where ρ_{ij} is the (i, j)-entry of the representation ρ. The group G itself can naturally be identified with a subgroup of $\tilde{G}$ in such a way that, for any ρ, one has $\rho = \tilde{\rho}|G$. Then one has $G = \{\tilde{g} \in \tilde{G} \mid \bar{\tilde{g}} = \tilde{g}\}$ (Tan'naka duality). If $\rho_0 : G \to GL_n(\boldsymbol{R})$ is a faithful representation, then the matrix entries ρ_{0ij} $(1 \leq i, j \leq n)$ generate $\mathcal{R}$ over $\boldsymbol{C}$, so that the corresponding representation $\tilde{\rho}_0$ of $\tilde{G}$ is also faithful. Since G is compact, ρ_0 and hence $\tilde{\rho}_0$ is fully reducible. Thus $\tilde{G}$ is a reductive $\boldsymbol{C}$-group in $GL_n(\boldsymbol{C})$. Replacing ρ_0 by an equivalent representation if necessary, we may assume that ρ_0 is "orthogonal", i. e., $\rho_0(G)\subset O_n(\boldsymbol{R})$. Then one has $\tilde{\rho}_0(\tilde{G})\subset O_n(\boldsymbol{C})$ and $\overline{\tilde{\rho}_0(\tilde{G})}=\tilde{\rho}_0(\tilde{G})$. It follows from Corollary 3. 2 that

$$(*)\qquad \begin{aligned} \tilde{\rho}_0(\tilde{G}) &= (\tilde{\rho}_0(\tilde{G})\cap U_n(\boldsymbol{C}))\cdot(\tilde{\rho}_0(\tilde{G})\cap \mathcal{P}_n(\boldsymbol{C})),\\ \rho_0(G) &= \tilde{\rho}_0(\tilde{G})\cap GL_n(\boldsymbol{R}) = \tilde{\rho}_0(\tilde{G})\cap U_n(\boldsymbol{C}). \end{aligned}$$

Thus $\rho_0(G)$ is a reductive $\boldsymbol{R}$-group, and for this structure of $\boldsymbol{R}$-group one has $G_{\boldsymbol{C}}=\tilde{G}$. Finally, since one has $\mathcal{R}=\boldsymbol{C}[\tilde{\rho}_0(\tilde{G})]$, all (continuous) representations of G are $\boldsymbol{R}$-homomorphisms, q. e. d.

Corollary 3. 4. *Let G be a compact Lie group. Then the identity connected component G° in the usual topology coincides with the identity connected component G^z of G in Zariski topology. Moreover, $(G^\circ)_{\boldsymbol{C}}$ coincides with the identity connected component of $G_{\boldsymbol{C}}$.*

This follows from the fact that, in the above setting, $\mathcal{R}$ is an integral domain if and only if G is connected in the usual topology. We remark that from the decompositions (*) it can be shown easily that G is a maximal compact subgroup of the $\boldsymbol{C}$-group $G_{\boldsymbol{C}}$. Moreover, any maximal compact subgroup of $G_{\boldsymbol{C}}$ is conjugate to G (see e. g., Matsumoto [5]; cf. also Prop. 8. 4).

As an example, let us consider a torus group $G=V/L$ in the sense of Lie groups, where V is an n-dimensional real vector space and L is a lattice in V, i. e., a discrete submodule of rank n. Thus $V\cong\boldsymbol{R}^n$ and $L\cong\boldsymbol{Z}^n$. The Lie algebra $\mathfrak{g}$ of G can naturally be identified with V by the relation $\exp x=(x \bmod L)$ for $x\in\mathfrak{g}=V$. Let L^* be the "dual lattice" of L in the dual space $\mathfrak{g}^*=V^*$, i. e., $L^*=\{\xi\in V^*\mid\langle\xi, x\rangle\in\boldsymbol{Z}$ for all $x\in L\}$. Then G, viewed as an $\boldsymbol{R}$-group, is an $\boldsymbol{R}$-torus and the character module $\boldsymbol{X}=(G_{\boldsymbol{C}})^\wedge$ is identified with $2\pi iL^*\subset\mathfrak{g}_{\boldsymbol{C}}^*$, where a $\boldsymbol{C}$-character χ is identified with $2\pi i\xi$ $(\xi\in L^*)$ by the relation

$$(3.6)\qquad \chi(\exp x) = \mathbf{e}(\langle\xi, x\rangle) \qquad (x\in\mathfrak{g}).$$

An analytic group (i. e., a connected Lie group) G is called (analytically) *reductive,* if $\mathfrak{g}=\operatorname{Lie} G$ is reductive. For a reductive analytic group G, the analytic subgroup G^a (resp. G^s) of G corresponding to $\mathfrak{g}^a$ (resp. $\mathfrak{g}^s$) will be called the *abelian* (resp. *semi-simple*) *part* of G. The relation between algebraic and analytic reductivities will be given by the following propositions.

Proposition 3. 5. *Let $\tilde{G}$ be an (analytically) reductive complex analytic subgroup of $GL_n(\boldsymbol{C})$ and suppose that the abelian part $\tilde{G}^a$ is isomorphic to a $\boldsymbol{C}$-torus. Then $\tilde{G}$ is a (Zariski connected) reductive $\boldsymbol{C}$-subgroup of $GL_n(\boldsymbol{C})$. Any holomorphic representation of $\tilde{G}$ is a $\boldsymbol{C}$-homomorphism with respect to this structure of $\boldsymbol{C}$-group; in particular, the $\boldsymbol{C}$-group structure of $\tilde{G}$ is uniquely determined. Moreover, there exists a compact $\boldsymbol{R}$-form U of $\tilde{G}$ which is a maximal compact subgroup of $\tilde{G}$, and every maximal compact subgroup of $\tilde{G}$ is conjugate to U.*

Proof. Let $\tilde{\mathfrak{g}}^a$ (resp. $\tilde{\mathfrak{g}}^s$) be the abelian (resp. semi-simple) part of $\tilde{\mathfrak{g}}=\text{Lie }\tilde{G}$. By the assumption, one has $\tilde{G}^a\cong(\boldsymbol{C}^\times)^l$, so that there exists a unique maximal compact subgroup U^a of $\tilde{G}^a$, for which one has $U^a\cong(\boldsymbol{C}^{(1)})^l$. On the other hand, if $\mathfrak{u}^s$ is a compact real form of $\tilde{\mathfrak{g}}^s$, the corresponding (real) analytic subgroup U^s of $\tilde{G}^s$ is compact (Weyl). Thus $U=U^aU^s$ is compact and $\mathfrak{u}=\text{Lie }U$ is a real form of $\tilde{\mathfrak{g}}$. By Proposition 3. 3, U may be viewed as a reductive $\boldsymbol{R}$-group (in $GL_{2n}(\boldsymbol{R})$). We claim that $\tilde{G}$ can be identified with $U_{\boldsymbol{C}}$, or more precisely, $\tilde{G}$ is a $\boldsymbol{C}$-group in $GL_n(\boldsymbol{C})$ canonically $\boldsymbol{C}$-isomorphic to $U_{\boldsymbol{C}}$ (in $GL_{2n}(\boldsymbol{C})$).

To prove this, we regard the complex vector space $V=\boldsymbol{C}^n$, on which $\tilde{G}$ is acting, as a $2n$-dimensional real vector space with a "complex structure" I (i. e., $I\in GL(V)$, $I^2=-1_V$, cf. II, § 3) and let

$$V_{\boldsymbol{C}}=V\otimes_{\boldsymbol{R}}\boldsymbol{C}=V_+\oplus V_-,$$

where $V_\pm$ are the $(\pm i)$-eigenspaces of I in $V_{\boldsymbol{C}}$. Let $p_\pm$ denote the canonical projections $V_{\boldsymbol{C}}\to V_\pm$. Then, the restriction of p_+ (resp. p_-) on V gives a $\boldsymbol{C}$-linear (resp. $\boldsymbol{C}$-antilinear) isomorphism of complex vector spaces: $(V, \pm I)\cong V_\pm$. We set

$$GL(V_{\boldsymbol{C}}, I)=\{g\in GL(V_{\boldsymbol{C}})\,|\,gI=Ig\}.$$

Then the correspondence $g\mapsto(g|V_+, g|V_-)$ gives a $\boldsymbol{C}$-isomorphism $GL(V_{\boldsymbol{C}}, I)\cong GL(V_+)\times GL(V_-)$, and $GL(V, I)=GL(V_{\boldsymbol{C}}, I)\cap GL(V)$ is the general linear group of the original complex vector space (V, I). We denote by ι and π the natural injection $GL(V, I)\to GL(V_{\boldsymbol{C}}, I)$ and the natural projection $GL(V_{\boldsymbol{C}}, I)\to GL(V_+)$, respectively. Then $\pi\circ\iota: GL(V, I)\to GL(V_+)$ is a $\boldsymbol{C}$-isomorphism, where $GL(V, I)$ is viewed as a $\boldsymbol{C}$-group.

Now, U is a (Zariski connected) reductive $\boldsymbol{R}$-subgroup of $GL(V, I)$ (viewed as an $\boldsymbol{R}$-group in $GL(V)\cong GL_{2n}(\boldsymbol{R})$), so that $U_{\boldsymbol{C}}$ is a (Zariski connected) reductive $\boldsymbol{C}$-subgroup of $GL(V_{\boldsymbol{C}}, I)$ (in $GL(V_{\boldsymbol{C}})\cong GL_{2n}(\boldsymbol{C})$). First we want to show that the restriction map π induces a $\boldsymbol{C}$-isomorphism of $U_{\boldsymbol{C}}$ onto its image in $GL(V_+)$. It suffices to show that $\pi|U_{\boldsymbol{C}}$ is injective. If $X, Y\in\mathfrak{u}$, one has $X+iY|V_+=0\Rightarrow X+IY=0\Rightarrow X=Y=0$, because $\mathfrak{u}$ is a real form of $\tilde{\mathfrak{g}}$. Hence the kernel of $\pi|U_{\boldsymbol{C}}$ is a finite central subgroup of $U_{\boldsymbol{C}}$. But, since U is a maximal compact subgroup of $U_{\boldsymbol{C}}$ (as remarked after Corollary 3. 4), this kernel should be contained in U, and hence reduces to $\{1\}$, as desired, for $\pi|U$ is injective. On the other hand, it is clear that on the Lie algebra level one has $\pi(\mathfrak{u}_{\boldsymbol{C}})=\tilde{\mathfrak{g}}|V_+=\pi\iota(\tilde{\mathfrak{g}})$. Hence $\pi(U_{\boldsymbol{C}})=\pi\iota(\tilde{G})$. Thus $\pi\iota$ induces an isomorphism of $\tilde{G}$ onto $\pi(U_{\boldsymbol{C}})$, which proves our assertion. The remaining part of the Proposition follows immediately from Proposition 3. 3 and the remark following Corollary 3. 4, q. e. d.

Proposition 3. 6. *Let G_0 be an (analytically) reductive real analytic subgroup of $GL_n(\boldsymbol{R})$ and suppose that the abelian part G_0^a is compact. Then, there exists a (Zariski connected) reductive $\boldsymbol{R}$-group G in $GL_n(\boldsymbol{R})$ such that the identity connected component of G (in the usual topology) coincides with G_0.*

Proof. Let $\tilde{G}$ be the complex analytic subgroup of $GL_n(\boldsymbol{C})$ corresponding to the complexification $\tilde{\mathfrak{g}}=\mathfrak{g}_C$. Then $\tilde{G}$ is (analytically) reductive and the abelian part $\tilde{G}^a$ is isomorphic to a $\boldsymbol{C}$-torus. Hence, by Proposition 3. 5, $\tilde{G}$ has a (uniquely determined) $\boldsymbol{C}$-group structure. Since $\mathfrak{g}$ is an $\boldsymbol{R}$-form of $\tilde{\mathfrak{g}}=\mathrm{Lie}\,\tilde{G}$, $G=\tilde{G}\cap GL_n(\boldsymbol{R})$ is an $\boldsymbol{R}$-form of $\tilde{G}$ (which is Zariski connected), and the identity connected component of G coincides with the (real) analytic subgroup corresponding to $\mathfrak{g}$, i. e., G_0, q. e. d.

Remark. Without the compactness assumption on G_0^a, Proposition 3. 6 is false. For instance, let $G_0=\boldsymbol{R}_+^\times\subset GL_1(\boldsymbol{R})=\boldsymbol{R}^\times$. Then, for $\alpha\in\boldsymbol{R}$, the analytic endomorphism $e^t\mapsto e^{\alpha t}$ of G_0 is extendible to an $\boldsymbol{R}$-endomorphism of $\boldsymbol{R}^\times$ if and only if α is an integer. It follows that, for an irrational α, the analytic group $G_0'=\left\{\begin{pmatrix} e^t & 0 \\ 0 & e^{\alpha t}\end{pmatrix}(t\in\boldsymbol{R})\right\}\subset GL_2(\boldsymbol{R})$ can not be the identity connected component of any $\boldsymbol{R}$-group in $GL_2(\boldsymbol{R})$, although G_0' is fully reducible and closed. Even under the assumption of Proposition 3. 6, the analogue of Proposition 3. 3 or 3. 5 concerning continuous representations is not true in general. [For instance, $ad: SL_3(\boldsymbol{R})\to\mathrm{Ad}(\mathfrak{sl}_3(\boldsymbol{R}))$ is an analytic isomorphism, but not an $\boldsymbol{R}$-isomorphism.] However, in the notation of the above proof, when $\tilde{G}^s(=G_C^s)$ is simply connected, one can see easily that any continuous representation of G_0 can uniquely be extended to an $\boldsymbol{R}$-representation of G. Thus the $\boldsymbol{R}$-isomorphism class of G for which G_C^s is simply connected is uniquely determined by the analytic isomorphism class of G_0. (In general, the analytic isomorphism class of G_0 determines only the bijective $\boldsymbol{R}$-isogeny class of G.)

§ 4. Cartan involutions of reductive $\boldsymbol{R}$-groups.

Let $\mathfrak{g}$ be a Lie algebra over F and let θ be an "involution" of $\mathfrak{g}$ (i. e., $\theta\in\mathrm{Aut}\,\mathfrak{g}$, $\theta^2=id$). We set

(4. 1) $$\mathfrak{k}=\mathfrak{g}(\theta\,;1),\qquad \mathfrak{p}=\mathfrak{g}(\theta\,;-1),$$

where $\mathfrak{g}(\theta\,;\pm1)=\{X\in\mathfrak{g}\,|\,\theta X=\pm X\}$. Then it is clear that

(4. 2) $$\begin{aligned} &\mathfrak{g}=\mathfrak{k}+\mathfrak{p}, \quad \mathfrak{k}\cap\mathfrak{p}=\{0\},\\ &[\mathfrak{k},\mathfrak{k}]\subset\mathfrak{k},\quad [\mathfrak{k},\mathfrak{p}]\subset\mathfrak{p},\quad [\mathfrak{p},\mathfrak{p}]\subset\mathfrak{k}.\end{aligned}$$

It follows that $\mathfrak{k}$ and $\mathfrak{p}$ are orthogonal with respect to the Killing form $B^{\mathfrak{g}}$. (If $\mathfrak{g}$ is semi-simple, one has $\mathfrak{p}=\mathfrak{k}^\perp$.) Conversely, the decomposition of $\mathfrak{g}$ satisfying the condition (4. 2) determines an involution θ by the relation (4. 1).

Now let G be a Zariski connected,* reductive $\boldsymbol{R}$-group and let $\mathfrak{g}=\mathrm{Lie}\,G$. An involution θ of $\mathfrak{g}$ is called a *Cartan involution* of $\mathfrak{g}$ (relative to G), if $\mathfrak{u}=\mathfrak{k}+i\mathfrak{p}$ is a "compact" real form of $\mathfrak{g}_C$ (relative to G_C), i. e., if the (real) analytic subgroup U of G_C corresponding to $\mathfrak{u}$ is compact. (By the proof of Proposition 3. 5, U is then a compact real form of G_C.) If we set

(4. 3) $$\mathfrak{k}^{(a)}=\mathfrak{k}\cap\mathfrak{g}^a,\quad \mathfrak{k}^{(s)}=\mathfrak{k}\cap\mathfrak{g}^s,\quad \mathfrak{p}^{(a)}=\mathfrak{p}\cap\mathfrak{g}^a,\quad \mathfrak{p}^{(s)}=\mathfrak{p}\cap\mathfrak{g}^s,$$

then it is easy to see that the above condition on $\mathfrak{u}$ is equivalent to the following two conditions:

(i) $\mathfrak{k}^{(a)}=\mathrm{Lie}(G^a)^c$, $\mathfrak{p}^{(a)}=\mathrm{Lie}\,A$, where $(G^a)^c$ and A denote the $\boldsymbol{R}$-compact and $\boldsymbol{R}$-split parts of the $\boldsymbol{R}$-torus G^a (see § 2), and

(ii) $\mathfrak{u}^s=\mathfrak{k}^{(s)}+i\mathfrak{p}^{(s)}$ is a compact real form of $\mathfrak{g}_C^s$, i. e., $B^{\mathfrak{g}^s}(X,\theta Y)$ $(X, Y\in\mathfrak{g}^s)$ is negative definite.

(In particular, if $\mathfrak{g}$ is semi-simple, the above definition of Cartan involution is independent of the choice of $\boldsymbol{R}$-group G.) The decomposition $\mathfrak{g}=\mathfrak{k}+\mathfrak{p}$ corresponding to a Cartan involution is called a *Cartan decomposition.*

Example 1. For $\mathfrak{g}=\mathfrak{gl}_n(F)$ ($F=\boldsymbol{R}$ or $\boldsymbol{C}$), the "standard" Cartan involution is given by $X\mapsto -{}^t\bar{X}$, for which one has $\mathfrak{p}=\mathscr{H}_n(F)$.

Example 2. In the above notation, the complex conjugation of $\mathfrak{g}_C$ relative to the compact real form $\mathfrak{u}$ is a Cartan involution of $\mathfrak{g}_C$ (viewed as a real Lie algebra), and the corresponding Cartan decomposition is given by

$$\mathfrak{g}_C = \mathfrak{u}+i\mathfrak{u}.$$

Now, suppose there is given a Cartan involution θ of $\mathfrak{g}$ (relative to G), and let σ_0 and σ denote the complex conjugations of G_C with respect to the $\boldsymbol{R}$-forms G and U. Then clearly one has $\sigma_0\sigma=\sigma\sigma_0$. Hence $\sigma_0\sigma$ is an involutive $\boldsymbol{C}$-automorphism of G_C, leaving G and U stable, and $\sigma_0\sigma|G=\sigma|G$ is an involutive $\boldsymbol{R}$-automorphism of G, whose differential is equal to θ. Thus a Cartan involution θ of $\mathfrak{g}$ (relative to G) is always liftable to an $\boldsymbol{R}$-automorphism of G, called a (global) "Cartan involution" of G and denoted by the same letter θ. In the following, we distinguish the actions of θ on G and $\mathfrak{g}$ by writing them in the form g^θ ($g\in G$) and θX ($X\in\mathfrak{g}$), respectively. A similar convention will also apply to the complex conjugations σ_0 and σ. It is clear that, conversely, if we have a compact $\boldsymbol{R}$-form U of G_C with complex conjugation σ such that $\sigma_0\sigma=\sigma\sigma_0$, then $\theta=\sigma|G$ (or $\sigma|\mathfrak{g}$) is a Cartan involution of G (or $\mathfrak{g}$). The existence of Cartan involution will be shown in Corollary 4.3 below.

(When G_1 is an open subgroup of a not necessarily Zariski connected reductive $\boldsymbol{R}$-group G, an involutive analytic automorphism of G_1 whose differential is a Cartan involution of $\mathfrak{g}$ (relative to G^z) will also be called a Cartan involution of G_1. For the treatment of this non-connected case, see Mostow [2].)

Lemma 4.1. *Let G be a Zariski connected reductive $\boldsymbol{R}$-group, and let θ_1 and θ_2 be two Cartan involutions of G. If $\theta_1\theta_2=\theta_2\theta_1$, then $\theta_1=\theta_2$.*

Proof. Let $\mathfrak{g}=\mathrm{Lie}\ G=\mathfrak{g}^a\oplus\mathfrak{g}^s$. Since one always has $\theta_1=\theta_2$ on $\mathfrak{g}^a$, it is enough to show that $\theta_1=\theta_2$ on $\mathfrak{g}^s$. Hence, for the proof, we may assume that $\mathfrak{g}$ is semi-simple. Let

$$\mathfrak{g} = \mathfrak{k}_1+\mathfrak{p}_1 = \mathfrak{k}_2+\mathfrak{p}_2$$

be the Cartan decompositions of $\mathfrak{g}$ $(=\mathfrak{g}^s)$ corresponding to θ_1 and θ_2. Then, since $\theta_1\theta_2=\theta_2\theta_1$, one has

$$\mathfrak{g} = \mathfrak{k}_1\cap\mathfrak{k}_2+\mathfrak{k}_1\cap\mathfrak{p}_2+\mathfrak{p}_1\cap\mathfrak{k}_2+\mathfrak{p}_1\cap\mathfrak{p}_2.$$

But the restriction of the Killing form B on $\mathfrak{k}_1 \times \mathfrak{k}_1$ (resp. $\mathfrak{p}_2 \times \mathfrak{p}_2$) is negative (resp. positive) definite, so that one has $\mathfrak{k}_1 \cap \mathfrak{p}_2 = \{0\}$. Similarly, $\mathfrak{p}_1 \cap \mathfrak{k}_2 = \{0\}$. Hence one has $\mathfrak{k}_1 = \mathfrak{k}_2$ and $\mathfrak{p}_1 = \mathfrak{p}_2$, i. e., $\theta_1 = \theta_2$, q. e. d.

Theorem 4. 2. *Let G' be a Zariski connected, reductive $\boldsymbol{R}$-group and G a Zariski connected $\boldsymbol{R}$-subgroup of G'.*

(i) *If there exists a Cartan involution θ' of G' leaving G stable, then G is reductive and $\theta = \theta' | G$ is a Cartan involution of G.*

(ii) *If G is reductive, every Cartan involution θ of G is extendible to a Cartan involution of G'.*

(iii) *If θ_1' and θ_2' are two Cartan involutions of G' extending one and the same Cartan involution θ of G, there exists an element a' in $C_{G'}(G)$ (the centralizer of G in G') such that $\theta_2' = \nu_{a'} \theta_1' \nu_{a'}^{-1}$, where $\nu_{a'}$ denotes the inner automorphism of G' defined by a': $\nu_{a'}(g') = a' g' a'^{-1}$ $(g' \in G')$.*

Proof. (i) Suppose there is a Cartan involution θ' of G' leaving G stable. Let U' be the corresponding compact $\boldsymbol{R}$-form of G_C' and let σ_0', σ' be the complex conjugations of G_C' with respect to G' and U'. Then $\sigma_0' \sigma' = \sigma' \sigma_0'$ and $\sigma' | G' = \theta'$. Since θ' leaves G and $\mathfrak{g}$ stable, σ' leaves $\mathfrak{g}_C$ and G_C stable, and the same is trivially true for σ_0'. Hence, putting $\sigma_0 = \sigma_0' | G_C$, $\sigma = \sigma' | G_C$, we obtain complex conjugations of G_C with respect to the $\boldsymbol{R}$-forms G and $U = U' \cap G_C$. Since U is compact, and hence reductive, so are G_C and G. Since $\sigma_0 \sigma = \sigma \sigma_0$, $\sigma | G = \theta' | G = \theta$ is a Cartan involution of G.

(ii) Suppose G is reductive and θ is a Cartan involution of G. Let U be the corresponding compact $\boldsymbol{R}$-form of G_C and let σ_0 and σ be the complex conjugations of G_C with respect to G and U. On the other hand, let U' be a compact $\boldsymbol{R}$-form of G_C' and let σ_0' and σ' be the complex conjugations of G_C' with respect to G' and U'. By the conjugacy of maximal compact subgroups of G_C', we may assume without loss of generality that $U \subset U'$. Then, both σ_0' and σ' leave G stable. In fact, it is clear that $\sigma_0' | G = id$, $\sigma' | U = id$. Hence $\sigma' | \mathfrak{g} = \sigma | \mathfrak{g} = \theta$, and so $\sigma' | G = \theta$. Now, both $\sigma_0' \sigma'$ and $\sigma' \sigma_0'$ are $\boldsymbol{C}$-automorphisms of G_C' leaving $G_C'^a$ and $G_C'^s$ stable. It is clear that one has $\sigma_0' \sigma' = \sigma' \sigma_0'$ on $G_C'^a$. Actually, the restriction of $\sigma_0' \sigma' = \sigma' \sigma_0'$ to G'^a is the unique Cartan involution of G'^a. If $\sigma_0' \sigma' = \sigma' \sigma_0'$ holds also on $G_C'^s$, then σ' leaves G' stable and $\theta' = \sigma' | G'$ is a Cartan involution of G' extending θ. We shall now show that σ' can always be modified so that this condition is satisfied.

We denote by $\sigma_0'^{(s)}$, $\sigma'^{(s)}$ the restriction of σ_0', σ' on $G_C'^s$ or $\mathfrak{g}_C'^s$. Put $\mathscr{G}_1 = \mathrm{Aut}(\mathfrak{g}_C'^s)$, which is a $\boldsymbol{C}$-group in $GL(\mathfrak{g}_C'^s)$. We define a positive definite hermitian form h on $\mathfrak{g}_C'^s$ by

$$h(X, Y) = -B'(\sigma'^{(s)} X, Y) \qquad (X, Y \in \mathfrak{g}_C'^s),$$

where B' is the Killing form of $\mathfrak{g}_C'^s$, and denote by $*$ the adjoint with respect to h. Then, since the Killing form is invariant under any automorphism, one has for $\varphi \in \mathscr{G}_1$

$$\varphi^* = \sigma'^{(s)}\varphi^{-1}\sigma'^{(s)} \in \mathcal{G}_1.$$

In particular, if we put $\varphi_1 = \sigma'^{(s)}\sigma_0'^{(s)}$ (viewed as an element of $\mathcal{G}_1$), we have $\varphi_1^* = \varphi_1$. It follows from Lemma 3. 1 that $X = \log \varphi_1^2 \in \operatorname{Lie} \mathcal{G}_1 = ad(\mathfrak{g}_C'^s)$. Then one has

$$\varphi_1 = u \exp\frac{X}{2},$$

with $u \in \mathcal{G}_1$, $u^* = u^{-1} = u$, and $uX = Xu$. Since $X^* = X$, one can write $X = ad\, x_1$ with $x_1 \in i\mathfrak{u}'^s$. Then, since $\varphi_1 X \varphi_1^{-1} = ad(\varphi_1 x_1) = X$, one has $\varphi_1 x_1 = x_1$. Thus one has

(*) $$\sigma_0' x_1 = \sigma' x_1 = -x_1.$$

Let $g_1 = \exp x_1 \in G_C'^s$. Then, by the canonical homomorphism $G_C'^s \to \operatorname{Ad}(\mathfrak{g}_C'^s) \subset \mathcal{G}_1$, g_1 goes to $\exp X$. On the other hand, we have a natural inclusion map $\operatorname{Aut}(G_C'^s) \to \operatorname{Aut}(\mathfrak{g}_C'^s) = \mathcal{G}_1$, by which the inner automorphism $\nu_{g_1}^{(s)}$ of $G_C'^s$ defined by g_1 is identified with $\exp X$. Therefore, in $\operatorname{Aut}(G_C'^s)$, one has

$$(\sigma'^{(s)}\sigma_0'^{(s)})^2 = \exp X = \nu_{g_1}^{(s)},$$

which means that $(\sigma'\sigma_0')^2 = \nu_{g_1}$ on $G_C'^s$. But, it is clear that $(\sigma'\sigma_0')^2 = \nu_{g_1} = id$ on $G_C'^a$. Hence one has

(**) $$(\sigma'\sigma_0')^2 = \nu_{g_1}.$$

Since $(\sigma'\sigma_0')^2 | G_C = id$, one has $g_1 \in C_{G'_C}(G_C)$. If we take a matrix realization $G_C' \subset GL_n(\boldsymbol{C})$ in such a way that $U' \subset U_n(\boldsymbol{C})$, then x_1 is expressed by a hermitian matrix. Hence, by Lemma 3. 1, we have $x_1 \in \operatorname{Lie}(C_{G'_C}(G_C))$, i. e.,

(***) $$[x_1, \mathfrak{g}_C] = 0.$$

Now, put

$$\sigma'' = \sigma'\nu_{g_1^{1/2}} = \nu_{g_1^{1/2}}^{-1}\sigma' = \nu_{g_1^{1/4}}^{-1}\sigma'\nu_{g_1^{1/4}},$$

[where by definition $g_1^\lambda = \exp(\lambda x_1)$ for $\lambda \in \boldsymbol{R}$]. Then σ'' is a complex conjugation of G_C' with respect to a compact $\boldsymbol{R}$-form $g_1^{-1/4}U'g_1^{1/4}$, and by (*), (**), (***) one has $\sigma_0'\sigma'' = \sigma''\sigma_0'$, $\sigma''|G = \sigma'|G = \theta$. Hence $\theta' = \sigma''|G'$ is a Cartan involution of G' extending θ.

(iii) Let θ_1' and θ_2' be two Cartan involutions of G' extending the same Cartan involution θ of G. Let U_1' and U_2' be the corresponding compact $\boldsymbol{R}$-forms of G_C', and σ_1' and σ_2' the complex conjugations of G_C' with respect to U_1' and U_2'. Then one has $\sigma_0'\sigma_i' = \sigma_i'\sigma_0'$ and $\sigma_i'|G' = \theta_i'$ $(i = 1, 2)$. Applying the same argument as in the proof of (ii) to σ_1' and σ_2' instead of σ' and σ_0', we see that there exists an element x_2 in $\mathfrak{g}_C'^s$ satisfying the following three conditions:

$$\sigma_1' x_2 = \sigma_2' x_2 = -x_2,$$
$$(\sigma_1'\sigma_2')^2 = \nu_{g_2} \qquad (g_2 = \exp x_2 \in G_C'^s),$$
$$[x_2, \mathfrak{g}_C] = 0.$$

Since σ_1' and σ_2' commute with σ_0', so does ν_{g_2}. This means that $g_2^{\sigma_0'}g_2^{-1}$ is in the center of $G_C'^s$, which is finite. But, for a suitable matrix realization, one may assume that both g_2 and $g_2^{\sigma_0'}$ are positive definite hermitian matrices. Hence one can conclude that $g_2^{\sigma_0'} = g_2$ and so $g_2 \in G'^s$, $x_2 \in \mathfrak{g}'^s$. Putting $\sigma_1'' = \sigma_1'\nu_{g_2^{1/2}}$, one obtains a complex conjugation of G_C' with respect to $U_1'' = g_2^{-1/4}U_1'g_2^{1/4}$ such that σ_0', σ_1'', and σ_2' commute with one another. Therefore, by Lemma 4. 1, one has $\sigma_2'|G=\sigma_1''|G$ and so $\sigma_2'=\sigma_1''=\nu_{g_2^{-1/4}}\sigma'\nu_{g_2^{1/4}}$. Thus $\theta_2'=\nu_{a'}\theta_1'\nu_{a'}^{-1}$ where $a'=g_2^{-1/4}\in C_{G'}(G)$. (This proof shows that a' can be chosen so that $a'=\exp x'$ with $x'\in \mathfrak{g}'^{s}$, $\theta_1'x'=\theta_2'x'=-x'$, $[x', \mathfrak{g}]=0$.) q. e. d.

Corollary 4. 3. *Any Zariski connected reductive $\boldsymbol{R}$-group G (and hence also $\mathfrak{g}=\mathrm{Lie}\ G$) has a Cartan involution. Two Cartan involutions of G are conjugate to each other by an inner automorphism of G.*

This follows from Theorem 4. 2 (ii), (iii) applied to the case $G=\{1\}$.

Corollary 4. 4 (*Mostow*). *Let G be a Zariski connected $\boldsymbol{R}$-group in $GL_n(\boldsymbol{R})$. Then G is reductive if and only if one has ${}^tG=G$ for a suitable choice of basis and, when this is the case, the map $g\mapsto {}^tg^{-1}$ $(g\in G)$ is a Cartan involution of G. All Cartan involutions θ of a reductive $\boldsymbol{R}$-group G in $GL_n(\boldsymbol{R})$ can be obtained in this way (for suitable bases of $\boldsymbol{R}^n$).*

This follows from Theorem 4. 2 applied to the case $G'=GL_n(\boldsymbol{R})$. One can obtain a similar result for $\boldsymbol{R}$-groups in $GL_n(\boldsymbol{C})$ by replacing ${}^tg$ by ${}^t\bar{g}$. In fact, by Theorem 4. 2 (iii), any Cartan involution θ of $GL_n(F)$ $(F=\boldsymbol{R}, \boldsymbol{C})$ is of the form $g^\theta=a{}^t\bar{g}^{-1}a^{-1}$ with $a\in\mathcal{P}_n(F)$ and so, after a change of coordinates $x\mapsto a^{1/2}x$ $(x\in F^n)$, θ is transformed into the standard form $\theta: g\mapsto {}^t\bar{g}^{-1}$. [If $\mathfrak{gl}_n(F)=\mathcal{M}_n(F)$ is viewed as an associative algebra over F, then $-\theta$ is a "positive involution" in the sense of Appendix, § 3.]

Corollary 4. 5. *Let G be a Zariski connected reductive $\boldsymbol{R}$-group in $GL_n(\boldsymbol{R})$ with Cartan involution θ and set*

$$K=\{g\in G\,|\,g^\theta=g\}. \tag{4. 4}$$

Then, K is a maximal compact subgroup of G and every maximal compact subgroup of G is conjugate to K. Every element g in G can be written uniquely in the form

$$g=k\cdot\exp X \quad \textit{with} \quad k\in K,\ X\in\mathfrak{p}, \tag{4. 5}$$

and the natural map $(k, X)\mapsto k\cdot\exp X$ gives a homeomorphism $K\times\mathfrak{p}\to G$. [The topological decomposition $G=KP$, $P=\exp\mathfrak{p}$, is called a "global Cartan decomposition" of G corresponding to the Cartan involution θ. E. g., $GL_n(\boldsymbol{C})=U_n(\boldsymbol{C})\cdot\mathcal{P}_n(\boldsymbol{C})$.]

Proof. Let K_1 be any compact subgroup of G. Then K_1 is a reductive $\boldsymbol{R}$-subgroup of G with id as a Cartan involution. By a similar argument as in the proof of Theorem 4. 2 (ii), it can be shown that there exists a Cartan involution θ_1 of G such that $\theta_1|K_1=id$. Then, by Theorem 4. 2 (iii), a conjugate of K_1 is contained in K. Therefore K is maximal compact and every maximal compact subgroup is conjugate to K. For the proof of the remaining part, we may assume by Corollary 4. 4 that θ is the standard Cartan involution $g\mapsto {}^tg^{-1}$. Then our assertion follows from Corollary 3. 2, q. e. d.

Remark. It follows from Corollary 4. 5 that K is connected (either in Zariski topology or in the usual topology) if and only if G is connected in the usual topology. In general, the number of connected components of G (in the usual topology) is equal to that of K (in either sense).

§ 5. Relative roots and parabolic subgroups.

In this section, we summarize basic facts on parabolic subgroups which play an important role in the theory of reductive algebraic groups.

Let G be a Zariski connected F-group with Lie algebra $\mathfrak{g}$. A *maximal F-split torus* A in G is a maximal element in the set of all F-split tori contained in G. It is known that all maximal F-split tori in G are mutually conjugate, so that their dimensions are equal. The common dimension of maximal F-split tori in G is called the *(F-)rank* of G (or of $\mathfrak{g}$) and denoted by F-rank G.

Let A be a maximal F-split torus in G. We denote the F-character module of A by $\boldsymbol{Y}$ and the Lie algebra of A by $\mathfrak{a}$. As explained in § 2, $\boldsymbol{Y}$ can naturally be embedded in the dual space $\mathfrak{a}^*$ of $\mathfrak{a}$. For $\chi \in \boldsymbol{Y}$, we put

$$\mathfrak{g}_\chi = \{x \in \mathfrak{g} \mid (ad\ a)x = \chi(a)x \text{ for } a \in A\}$$

and define the *F-root system*

$$\mathfrak{r} = \{\chi \in \boldsymbol{Y} \mid \chi \neq 0,\ \mathfrak{g}_\chi \neq \{0\}\}.$$

Then, as the restriction to A of the adjoint representation of G is fully reducible, one has a direct sum decomposition

$$\mathfrak{g} = \mathfrak{c}(\mathfrak{a}) + \sum_{\alpha \in \mathfrak{r}} \mathfrak{g}_\alpha, \tag{5. 1}$$

where $\mathfrak{c}(\mathfrak{a})$ denotes the centralizer of $\mathfrak{a}$ in $\mathfrak{g}$. $\alpha \in \mathfrak{r}$ is called an *F-root* of G (or $\mathfrak{g}$) relative to A (or $\mathfrak{a}$). Through the natural inclusion $\boldsymbol{Y} \subset \mathfrak{a}^*$, an F-root α is often identified with the corresponding linear functional $d\alpha/\alpha$ on $\mathfrak{a}$.

Now assume that G is reductive. Then it is known that $\mathfrak{r}$ has the property of a *root system in a wider sense*, i. e.,

(R 1) $0 \notin \mathfrak{r}$; and $\alpha \in \mathfrak{r}$ implies $-\alpha \in \mathfrak{r}$.

(R 2) There exists a (positive definite) inner product $\langle\ \rangle$ on $\boldsymbol{Y}_{\boldsymbol{Q}}$ such that for any $\alpha, \beta \in \mathfrak{r}$ one has $c_{\alpha\beta} = 2\langle\alpha, \beta\rangle / \langle\alpha, \alpha\rangle \in \boldsymbol{Z}$ and

$$s_\alpha \beta = \beta - c_{\alpha\beta}\alpha \in \mathfrak{r},$$

s_α denoting the symmetry of $\boldsymbol{Y}_{\boldsymbol{Q}}$ with respect to α. (Actually, the $c_{\alpha\beta}$ and hence the s_α are determined independently of the choice of the inner product.)

The (finite) subgroup W of Aut($\boldsymbol{Y}$) generated by $\{s_\alpha (\alpha \in \mathfrak{r})\}$ is called the *Weyl group* of $\mathfrak{r}$. By the duality of a torus, W can naturally be identified with a subgroup of Aut(A). Let $N(A)$ (resp. $C(A)$) denote the normalizer (resp. the centralizer) of A in G. Then it is known that $C(A)$ coincides with the Zariski connected component of $N(A)$ and the factor group $N(A)/C(A)$ is canonically isomorphic to the Weyl group $W(\subset \mathrm{Aut}(A))$. We recall that a root system *in the ordinary sense* is the one which is "reduced", i. e., the one satisfying the following additional condition:

(R 1a) If $\alpha \in \mathfrak{r}$ and $m\alpha \in \mathfrak{r}$ with $m \in \boldsymbol{Z}$, then $m = \pm 1$.
According to the classification theory of root systems, the only "irreducible" root systems in a wider sense, which are not reduced, are of type (BC_r) (cf. II, § 4).

If G^a (resp. G^s) denotes the abelian (resp. semi-simple) part of G, then A is an almost direct product of $A^{(a)}$ and $A^{(s)}$ where $A^{(a)}$ (resp. $A^{(s)}$) denotes a maximal F-split torus of G^a (resp. G^s). The subspace of $\boldsymbol{Y}_{\boldsymbol{Q}}$ generated by $\mathfrak{r}$ coincides with the annihilator of $A^{(a)}$ in $\boldsymbol{Y}_{\boldsymbol{Q}}$, so that $\mathfrak{r}$ can be identified with the root system of G^s relative to $A^{(s)}$.

We introduce a linear order (compatible with the vector operations) in $\boldsymbol{Y}_{\boldsymbol{Q}}$, and define the notion of "positive" roots in the usual manner. A positive root α is called *simple* if it can not be written as a sum of two positive roots. The set of all simple roots Δ is called the *fundamental system* of $\mathfrak{r}$. It is known that, if $r =$ F-rank $G^s (= \dim A^{(s)})$, then Δ consists of r linearly independent roots $\gamma_1, \cdots, \gamma_r$ and every F-root $\alpha \in \mathfrak{r}$ can be expressed uniquely in the form

$$\alpha = m_1\gamma_1 + \cdots + m_r\gamma_r,$$

where m_i's are all ≥ 0 or all ≤ 0, according as α is positive or negative. Moreover, all fundamental systems of $\mathfrak{r}$ are mutually conjugate with respect to the Weyl group W. Therefore, the pair (A, Δ) is uniquely determined up to an inner automorphism of G.

For a fixed pair (A, Δ), we put

(5.2) $$\mathfrak{n} = \sum_{\substack{\alpha \in \mathfrak{r} \\ \alpha > 0}} \mathfrak{g}_\alpha.$$

Then $\mathfrak{n}$ is the Lie algebra of a maximal unipotent F-subgroup N of G. The semi-direct product $B = AN$ is called an *F-Borel subgroup* of G. If $\mathcal{G}$ is the (reductive) algebraic group associated with G (i. e., $\mathcal{G} = G_{\boldsymbol{F}}$), then an (absolute) *Borel subgroup* (i. e., an $\boldsymbol{F}$-Borel subgroup in the above sense) of $\mathcal{G}$ is a maximal Zariski connected solvable algebraic subgroup of $\mathcal{G}$. It is known that all Borel subgroups of $\mathcal{G}$ are mutually conjugate. An F-subgroup H of G is called *parabolic* if $\mathcal{H} = H_{\boldsymbol{F}}$ contains a Borel subgroup of $\mathcal{G}$, or equivalently, if $\mathcal{G}/\mathcal{H}$ is projective. It is known that any parabolic subgroup H coincides with its own normalizer : $N(H) = H$, which implies, in particular, that H is Zariski connected. Standard parabolic subgroups of G are obtained in the following manner. For the fixed pair (A, Δ), let Γ be any subset of Δ, and let $(\Gamma)_{\boldsymbol{Q}}$ be the subspace of $\boldsymbol{Y}_{\boldsymbol{Q}}$ spanned by Γ. Let A_Γ denote the annihilator of $(\Gamma)_{\boldsymbol{Q}} \cap \boldsymbol{Y}$ in A ; then $\mathfrak{a}_\Gamma = \operatorname{Lie} A_\Gamma$ is the annihilator of $(\Gamma)_F (\subset \mathfrak{a}^*)$ in $\mathfrak{a}$. We put

$$\mathfrak{r}_\Gamma = \mathfrak{r} \cap (\Gamma)_{\boldsymbol{Z}} = \{\alpha \in \mathfrak{r} \mid \alpha|\mathfrak{a}_\Gamma = 0\},$$

and

(5.3) $$\mathfrak{b}_\Gamma = \mathfrak{c}(\mathfrak{a}_\Gamma) + \sum_{\substack{\alpha > 0 \\ \alpha \notin \mathfrak{r}_\Gamma}} \mathfrak{g}_\alpha.$$

Then there is a (unique) parabolic subgroup B_Γ of G with $\operatorname{Lie} B_\Gamma = \mathfrak{b}_\Gamma$. We call B_Γ (resp. $\mathfrak{b}_\Gamma$) a *parabolic subgroup* (resp. *subalgebra*) *belonging to* (A, Δ) (or $(\mathfrak{a}, \Delta)$).

Clearly the reductive part of B_Γ is given by $C(A_\Gamma)$ and the F-root system of $C(A_\Gamma)$ relative to A can be identified with $\mathfrak{r}_\Gamma$. It is known that any parabolic subgroup of G is conjugate to one of the standard parabolic subgroups constructed above. Thus the set $\{B_\Gamma(\Gamma\subset\Delta)\}$ constitutes a full set of representatives of the conjugacy classes of parabolic subgroups of G. It is clear that, for two subsets Γ, Γ' of Δ, one has

$$\Gamma\subset\Gamma' \Longleftrightarrow \mathfrak{a}_\Gamma\supset\mathfrak{a}_{\Gamma'} \Longleftrightarrow \mathfrak{b}_\Gamma\subset\mathfrak{b}_{\Gamma'}. \tag{5.4}$$

In particular,

$$\mathfrak{b}_\phi = \mathfrak{c}(\mathfrak{a})+\mathfrak{n}$$

is a minimal parabolic subalgebra of $\mathfrak{g}$. From the theory of algebraic Lie algebras (Chevalley [2], [4]), it is easy to see that all Lie subalgebras of $\mathfrak{g}$ containing $\mathfrak{b}_\phi$ are algebraic and hence of the form $\mathfrak{b}_\Gamma$ for some $\Gamma\subset\Delta$.

It is known that any algebraic group defined over F contains an (absolute) maximal torus $\mathcal{T}$ defined over F. It follows that there is an F-torus T in $C(A)$ such that $\mathcal{T}=T_F$ is maximal in $C(A)_F$. Then T is (absolutely) maximal in G and A coincides with the F-split part of T. Let $\boldsymbol{X}$ be the (absolute) character module of $\mathcal{T}$ and $\boldsymbol{X}_0$ the annihilator of A in $\boldsymbol{X}$. Then one has an exact sequence

$$0\longrightarrow \boldsymbol{X}_0\longrightarrow \boldsymbol{X}\overset{\pi}{\longrightarrow} \boldsymbol{Y}\longrightarrow 0,$$

π denoting the restriction map: $\pi(\chi)=\chi|A$. If we denote by $\tilde{\mathfrak{r}}$ the (absolute) root system of $\mathcal{G}$ relative to $\mathcal{T}$, then from the definition it is clear that the F-root system $\mathfrak{r}$ of G relative to A is given by $\mathfrak{r}=\pi(\tilde{\mathfrak{r}})-\{0\}$. (In this sense, F-roots are also called the "restricted roots".) We denote by $\tilde{W}$ the Weyl group of $\tilde{\mathfrak{r}}$.

Given a linear order in $\boldsymbol{Y}_Q$, there exists a linear order in $\boldsymbol{X}_Q$ such that for $\chi\in\boldsymbol{X}_Q$ one has

$$\chi>0,\ \chi\notin\boldsymbol{X}_{0Q}\Longleftrightarrow\pi(\chi)>0.$$

Let $\tilde{\Delta}=\{\alpha_1,\cdots,\alpha_l\}$ be the fundamental system of $\tilde{\mathfrak{r}}$ with respect to such an order and put $\Delta_0=\tilde{\Delta}\cap\boldsymbol{X}_0$. Then Δ_0 is a fundamental system of $\mathfrak{r}_0=\tilde{\mathfrak{r}}\cap\boldsymbol{X}_0$ and one has $\Delta=\pi(\tilde{\Delta}-\Delta_0)$. For each $\sigma\in\mathrm{Gal}(\bar{F}/F)$, there exists a unique element w_σ in $\tilde{W}$ such that $\tilde{\Delta}^\sigma=w_\sigma\tilde{\Delta}$. One defines a new action of σ on $\tilde{\Delta}$ by $\alpha_i^{[\sigma]}=w_\sigma^{-1}\alpha_i^\sigma$. Then, for $\alpha_i,\ \alpha_j\in\tilde{\Delta}-\Delta_0$, one has $\pi(\alpha_i)=\pi(\alpha_j)$ if and only if $\alpha_i^{[\sigma]}=\alpha_j$ for some $\sigma\in\mathrm{Gal}(\bar{F}/F)$ (cf. Satake [10], pp. 71–72). It is convenient to illustrate this situation by a Dynkin diagram of $\tilde{\Delta}$ with additional marks ("Γ-diagram", loc. cit.), in which simple roots in Δ_0 (resp. $\tilde{\Delta}-\Delta_0$) are represented by black (resp. white) vertices and the action of $[\sigma]$ is indicated by an arrow. Examples of such diagrams for simple $\boldsymbol{R}$-groups of hermitian type will appear in III, § 4.

When $F=\boldsymbol{R}$ and $\boldsymbol{F}=\boldsymbol{C}$, if A is a maximal $\boldsymbol{R}$-split torus in a reductive $\boldsymbol{R}$-group G, one has $A\cong(\boldsymbol{R}^\times)^r$ ($r=\boldsymbol{R}$-rank G) and so $A^\circ\cong(\boldsymbol{R}_+^\times)^r$ is a vector group. Let T be a maximal $\boldsymbol{R}$-torus containing A. Then, by Theorem 4.2, there exists a global Cartan decomposition $G=KP$ such that

$$T=(T\cap K)\cdot(T\cap P). \tag{5.5}$$

Consequently one has $A^\circ = T \cap P$ and so $\mathfrak{a} \subset \mathfrak{p}$. (Conversely, for any Cartan decomposition such that $\mathfrak{a} \subset \mathfrak{p}$, one can find a maximal $\boldsymbol{R}$-torus T containing A and satisfying (5. 5).) Thus $\mathfrak{a} = \mathrm{Lie}\, A$ can be characterized as a maximal element in the set of abelian subalgebras contained in the $\mathfrak{p}$-part of some Cartan decomposition of $\mathfrak{g}$. When $\mathfrak{a}$ and $\mathfrak{p}$ are as above, every element in $\mathfrak{p}$ is conjugate to an element in $\mathfrak{a}$ under the adjoint action of K°. Thus one has $\mathfrak{p} = ad(K^\circ)\mathfrak{a}$ and so $G^\circ = K^\circ A^\circ K^\circ$. Moreover, it is known that every coset in $N(A)/C(A)$ has a representative in K°, so that one has

(5. 6) $$W \cong N(A) \cap K^\circ / C(A) \cap K^\circ.$$

A fundamental domain of W in $\mathfrak{a}$ is given by a (closed) "Weyl chamber"

$$\mathfrak{a}^+ = \{x \in \mathfrak{a} \mid \gamma_i(x) \geq 0 \text{ for all } \gamma_i \in \Delta\}.$$

Therefore we have $\mathfrak{p} = ad(K^\circ)\mathfrak{a}^+$, and so

(5. 7) $$P = \bigcup_{k \in K^\circ} k(\exp \mathfrak{a}^+)k^{-1},$$

(5. 8) $$G^\circ = K^\circ P = K^\circ(\exp \mathfrak{a}^+)K^\circ.$$

It is also known that one has

(5. 9) $$G = K \cdot A^\circ N$$

which is a topological direct product decomposition. The decomposition (5. 9) (which is also valid for any $\boldsymbol{R}$-group, not necessarily reductive) is called an *Iwasawa decomposition* of G (cf. Helgason [1], VI).

§ 6. The structure group of a (non-degenerate) JTS.

Let V be a finite-dimensional vector space over a field F of characteristic zero. When there is given an F-trilinear map $\{\ \}: V \times V \times V \to V$ satisfying the following two conditions (JT 1), (JT 2), V is called a *Jordan triple system* (abbreviated as JTS) over F:

(JT 1) $$\{x, y, z\} = \{z, y, x\},$$

(JT 2) $$\{a, b, \{x, y, z\}\} = \{\{a, b, x\}, y, z\} - \{x, \{b, a, y\}, z\} + \{x, y, \{a, b, z\}\}$$
for all $a, b, x, y, z \in V$.

For $x, y \in V$, we define $x \square y \in \mathrm{End}(V)$ by

(6. 1) $$(x \square y)z = \{x, y, z\}.$$

Then (JT 2) can be rewritten as

(JT 2′) $$[a \square b, x \square y] = ((a \square b)x) \square y - x \square ((b \square a)y).$$

It follows that $V \square V$, the subspace of $\mathrm{End}(V)$ spanned by $a \square b$ $(a, b \in V)$, is a Lie subalgebra of $\mathfrak{gl}(V)$.

One can define a JTS structure on any Jordan algebra. First let us recall some definitions and basic identities. A finite dimensional (non-associative) algebra A over F is called a *Jordan algebra* if the following two conditions are satisfied:

(J 1) $$xy = yx,$$

(J 2) $$x^2(xy) = x(x^2y) \qquad \text{for all} \quad x, y \in A.$$

For $a \in A$, we define $T_a \in \mathrm{End}(V)$ by $T_a x = ax$, where V is the underlying vector space of A.

Polarizing the identity (J 2), one obtains

$$(ab)(cd) + (bc)(ad) + (ca)(bd) = a((bc)d) + b((ca)d) + c((ab)d) \qquad (a, b, c, d \in A).$$

This can be rewritten as

(6. 2) $$[T_a, T_{bc}] + [T_b, T_{ca}] + [T_c, T_{ab}] = 0,$$

or

(6. 2′) $$T_{cd}T_b + T_{bc}T_d + T_{bd}T_c = T_{(bc)d} + T_b T_d T_c + T_c T_d T_b.$$

Interchanging b and d in (6. 2′) and subtracting, one obtains

$$T_{d(bc)} - T_{b(dc)} = [[T_d, T_b], T_c],$$

or, changing notation,

(6. 3) $$T_{a(bc) - b(ac)} = [[T_a, T_b], T_c].$$

This means that $[T_a, T_b]$ is a derivation of the Jordan algebra A. In general, a derivation of A which is a linear combination of $[T_a, T_b]$ $(a, b \in A)$ is called an "inner" derivation.

Now we define a trilinear product $\{\ \}$ on $V(=A)$ by

(6. 4) $$\{x, y, z\} = (xy)z + x(yz) - y(xz),$$

or equivalently, by

(6. 4′) $$x \square y = T_{xy} + [T_x, T_y].$$

Then (J 1) implies (JT 1). Since $D = [T_a, T_b]$ is a derivation of the Jordan algebra A, D is also a derivation for the triple product, i. e.,

$$D\{x, y, z\} = \{Dx, y, z\} + \{x, Dy, z\} + \{x, y, Dz\}.$$

Hence, to prove (JT 2), it suffices to prove the relation

$$T_a\{x, y, z\} = \{T_a x, y, z\} - \{x, T_a y, z\} + \{x, y, T_a z\},$$

which follows immediately from (6. 2) and (6. 3). Thus the Jordan algebra A becomes a JTS with the trilinear product (6. 4). When A has a unit element e, one can recover the Jordan product from $\{\ \}$ by

(6. 5) $$\{x, y, e\} = \{x, e, y\} = \{e, x, y\} = xy.$$

Returning to an (arbitrary) JTS $(V, \{\ \})$, we define the *trace form* by

(6. 6) $$\tau(x, y) = \mathrm{tr}(x \square y),$$

and from now on assume that $(V, \{\ \})$ is "non-degenerate" (or "semi-simple"), that is,

(JT 3) *τ is non-degenerate.*

Under this condition, we define the adjoint * with respect to τ by

$$\tau(Tx, y) = \tau(x, T^*y) \qquad \text{for} \quad T \in \mathrm{End}(V).$$

Then from (JT 2′) one has

$$\tau((a\square b)x, y) = \tau(x, (b\square a)y),$$

i. e.,

(6. 7) $$(a\square b)^* = b\square a.$$

It follows that

(6. 8) $$\tau(a, b) = \tau(b, a),$$

i. e., τ is symmetric, (which legitimates the use of the adjoint *).

For a JTS coming from a Jordan algebra A, one has by (6. 4′)

(6. 9) $$\tau(x, y) = \mathrm{tr}(T_{xy}).$$

Hence, from (6. 3) one obtains

$$\tau(ax, y) = \tau(x, ay),$$

i. e.,

(6. 10) $$T_a^* = T_a.$$

It is known (Albert [6], [7], Jacobson [7], Braun-Koecher [1]) that in the Jordan algebra case (JT 3) is equivalent to the "semi-simplicity" of A. By definition a Jordan algebra A is *semi-simple,* if the "radical" (the unique maximal solvable ideal) of A reduces to $\{0\}$, or equivalently, if $\{T_a (a \in A)\}$ is fully reducible and if $T_a = 0$ implies $a = 0$. As we shall show below (Lem. 6. 1), a semi-simple Jordan algebra A has always a unit element.

Example. The space $V = \mathcal{M}_{p,q}(F)$ of all $p \times q$ matrices with entries in F becomes a JTS with respect to the triple product

(6. 11 a) $$\{x, y, z\} = \frac{1}{2}(x{}^t yz + z{}^t yx).$$

The trace form is given by

$$\tau(x, y) = \frac{p+q}{2}\mathrm{tr}(x{}^t y),$$

which is non-degenerate. When $p = q$, V has another (non-degenerate) triple product

(6. 11 b) $$\{x, y, z\} = \frac{1}{2}(xyz + zyx)$$

which comes from a Jordan product

(6. 11 c) $$x \circ y = \frac{1}{2}(xy + yx).$$

Lemma 6. 1. *Let $(V, \{\ \})$ be a non-degenerate JTS and let (c_i), (c'_i) be mutually dual bases of V with respect to the trace form τ. Then one has*

(6. 12 a) $$\sum_i c'_i \Box c_i = 1_V.$$

In the Jordan algebra case, one has

(6. 12 b) $$\sum_i c'_i c_i = e \text{ (the unit element)},$$

(6. 12 c) $$\sum_i [T_{c'_i}, T_{c_i}] = 0.$$

Proof. One has by (6. 7)

$$\tau(x, (\sum c'_i \Box c_i) y) = \sum \tau((x \Box y) c_i, c'_i) = \mathrm{tr}(x \Box y) = \tau(x, y)$$

for all $x, y \in V$, which proves (6. 12 a). In the Jordan algebra case, one has by (6. 10)

$$\tau(x, (\sum c'_i \cdot c_i) y) = \sum \tau((xy) c_i, c'_i) = \mathrm{tr}(T_{xy}) = \tau(x, y),$$

which proves that $\sum c'_i c_i$ is the unit element. (6. 12 c) follows from (6. 4′) and (6. 12 a, b), q. e. d.

The *structure group* $\Gamma(V, \{\ \})$ is by definition the group of all invertible linear transformations g of V satisfying the condition

(6. 13) $$g\{x, y, z\} = \{gx, g^{*-1}y, gz\} \qquad (x, y, z \in V),$$

or equivalently,

(6. 13′) $$g(x \Box y) g^{-1} = (gx) \Box (g^{*-1}y) \qquad (x, y \in V).$$

$\Gamma(V, \{\ \})$ is an F-group in $GL(V)$. It is clear from (6. 7) and (6. 13′) that $\Gamma(V, \{\ \})$ is self-adjoint, i. e., closed under the transposition $g \mapsto g^*$.

Lemma 6. 2. *Let $g \in GL(V)$ and $h \in \mathrm{End}(V)$. Then, one has the relation*

(6. 14) $$g\{x, y, z\} = \{gx, hy, gz\} \qquad \text{for all } x, y, z \in V,$$

if and only if $g \in \Gamma(V, \{\ \})$ and $h = g^{*-1}$.

Proof. The "if" part is trivial. If (6. 14) holds, then one has $g(x \Box y) g^{-1} = (gx) \Box (hy)$, and so

$$\tau(gx, hy) = \tau(x, y) \qquad (x, y \in V).$$

Hence one has $h = g^{*-1}$ and $g \in \Gamma(V, \{\ \})$, q. e. d.

Corollary 6. 3. *$g \in GL(V)$ is an automorphism of the JTS $(V, \{\ \})$ (i. e., (6. 14) holds with $g = h$) if and only if $g \in \Gamma(V, \{\ \})$ and $g = g^{*-1}$.*

The Lie algebra of $\Gamma(V, \{\ \})$ is called the *structure algebra.* This algebra consists of all $T \in \mathfrak{gl}(V)$ satisfying the condition

$$(6.15)\qquad [T, x\square y] = (Tx)\square y - x\square (T^*y) \qquad (x, y \in V).$$

It is clear that $1_V \in \operatorname{Lie} \Gamma(V, \{\ \})$. From (JT 2′) and (6.7) we see that $a\square b \in \operatorname{Lie} \Gamma(V, \{\ \})$ for all $a, b \in V$. [Actually, we shall see in the next section that $\operatorname{Lie} \Gamma(V, \{\ \}) = V\square V$.] By Corollary 6.3 (and § 1, Exerc. 3), the Lie algebra of $\operatorname{Aut}(V, \{\ \})$ is given by

$$(6.16)\qquad \operatorname{Der}(V, \{\ \}) = \{T \in \operatorname{Lie} \Gamma(V, \{\ \}) \mid T^* = -T\}.$$

Exercises

In the following exercises, $(V, \{\ \})$ is a JTS, not necessarily satisfying the condition (JT 3). For $a \in V$, we define the "quadratic map" $P(a) \in \operatorname{End}(V)$ by $P(a)x = \{a, x, a\}$.

1. From (JT 1) and (JT 2) prove the following formulas for $a, x, y \in V$:

$$(6.17)\qquad P(a)(x\square a) = (a\square x)P(a),$$

$$(6.17')\qquad P(a)\{x, a, y\} = \{P(a)x, y, a\} = \{a, x, P(a)y\},$$

$$(6.18)\qquad (P(x)a)\square a = x\square (P(a)x) = 2(x\square a)^2 - P(x)P(a),$$

$$(6.18')\qquad 2\{x, a, y\}\square a = x\square (P(a)y) + y\square (P(a)x),$$

$$(6.19)\qquad (P(x)P(a)x)\square a = (P(x)a)\square (P(a)x) = x\square (P(a)P(x)a).$$

Hint. To obtain (6.17) and (6.18), compute $\{a, y, \{a, x, a\}\}$, $\{x, a, \{x, a, y\}\}$ and $\{y, a, \{x, a, x\}\}$.

Remark. The so-called "fundamental formula"

$$(6.20)\qquad P(P(a)x) = P(a)P(x)P(a)$$

can be derived from these formulas as follows (Meyberg [1]):

$$\begin{aligned} P(P(a)x)y &= 2\{a, x, \{a, y, P(a)x\}\} - P(a)\{x, P(a)x, y\} && \text{(by (JT 2))} \\ &= 2P(a)\{x, a, \{y, a, x\}\} - P(a)\{x, P(a)x, y\} && \text{(by (6.17))} \\ &= P(a)P(x)P(a)y \qquad \text{(by (6.18′))}. \end{aligned}$$

Conversely, if one defines the triple product $\{\ \}$ by

$$\{x, y, z\} = \frac{1}{2}(P(x+z) - P(x) - P(z))y,$$

then the condition (JT 2) follows from (6.17), (6.18) and (6.20) (cf. Loos [8]).

2. Assuming (JT 3), show that $P(a)^* = P(a)$.

3. $a \in V$ is called *invertible* if $P(a)$ is non-singular; in that case, we put $a^{-1} = P(a)^{-1}a$. Prove the following relations:

$$(6.21)\qquad \begin{gathered} a\square a^{-1} = a^{-1}\square a = 1_V, \\ P(a^{-1}) = P(a)^{-1}, \qquad (a^{-1})^{-1} = a. \end{gathered}$$

In the case of semi-simple Jordan algebra, one has $a \cdot a^{-1} = e$ and $[T_a, T_{a^{-1}}] = 0$.

Remark. A Jordan algebra is "power-associative" in the sense that, if one defines x^n inductively by $x^n = x^{n-1} \cdot x$, then the rule $x^m x^n = x^{m+n}$ holds for all positive integers m, n. For a Jordan algebra A with unit element over $\boldsymbol{R}$ or $\boldsymbol{C}$, the "exponential map" is defined as usual by $\exp x = \sum_{n=0}^{\infty} \frac{1}{n!} x^n = (\exp T_x)e$. Then one has the relation

$$P(\exp x) = \exp(2T_x)$$

(cf. Koecher [5], IV, § 4 ; Braun-Koecher [1], XI, Satz 2. 2), which implies $(\exp x)^{-1}=\exp(-x)$.

4. For $a\in V$, we put

$$x \underset{a}{\cdot} y = \{x, a, y\}, \qquad (6.22)$$
$$\{x, y, z\}_a = \{x, P(a)y, z\}, \qquad x \underset{a}{\square} y = x\square P(a)y.$$

Prove the following:

4. 1) With the product $\underset{a}{\cdot}$, V becomes a Jordan algebra. (This is almost equivalent to (6. 19). In a more general context, this result is due to Meyberg. Cf. Koecher [9], p. 19 and p. 50.)

4. 2) One has

$$x \underset{a}{\square} y = (x \underset{a}{\cdot} y)\square a + [x\square a, y\square a]. \qquad (6.23)$$

In other words, $\{\ \}_a$ is a Jordan triple product coming from the Jordan product $\underset{a}{\cdot}$. [This construction of a new JTS $(V, \{\ \}_a)$ from the given one is called a "mutation" by a.]

4. 3) The Jordan algebra V with the product $\underset{a}{\cdot}$ has a unit element if and only if a is invertible, and in that case the unit element is given by a^{-1}. In particular, the triple product $\{\ \}$ itself comes from a Jordan algebra with unit element e, if and only if $P(e)=1_V$.

Remark. Under the additional assumption (JT 3), it follows from Lemma 6. 1 that an endomorphism g of V gives a JTS isomorphism $(V, \{\ \})\to(V, \{\ \}_a)$ if and only if $g\in\Gamma(V, \{\ \})$ and $(gg^*)^{-1}=P(a)$, i. e., $P(g^*a)=1$. Therefore, if such an isomorphism g exists, then a is invertible, $\{\ \}$ comes from a Jordan algebra structure with the unit element $e=g^*a=g^{-1}a^{-1}$, and g is a Jordan algebra isomorphism for these Jordan algebra structures, i. e., $g(xy)=gx\underset{a}{\cdot}gy$ $(x, y\in V)$.

5. Put

$$K(a, b) = 1_V-2a\square b+P(a)P(b) \qquad (6.24)$$

and prove

$$K(a, b)P(a) = P(a-P(a)b). \qquad (6.25)$$

(This relation is equivalent to the fundamental formula. Cf. Koecher [9], p. 53.) When $1-a\square b$ is non-singular, prove

$$(1-a\square b)^{-1}K(a, b)(1-a\square b)^{-1}a = a. \qquad (6.26)$$

Hint. Using (6. 17), one obtains $P(a)P(b)(a\square b)^k a=(a\square b)^{k+2}a$ $(k\geq 0)$.

6. When there exist invertible elements, define rational mappings

$$t_b(x) = x+b, \qquad j(x) = -x^{-1}, \qquad (6.27)$$
$$\tilde{t}_b(x) = (1-x\square b)^{-1}x,$$

and prove

$$\tilde{t}_b = j\circ t_b\circ j. \qquad (6.28)$$

From this, deduce the formula

$$\frac{d}{d\lambda}(x+\lambda b)^{-1}\Big|_{\lambda=0} = -P(x)^{-1}b. \qquad (6.29)$$

Remark. In Koecher [9], the group of rational automorphisms of V generated by $\{t_a, \tilde{t}_b\ (a, b\in V)\}$ and by $\Gamma(V, \{\ \})$, called the group of "essential automorphisms", is studied in detail. Its Zariski closure is an F-group whose Lie algebra is the symmetric Lie algebra $\mathfrak{G}(V, \{\ \})$ to be considered in the next section. Cf. also Koecher [6], [8], Springer [5], and Loos [10].

§ 7. The symmetric Lie algebra associated with a JTS.

Let V be an n-dimensional vector space over F. For a non-negative integer ν, we denote by $\mathfrak{P}_\nu$ the space of all homogeneous polynomial maps $V \to V$ of degree ν, and put $\mathfrak{P} = \bigoplus_{\nu=0}^{\infty} \mathfrak{P}_\nu$. In the following, we make the natural identifications $\mathfrak{P}_0 = V$, $\mathfrak{P}_1 = \mathrm{End}(V)$. Fixing a basis of V, we denote by a^i the i-th component of $a \in V$ and write $a = (a^i)$. Similarly, $p \in \mathfrak{P}$ is written as $p(x) = (p^i(x))$, where p^i's are polynomial functions on V. To each $p \in \mathfrak{P}$, we associate a (formal) polynomial vector field $p\frac{\partial}{\partial x}$ on V defined by

$$p\frac{\partial}{\partial x} = \sum_{i=1}^{n} p^i(x)\frac{\partial}{\partial x^i}. \tag{7.1}$$

Note that this definition is independent of the choice of basis. If we define the bracket product $[p, q]$ for $p, q \in \mathfrak{P}$ by

$$[p, q]^i = \sum_{j=1}^{n}\left(\frac{\partial p^i}{\partial x^j} q^j - p^j \frac{\partial q^i}{\partial x^j}\right) \qquad (1 \leq i \leq n), \tag{7.2}$$

then we have

$$-[p, q]\frac{\partial}{\partial x} = \left[p\frac{\partial}{\partial x}, q\frac{\partial}{\partial x}\right], \tag{7.2'}$$

where the bracket on the right-hand side is the usual Poisson bracket. Thus $\mathfrak{P}$ is an (infinite-dimensional) Lie algebra over F isomorphic to the Lie algebra of polynomial vector fields on V by the correspondence $p \leftrightarrow -p\frac{\partial}{\partial x}$. Clearly one has

$$[\mathfrak{P}_\nu, \mathfrak{P}_\mu] \subset \mathfrak{P}_{\nu+\mu-1}. \tag{7.3}$$

In particular, $\mathfrak{P}_1$ is a Lie subalgebra, identical to $\mathfrak{gl}(V)$. Also, for $A \in \mathfrak{gl}(V) = \mathfrak{P}_1$ and $b \in V = \mathfrak{P}_0$, one has $[A, b] = Ab$.

Now let $(V, \{\ \})$ be a non-degenerate JTS. For $b \in V$, we set

$$p_b(x) = \{x, b, x\} \ (= P(x)b) \tag{7.4}$$

and define $\mathfrak{G}(V, \{\ \}) = \mathfrak{G}_{-1} + \mathfrak{G}_0 + \mathfrak{G}_1$ by

$$\begin{cases} \mathfrak{G}_{-1} = V = \mathfrak{P}_0, \\ \mathfrak{G}_0 = V \square V \subset \mathfrak{P}_1, \\ \mathfrak{G}_1 = \{p_b \mid b \in V\} \subset \mathfrak{P}_2. \end{cases} \tag{7.5}$$

We write (a, T, b) for $a + T + p_b$ $(a, b \in V,\ T \in V \square V)$. Then we obtain the following

Proposition 7.1 (*Koecher*). 1) *$\mathfrak{G}(V, \{\ \})$ is a (graded) Lie subalgebra of $\mathfrak{P}$. For*

$$X = (a, T, b) \text{ and } X' = (a', T', b') \in \mathfrak{G}(V, \{\ \}),$$

one has

$$[X, X'] = (Ta' - T'a,\ 2a' \square b + [T, T'] - 2a \square b',\ T'^{*}b - T^{*}b'). \tag{7.6}$$

2) *The map*

$$\theta : X = (a, T, b) \longmapsto (b, -T^*, a) \tag{7.7}$$

is an involutive automorphism of $\mathfrak{G}$ *with* $\theta\mathfrak{G}_\nu = \mathfrak{G}_{-\nu}$.

3) $\mathfrak{G}$ *is semi-simple.*

The verifications of 1), 2) are straightforward (cf. Koecher [9], II, § 3). To prove 3), we need the following formula for the Killing form B of the Lie algebra $\mathfrak{G}$ (Koecher [9], II, § 4):

$$B(X, X') = B^{\mathfrak{G}_0}(T, T') + 2\,\mathrm{tr}_V(TT') - 4\tau(a, b') - 4\tau(b, a'). \tag{7.8}$$

Clearly, one has $B(\mathfrak{G}_\nu, \mathfrak{G}_\mu) = 0$ unless $\nu + \mu = 0$. Hence, to establish (7. 8), it is enough to prove the following formulas:

$$B(T, T') = B^{\mathfrak{G}_0}(T, T') + 2\,\mathrm{tr}_V(TT'), \tag{7.9 a}$$

$$B(a, p_{b'}) = -4\tau(a, b'). \tag{7.9 b}$$

(7. 9 a) follows immediately from the definitions and (7. 6). Since $1_V \in \mathfrak{G}_0$ (Lem. 6. 1), the left-hand side of (7. 9 b) can be transformed as follows:

$$B(a, p_{b'}) = B([1_V, a], p_{b'}) = B(1_V, [a, p_{b'}]).$$

By (7. 6) and (7. 9 a), this is equal to

$$\begin{aligned} &= -2B(1_V, a \square b') \\ &= -4\,\mathrm{tr}(a \square b') = -4\tau(a, b'), \end{aligned}$$

which proves (7. 9 b). By a similar computation, one also obtains

$$B(T, x \square y) = 2\tau(Tx, y) \qquad (x, y \in V,\ T \in \mathfrak{G}_0). \tag{7.10}$$

From (7. 8), (7. 10) and (JT 3) we can conclude that the Killing form B is non-degenerate, which proves 3).

We note that by (7. 6), (7. 7) one has

$$x \square y = -\frac{1}{2}[x, \theta y], \tag{7.11}$$

or

$$\{x, y, z\} = -\frac{1}{2}[[x, \theta y], z]. \tag{7.11'}$$

We also note that the gradation of $\mathfrak{G}$ coincides with its eigenspace decomposition with respect to $ad(-1_V)$, i. e.,

$$\mathfrak{G}_\nu = \mathfrak{G}(ad(-1_V); \nu) \qquad (\nu = 0, \pm 1). \tag{7.12}$$

In general, a pair $(\mathfrak{G}, \theta)$ formed of a graded Lie algebra $\mathfrak{G} = \mathfrak{G}_{-1} + \mathfrak{G}_0 + \mathfrak{G}_1$ and an involution θ of $\mathfrak{G}$ such that $\theta\mathfrak{G}_\nu = \mathfrak{G}_{-\nu}$ is called a (non-degenerate) *symmetric Lie algebra* if $\mathfrak{G}$ is semi-simple and $\mathfrak{G}_0 = [\mathfrak{G}_1, \mathfrak{G}_{-1}]$. Proposition 7. 1 and (7. 11) show that the pair $(\mathfrak{G}(V, \{\ \}), \theta)$ is a symmetric Lie algebra. Conversely, it is known (Meyberg, U. Hirzebruch) that, for any (non-degenerate) symmetric Lie algebra $(\mathfrak{G}, \theta)$, $V = \mathfrak{G}_{-1}$ with the trilinear product $\{\ \}$ defined by (7. 11') is a non-degenerate JTS and one has $\mathfrak{G}=\mathfrak{G}(V, \{\ \})$ (Koecher [9], II, § 5). Thus there is a one-to-one correspondence between non-degenerate JTS's and symmetric Lie algebras.

Proposition 7. 2. *Let $\mathfrak{G}=\mathfrak{G}(V, \{\ \})$. Then $g\in GL(V)$ extends to an automorphism of the graded Lie algebra $\mathfrak{G}$ (i. e., an automorphism of $\mathfrak{G}$ preserving the gradation) if and only if $g\in\Gamma(V, \{\ \})$. In that case, the extension of g is unique and is given by*

$$\tilde{g}: X=(a, T, b) \longmapsto (ga, gTg^{-1}, g^{*-1}b). \tag{7. 13}$$

Every automorphism of the graded Lie algebra $\mathfrak{G}$ is obtained in this way. $\tilde{g}$ is an automorphism of the symmetric Lie algebra $(\mathfrak{G}, \theta)$, i. e., $\tilde{g}$ commutes with θ, if and only if $g\in$ $\mathrm{Aut}(V, \{\ \})$.

The proof is again straightforward, and so left for an exercise of the reader. (The second part will be generalized in Proposition 9. 1.)

Proposition 7. 2 implies that $\Gamma(V, \{\ \})$ is canonically F-isomorphic to the automorphism group of the graded Lie algebra $\mathfrak{G}$, i. e., the centralizer of $ad(1_V)$ in $\mathrm{Aut}(\mathfrak{G})$. Thus $\Gamma(V, \{\ \})$ is reductive, and $\mathrm{Lie}\,\Gamma(V, \{\ \})$ may be identified with the derivation algebra of the graded Lie algebra $\mathfrak{G}$. But, since $\mathfrak{G}$ is semi-simple, one has $\mathrm{Der}(\mathfrak{G})=ad(\mathfrak{G})\cong\mathfrak{G}$. Therefore a derivation of $\mathfrak{G}$ preserving the gradation must be of the form $D=ad\,T$ with $T\in\mathfrak{G}_0=V\,\square\,V$. Thus one has

Proposition 7. 3. *The structure group $\Gamma(V, \{\ \})$ of a non-degenerate JTS is reductive, and* $\mathrm{Lie}\,\Gamma(V, \{\ \})$ *coincides with $V\,\square\,V$, i. e., in the above notation, one has*

$$\mathfrak{G}_0 = \mathrm{Lie}\,\Gamma(V, \{\ \}). \tag{7. 14}$$

If we set

$$\mathfrak{G}_0^{\pm} = \{T\in\mathfrak{G}_0 \mid T^*=\pm T\}, \tag{7. 15}$$

then $\mathfrak{G}_0=\mathfrak{G}_0^+ + \mathfrak{G}_0^-$ and by (6. 16) (or Prop. 7. 2) and (7. 14) one has

$$\mathfrak{G}_0^- = \mathrm{Der}(V, \{\ \}). \tag{7. 16}$$

In the Jordan algebra case, one has by (6. 4′), (6. 10)

$$\mathfrak{G}_0^+ = \{T_a (a\in A)\} \quad \text{and} \quad \mathfrak{G}_0^- = [\mathfrak{G}_0^+, \mathfrak{G}_0^+], \tag{7. 17}$$

that is, $\mathfrak{G}_0^-$ is the space of inner derivations. From these we obtain

Proposition 7. 4. *Let A be a semi-simple Jordan algebra with unit element e. Then, for $T\in\mathrm{Lie}\,\Gamma(A, \{\ \})$, the following conditions are all equivalent:*

(a) $T\in \mathrm{Der}_{\mathrm{JTS}}(A)$,

(b) $T\in \mathrm{Der}_{\mathrm{J.alg.}}(A)$,

(b′) *T is an inner derivation of the Jordan algebra A,*

(c) $T^* = -T$,

(d) $Te = 0$.

Also, for $a\in A$, the linear transformation $T=T_a$ is characterized as the unique element in

Lie $\Gamma(A, \{\ \})$ *satisfying the conditions*

$$T^* = T \quad and \quad Te = a. \tag{7.18}$$

In fact, the equivalence of the conditions (a), (b′), (c) is shown above. Since (b′)⇒(b)⇒(a), these three conditions are also equivalent. Finally, it is clear that $T_a e = a$ and $[T_a, T_b]e = 0$ for all $a, b \in A$, which together with (7.17) implies the equivalence (c)⇔(d) and the last assertion of the Proposition.

Thus, in the Jordan algebra case, identifying $\mathfrak{G}_0^+$ with A, one obtains the expression

$$\mathrm{Lie}\, \Gamma(A, \{\ \}) = A \oplus \mathrm{Der}(A) \tag{7.19}$$

with the bracket product defined in a natural manner (Exerc. 2).

Remark 1. The fact that all derivations of a semi-simple Jordan algebra A are inner is due to Jacobson [1].

Remark 2. Simple symmetric Lie algebras $\mathfrak{G}$, or equivalently, simple (non-degenerate) JTS's V, over an algebraically closed field F have been classified by Meyberg: $\mathfrak{G}$ is of one of the types A, B, C, D, E_6, and E_7 (cf. Loos [8]). The case where $F = \boldsymbol{R}$ and V is hermitian positive definite will appear in Ch. II in connection with symmetric domains, see also V, §§ 5, 6. For the case where V is a formally real Jordan algebra, see §§ 8, 9. More generally, the case where $F = \boldsymbol{R}$ and V is positive definite has been treated by Kobayashi-Nagano [1] in connection with "symmetric R-spaces"; cf. also Loos [2]. A complete classification of simple JTS over $\boldsymbol{R}$ was given recently by Neher [1].

Exercises

1 (Kaup [3]). Let V be a vector space and let $p(x) \in \mathfrak{P}_2$. Define a (commutative) bilinear product $x \cdot y$ in V by

$$x \cdot y = \frac{1}{2}(p(x+y) - p(x) - p(y)).$$

Prove that the following two conditions are equivalent.

(a) The product $x \cdot y$ satisfies (J 2) (hence defines a Jordan algebra structure on V).

(b) $(ad\, p(x))^3 \mathfrak{P}_0 = \{0\}$.

Hint. (b) is equivalent to

$$x^3 y - x^2(xy) - 2x(x^2 y) + 2x(x(xy)) = 0 \qquad (x, y \in V),$$

which implies $x^4 = (x^2)^2$. Substitute $x+y$ for x in the last identity and compare the linear terms in y to obtain another identity

$$x^3 y - 4x^2(xy) + x(x^2 y) + 2x(x(xy)) = 0.$$

(J 2) follows from these two identities.

2. Let A be a semi-simple Jordan algebra. Show that, in the expression (7.19), one has

$$[a+D, a'+D'] = (Da' - D'a) + ([T_a, T_{a'}] + [D, D']) \qquad (a, a' \in A,\ D, D' \in \mathrm{Der}(A)). \tag{7.20}$$

Verify directly that, for any Jordan algebra A (not necessarily semi-simple), (7.20) defines a Lie algebra structure on $A \oplus \mathrm{Der}(A)$.

§ 8. Formally real Jordan algebras and self-dual homogeneous cones.

In the remaining of this chapter (§§ 8, 9), we assume that $F=\boldsymbol{R}$. A Jordan algebra A over $\boldsymbol{R}$ is called *formally real* (or *compact*) if the trace form τ is positive definite. It is known that this condition is equivalent to saying that $x^2+y^2=0$ $(x, y \in A)$ implies $x=y=0$. (It is clear that the first condition implies the second. For the converse, see V, § 6, Exerc. 4.) To explain the connection between formally real Jordan algebras and self-dual homogeneous cones, we start with giving some basic definitions and results on open convex cones.

Let U be a vector space over $\boldsymbol{R}$ of dimension m. By a *non-degenerate* (or *regular*) *open convex cone*, or simply a "cone", in U we mean a (non-empty) open subset Ω of U satisfying the condition:

$$x, y \in \Omega \Longrightarrow \lambda x+\mu y \in \Omega \qquad \text{for any positive real numbers } \lambda, \mu$$

and, in addition, not containing any straight line. For a cone Ω, we define the (linear) automorphism group of Ω by

$$G(\Omega) = \{g \in GL(U) \mid g\Omega=\Omega\}. \tag{8.1}$$

Clearly $G(\Omega)$ is a closed subgroup of $GL(U)$ and hence a Lie group. Ω is called *homogeneous* if $G(\Omega)$ is transitive on Ω. Fixing a (positive definite) inner product $\langle\ \rangle$ on U, we define the "dual" of Ω by

$$\Omega^* = \{x \in U \mid \langle x, y\rangle >0 \text{ for all } y \in \bar{\Omega}-\{0\}\}, \tag{8.2}$$

which is easily seen to be also a (non-degenerate, open convex) cone in U. It is also easy to see that $\Omega^{**}=\Omega$ and

$$G(\Omega^*) = {}^tG(\Omega), \tag{8.3}$$

t denoting always the transpose with respect to $\langle\ \rangle$. Ω is called *self-dual* if $\Omega^*=\Omega$.

We need the following results.

Lemma 8.1. *Let Ω be a cone in U and Ω^* its dual. Then there exists a bijection $\Omega \ni x \mapsto x^* \in \Omega^*$ satisfying the relation $(gx)^*={}^tg^{-1}x^*$ for all $g \in G(\Omega)$, $x \in \Omega$.*

For the proof, see Vinberg [5], I, § 4, or Koecher [5], I, § 8. x^* is given by

$$x^* = \int_{\Omega^* \cap H_m(x)} y\, d^{(m)}y \Big/ \int_{\Omega^* \cap H_m(x)} d^{(m)}y, \tag{8.4}$$

where $H_m(x)$ is the hyperplane $\{y \in U \mid \langle x, y\rangle =m\}$ $(m=\dim U)$ and $d^{(m)}y$ is the Lebesgue measure on $H_m(x)$. In other words, x^* is the center of gravity of $\Omega^* \cap H_m(x)$ (cf. Exerc. 3).

Lemma 8.2 (*Ochiai* [1]). *One has $\Omega \cap \Omega^* \neq \phi$.*

It follows from Lemma 8.1 and (8.3) that, if Ω is homogeneous, so is Ω^*. (8.3) also implies that, if Ω is self-dual, then $G(\Omega)$ is "self-adjoint", i. e., ${}^tG(\Omega)=G(\Omega)$. For a homogeneous cone, the converse of this is also true. For later use, we state this in a slightly more general form.

Lemma 8. 3. *Let Ω be a cone in U and suppose that there exists an analytic subgroup G_1 of $G(\Omega)$ which is transitive on Ω and self-adjoint. Then Ω is self-dual and $G_1=G(\Omega)^\circ$.*

By Lemmas 8. 1, 8. 2, taking $x_0 \in \Omega$ such that $x_0^* \in \Omega$, one has from the assumptions

$$\Omega^* = (G_1 x_0)^* = {}^tG_1 x_0^* = G_1 x_0^* = \Omega.$$

Thus Ω is self-dual. The proof of $G_1=G(\Omega)^\circ$ will be given later (p. 33).

Proposition 8. 4 (*Vinberg*). *Let Ω be a homogeneous cone in U. Then there exists an $\boldsymbol{R}$-group G in $GL(U)$ such that $G^\circ \subset G(\Omega) \subset G$. For any $x_0 \in \Omega$, the stabilizer $K=G(\Omega)_{x_0}$ is a maximal compact subgroup of $G(\Omega)$, and every maximal compact subgroup of $G(\Omega)$ is conjugate to K.*

Proof. Let G be the normalizer of $\mathfrak{g}(\Omega)=\text{Lie } G(\Omega)$ in $GL(U)$, i. e.,

$$G = \{g \in GL(U) \mid g(\mathfrak{g}(\Omega))g^{-1}=\mathfrak{g}(\Omega)\}.$$

Then clearly G is an $\boldsymbol{R}$-group containing $G(\Omega)$. Let $g \in G$ and $x_0 \in \Omega$. Then, since Ω is homogeneous, one has

$$g\Omega = gG(\Omega)^\circ x_0 = G(\Omega)^\circ(gx_0).$$

Thus $g\Omega$ is an (open) $G(\Omega)^\circ$-orbit in U. Let Ξ denote the union of all open $G(\Omega)^\circ$-orbits other than Ω. Then, Ξ is open, $\Xi \cap \Omega=\phi$, and for every $g \in G$ one has either $g\Omega=\Omega$ or $g\Omega \subset \Xi$. Therefore $G(\Omega)$ is open in G and one has $G^\circ \subset G(\Omega)$. Next, let $K=G(\Omega)_{x_0}$. Then, since K leaves stable a bounded open set $\Omega \cap (x_0-\Omega)$, K is compact. On the other hand, since $\Omega=G(\Omega)x_0$ is convex, it is easy to see that any compact subgroup K' of $G(\Omega)$ has a fixed point of the form g_1x_0 with $g_1 \in G(\Omega)$; then $K' \subset g_1Kg_1^{-1}$. It follows that K is maximal compact, and every maximal compact subgroup of $G(\Omega)$ is conjugate to K, q. e. d.

Now let Ω be a self-dual homogeneous cone in U. Then, in the notation of Proposition 8. 4, we have from the results of § 4 that G^z is a reductive $\boldsymbol{R}$-group with Cartan involution $g \mapsto {}^tg^{-1}$. Therefore, (identifying U with $\boldsymbol{R}^m$ with respect to an orthonormal basis and) putting

$$K' = G(\Omega) \cap O_m(\boldsymbol{R}), \qquad P = G(\Omega) \cap \mathcal{P}_m(\boldsymbol{R}),$$

one has a global decomposition $G(\Omega)=K'P$, where K' is also a maximal compact subgroup of $G(\Omega)$. By the conjugacy of maximal compact subgroups of $G(\Omega)$, one may assume that $K'=K=G(\Omega)_{x_0}$. When this equality holds, we say that the reference point x_0 is "compatible" with $\langle\ \rangle$. In the following, we fix such an x_0 once and for all and denote it by e.

Let

$$\mathfrak{g}(\Omega) = \mathfrak{k} + \mathfrak{p}$$

be the corresponding Cartan decomposition of $\mathfrak{g}(\Omega) = \text{Lie } G(\Omega)$. Then, for $T \in \mathfrak{g}(\Omega)$, one has

$$T \in \mathfrak{k} \Longleftrightarrow {}^tT = -T \Longleftrightarrow Te = 0. \tag{8.5}$$

Hence, for each $a \in U$, there exists a unique element $T_a \in \mathfrak{p}$ such that $T_a e = a$, and the map $a \mapsto T_a$ gives a linear isomorphism $U \cong \mathfrak{p}$. We note that, since $G(\Omega)$ acts effectively on Ω, one has $\mathfrak{g}(\Omega)^a \cap \mathfrak{k} = \{0\}$, which implies

$$\mathfrak{k} = [\mathfrak{p}, \mathfrak{p}] \quad \text{and} \quad \mathfrak{p} = \mathfrak{k}^\perp,$$

($\mathfrak{k}^\perp$ denoting the orthogonal space of $\mathfrak{k}$ with respect to the Killing form of $\mathfrak{g}(\Omega)$). Thus the above Cartan decomposition and the T_a ($a \in U$) are uniquely determined only by (Ω, e), independently of the choice of the (compatible) inner product $\langle\ \rangle$.

We can now prove the remaining part of Lemma 8. 3. Let $p \in P (= \exp \mathfrak{p})$. Then, since G_1 is transitive on Ω, there exists $g_1 \in G_1$ such that $p^{1/2}e = g_1 e$. Then one has $g_1^{-1} p^{1/2} \in K$ and so $g_1 {}^t g_1 = p$, which shows that $P \subset G_1$. Since $G(\Omega)^\circ$ is generated by P, one has $G_1 = G(\Omega)^\circ$, completing the proof of Lemma 8. 3.

The following theorem is fundamental:

Theorem 8. 5 (*Koecher, Vinberg*). *Let Ω be a self-dual homogeneous cone in U. Then U endowed with a product*

$$xy = T_x y \qquad (x, y \in U) \tag{8.6}$$

becomes a formally real Jordan algebra with unit element e, and one has

$$\begin{aligned} \Omega &= \exp U \\ &(= \textit{Interior of } \{x^2 \mid x \in U\} = \{x^2 \mid x \in U^\times\}), \end{aligned} \tag{8.7}$$

(*where $U^\times$ denotes the set of invertible elements in U*). *Conversely, for any formally real Jordan algebra U with unit element e, the set Ω defined by* (8. 7) *is a homogeneous open convex cone in U, which is self-dual with respect to the trace form τ. Moreover, one has* Lie $\Gamma(U, \{\ \})= \mathfrak{g}(\Omega)$ *and the operator T_x for the Jordan algebra U defined by* (8. 6) *coincides with the operator T_x for Ω defined above.*

We give a proof after Vinberg [1] and Ash et al. [1]. (Cf. also Koecher [5] and Braun-Koecher [1].) The essential step is contained in the following

Lemma 8. 6. *Let $\mathfrak{g}$ be an algebraic Lie subalgebra of $\mathfrak{gl}(U)$ and suppose that there exist $e \in U$ and an inner product $\langle\ \rangle$ on U satisfying the following conditions:*

(i) ${}^t\mathfrak{g} = \mathfrak{g}$,

(ii) *one has* ${}^tX = -X \Longleftrightarrow Xe = 0$ *for* $X \in \mathfrak{g}$,

(iii) $\mathfrak{g}e = U$.

Then: 0) *For each $a \in U$, there exists a unique element T_a in $\mathfrak{g}$ such that ${}^tT_a = T_a$ and $T_a e = a$.*

1) *The space U, endowed with the product $ab = T_a b$, becomes a formally real Jordan algebra with unit element e.*

2) *Let G_1 be the analytic subgroup of $GL(U)$ corresponding to $\mathfrak{g}$. Then the orbit $\Omega = G_1 e$ is a self-dual homogeneous cone in U.*

Proof. 0) is trivial. Let $\mathfrak{g} = \mathfrak{k} + \mathfrak{p}$ be the Cartan decomposition corresponding to the involution $X \mapsto -{}^t X$. Then one has $\mathfrak{p} = \{T_a (a \in U)\}$ and (8. 5) holds for $T \in \mathfrak{g}$. To prove 1), let $a, b, c \in U$. Then, first, from $[T_a, T_b] \in \mathfrak{k}$, one has

$$0 = [T_a, T_b]e = T_a b - T_b a = ab - ba.$$

Thus the product is commutative. Next, from $[[T_a, T_b], T_c] \in \mathfrak{p}$ and

$$[[T_a, T_b], T_c]e = [T_a, T_b]c = a(bc) - b(ac),$$

one has

$$[[T_a, T_b], T_c] = T_{a(bc)-b(ac)},$$

i. e., the relation (6. 3). Denoting the "associater" $a(cb) - (ac)b$ by $[a, c, b]$, we can rewrite this as

$$[a, c, b]d = [a, cd, b] - c[a, d, b].$$

Putting $c = d$, one has

$$(*) \qquad 2[a, c, b]c = [a, c^2, b].$$

Now, for any $a, b, x \in U$, one has by (*) and the relation $\langle ab, c \rangle = \langle a, bc \rangle$

$$\begin{aligned} \langle [a^2, b, a], x \rangle &= \langle a^2, [x, a, b] \rangle = \langle [b, a^2, x], a \rangle \\ &= 2\langle [b, a, x]a, a \rangle = 2\langle [b, a, x], a^2 \rangle. \end{aligned}$$

Since $[x, a, b] = -[b, a, x]$, this must be equal to zero, whence follows $[a^2, b, a] = 0$, i. e., (J 2). If $x^2 + y^2 = 0$, one has

$$0 = \langle x^2 + y^2, e \rangle = \langle x, x \rangle + \langle y, y \rangle,$$

so that $x = y = 0$. Therefore U is formally real.

2) First by (iii) Ω is open. Define Ω^* by (8. 2). Then it is clear that Ω^* is a (non-degenerate, open convex) cone. To see that Ω^* is non-empty, we show that $\Omega \subset \Omega^*$. Let $G_1 = \exp \mathfrak{p} \cdot K_1$ be the global Cartan decomposition. Then, for $g_1, g_2 \in G_1$, if one puts $g_1^{-1} g_2 = pk$ with $p \in \exp \mathfrak{p}$, $k \in K_1$, then one has

$$\langle g_1 e, g_2 e \rangle = \langle p^{\frac{1}{2}} e, p^{\frac{1}{2}} e \rangle > 0.$$

Hence one has $\Omega \subset \bar{\Omega}^*$ and so $\Omega \subset \Omega^*$. One has also $G_1 \subset G(\Omega^*)$. Therefore, on Ω^*, there exists a G_1-invariant Riemannian metric q (cf. Exerc. 4). Since Ω is open and $e \in \Omega \subset \Omega^*$, there exists $\rho > 0$ such that a geodesic ball about e with radius ρ is contained in Ω. For any $x \in \Omega^*$, there exists a sequence $\{x_i (0 \leq i \leq s)\}$ in Ω^* such that $x_0 = e$, $x_s = x$ and the geodesic distance (with respect to q) of x_{i-1} and x_i is $< \rho$ for $1 \leq i \leq s$. Then one has $x_i = g_i x_{i-1}$ for some $g_i \in G_1$ and hence $x = g_s g_{s-1} \cdots g_1 e \in \Omega$. Thus $\Omega = \Omega^*$, q. e. d.

Proof of Theorem 8. 5. Let Ω be a self-dual homogeneous cone in U with a reference point e. We know that $\mathfrak{g}=\mathfrak{g}(\Omega)$ is algebraic and satisfies the conditions (i)$\sim$(iii) in Lemma 8. 6 (with respect to any compatible e and $\langle\ \rangle$). Hence, by 1), U becomes a formally real Jordan algebra with respect to the product (8. 6). Moreover, in the above notation, one has

$$(**) \qquad \Omega = G_1 e = (\exp \mathfrak{p})e = \exp U$$

(cf. Remark following § 6, Exerc. 3). [Other equalities in (8. 7) follow from this and V, § 6, Exerc. 4.] Conversely, let U be a formally real Jordan algebra and define T_x by (8. 6). Put $\mathfrak{p}=\{T_x(x\in U)\}$, $\mathfrak{k}=[\mathfrak{p},\mathfrak{p}]$, and $\mathfrak{g}=\mathfrak{k}+\mathfrak{p}$. Then, by Proposition 7. 4 and (7. 19), $\mathfrak{g}=\mathrm{Lie}\,\Gamma(U,\{\ \})$ and the conditions (i)$\sim$(iii) are again satisfied (with respect to the trace form τ). Hence, by Lemma 8. 6, 2), $\Omega=G_1e$ is a self-dual homogeneous cone and one has (**). Moreover, by Lemma 8. 3, one has $G_1=G(\Omega)^\circ$ and hence $\mathfrak{g}=\mathfrak{g}(\Omega)$; therefore T_x coincides with the operator defined for Ω, q. e. d.

As was noted in the above proof, in the notation of § 7 (with $V=U$), one has by (7. 14), (7. 17) and Proposition 7. 4

$$(8.\,8) \qquad \begin{aligned} &\mathfrak{p} = \mathfrak{S}_0^+, \qquad \mathfrak{k} = \mathfrak{S}_0^- = \mathrm{Der}_{\mathrm{J.alg.}}(U), \\ &\mathfrak{g}(\Omega) = \mathfrak{S}_0 = \mathrm{Lie}\,\Gamma(U,\{\ \}). \end{aligned}$$

Thus the Cartan involution of $\mathfrak{g}(\Omega)$ at e coincides with $\theta|\mathfrak{S}_0$. We also note that the reference point $e\in\Omega$ and the inner product $\langle\ \rangle$, which are "compatible" (i. e., corresponding to the same maximal compact subgroup of $G(\Omega)$), do *not* determine each other uniquely. However, when one of them is given, the other can uniquely be normalized by the condition

$$(8.\,9) \qquad \tau(x,y) = \langle x,y\rangle \qquad (x,y\in U).$$

In view of (8. 8) we see that, even if $\langle\ \rangle$ is not normalized, the adjoints of $T\in\mathfrak{g}(\Omega)$ with respect to τ and $\langle\ \rangle$ are always the same.

Theorem 8. 5 gives a one-to-one correspondence between formally real Jordan algebras and the self-adjoint homogeneous cones Ω with a "reference point" e. In the next section, we shall show that this correspondence is actually an equivalence of two categories.

An (open convex) cone Ω is called *reducible* or *decomposable* if there is a direct sum decomposition of the ambient space $U=U_1\oplus U_2$, $U_i\neq\{0\}$, such that $\Omega=\Omega_1\times\Omega_2$ with $\Omega_i\subset U_i$; and if there is no such decomposition Ω is called *irreducible*. By Theorem 8. 5, it is clear that a self-dual homogeneous cone Ω is irreducible if and only if the corresponding Jordan algebra U is simple. [In that case, U is central and one has $\mathfrak{g}(\Omega)^a=\{1_U\}_R$.] In general, any formally real Jordan algebra U is uniquely decomposed into the direct sum of simple formally real Jordan algebras U_i and hence the corresponding cone Ω is also uniquely decomposed into the direct product of the irreducible self-dual homogeneous cones Ω_i corresponding to U_i:

$$(8.\,10) \qquad U=U_1\oplus\cdots\oplus U_s, \qquad \Omega=\Omega_1\times\cdots\times\Omega_s, \qquad \Omega_i\subset U_i \quad (1\leq i\leq s).$$

The simple formally real Jordan algebras, or equivalently, the irreducible self-dual homogeneous cones, are classified completely (cf. e. g., Braun-Koecher [1]). There are the following five types of irreducible self-dual homogeneous cones:

(I–II–III) *(classical cones)* $U=\mathcal{H}_n(\mathcal{K})$ $(\mathcal{K}=\boldsymbol{R}, \boldsymbol{C}, \boldsymbol{H})$, $\Omega=\mathcal{P}_n(\mathcal{K})$ (cf. Appendix, § 3). The Jordan algebra structure on U is given by

$$x\circ y=\frac{1}{2}(xy+yx), \qquad e=1_n. \tag{8.11}$$

Then one has

$$\{x, y, z\}=\frac{1}{2}(xyz+zyx), \tag{8.12}$$

$$\tau(x, y)=c\operatorname{tr}(x\circ y), \tag{8.13}$$

where tr is the "reduced trace" of $\mathcal{M}_n(\mathcal{K})$ over its center and $c=(n+1)/2, n, n-1/2$, according as $\mathcal{K}=\boldsymbol{R}, \boldsymbol{C}, \boldsymbol{H}$. Since, for $n=1$, all $\mathcal{P}_1(\mathcal{K})$ $(\mathcal{K}=\boldsymbol{R}, \boldsymbol{C}, \boldsymbol{H})$ coincide, we assume from now on, unless otherwise specified, that $n\geq 2$ when $\mathcal{K}=\boldsymbol{C}$ or $\boldsymbol{H}$. Then $GL_n(\mathcal{K})$ acts linearly on $\mathcal{P}_n(\mathcal{K})$ by

$$GL_n(\mathcal{K})\times\mathcal{P}_n(\mathcal{K})\ni(g, p)\longmapsto gp{}^t\bar{g}\in\mathcal{P}_n(\mathcal{K}), \tag{8.14}$$

and one has a surjective homomorphism $GL_n(\mathcal{K})\rightarrow G(\Omega)^\circ$, of which the kernel is $\{\zeta 1_n | \zeta\in\operatorname{Cent}(\mathcal{K}), \zeta\bar{\zeta}=1\}$. Therefore, $\mathfrak{g}(\Omega)$ may be identified with

$$\mathfrak{gl}_n^0(\mathcal{K})=\{X\in\mathfrak{gl}_n(\mathcal{K}) | \operatorname{tr} X\in\boldsymbol{R}\}\cong\{1_n\}_{\boldsymbol{R}}\oplus\mathfrak{sl}_n(\mathcal{K}). \tag{8.15}$$

(IV) Let U be an m-dimensional vector space over F (any field of characteristic zero), and let S be a non-degenerate symmetric bilinear form on $U\times U$. Replacing S by a scalar multiple if necessary, we assume that there exists an element e in U with $S(e, e)=1$. Then, for $m\geq 3$, U becomes a simple Jordan algebra with unit element e with respect to the product

$$xy=S(x, e)y+S(y, e)x-S(x, y)e. \tag{8.16}$$

We denote this Jordan algebra by $J(U, S, e)$. The corresponding Jordan triple product $\{\ \}$ is given by

$$\{x, y, z\}=S(x, z)\hat{y}-S(x, \hat{y})z-S(z, \hat{y})x, \tag{8.17}$$

where $\hat{y}=y-2S(y, e)e$, and the trace form is given by

$$\tau(x, y)=-mS(x, \hat{y}), \tag{8.18}$$

which is non-degenerate. When $F=\boldsymbol{R}$, the isomorphism class of $J(U, S, e)$ depends only on the signature (p, q) of S. From (8. 18) it is easy to see that τ is positive definite if and only if S is of signature $(1, m-1)$. The corresponding self-dual homogeneous cone Ω (called a "circular cone") is then given by

$$\Omega=\{x\in U | S(x, x)>0,\ S(x, e)>0\}. \tag{8.19}$$

The typical Ω with $U=\boldsymbol{R}^m$, $S=\operatorname{diag}(1, -1_{m-1})$, $e={}^t(1, 0, \cdots, 0)$ will be denoted by $\mathcal{P}(1, m-1)$. Clearly $\mathfrak{g}(\mathcal{P}(1, m-1))^s=\mathfrak{o}(1, m-1)$. There are the following isomorphisms:

$$\mathcal{P}(1, 2)\cong\mathcal{P}_2(\boldsymbol{R}), \qquad \mathcal{P}(1, 3)\cong\mathcal{P}_2(\boldsymbol{C}), \qquad \mathcal{P}(1, 5)\cong\mathcal{P}_2(\boldsymbol{H}).$$

(*V*) (*exceptional cone*) $U=\mathcal{H}_3(\boldsymbol{O})$, $\Omega=\mathcal{P}_3(\boldsymbol{O})$, where $\boldsymbol{O}$ is the (non-associative) algebra of "Cayley numbers" (or "octonions"). In this case, $\mathfrak{g}(\Omega)^s$ is an exceptional simple Lie algebra of type (E_6) and of $\boldsymbol{R}$-rank 2.

For convenience of the reader, we give a list of the relevant data on Lie algebras. [The notation $\mathfrak{g}=(X_{l,r})$ indicates that $\mathfrak{g}$ is of type X, of absolute rank l and of $\boldsymbol{R}$-rank r. We put $r=\boldsymbol{R}$-rank $\mathfrak{g}(\Omega)$; for the meaning of d, see the Remark below.]

Ω	$\mathfrak{k}=\mathrm{Der}\,U$	$\mathfrak{g}(\Omega)^s=\mathfrak{G}_0^s$	$\mathfrak{G}(U,\{\ \})$	$\dim U$	$\dim\mathfrak{k}$	r	d
$\mathcal{P}_n(\boldsymbol{R})$	$\mathfrak{o}(n)$	$\mathfrak{sl}_n(\boldsymbol{R})$	$\mathfrak{sp}_{2n}(\boldsymbol{R})$	$\frac{1}{2}n(n+1)$	$\frac{1}{2}n(n-1)$	n	1
$\mathcal{P}_n(\boldsymbol{C})$	$\mathfrak{su}(n)$	$\mathfrak{sl}_n(\boldsymbol{C})$	$\mathfrak{su}(n,n)$	n^2	n^2-1	n	2
$\mathcal{P}_n(\boldsymbol{H})$	$\mathfrak{su}_n(\boldsymbol{H})$	$\mathfrak{sl}_n(\boldsymbol{H})$	$\mathfrak{su}^-_{2n}(\boldsymbol{H})$	$n(2n-1)$	$(n+1)(2n-1)$	n	4
$\mathcal{P}(1,n-1)$	$\mathfrak{o}(n-1)$	$\mathfrak{o}(1,n-1)$	$\mathfrak{o}(2,n)$	n	$\frac{1}{2}(n-1)(n-2)$	2	$n-2$
$\mathcal{P}_3(\boldsymbol{O})$	(F_4)	$(E_{6,2})$	$(E_{7,3})$	27	52	3	8

We note that one has the following relations:

$$\begin{aligned} \dim\mathfrak{g}(\Omega)^s &= \dim U+\dim\mathfrak{k}-1, \\ \dim\mathfrak{G}(U,\{\ \}) &= 3\dim U+\dim\mathfrak{k}, \\ \boldsymbol{R}\text{-rank}\,\mathfrak{G}(U,\{\ \}) &= r \qquad \text{(cf. § 9, Exerc. 5)}, \\ \dim U &= r+\frac{1}{2}r(r-1)d \qquad \text{(cf. Remark below)}. \end{aligned} \tag{8.20}$$

Remark. The classification of formally real Jordan algebras was obtaind as early as in 1934 by P. Jordan, J. von Neumann, and Wigner [1]. As a result of the classification, we see that, for an irreducible self-dual homogeneous cone Ω, the system of $\boldsymbol{R}$-roots of $\mathfrak{g}(\Omega)^s$ is always of type (A_{r-1}), and a complete invariant for the linear isomorphism class of Ω is given by a pair (r, d), where $r=\boldsymbol{R}$-rank $\mathfrak{g}(\Omega)$ and d denotes the common multiplicity of $\boldsymbol{R}$-roots ; when $r=1$, i. e., $\Omega=\boldsymbol{R}_+^\times$, we put $d=1$. (Cf. Vinberg [6], II, § 2 ; Loos [1], Vol. II, pp. 170–171. Cf. also V, § 6, Exerc. 9.) For the relation between Jordan algebras and exceptional Lie algebras, cf. e. g., Schafer [1], Springer [4], Tits [3], [4], and Jacobson [7].

Exercises

Let Ω be a cone in U and Ω^* its dual. Consider the integral

$$\phi(x) = \int_{\Omega^*}\exp(-\langle x,y\rangle)dy \qquad (x\in U) \tag{8.21}$$

and prove the following properties (Vinberg [5]).

1. For any $\lambda>0$, let $H_\lambda(x)=\{y\in U\mid\langle x,y\rangle=\lambda\}$ and let $d^{(\lambda)}y$ denote the Lebesgue measure on $H_\lambda(x)$. Then one has

$$\phi(x) = \frac{(m-1)!}{\lambda^{m-1}}\int_{\Omega^*\cap H_\lambda(x)} d^{(\lambda)}y. \tag{8.22}$$

Hence the integral is convergent for all $x\in\Omega$. When x converges to a boundary point of Ω, $\phi(x)$ tends to infinity. (ϕ is called the "characteristic" of Ω.)

2. One has

$$\phi(gx) = |\det(g)|^{-1}\phi(x) \tag{8.23}$$

for all $x \in \Omega$, $g \in G(\Omega)$. Thus $\phi(x)dx$ is an invariant measure on Ω.

3. Writing $\partial_a = a\frac{\partial}{\partial x} = \sum_{i=1}^{m} a^i \frac{\partial}{\partial x^i}$, one has

$$\begin{aligned} \partial_a \log \phi(x) &= -\phi(x)^{-1}\int_{\Omega^*} \langle a, y\rangle \exp(-\langle x, y\rangle)\, dy \\ &= -\langle a, x^*\rangle \qquad (a \in U,\ x \in \Omega), \end{aligned} \tag{8.24}$$

where x^* is defined by (8. 4). The tangent space to the hypersurface $\{y \in U \mid \phi(y) = \phi(x)\}$ at $x \in \Omega$ is given by $\{y \in U \mid \langle x^*, y\rangle = m\}$.

4. For $x \in \Omega$, the bilinear form

$$q_x(a, b) = \partial_a \partial_b \log \phi(x) \qquad (a, b \in U)$$

is symmetric and positive definite, so that it defines an invariant Riemannian metric on Ω.

5. Let Ω be homogeneous and self-dual. Then

5. 1) One has

$$\begin{aligned} (\partial_a \partial_b \log \phi)(e) &= \operatorname{tr}(T_{ab}) \qquad (\text{i. e., } q_e = \tau), \\ (\partial_a \partial_b \partial_c \log \phi)(e) &= -2\operatorname{tr}(T_{(ab)c}). \end{aligned} \tag{8.25}$$

Hint. To prove (8. 25), first deduce from (8. 23)

$$\log \phi((\exp T_a)e) = \log \phi(e) - \operatorname{tr}(T_a).$$

Then, substitute λa ($\lambda \in \boldsymbol{R}$) for a, expand the left-hand side in Taylor series in λ, and compare the terms of the first, second, and third order in λ to obtain the relations

$$\begin{aligned} (\partial_a \log \phi)(e) &= -\operatorname{tr}(T_a), \\ (\partial_a^2 \log \phi)(e) &= \operatorname{tr}(T_{a^2}), \\ (\partial_a^3 \log \phi)(e) &= -2\operatorname{tr}(T_{a^3}). \end{aligned}$$

5. 2) When $\langle\ \rangle$ is normalized by (8. 9), one has $e^* = e$ and $x^* = x^{-1}$ for $x \in \Omega$.

6. (Generalization of the Gamma function) Let Ω be a self-dual homogeneous cone. Then there exists a holomorphic function $\Gamma_\Omega(s)$ ($s \in \boldsymbol{C}$, Re $s > 1$) such that one has

$$(\det P(x))^{-\frac{s}{2}}\Gamma_\Omega(s) = \int_\Omega (\det P(y))^{\frac{s-1}{2}} \exp(-\langle x, y\rangle)\, dy \qquad (x \in \Omega,\ \operatorname{Re} s > 1) \tag{8.26}$$

(cf. Koecher [1], Resnikoff [2]).

§ 9. Morphisms of JTS's and self-dual homogeneous cones.

A JTS $(V, \{\ \})$ over $\boldsymbol{R}$ is called *positive definite* if the trace form τ is positive definite. In that case, since the structure group $\Gamma(V, \{\ \})$ is stable under $*$, its Zariski connected component $\Gamma(V, \{\ \})^z$ is a reductive $\boldsymbol{R}$-group with a Cartan involution $\theta : g \mapsto g^{*-1}$. It follows by (7. 8) that

$$B(X, \theta X') = -B^{\mathfrak{G}_0}(T, T'^*) - 2\operatorname{tr}_V(TT'^*) - 4\tau(a, a') - 4\tau(b, b')$$

is negative definite. Therefore θ is a Cartan involution of the semi-simple Lie algebra $\mathfrak{G} = \mathfrak{G}(V, \{\ \})$.

Let $(V, \{\ \})$ and $(V', \{\ \}')$ be positive definite JTS's over $\boldsymbol{R}$ and let $(\mathfrak{G}, \theta)$ and $(\mathfrak{G}', \theta')$ be the corresponding symmetric Lie algebras. A linear map $\varphi: V \to V'$ is called a *JTS homomorphism* if one has

$$(9.1) \qquad \varphi\{x, y, z\} = \{\varphi x, \varphi y, \varphi z\}' \qquad \text{for all } x, y, z \in V.$$

A Lie algebra homomorphism $\rho: \mathfrak{G} \to \mathfrak{G}'$ is called a *symmetric Lie algebra homomorphism* if one has

$$(9.2) \qquad \rho(\mathfrak{G}_\nu) \subset \mathfrak{G}'_\nu \quad (\nu = 0, \pm 1), \quad \text{and}$$

$$(9.3) \qquad \rho \circ \theta = \theta' \circ \rho.$$

ρ is called an *extension* of a linear map φ if one has $\rho | \mathfrak{G}_{-1} = \varphi$ after the identification $\mathfrak{G}_{-1} = V$.

Proposition 9.1. *The notation being as above, a linear map $\varphi: V \to V'$ is (uniquely) extendible to a symmetric Lie algebra homomorphism $\rho: \mathfrak{G} \to \mathfrak{G}'$ if and only if φ is a JTS homomorphism.*

Proof. The "only if" part is clear from (7.11'). To prove the "if" part, let φ be a JTS homomorphism and put

$$\bar{\mathfrak{G}} = \varphi(V) + \varphi(V) \square \varphi(V) + \theta' \varphi(V) \ (\subset \mathfrak{G}').$$

Then, as is easily seen, $\bar{\mathfrak{G}}$ is a Lie subalgebra of $\mathfrak{G}'$ stable under θ'. We shall show, for instance, that $\varphi(V) \square \varphi(V)$ is a subalgebra stable under θ'. Let $a, b, x, y \in V$. Then one has by (JT 2')

$$\begin{aligned}[][\varphi a \square \varphi b, \varphi x \square \varphi y] &= \{\varphi a, \varphi b, \varphi x\}' \square \varphi y - \varphi x \square \{\varphi b, \varphi a, \varphi y\}' \\ &= \varphi\{a, b, x\} \square \varphi y - \varphi x \square \varphi\{b, a, y\}\end{aligned}$$

and by (7.7) and (6.7)

$$\theta'(\varphi x \square \varphi y) = -(\varphi x \square \varphi y)^* = -\varphi y \square \varphi x,$$

which proves our assertion. It follows that $\bar{\mathfrak{G}}$ is reductive, and hence one has $\bar{\mathfrak{G}}^s = [\bar{\mathfrak{G}}, \bar{\mathfrak{G}}]$. Therefore, for $T' \in \varphi(V) \square \varphi(V) \subset \bar{\mathfrak{G}}^s$, one has by (7.6)

$$(*) \qquad T'\varphi(V) = T'^*\varphi(V) = 0 \Longrightarrow [T', \bar{\mathfrak{G}}] = 0 \Longrightarrow T' = 0.$$

Now we define $\rho: \mathfrak{G} \to \bar{\mathfrak{G}} \subset \mathfrak{G}'$ by

$$(9.4) \qquad \begin{cases} \rho(x) = \varphi(x), \\ \rho(x \square y) = \varphi(x) \square \varphi(y), \\ \rho(\theta x) = \theta' \varphi(x) \qquad (x, y \in V). \end{cases}$$

To see that ρ is well-defined, we have to check that, for $x_i, y_i \in V$, $\sum x_i \square y_i = 0$ implies $\sum \varphi x_i \square \varphi y_i = 0$. If we put $T' = \sum \varphi x_i \square \varphi y_i$, then from $\sum x_i \square y_i = 0$ one has

$$T'(\varphi z) = \sum \{\varphi x_i, \varphi y_i, \varphi z\}' = \varphi(\sum \{x_i, y_i, z\}) = 0 \qquad (z \in V),$$

and similarly, from $\sum y_i \square x_i = (\sum x_i \square y_i)^* = 0$, one has $T'^*(\varphi z) = 0$ for all $z \in V$. Hence by (*) one has $T' = 0$, as desired. It is then easy to see that ρ is a Lie algebra homomorphism, and it is clear from the definition that ρ is the unique extension of φ satisfying the conditions (9. 2), (9. 3), q. e. d.

Proposition 9. 1 shows that the correspondence between positive definite JTS's $(V, \{\ \})$ and symmetric Lie algebras $(\mathfrak{G}, \theta)$ with Cartan involution θ given in § 7 is actually an equivalence of the two categories*).

Now let Ω, Ω' be self-dual homogeneous cones with reference points e, e' in real vector spaces U, U' with (compatible but not necessarily normalized) inner products $\langle\ \rangle$, $\langle\ \rangle'$. As explained in § 8, U, U' have (uniquely determined) structures of formally real Jordan algebras with unit elements e, e'. In what follows, the objects relative to (U', e') will be denoted by the corresponding symbols with prime.

Proposition 9. 2. *The notation being as above, a linear map $\varphi: U \rightarrow U'$ with $\varphi(e)=e'$ is a (unital) Jordan algebra homomorphism if and only if there exists a Lie algebra homomorphism $\rho_0: \mathfrak{g}(\Omega) \rightarrow \mathfrak{g}(\Omega')$ satisfying the following conditions:*

$$\varphi(Tx) = \rho_0(T)\varphi(x), \tag{9.5}$$

$$\rho_0({}^tT) = {}^t\rho_0(T) \qquad (T \in \mathfrak{g}(\Omega),\ x \in U), \tag{9.6}$$

where t denotes the adjoint with respect to the given inner products in U and U'. When these conditions are satisfied, φ and ρ_0 determine each other uniquely, and one has $\varphi(\Omega) \subset \Omega'$.

Proof. First suppose φ is a Jordan algebra homomorphism. Then, φ, being also a JTS homomorphism, can uniquely be extended to a symmetric Lie algebra homomorphism $\rho: \mathfrak{G} \rightarrow \mathfrak{G}'$ by Proposition 9. 1. Identifying $\mathfrak{G}_0$ and $\mathfrak{G}_0'$ with $\mathfrak{g}(\Omega)$ and $\mathfrak{g}(\Omega')$, respectively, put $\rho_0 = \rho|\mathfrak{G}_0$. Then it is clear from (9. 2) and (9. 3) that ρ_0 is a Lie algebra homomorphism of $\mathfrak{g}(\Omega)$ into $\mathfrak{g}(\Omega')$ satisfying the conditions (9. 5), (9. 6).

Conversely, suppose there is given a Lie algebra homomorphism ρ_0 satisfying these conditions. Then one has

$$\rho_0(T_a)e' = \varphi(T_a e) = \varphi(a),$$
$${}^t(\rho_0(T_a)) = \rho_0(T_a).$$

Hence, by the characterization of the operator $T'_{a'}$ $(a' \in U')$, one has

$$\rho_0(T_a) = T'_{\varphi(a)}. \tag{9.7}$$

Since $\mathfrak{p} = \{T_a (a \in U)\}$ generates $\mathfrak{g}(\Omega)$, this implies that φ and ρ_0 determine each other uniquely. Moreover one has

$$\begin{aligned}\varphi(xy) &= \varphi(T_x y) = \rho_0(T_x)\varphi(y)\\ &= T'_{\varphi(x)}\varphi(y) = \varphi(x)\varphi(y),\end{aligned}$$

i. e., φ is a Jordan algebra homomorphism. Finally, since $\Omega = (\exp \mathfrak{p})e$, one has

$$\varphi(\Omega) = (\exp \rho_0(\mathfrak{p}))e' \subset (\exp \mathfrak{p}')e' = \Omega', \qquad \text{q. e. d.}$$

---

*) In a written communication with the author (November, 1975), Prof. N. Jacobson has shown that this result is also true for any semi-simple Jordan algebra over F. The case of semi-simple JTS is treated in the thesis of E. Neher [1].

We call a pair (ρ_0, φ) satisfying the conditions (9. 5), (9. 6) in Proposition 9. 2 an *equivariant pair*. Sometimes a linear map $\varphi: U \to U'$ with $\varphi(e)=e'$, or a Lie algebra homomorphism $\rho_0: \mathfrak{g}(\Omega) \to \mathfrak{g}(\Omega')$ alone is called "equivariant" if it belongs to an equivariant pair. Combining Theorem 8. 5 with Proposition 9. 2 we can conclude that the following three categories are equivalent:

($\mathcal{J}$) The category of formally real Jordan algebras U, the morphisms being unital Jordan algebra homomorphisms.

($\mathcal{C}$) The category of self-dual homogeneous cones Ω with reference point e, the morphisms being equivariant linear maps φ of the ambient spaces U.

($\mathcal{RL}$) The category of reductive Lie algebras $\mathfrak{g}(\Omega)$ (corresponding to self-dual homogeneous cones Ω) with Cartan involution θ_0, the morphisms being equivariant Lie algebra homomorphisms.

Note that the Lie algebras $\mathfrak{g}=\mathfrak{g}(\Omega)$ corresponding to self-dual homogeneous cones are characterized by the properties stated in Lemma 8. 6. As a consequence of the above equivalences, we obtain

Corollary 9. 3. *The notation being as in Proposition* 9. 2, *let φ be a linear isomorphism $U \to U'$ with $\varphi(e)=e'$. Then the following three conditions are equivalent:*

(a) *φ is a Jordan algebra isomorphism.*

(b) $\varphi(\Omega) = \Omega'$.

(c) $\varphi \circ \mathfrak{g}(\Omega) \circ \varphi^{-1} = \mathfrak{g}(\Omega')$.

Proof. (a)⇒(b)⇒(c) is trivial. Suppose (c) is satisfied. Then one has $\varphi G(\Omega)^\circ \varphi^{-1} = G(\Omega')^\circ$ and hence

$$\Omega' = G(\Omega')^\circ e' = \varphi(G(\Omega)^\circ e) = \varphi(\Omega),$$

i. e., (b). Define $\rho_0: \mathfrak{g}(\Omega) \to \mathfrak{g}(\Omega')$ by $\rho_0(T) = \varphi T \varphi^{-1}$ for $T \in \mathfrak{g}(\Omega)$. Then ρ_0 is clearly a Lie algebra isomorphism satisfying (9. 5). Moreover, since $\varphi(\Omega)=\Omega'$, $\varphi(e)=e'$, Ω is self-dual with respect to the inner products $\langle x, y\rangle$ and $\langle \varphi x, \varphi y\rangle'$ $(x, y \in U)$, and these inner products correspond to the same maximal compact subgroup of $G(\Omega)$. Hence the condition (9. 6) is also satisfied. Therefore by Proposition 9. 2 φ is a Jordan algebra isomorphism, q. e. d.

Proposition 9. 4. *Let U be a formally real Jordan algebra corresponding to a self-dual homogeneous cone Ω with a reference point e. Then one has*

$$\Gamma(U, \{\ \})^\circ \subset G(\Omega) \subset \Gamma(U, \{\ \}), \tag{9. 8}$$

$$K = \{g \in G(\Omega) \mid ge=e\} = \mathrm{Aut}_{\mathrm{J.alg.}}(U), \tag{9. 9}$$

and the Zariski closure of $G(\Omega)$ coincides with $\Gamma(U, \{\ \})$. When Ω is irreducible, one has

$$\Gamma(U, \{\ \}) = G(\Omega) \times \{\pm 1_U\} \qquad (\textit{as Lie group}). \tag{9. 10}$$

Proof. In this proof we write Γ for $\Gamma(U, \{\ \})$. (9. 9) follows from Corollary 9. 3 (or directly from Theorem 8. 5). On the other hand, by Corollary 6. 3 one has

$$\{g \in \Gamma \mid {}^t g^{-1} = g\} = \mathrm{Aut}_{\mathrm{JTS}}(U),$$

and clearly $\mathrm{Aut}_{\mathrm{J.alg.}}(U) \subset \mathrm{Aut}_{\mathrm{JTS}}(U)$. Since $\mathrm{Lie}\,\Gamma = \mathfrak{g}(\Omega)$ (Th. 8. 5), it follows that $G(\Omega) = K \cdot \exp \mathfrak{p} \subset \Gamma$ and $G(\Omega)^\circ = \Gamma^\circ$, which proves (9. 8). Now, when Ω is irreducible, $g \in \mathrm{Aut}_{\mathrm{JTS}}(U)$ implies $ge = \pm e$ (Exerc. 1), so that one has

$$(\Gamma : G(\Omega)) = (\mathrm{Aut}_{\mathrm{JTS}}(U) : \mathrm{Aut}_{\mathrm{J.alg.}}(U)) = 2.$$

Clearly $\boldsymbol{R}^\times \subset \Gamma$, but $-1 \notin G(\Omega)$. Hence one has (9. 10) and the Zariski closure of $G(\Omega)$ coincides with Γ. This last assertion can easily be extended to the reducible case, q. e. d.

For $g \in \Gamma = \Gamma(U, \{\ \})$ and $x \in \mathfrak{g}(\Omega)^s$, put $\alpha(g)(x) = gxg^{-1}$. Then α is an $\boldsymbol{R}$-homomorphism of Γ into $\mathrm{Aut}\,\mathfrak{g}(\Omega)^s$. We assume Ω to be irreducible. Then, since $\mathfrak{g}(\Omega)^s$ is absolutely irreducible in $\mathrm{End}\, U$ (or by the uniqueness mentioned in Proposition 9. 2), one has $\mathrm{Ker}\,\alpha = \boldsymbol{R}^\times$. It follows (§ 1, Exerc. 1) that the image $\alpha(\Gamma)$ is Zariski closed and one has an $\boldsymbol{R}$-isomorphism

(9. 11) $$\Gamma/\boldsymbol{R}^\times \cong \alpha(\Gamma) \ (\subset \mathrm{Aut}\,\mathfrak{g}(\Omega)^s),$$

which also induces a Lie group isomorphism $G(\Omega)/\boldsymbol{R}_+^\times \cong \alpha(G(\Omega)) = \alpha(\Gamma)$. Thus one has

$$\begin{aligned}(\Gamma : \Gamma^z) &= (\alpha(\Gamma) : \mathrm{Ad}(\mathfrak{g}(\Omega)^s)) \\ &\leq (\alpha(\Gamma) : \mathrm{Inn}(\mathfrak{g}(\Omega)^s)) = (G(\Omega) : G(\Omega)^\circ).\end{aligned}$$

Since, for each real simple Lie algebra $\mathfrak{g}$, the structure of $\mathrm{Aut}\,\mathfrak{g}/\mathrm{Inn}\,\mathfrak{g}$ is well-known (see e. g., Takeuchi [1]), these indices can easily be determined. We list the result in the following table (where 2×2 means that the factor group is an abelian group of type (2, 2)).

Ω	$\mathfrak{g}(\Omega)^s$	$\mathrm{Aut}/\mathrm{Inn}(\mathfrak{g}(\Omega)^s)$	$(\Gamma : \Gamma^z)$	$(G(\Omega) : G(\Omega)^\circ)$
$\mathcal{P}_n(\boldsymbol{R})$ $(n\geq 3)$	$\mathfrak{sl}_n(\boldsymbol{R})$	$\begin{cases} 2\times 2 & (n \text{ even}) \\ 2 & (n \text{ odd})\end{cases}$	1	$\begin{cases} 2 & (n \text{ even}) \\ 1 & (n \text{ odd})\end{cases}$
$\mathcal{P}_n(\boldsymbol{C})$ $(n\geq 3)$	$\mathfrak{sl}_n(\boldsymbol{C})$	2×2	2	2
$\mathcal{P}_n(\boldsymbol{H})$ $(n\geq 3)$	$\mathfrak{sl}_n(\boldsymbol{H})$	2	1	1
$\mathcal{P}(1, n-1)$ $(n\geq 3)$	$\mathfrak{o}(1, n-1)$	2	$\begin{cases} 2 & (n \text{ even}) \\ 1 & (n \text{ odd})\end{cases}$	2
$\mathcal{P}_3(\boldsymbol{O})$	$(E_{6,2})$	2	1	1

A representative g_1 for the non-trivial class in $G(\Omega)/G(\Omega)^\circ$ is given as follows:
For $\Omega = \mathcal{P}_n(\boldsymbol{R})$ $(n \geq 2$, even$)$, $g_1 : x \mapsto E_1 x E_1$, $E_1 = \mathrm{diag}(-1, 1, \cdots, 1)$.
For $\Omega = \mathcal{P}_n(\boldsymbol{C})$ $(n\geq 2)$ and $\mathcal{P}_2(\boldsymbol{H})$, $g_1 : x \mapsto \bar{x} = {}^t x$.
For $\Omega = \mathcal{P}(1, n-1)$, $g_1 : x \mapsto E_2 x$, $E_2 = \mathrm{diag}(1, -1, 1, \cdots, 1)$.

Note that $\mathcal{P}_n(\boldsymbol{H})$ $(n\geq 3)$ is not stable under $x \mapsto \bar{x}$. For instance,

$$\begin{pmatrix} 1 & \boldsymbol{k} & \boldsymbol{j} \\ -\boldsymbol{k} & 2 & \boldsymbol{i} \\ -\boldsymbol{j} & -\boldsymbol{i} & 2 \end{pmatrix} \in \mathcal{P}_3(\boldsymbol{H}), \quad \text{but} \quad \begin{pmatrix} 1 & -\boldsymbol{k} & -\boldsymbol{j} \\ \boldsymbol{k} & 2 & -\boldsymbol{i} \\ \boldsymbol{j} & \boldsymbol{i} & 2 \end{pmatrix} \notin \mathcal{P}_3(\boldsymbol{H}).$$

Remark 1. It can be seen that the Cartan involution θ of $\mathfrak{g}(\Omega)^s$ belongs to $\alpha(\Gamma)$ [or equivalently, there exists $g_2 \in G(\Omega)$ such that $x^{-1} = \lambda(x) g_2 x$ for all $x \in \Omega$, where $\lambda(x) \in \boldsymbol{R}_+^\times$], if and only if $r=2$ (cf. Exerc. 3, 4). When $r>2$, Aut $\mathfrak{g}(\Omega)^s$ is generated by $\alpha(\Gamma)$ and θ, so that one has (Aut $\mathfrak{g}(\Omega)^s$: $\alpha(\Gamma))=2$.

Remark 2. We shall see later on (V, § 1, (A) ; Cor. V. 3. 7) that, for the tube domain $\mathcal{D}=U+i\Omega$ corresponding to a self-dual homogeneous cone Ω, the Lie algebra of Hol$(\mathcal{D})$ can be identified with the symmetric Lie algebra $\mathfrak{G}(U, \{\ \})$. Moreover, the group of affine automorphisms Aff$(\mathcal{D})$ is given by $\{z \mapsto gz+a \mid g \in G(\Omega), a \in U\}$ and Hol$(\mathcal{D})$ is generated by Aff$(\mathcal{D})$ and the symmetry of $\mathcal{D}$ at ie (given by the rational map $x \mapsto -x^{-1}$).

Exercises

1. Let A be a semi-simple Jordan algebra with unit element e, and let e' be any element such that $P(e')=1$, or equivalently, $e'^2=e$ and $T_{e'}^2=1$. Show that e' is in the "center" of A, i. e., $[T_x, T_{e'}]=0$ for all $x \in A$. (*Hint.* Use (6. 17′) to obtain $x \square e' = e' \square x$.)

Remark. In particular, if A is simple, this implies that $e'=\pm e$. In general, if $A=\bigoplus_{i=1}^{s} A_i$ is the decomposition of A into simple components and if $e=\sum e_i$ with $e_i \in A_i$, then one has $e'=\sum \pm e_i$. From this, we can conclude that the number of distinct Jordan algebra structures (with unit element) compatible with the given JTS structure is 2^s. Otherwise expressed, one has

$$(\mathrm{Aut}_{\mathrm{JTS}}(A) : \mathrm{Aut}_{\mathrm{J.alg.}}(A)) = 2^s \tag{9. 2}$$

(Braun-Koecher [1], IV, § 6).

2. Let A be a real simple Jordan algebra of dimension m and let $\rho: A \to \mathcal{H}_n(\boldsymbol{C})$ be a unital Jordan algebra homomorphism. Show that

$$\mathrm{tr}\, \rho(x) = \frac{n}{m} \mathrm{tr}\,(T_x).$$

3. Let $A=J(U, S, e)$ be a simple Jordan algebra of type (IV). For $x \in A$, show that x is invertible if and only if $S(x, x) \neq 0$ and that if this condition is satisfied, one has

$$x^{-1} = -S(x, x)^{-1} \hat{x} \qquad (\hat{x} = x - 2S(x, e)e).$$

4. Let $\Omega = \mathcal{P}_2(\boldsymbol{H})$ and $x=\begin{pmatrix} x_1 & x_{12} \\ \bar{x}_{12} & x_2 \end{pmatrix} \in \Omega$. Show that the reduced norm of x is given by $n(x)=(x_1x_2 - x_{12}\bar{x}_{12})^2$ and

$$x^{-1} = n(x)^{-\frac{1}{2}} \begin{pmatrix} 0 & -1 \\ 1 & 0 \end{pmatrix} \bar{x} \begin{pmatrix} 0 & 1 \\ -1 & 0 \end{pmatrix}.$$

5. Let $(V, \{\ \})$ be a positive definite JTS over $\boldsymbol{R}$, and $\mathfrak{G}=\mathfrak{G}(V, \{\ \})$ the associated symmetric Lie algebra. Show that $\boldsymbol{R}$-rank $\mathfrak{G} = \boldsymbol{R}$-rank $\mathfrak{G}_0$. (*Hint.* Let $\mathfrak{a}$ be an abelian subalgebra of $\mathfrak{G}_0$ contained in its $\mathfrak{p}$-part and maximal with respect to this property. Then, one has $1_V \in \mathfrak{a}$ and so $\mathfrak{c}(\mathfrak{a}) \subset \mathfrak{G}_0$, which shows that $\mathfrak{a}$ is also maximal in $\mathfrak{G}$ with respect to the same property.)

Chapter II

Basic Concepts on Symmetric Domains

§ 1. Riemannian Symmetric Spaces.

In this section, we summarize some basic concepts and results on (globally) symmetric Riemannian manifolds. The main references will be Kobayashi-Nomizu [1], abbreviated as K-N, and Helgason [1], [1a]. All manifolds and mappings to be considered in this chapter are supposed to be C^∞.

Let M be a connected Riemannian manifold with a (C^∞) Riemannian metric q. We denote by $I(M)$ the group of isometries of M and by $I(M)_x$ the isotropy subgroup at $x \in M$ (i. e., the stabilizer of x in $I(M)$). It is known (K-N, VI, Th. 3. 4) that $I(M)$ with the compact-open topology becomes a Lie group and $I(M)_x$ is a compact subgroup. For $x \in M$, let $T_x(M)$ denote the tangent space at x to M and q_x the (positive definite) inner product on $T_x(M)$ defining q. For $\varphi \in I(M)_x$ let $d_x\varphi$ denote the differential of φ at x. Then one has $d_x\varphi \in O(T_x(M), q_x)$ and the linear representation

$$I(M)_x \ni \varphi \longmapsto d_x\varphi \in O(T_x(M), q_x) \tag{1. 1}$$

is faithful, so that the compact Lie group $I(M)_x$ is isomorphic to its image in $O(T_x(M), q_x)$, called the linear isotropy subgroup at x. (This follows immediately from the existence of "normal coordinates". Cf. K-N, IV, § 3; Helgason [1], I, Lem. 11. 2.)

An isometry $s_x \in I(M)_x$ is called a *symmetry* at x if $s_x^2 = id$ and x is an isolated fixed point of s_x. These conditions are equivalent to saying that $d_x s_x = -1_{T_x}$. Hence the symmetry s_x (if it exists) is uniquely determined by x. A Riemannian manifold M is called (globally) *symmetric*, or we say M is a *Riemannian symmetric space*, if there exists a symmetry s_x at every point $x \in M$. A Riemannian symmetric space is always "homogeneous" (i. e., $I(M)$ is transitive on M) and hence complete (Helgason [1], p. 170).

Let M be a Riemannian symmetric space, $x_0 \in M$, and set $G = I(M)$, $K = I(M)_{x_0}$, $s = s_{x_0}$. Then the correspondence $gx_0 \leftrightarrow gK$ gives a G-equivariant diffeomorphism $M \approx G/K$. The Lie algebras $\mathfrak{g}$ and $\mathfrak{k}$ of G and K are canonically identified with the tangent spaces $T_e(G)$ and $T_e(K)$ at the identity element $e = id_M$. The natural projection map $\pi: G \to M$ defined by $\pi(g) = gx_0$ gives rise to an exact sequence of linear mappings

$$0 \longrightarrow \mathfrak{k} \longrightarrow \mathfrak{g} \xrightarrow{d\pi} T_{x_0}(M) \longrightarrow 0,$$

so that $T_{x_0}(M)$ may also be identified with $\mathfrak{g}/\mathfrak{k}$ and the isotropy representation (1. 1) of K coincides with the one induced on $\mathfrak{g}/\mathfrak{k}$ by the adjoint representation. For $g \in G$, one has

$$g \in K \Longleftrightarrow gx_0 = x_0 \Longrightarrow gsg^{-1} = s_{gx_0} = s. \tag{1.2}$$

The last relation can also be expressed as $\nu_g s = s$, or equivalently $\nu_s g = g$, where ν_g denotes the inner automorphism of G defined by g. If we put $\theta = ad\, s\, (= d_e \nu_s)$, then θ is an involutive automorphism of $\mathfrak{g}$, and by (1.2) one has $\theta|\mathfrak{k} = id$. On the other hand, θ induces $d_{x_0} s = -id$ on $T_{x_0}(M) = \mathfrak{g}/\mathfrak{k}$. Thus one has $\mathfrak{k} = \mathfrak{g}(\theta\,;1)$, which implies

$$C_G(s)^\circ \subset K \subset C_G(s), \tag{1.3}$$

where $C_G(s)$ denotes the centralizer of s in G.

We put $\mathfrak{p} = \mathfrak{g}(\theta\,;-1)$. Then clearly one has

$$\begin{gathered} \mathfrak{g} = \mathfrak{k} + \mathfrak{p}, \qquad \mathfrak{k} \cap \mathfrak{p} = \{0\}, \\ [\mathfrak{k}, \mathfrak{k}] \subset \mathfrak{k}, \qquad [\mathfrak{k}, \mathfrak{p}] \subset \mathfrak{p}, \qquad [\mathfrak{p}, \mathfrak{p}] \subset \mathfrak{k}. \end{gathered} \tag{1.4}$$

In the following, we always make a natural identification $T_{x_0}(M) = \mathfrak{p}$. Then $X \in \mathfrak{p}$ is tangent to the (geodesic) curve $(\exp tX)x_0$ $(t \in \boldsymbol{R})$. For $k \in K$, $d_{x_0} k$ is identified with the restriction of $ad\, k = d_e \nu_k$ on $\mathfrak{p}$, which we denote by $ad_{\mathfrak{p}}\, k$. If moreover $k = \exp X$ $(X \in \mathfrak{k})$, the latter is equal to $\exp(ad\, X)|\mathfrak{p}$. After these identifications, it is known (Helgason [1], IV, Th. 4. 2 ; K-N, XI, Th. 3. 2) that the "curvature tensor" R of M is given by

$$R_{x_0}(X, Y)Z = -[[X, Y], Z] \qquad (X, Y, Z \in \mathfrak{p}). \tag{1.5}$$

One then has $\nabla R = 0$ and the "holonomy algebra" at x_0 is given by $ad_{\mathfrak{p}}\, [\mathfrak{p}, \mathfrak{p}]$. Also, the "sectional curvature" at x_0 in the direction of a plane $\{X, Y\}_{\boldsymbol{R}}$ $(X, Y \in \mathfrak{p})$ is given by

$$-q_{x_0}(R_{x_0}(X, Y)X, Y) = -q_{x_0}((ad\, X)^2 Y, Y), \tag{1.6}$$

where X, Y are taken to be orthonormal with respect to q_{x_0}.

Now let M be a simply connected Riemannian symmetric space. Then (K-N, XI, Th. 6. 6 ; Helgason [1], V) M is uniquely decomposed into the direct product

$$M = M_0 \times M_1 \times \cdots \times M_r, \tag{1.7}$$

called the "De Rham decomposition", where M_0 is a Euclidean space and the M_i $(1 \leq i \leq r)$ are irreducible (i. e., indecomposable) Riemannian symmetric spaces ; moreover one has

$$I(M)^\circ = I(M_0)^\circ \times I(M_1)^\circ \times \cdots \times I(M_r)^\circ, \tag{1.8}$$

where the $I(M_i)^\circ$ $(1 \leq i \leq r)$ are simple and semi-simple.

Let M be an irreducible (simply connected, non-Euclidean) Riemannian symmetric space. Then the adjoint representation of $\mathfrak{k}$ (hence that of K°) on $\mathfrak{p}$ is irreducible, so that a K°-invariant (positive definite) inner product on $\mathfrak{p}$ is uniquely determined up to a positive constant. Hence one has

$$q_{x_0}(X, Y) = cB_{\mathfrak{p}}(X, Y) \qquad (X, Y \in \mathfrak{p}) \tag{1.9}$$

with a non-zero constant c, $B_{\mathfrak{p}}$ denoting the restriction of the Killing form B on $\mathfrak{p}$. Clearly one has

(1. 10) $$M \text{ compact} \Longleftrightarrow I(M) \text{ compact} \Longleftrightarrow c < 0.$$

Therefore, if M is non-compact, one has $c>0$ in (1. 9), which implies that the decomposition (1. 4) is a Cartan decomposition of $\mathfrak{g}$ (cf. I, § 4). In general, as is seen from (1. 6) and (1. 9), a simply connected Riemannian symmetric space M has a non-positive sectional curvature if and only if there is no compact factor in the De Rham decomposition (1. 7).

A Riemannian symmetric space M is called *of non-compact type* if, in the decomposition (1. 7) of the universal covering space of M, every irreducible component M_i $(1 \leq i \leq r)$ is non-compact and M_0 reduces to a point. G° is then a semi-simple Lie group and K° is a maximal compact subgroup of G°. Moreover one has a (topological) direct product decomposition $G^\circ = K^\circ \cdot \exp \mathfrak{p}$, which gives rise to a diffeomorphism $M \approx \exp \mathfrak{p} \approx \mathfrak{p}$. Hence, in this case, M itself is simply connected and the center of G° is finite. (It will be shown in § 2 that the center of G° reduces to the identity.)

Conversely, let G° be any connected semi-simple Lie group with a finite center and K° a maximal compact subgroup of G°. Then, starting with any K°-invariant (positive definite) inner product on $\mathfrak{p}$, one can define a G°-invariant Riemannian metric q on the coset space $M = G^\circ/K^\circ$. With this Riemannian metric, M becomes a Riemannian symmetric space of non-compact type, and one has a natural surjective homomorphism $G^\circ \to I(M)^\circ$, of which the kernel is compact. (The surjectivity follows from the fact that $I(M)^\circ$ is generated by $\exp \mathfrak{p}$.) The G°-invariant Riemannian metric q on M is not unique, but is determined up to a positive constant on each irreducible component M_i of M. We say q is *normalized* if the relation (1. 9) holds with $c = 1/2$.

Let M be a Riemannian symmetric space of non-compact type. Then, the exponential map $\mathfrak{p} \to \exp \mathfrak{p}$ being bijective, every point $x \in M$ can be expressed uniquely in the form

(1. 11) $$x = (\exp X)x_0 \qquad \text{with} \quad X \in \mathfrak{p},$$

and $\{(\exp tX)x_0 \ (t \in \boldsymbol{R})\}$ is a geodesic passing through the "origin" x_0. Thus, for every $x_1 \in M$, $x_0 \neq x_1$, there exists a unique geodesic passing through x_0 and x_1. Since the symmetry $s = s_{x_0}$ is given by

$$(\exp X)x_0 \longmapsto (\exp(-X))x_0,$$

it follows that s has no fixed point other than x_0. Therefore in (1. 2) one has

(1. 12) $$gx_0 = x_0 \Longleftrightarrow \nu_g s = s,$$

which implies that $K = C_G(s)$ in (1. 3).

Exercise

1. Let Ω be a self-dual homogeneous cone with reference point e (I, § 8). Show that Ω endowed with the metric defined in I, § 8, Exerc. 4 is a Riemannian symmetric space and that the symmetry at e is given by $x \mapsto x^{-1}$.

§ 2. Equivariant maps of Riemannian symmetric spaces.

Let M and M' be Riemannian symmetric spaces of non-compact type and let $G=I(M)$, $G'=I(M')$, $\mathfrak{g}=\text{Lie } G$, $\mathfrak{g}'=\text{Lie } G'$. A pair (ρ, φ) formed of a Lie algebra homomorphism $\rho: \mathfrak{g}\to\mathfrak{g}'$ and a C^∞-map $\varphi: M\to M'$ is called *weakly equivariant* if one has

$$\varphi((\exp X)x) = (\exp \rho(X))\varphi(x) \qquad \text{for all } X\in\mathfrak{g},\ x\in M. \tag{2. 1}$$

In the following, we fix the origins $x_0\in M$ and $x_0'\in M'$ once and for all and set $K=G_{x_0}$, $K'=G'_{x_0'}$, $\theta=ad\, s_{x_0}$, $\theta'=ad\, s_{x_0'}$. $\mathfrak{g}=\mathfrak{k}+\mathfrak{p}$ and $\mathfrak{g}'=\mathfrak{k}'+\mathfrak{p}'$ are the corresponding Cartan decompositions of $\mathfrak{g}$ and $\mathfrak{g}'$.

Lemma 2. 1. *If (ρ, φ) is a weakly equivariant pair and $\varphi(x_0)=x_0'$, then one has $\rho(\mathfrak{k})\subset\mathfrak{k}'$. Conversely, if ρ is a Lie algebra homomorphism of $\mathfrak{g}$ into $\mathfrak{g}'$ with $\rho(\mathfrak{k})\subset\mathfrak{k}'$, then there exists a unique C^∞-map $\varphi: M\to M'$ such that $\varphi(x_0)=x_0'$ and (ρ, φ) is weakly equivariant.*

Proof. The first assertion is obvious. To prove the second, let ρ be a Lie algebra homomorphism $\mathfrak{g}\to\mathfrak{g}'$ with $\rho(\mathfrak{k})\subset\mathfrak{k}'$. We recall that one has $K^\circ=K\cap G^\circ$ (resp. $K'^\circ=K'\cap G'^\circ$) so that one has a natural identification $M=G^\circ/K^\circ$ (resp. $M'=G'^\circ/K'^\circ$). The Lie algebra homomorphism ρ can be lifted to a "local" homomorphism, denoted again by ρ, of G° into G'°. One can then construct a covering group $(\tilde{G}^\circ, \pi)$ of G° and a (global) homomorphism $\tilde{\rho}: \tilde{G}^\circ\to G'^\circ$ such that $\tilde{\rho}$ coincides with $\rho\circ\pi$ in a neighbourhood of the identity of $\tilde{G}^\circ$. Putting $\tilde{K}^\circ=\pi^{-1}(K^\circ)$, one has $\tilde{G}^\circ/\tilde{K}^\circ\approx G^\circ/K^\circ=M$. Since $\rho(\mathfrak{k})\subset\mathfrak{k}'$, one has $\tilde{\rho}(\tilde{K}^\circ)\subset K'^\circ$, so that $\tilde{\rho}$ induces a natural C^∞-map

$$\varphi:\ M=\tilde{G}^\circ/\tilde{K}^\circ \longrightarrow M'=G'^\circ/K'^\circ,$$

which clearly satisfies the conditions $\varphi(x_0)=x_0'$ and

$$\varphi(\tilde{g}x) = \tilde{\rho}(\tilde{g})\varphi(x) \qquad \text{for all } \tilde{g}\in\tilde{G}^\circ,\ x\in M. \tag{2. 1'}$$

The condition (2. 1) follows from (2. 1'). Thus (ρ, φ) is weakly equivariant. Writing $x\in M$ in the form $x=(\exp X)x_0$ with $X\in\mathfrak{p}$, one has by (2. 1)

$$\varphi(x) = (\exp \rho(X))x_0', \tag{2. 2}$$

which proves the uniqueness of φ, q. e. d.

A weakly equivariant pair (ρ, φ) with $\varphi(x_0)=x_0'$ is said to be (*strongly*) *equivariant* (*at x_0*) if one has

$$\rho\circ\theta = \theta'\circ\rho. \tag{2. 3}$$

When this condition is satisfied, ρ is also called a "homomorphism" of $(\mathfrak{g}, \theta)$ into $(\mathfrak{g}', \theta')$; we say also that ρ and φ are *associated* with each other. Clearly the condition (2. 3) is equivalent to saying that

$$\rho(\mathfrak{k})\subset\mathfrak{k}', \qquad \rho(\mathfrak{p})\subset\mathfrak{p}'. \tag{2. 3'}$$

Hence, if (ρ, φ) is equivariant, one has $\rho(X)\in\mathfrak{p}'$ in (2. 2), and the following diagram is commutative:

$$\begin{array}{ccc} \mathfrak{p} & \xrightarrow{\approx} & M \\ {\scriptstyle\rho}\downarrow & & \downarrow{\scriptstyle\varphi} \\ \mathfrak{p}' & \xrightarrow{\approx} & M' \end{array} \tag{2.4}$$

Since the homomorphism ρ is uniquely determined by its restriction to $\mathfrak{p}$, we see that ρ and φ determine each other uniquely. Also, in this situation, it is clear that φ is surjective (resp. injective) if and only if ρ is so.

The following lemma shows that the notion of equivariance is independent of the choice of the "origin" x_0.

Lemma 2. 2. *If (ρ, φ) is (strongly) equivariant at x_0, it is so everywhere, i. e., for any $x_1 \in M$ one has*

$$\rho \circ \theta_{x_1} = \theta'_{\varphi(x_1)} \circ \rho, \tag{2.5}$$

where θ_{x_1} and $\theta'_{\varphi(x_1)}$ are Cartan involutions of $\mathfrak{g}$ and $\mathfrak{g}'$ at x_1 and $\varphi(x_1)$.

Proof. Take $X_1 \in \mathfrak{p}$ such that one has $x_1 = (\exp X_1)x_0$. Then by (2. 2) one has $\varphi(x_1) = (\exp \rho(X_1))x'_0$. Since the symmetry s_{x_1} at x_1 is given by

$$s_{x_1} = (\exp X_1)s(\exp X_1)^{-1},$$

one has in Aut $\mathfrak{g}$

$$\theta_{x_1} = ad\, s_{x_1} = (\exp(ad\, X_1))\theta(\exp(ad\, X_1))^{-1}.$$

Similarly,

$$\theta'_{\varphi(x_1)} = (\exp(ad\, \rho(X_1)))\theta'(\exp(ad\, \rho(X_1)))^{-1}.$$

Hence by (2. 3) one has

$$\begin{aligned} \rho\theta_{x_1} &= \rho(\exp ad\, X_1)\theta(\exp ad\, X_1)^{-1} \\ &= (\exp ad\, \rho(X_1))\rho\theta(\exp ad\, X_1)^{-1} \\ &= (\exp ad\, \rho(X_1))\theta'\rho(\exp ad\, X_1)^{-1} \\ &= (\exp ad\, \rho(X_1))\theta'(\exp ad\, \rho(X_1))^{-1}\rho \\ &= \theta'_{\varphi(x_1)}\rho, \end{aligned}$$

which proves our assertion, q. e. d.

Now we want to obtain a geometric characterization of C^∞-maps $\varphi: M \to M'$ associated to homomorphisms $\rho: (\mathfrak{g}, \theta) \to (\mathfrak{g}', \theta')$. Let (ρ, φ) be an equivariant pair. Then one has

$$\varphi((\exp X)x_0) = (\exp \rho(X))x'_0,$$

where $X \in \mathfrak{p}$ and $\rho(X) \in \mathfrak{p}'$. This implies that one has $(d_{x_0}\varphi)(X) = \rho(X)$ for $X \in \mathfrak{p}$. More generally, for any $x_1 \in M$ and for any $Y \in \mathfrak{p}_{x_1} = T_{x_1}(M)$ one has

$$\varphi((\exp Y)x_1) = (\exp \rho(Y))\varphi(x_1), \tag{2.6}$$

where one has $\rho(Y) \in \mathfrak{p}'_{\varphi(x_1)} = T_{\varphi(x_1)}(M')$ by Lemma 2. 2. Thus one has

$$d_{x_1}\varphi = \rho|\mathfrak{p}_{x_1}, \tag{2.7}$$

where $\mathfrak{p}_{x_1}$ is the $\mathfrak{p}$-part of the Cartan decomposition of $\mathfrak{g}$ at x_1. When ρ is injective, we can replace the Riemannian metric on M by the "induced" metric

$$q_{x_1}(X, Y) = q'_{\varphi(x_1)}(\rho(X), \rho(Y)) \qquad (X, Y \in \mathfrak{p}_{x_1}).$$

Then φ becomes an isometry of M into M'. Moreover the image $\varphi(M)$ is "totally geodesic" in M', because for any $x_1 \in M$ the image of a geodesic $\{(\exp tY)x_1\}$ passing through x_1 is a geodesic $\{(\exp t\rho(Y))\varphi(x_1)\}$ in M' passing through $\varphi(x_1)$.

In general, when ρ is not injective, let $\mathfrak{g}^{(2)} = \operatorname{Ker} \rho$. Then, since $\mathfrak{g}$ is semi-simple, there exists a unique ideal $\mathfrak{g}^{(1)}$ such that $\mathfrak{g} = \mathfrak{g}^{(1)} \oplus \mathfrak{g}^{(2)}$. One also has the corresponding direct product decomposition $M = M^{(1)} \times M^{(2)}$ of the symmetric space. Let $p^{(i)}$ denote the projection $M \to M^{(i)}$. Then $\rho|\mathfrak{g}^{(1)}$ is an injective homomorphism of $\mathfrak{g}^{(1)}$ into $\mathfrak{g}'$ and hence (by Lem. 2. 1) there exists a unique injective C^∞-map $\varphi^{(1)}$: $M^{(1)} \to M'$ such that $\varphi^{(1)}(p^{(1)}(x_0)) = x'_0$ and the pair $(\rho|\mathfrak{g}^{(1)}, \varphi^{(1)})$ is (strongly) equivariant. By the uniqueness of φ, one then has $\varphi = \varphi^{(1)} \circ p^{(1)}$. By the result for the injective case, $\varphi(M) = \varphi^{(1)}(M^{(1)})$ is totally geodesic in M', and after a suitable adjustment of the Riemannian metric on $M^{(1)}$, $\varphi^{(1)}$ becomes an isometry. A C^∞-map $\varphi: M \to M'$ which can be factored in the way described above will be called a *totally geodesic map* of M into M'.

Conversely, let φ be a totally geodesic map of M into M' with $\varphi(x_0) = x'_0$. We shall show that there exists a (unique) homomorphism $\rho: (\mathfrak{g}, \theta) \to (\mathfrak{g}', \theta')$ associated with φ. By the definition, one has decompositions

$$\mathfrak{g} = \mathfrak{g}^{(1)} \oplus \mathfrak{g}^{(2)}, \qquad M = M^{(1)} \times M^{(2)}$$

and an isometry $\varphi^{(1)}$ of $M^{(1)}$ into M' (after a suitable adjustment of the Riemannian metric) such that $\varphi = \varphi^{(1)} \circ p^{(1)}$ and $\varphi^{(1)}(M^{(1)})$ is totally geodesic in M'. Let $x_0 = (x_0^{(1)}, x_0^{(2)})$ and $\mathfrak{k}^{(i)} = \mathfrak{k} \cap \mathfrak{g}^{(i)}$, $\mathfrak{p}^{(i)} = \mathfrak{p} \cap \mathfrak{g}^{(i)}$. Then, since $\varphi^{(1)}$ preserves the Riemannian connection, $\lambda = d_{x_0^{(1)}} \varphi^{(1)}$ is an injective linear map of $\mathfrak{p}^{(1)} = T_{x_0^{(1)}}(M^{(1)})$ into $\mathfrak{p}' = T_{x_0'}(M')$ satisfying the condition

$$\lambda([[X, Y], Z]) = [[\lambda(X), \lambda(Y)], \lambda(Z)]$$

for all $X, Y, Z \in \mathfrak{p}^{(1)}$. It follows from the Lemma 2. 3 below that such a λ can uniquely be extended to an (injective) Lie algebra homomorphism $\rho^{(1)}: \mathfrak{g}^{(1)} \to \mathfrak{g}'$. Composing $\rho^{(1)}$ with the natural projection $\mathfrak{g} \to \mathfrak{g}^{(1)}$, one obtains a homomorphism $\rho: (\mathfrak{g}, \theta) \to (\mathfrak{g}', \theta')$. Then for $X = X^{(1)} + X^{(2)}$, $X^{(i)} \in \mathfrak{p}^{(i)}$, one has

$$\begin{aligned} (\exp \rho(X))x'_0 &= (\exp \lambda(X^{(1)}))x'_0 \\ &= \varphi^{(1)}((\exp X^{(1)})x_0^{(1)}) \\ &= \varphi((\exp X)x_0), \end{aligned}$$

which proves that (ρ, φ) is equivariant.

Lemma 2. 3. *Let* $\mathfrak{g}$ *and* $\mathfrak{g}'$ *be real semi-simple Lie algebras and* $\mathfrak{g} = \mathfrak{k} + \mathfrak{p}$, $\mathfrak{g}' = \mathfrak{k}' + \mathfrak{p}'$ *be the Cartan decompositions of* $\mathfrak{g}$ *and* $\mathfrak{g}'$ *corresponding to* θ *and* θ'. *Let* $\lambda: \mathfrak{p} \to \mathfrak{p}'$ *be a linear map preserving "Lie triple products", i. e., satisfying*

(2. 8) $$\lambda([[X, Y], Z]) = [[\lambda(X), \lambda(Y)], \lambda(Z)]$$

for all $X, Y, Z \in \mathfrak{p}$. Then λ can uniquely be extended to a homomorphism $\rho : (\mathfrak{g}, \theta) \to (\mathfrak{g}', \theta')$.

Proof. Put $\mathfrak{g}_1 = [\lambda(\mathfrak{p}), \lambda(\mathfrak{p})] + \lambda(\mathfrak{p}) \subset \mathfrak{g}'$. Then by (2. 8) it is clear that $\mathfrak{g}_1$ is a Lie subalgebra of $\mathfrak{g}'$ stable under θ'. Hence $\mathfrak{g}_1$ is reductive and $[\mathfrak{g}_1, \mathfrak{g}_1]$ is semi-simple. Since $\mathfrak{g} = \mathfrak{k} + \mathfrak{p}$ and $\mathfrak{k} = [\mathfrak{p}, \mathfrak{p}]$, we can extend λ to a linear map $\rho : \mathfrak{g} \to \mathfrak{g}_1 \subset \mathfrak{g}'$ by setting

$$\rho(\Sigma[X_i, Y_i]) = \Sigma[\lambda(X_i), \lambda(Y_i)] \qquad (X_i, Y_i \in \mathfrak{p})$$

and $\rho | \mathfrak{p} = \lambda$. As in Proposition I. 9. 1, it can be shown that ρ is well-defined. It is then clear that ρ is a Lie algebra homomorphism of $\mathfrak{g}$ into $\mathfrak{g}'$ satisfying the condition (2. 3). The uniqueness of the extension ρ is also clear, q. e. d.

Remark. It is known that there exists a one-to-one correspondence between (complete) totally geodesic submanifolds N of M containing x_0 and linear subspaces $\mathfrak{q}$ of $\mathfrak{p}$ closed under Lie triple product (so-called "Lie triple systems") by the relation $N = (\exp \mathfrak{q}) x_0$ or equivalently $T_{x_0}(N) = \mathfrak{q}$ (cf. Helgason [1], IV, Th. 7. 2 ; K-N, XI, Th. 4. 3). A totally geodesic subspace N with the induced Riemannian metric is a Riemannian symmetric space with non-positive curvature (possibly with a flat part).

Summing up, we obtain the following

Theorem 2. 4. *Let M and M' be Riemannian symmetric spaces of non-compact type, and $\mathfrak{g} = \mathrm{Lie}\ I(M)$ and $\mathfrak{g}' = \mathrm{Lie}\ I(M')$. Let $x_0 \in M$, $x_0' \in M'$ and let θ and θ' be Cartan involutions of $\mathfrak{g}$ and $\mathfrak{g}'$ at x_0 and x_0', respectively. Then, there exists a one-to-one correspondence between homomorphisms $\rho : (\mathfrak{g}, \theta) \to (\mathfrak{g}', \theta')$ and totally geodesic maps $\varphi : M \to M'$ with $\varphi(x_0) = x_0'$ by the relation of (ρ, φ) being (strongly) equivariant. When (ρ, φ) is equivariant, φ is an isometry of M onto M' (with respect to the normalized Riemannian metrics) if and only if ρ is an isomorphism.*

Theorem 2. 5 (*É. Cartan*). *Let M be a Riemannian symmetric space of non-compact type and let $G = I(M)$, $\mathfrak{g} = \mathrm{Lie}\ G$. Then the adjoint representation*

$$\alpha (= ad^G) : G \ni g \longmapsto d_e \nu_g \in \mathrm{Aut}\, \mathfrak{g} \tag{2. 9}$$

is an injective homomorphism of G into Aut $\mathfrak{g}$. *If M is irreducible, or if the Riemannian metric of M is normalized, then α is an isomorphism.*

Proof. We keep the notation introduced above, and denote by ν_θ the inner automorphism of Aut $\mathfrak{g}$ defined by θ and by $ad\, \theta = d\nu_\theta$ the corresponding automorphism of Lie(Aut $\mathfrak{g}$) $= ad\, \mathfrak{g}$. Then clearly one has

$$(ad\, \theta) \circ (ad\, X) = \theta \circ (ad\, X) \circ \theta = ad(\theta X)$$

for all $X \in \mathfrak{g}$. It follows that the global Cartan decomposition of Aut $\mathfrak{g}$ corresponding to θ is given by

$$\mathrm{Aut}\, \mathfrak{g} = \mathrm{Aut}(\mathfrak{g}, \theta) \cdot \exp(ad\, \mathfrak{p}),$$

where $\mathrm{Aut}(\mathfrak{g}, \theta) = \{\varphi \in \mathrm{Aut}\,\mathfrak{g} \mid \varphi\theta = \theta\varphi\}$. The homomorphism α defined by (2. 9) maps $K = I(M)_{x_0}$ and $\exp \mathfrak{p}$ into $\mathrm{Aut}(\mathfrak{g}, \theta)$ and $\exp(ad\,\mathfrak{p})$, respectively, and induces a bijection $\exp \mathfrak{p} \approx \exp(ad\,\mathfrak{p})$. Hence, to prove the injectivity of α, it is enough to show that $\alpha|K$ is injective. Suppose $k \in K$ and $\alpha(k) = d_e\nu_k = id$. Then one has $d_{x_0}k = d_e\nu_k|\mathfrak{p} = id$, which implies $k = e$ and proves our assertion, since the linear representation of K on $T_{x_0}(M)$ is faithful. To prove the last statement of the Theorem, we shall show that, if the Riemannian metric of M is normalized, one has $\alpha(K) = \mathrm{Aut}(\mathfrak{g}, \theta)$. Let $\varphi \in \mathrm{Aut}(\mathfrak{g}, \theta)$. Then by Theorem 2. 4 there exists a (unique) isometry $k \in K$ such that (φ, k) is equivariant. Then one has $\varphi|\mathfrak{p} = d_{x_0}k = (d_e\nu_k)|\mathfrak{p} = \alpha(k)|\mathfrak{p}$. Hence $\alpha(k) = \varphi$, q. e. d.

In view of Proposition I. 3. 6, this theorem implies that $G = I(M)$ has a (unique) structure of semi-simple $\boldsymbol{R}$-group such that G^z is the (algebraic) adjoint group.

Corollary 2. 6. *The centralizer of $I(M)^\circ$ in $I(M)$ reduces to the identity. In particular, the centers of $I(M)$ and $I(M)^\circ$ are trivial.*

This follows immediately from Theorem 2. 5.

Corollary 2. 7. *If $\varphi \in \mathrm{Aut}\,\mathfrak{g}$ commutes with all Cartan involutions of $\mathfrak{g}$, then $\varphi = id$.*

Proof. Let $g \in I(M)$ (with respect to the normalized metric) and suppose that $\alpha(g) = \varphi$ commutes with all Cartan involutions. Then, since $\alpha(s_x) = \theta_x$ $(x \in M)$, the Theorem implies that g commutes with all symmetries s_x, i. e., $gs_x = s_xg$, which by (1. 12) is equivalent to $gx = x$ for all $x \in M$, i. e., $g = id$, q. e. d.

Corollary 2. 8. *Let G_1 be a Zariski connected semi-simple $\boldsymbol{R}$-group. Then the intersection of all maximal compact subgroups of G_1 coincides with the center of G_1.*

Proof. Let Lie $G_1 = \mathfrak{g}$ and let $\alpha_1 = ad^{G_1} : G_1 \to \mathrm{Ad}\,\mathfrak{g} = (\mathrm{Aut}\,\mathfrak{g})^\circ$ be the adjoint representation of G_1. Then, since Ker α_1 = Cent G_1 is finite, $g \in G_1$ belongs to the intersection of all maximal compact subgroups of G_1 if and only if $\alpha_1(g)$ belongs to the intersection of all maximal compact subgroups of $\mathrm{Ad}\,\mathfrak{g}$. By Corollary 2. 7, this last condition is equivalent to saying $\alpha_1(g) = id$, i. e., $g \in$ Cent G_1, q. e. d.

For the structure of $\mathrm{Aut}\,\mathfrak{g}/\mathrm{Inn}\,\mathfrak{g}$ for real simple Lie algebras $\mathfrak{g}$, see Takeuchi [1], Matsumoto [2]. (Cf. also É. Cartan [3], Murakami [1], Loos [1].) The case of symmetric domains will be treated in § 8.

Exercise

1. Let G and G' be Zariski connected reductive $\boldsymbol{R}$-groups with Lie algebras $\mathfrak{g}$ and $\mathfrak{g}'$ and suppose there are given a Cartan involution θ (resp. θ') of G (resp. G') and an $\boldsymbol{R}$-homomorphism $\rho : G \to G'$.

We say that a triple (ρ, θ, θ') is *compatible* if $\rho\theta=\theta'\rho$. Assuming that (ρ, θ, θ') is compatible, prove the following.

1. 1) For any $g\in G$ and $g'\in G'$, $(\nu_{g'}\rho\nu_g^{-1}, \nu_g\theta\nu_g^{-1}, \nu_{g'}\theta'\nu_{g'}^{-1})$ is compatible.

1. 2) For $g\in G$, $(\rho, \nu_g\theta\nu_g^{-1}, \theta')$ is compatible if and only if $g\in(\operatorname{Ker}\rho)G^aK$, where G^a is the abelian part of G and K is the maximal compact subgroup of G corresponding to θ.

1. 3) For $g'\in G'$, $(\rho, \theta, \nu_{g'}\theta'\nu_{g'}^{-1})$ is compatible if and only if $g'\in C_{G'}(\rho(G))K'$, where K' is the maximal compact subgroup of G' corresponding to θ' and $C_{G'}$ denotes the centralizer in G'.

Hint. 1. 1) and the "if" parts of 1. 2), 1. 3) are obvious. To prove the "only if" part of 1. 3), write $g'=p'k'$ with $k'\in K'$, $p'\in P'$ and show that $(\rho, \theta, \nu_{g'}\theta'\nu_{g'}^{-1})$ is compatible if and only if $p'^2\in C_{G'}(\rho(G))$, which implies $p'\in C_{G'}(\rho(G))$. Next, if $(\rho, \nu_g\theta\nu_g^{-1}, \theta')$ is compatible, so is $(\rho, \theta, \nu_{\rho(g)}^{-1}\theta'\nu_{\rho(g)})$. Hence, writing $g=pk$ with $p\in P$, $k\in K$, one has by the above that $\rho(p)$ is in the center of $\rho(G)$, i. e., $p\in(\operatorname{Ker}\rho)$ (Cent G), whence follows the "only if" part of 1. 2).

§ 3. Hermitian symmetric spaces and hermitian JTS's.

We start with the notion of a complex structure on a vector space. Let V be a (finite-dimensional) real vector space. A *complex structure* I on V is a linear transformation of V such that $I^2=-1_V$. Given a complex structure I, one can convert V into a complex vector space by defining complex scalar multiplication by

$$\alpha x = (\operatorname{Re}\alpha)x+(\operatorname{Im}\alpha)Ix$$

for $\alpha\in\boldsymbol{C}$ and $x\in V$. The complex vector space thus obtained will be denoted by (V, I). If the complex dimension of (V, I) is n, then the real dimension of V is $2n$. Now let V_C be the complexification of V and extend I to a $\boldsymbol{C}$-linear transformation of V_C, denoted also by I. Then, putting $V_\pm=V_C(I;\pm i)$, one obtains a direct sum decomposition

$$V_C = V_+\oplus V_-,\qquad \bar{V}_+ = V_-. \tag{3.1}$$

The projection maps $p_\pm : V_C\to V_\pm$ are given by

$$p_\pm = \frac{1}{2}(1_V\mp iI), \tag{3.2}$$

and the restriction of p_+ (resp. p_-) on V gives a $\boldsymbol{C}$-linear (resp. $\boldsymbol{C}$-antilinear) isomorphism of (V, I) onto $V_\pm$.

Conversely, when there is given a direct sum decomposition of V_C of the form (3. 1), we can define a $\boldsymbol{C}$-linear transformation I of V_C by

$$I = i1_{V_+}\oplus(-i)1_{V_-}.$$

Then I is "real" (i. e., $\bar{I}=I$) and satisfies $I^2=-1_V$. Hence I induces a complex structure on V, for which one has $V_\pm=V_C(I;\pm i)$. Thus, one has a one-to-one correspondence between complex structures on V and the direct sum decompositions of V_C of the form (3. 1), or what amounts to the same thing, n-dimensional complex subspaces V_+ of V_C such that $V_+\cap\bar{V}_+=\{0\}$. It will also be useful to observe that a complex structure I on V is determined by an $\boldsymbol{R}$-homomorphism ϕ of the $\boldsymbol{R}$-torus $\boldsymbol{C}^{(1)}=\{\zeta\in\boldsymbol{C}\mid|\zeta|=1\}$ into $GL(V)$ given by

$$\phi : e^{it} \longmapsto \exp(tI) = e^{it}1_{V_+} \oplus e^{-it}1_{V_-} \qquad (t \in \boldsymbol{R}).$$

Now, let M be a Riemannian symmetric space of non-compact type and let the notation $G=I(M)$, $K=I(M)_{x_0}$, etc. be as in § 2. For simplicity, we assume in this section that the Riemannian metric on M is normalized, i. e., $q_x=\frac{1}{2}B_{\mathfrak{p}}$. An *almost complex structure* J on M is a C^∞ tensor field of type (1, 1) which assigns to each $x \in M$ a complex structure J_x on the tangent space $T_x(M)$. A *hermitian symmetric space* (of non-compact type) is a pair (M, J) formed of a Riemannian symmetric space M (of non-compact type) and an almost complex structure J on M satisfying the following conditions:

(a) J is "G°-invariant", i. e., one has

$$J_{gx} = (d_x g) J_x (d_x g)^{-1} \qquad \text{for all } g \in G^\circ,\ x \in M; \quad \text{and}$$

(b) the Riemannian metric q is "hermitian" with respect to J, i. e., one has

$$q_x(J_x X, J_x Y) = q_x(X, Y) \qquad \text{for all } X, Y \in T_x(M),\ x \in M.$$

Under these conditions, one can define a G°-invariant hermitian metric h on M by

(3. 3) $$h_x(X, Y) = q_x(X, Y) - iq_x(X, J_x Y) \qquad (X, Y \in T_x(M))$$

(which is $\boldsymbol{C}$-linear in the second variable Y). We shall see later on (§§ 4, 6) that the almost complex manifold (M, J) can be embedded in $\boldsymbol{C}^N$ as a bounded domain, so that J is actually a complex structure (i. e., satisfies the "integrability condition") and the hermitian metric h coincides (up to a homothety) with the "Bergman metric", which is Kählerian.

Now, since G° acts transitively on M, the conditions (a), (b) are equivalent to the following conditions concerning the complex structure $J_0=J_{x_0}$ on $\mathfrak{p}=T_{x_0}(M)$:

(3. 4) $$J_0 \circ ad_{\mathfrak{p}}(k) = ad_{\mathfrak{p}}(k) \circ J_0 \qquad \text{for all } k \in K^\circ,$$

(3. 5) $$B(J_0 X, J_0 Y) = B(X, Y) \qquad \text{for all } X, Y \in \mathfrak{p}.$$

The condition (3. 4) is clearly equivalent to

(3. 4') $$J_0[X, Y] = [X, J_0 Y] \qquad \text{for all } X \in \mathfrak{k},\ Y \in \mathfrak{p}.$$

We also consider the following condition:

(3. 6) $$[J_0 X, J_0 Y] = [X, Y] \quad , \qquad \text{for all } X, Y \in \mathfrak{p}.$$

Lemma 3. 1. *Any two of the conditions* (3. 4'), (3. 5), *and* (3. 6) *imply the remaining one.*

Proof. Under the condition (3. 5) the equivalence of (3. 4') and (3. 6) follows from the following equation

$$\begin{aligned} B([X, Y]-[J_0X, J_0Y], Z) &= B(Y, [Z, X]) - B(J_0Y, [Z, J_0X]) \\ &= B(J_0Y, J_0[Z, X] - [Z, J_0X]) \\ &\qquad (X, Y \in \mathfrak{p},\ Z \in \mathfrak{k}), \end{aligned}$$

since the restrictions of the Killing form B on $\mathfrak{k}$ and $\mathfrak{p}$ are both non-degenerate.

Next, assuming (3. 4′) and (3. 6), one has

$$ad(J_0Y) = \begin{cases} J_0 \circ ad\, Y & \text{on } \mathfrak{k} \\ -(ad\, Y) \circ J_0 & \text{on } \mathfrak{p} \end{cases}$$

for $Y \in \mathfrak{p}$. Hence

$$ad(J_0X)ad(J_0Y) = \begin{cases} ad\, X \circ ad\, Y & \text{on } \mathfrak{k} \\ J_0 \circ ad\, X \circ ad\, Y \circ J_0^{-1} & \text{on } \mathfrak{p} \end{cases}$$

for $X, Y \in \mathfrak{p}$, which implies (3. 5), q. e. d.

We put

$$\tilde{J} = \begin{cases} 0 & \text{on } \mathfrak{k} \\ J_0 & \text{on } \mathfrak{p}. \end{cases}$$

Then it is clear that the conditions (3. 4′), (3. 6) taken together are equivalent to saying that $\tilde{J}$ is a derivation of the Lie algebra $\mathfrak{g}$. When this is the case, since $\mathfrak{g}$ is semi-simple, the derivation $\tilde{J}$ can be expressed uniquely as $\tilde{J} = ad(H_0)$ with $H_0 \in \mathfrak{g}$. Then, since $\theta\tilde{J} = \tilde{J}\theta$, one has $\theta(H_0) = H_0$, i. e., $H_0 \in \mathfrak{k}$. Thus H_0 is in the center of $\mathfrak{k}$. Conversely, it is clear that, if H_0 is an element in the center of $\mathfrak{k}$ such that

(3. 7) $$J_0 = ad_{\mathfrak{p}}(H_0)$$

is a complex structure on $\mathfrak{p}$, then J_0 satisfies the conditions (3. 4′) and (3. 6), and hence gives rise to an almost complex structure J on M satisfying the conditions (a), (b). In general, a pair $(\mathfrak{g}, H_0)$ formed of a real semi-simple Lie algebra $\mathfrak{g}$ (with a Cartan decomposition $\mathfrak{g} = \mathfrak{k} + \mathfrak{p}$) and an element H_0 in the center of $\mathfrak{k}$ such that $(ad_{\mathfrak{p}}(H_0))^2 = -1_{\mathfrak{p}}$ will be called a semi-simple Lie algebra *of hermitian type.* Sometimes $\mathfrak{g}$ alone is called of hermitian type if there exists such an element H_0, and the element H_0 is referred to as an "element defining a hermitian structure" of $\mathfrak{g}$, or simply an *H-element of* $\mathfrak{g}$ (relative to $\mathfrak{k}$).

We have thus proved

Proposition 3. 2. *A Riemannian symmetric space M of non-compact type has a structure of a hermitian symmetric space if and only if the corresponding Lie algebra* $\mathfrak{g} = \mathrm{Lie}\ I(M)$ *is of hermitian type. For a fixed* $x_0 \in M$, *there exists a one-to-one correspondence between almost complex structures J on M satisfying the conditions* (a), (b) *and the H-elements* H_0 *of* $\mathfrak{g}$ *relative to* $\mathfrak{k} = \mathrm{Lie}\ I(M)_{x_0}$.

Next we explain the connection with hermitian JTS's. Let $(\mathfrak{g}, H_0)$ be a semi-simple Lie algebra of hermitian type, and put

(3. 8) $$\mathfrak{p}_\pm = \mathfrak{p}_C(J_0\,;\, \pm i),$$

where $J_0 = ad_{\mathfrak{p}}(H_0)$. Then one has the direct sum decomposition

(3. 9) $$\mathfrak{g}_C = \mathfrak{p}_+ + \mathfrak{k}_C + \mathfrak{p}_-, \qquad \bar{\mathfrak{p}}_+ = \mathfrak{p}_-$$

satisfying the following relations by (3. 4′), (3. 6) :

$$\text{(3.10)} \qquad [\mathfrak{k}_C, \mathfrak{p}_\pm] \subset \mathfrak{p}_\pm, \quad [\mathfrak{p}_+, \mathfrak{p}_-] \subset \mathfrak{k}_C, \quad [\mathfrak{p}_+, \mathfrak{p}_+] = [\mathfrak{p}_-, \mathfrak{p}_-] = 0.$$

Thus $\mathfrak{p}_\pm$ are abelian subalgebras of $\mathfrak{g}_C$. The decomposition (3. 9) will be called a *canonical decomposition* of $\mathfrak{g}_C$ (relative to H_0). The condition (3. 10) is equivalent to saying that the complex Lie algebra $\mathfrak{G}=\mathfrak{g}_C$ has a gradation given by $\mathfrak{G}_0=\mathfrak{k}_C$, $\mathfrak{G}_{\pm 1}=\mathfrak{p}_\mp$. Notice that in (3. 10) one has always the equation $[\mathfrak{k}_C, \mathfrak{p}_\pm]=\mathfrak{p}_\pm$, and the equation $[\mathfrak{p}_+, \mathfrak{p}_-]=\mathfrak{k}_C$ holds if and only if $\mathfrak{g}$ has no compact factor. Put

$$\text{(3.11)} \qquad \sigma X = \theta \bar{X} \quad \text{for} \quad X \in \mathfrak{g}_C.$$

Then σ is a complex conjugation of $\mathfrak{G}=\mathfrak{g}_C$ with respect to the compact real form $\mathfrak{u}=\mathfrak{k}+i\mathfrak{p}$, and as such σ is a Cartan involution of $\mathfrak{G}$ viewed as a real Lie algebra. Thus, if $\mathfrak{g}$ has no compact factor, the graded Lie algebra $\mathfrak{G}$ together with the involution σ becomes a (real) "symmetric Lie algebra" in the sense of I, § 7. The corresponding JTS (over $\boldsymbol{R}$) is $\mathfrak{G}_{-1}=\mathfrak{p}_+$ (viewed as a real vector space) with the triple product

$$\text{(3.12)} \qquad \{X, Y, Z\} = -\frac{1}{2}[[X, \sigma Y], Z] = \frac{1}{2}[[X, \bar{Y}], Z],$$

which is "positive definite" in the sense of I, § 9, because one has $\tau(X, Y)=-\frac{1}{2}\operatorname{Re} B(X, \sigma Y)$ by (I. 7. 8), where B is the Killing form of $\mathfrak{G}$ (over $\boldsymbol{C}$).

In general, a (real) JTS $(V, \{\ \})$ is called *hermitian*, if V has a structure of complex vector space and the triple product $\{x, y, z\}$ is $\boldsymbol{C}$-linear in x, z and $\boldsymbol{C}$-antilinear in y. In that case, if we denote by $\tau'(x, y)$ the trace (taken over $\boldsymbol{C}$) of the $\boldsymbol{C}$-linear transformation $y \square x$, then τ' is a hermitian form on V ($\boldsymbol{C}$-linear in y) and the usual trace form τ (over $\boldsymbol{R}$) is given by $\tau=2\operatorname{Re}\tau'$. Hence τ' is positive definite if and only if the JTS $(V, \{\ \})$ is positive definite in the previous sense; when this is the case, the hermitian JTS $(V, \{\ \})$ is also called *positive definite*. We have shown above that for any semi-simple Lie algebra $\mathfrak{g}$ of hermitian type the complex vector space $\mathfrak{p}_+$ with the triple product (3. 12) is a positive definite hermitian JTS.

Conversely, let $(V, \{\ \})$ be a positive definite hermitian JTS and let $(\mathfrak{G}, \sigma)$ be the corresponding symmetric Lie algebra. Then it is clear that $\mathfrak{G}$ has a natural structure of complex Lie algebra, for which σ is a $\boldsymbol{C}$-antilinear involution (i. e., a complex conjugation). Since τ is positive definite, σ is a Cartan involution of $\mathfrak{G}$ viewed as a real semi-simple Lie algebra (I, § 9), and hence the real form of $\mathfrak{G}$ defined by σ is compact. On the other hand, let us define a $\boldsymbol{C}$-linear involution θ of $\mathfrak{G}$ by

$$\text{(3.13)} \qquad \theta(X) = (-1)^\nu X \quad \text{for} \quad X \in \mathfrak{G}_\nu,$$

and set $\bar{X}=\theta\sigma X$ for $X \in \mathfrak{G}$. Then the complex conjugation $X \mapsto \bar{X}$ of $\mathfrak{G}$ determines another real form $\mathfrak{g}$, on which θ induces a Cartan involution (I, § 4). Thus, in the previous notation, one has $\mathfrak{g}=\mathfrak{k}+\mathfrak{p}$, $\mathfrak{k}_C=\mathfrak{G}_0$, $\mathfrak{p}_C=\mathfrak{G}_1+\mathfrak{G}_{-1}$, and $\bar{\mathfrak{G}}_1=\mathfrak{G}_{-1}$. Therefore, there is a unique H-element H_0 of $\mathfrak{g}$ relative to $\mathfrak{k}$ such that $(\mathfrak{g}, H_0)$ is a semi-simple Lie algebra of hermitian type and $\mathfrak{p}_\pm=\mathfrak{G}_{\mp 1}$.

We have thus proved

Proposition 3. 3. *Let* $(\mathfrak{g}, H_0)$ *be a semi-simple Lie algebra of hermitian type and let* $\mathfrak{p}_+=\mathfrak{g}_C(ad\,H_0\,;i)$. *Then* $\mathfrak{p}_+$ *with the triple product* (3. 12) *is a positive definite hermitian JTS. The corresponding symmetric Lie algebra* $(\mathfrak{G}, \sigma)$ *can be identified with the complexification of the non-compact part of* $\mathfrak{g}$ *with the gradation* $\mathfrak{G}_{\pm 1}=\mathfrak{p}_{\mp}$, $\mathfrak{G}_0=[\mathfrak{p}_+, \mathfrak{p}_-]$ *and the involution* σ *given by* (3. 11). *All positive definite hermitian JTS's are obtained in this manner from semi-simple Lie algebras of hermitian type.*

We know that, for a non-compact simple Lie algebra $\mathfrak{g}$, the restriction of the adjoint representation of $\mathfrak{k}$ on $\mathfrak{p}$ is faithful and irreducible. Hence the dimension of the center of $\mathfrak{k}$ is at most one, and it is equal to one if and only if $\mathfrak{g}$ is of hermitian type. Thus for a non-compact simple Lie algebra $\mathfrak{g}$ of hermitian type there are exactly two H-elements of $\mathfrak{g}$ relative to $\mathfrak{k}$. In general, if $\mathfrak{g}$ is semi-simple and of hermitian type and if

$$\mathfrak{g} = \mathfrak{g}_0 \oplus \mathfrak{g}_1 \oplus \cdots \oplus \mathfrak{g}_r \tag{3. 14}$$

is its direct sum decomposition such that $\mathfrak{g}_0$ is compact and the $\mathfrak{g}_i$ $(1 \leq i \leq r)$ are simple and non-compact, then each $\mathfrak{g}_i$ is of hermitian type and the number of H-elements of $\mathfrak{g}$ relative to (a fixed) $\mathfrak{k}$ is 2^r.

We also note that, for a hermitian symmetric space M, one has an $\boldsymbol{R}$-homomorphism

$$\phi:\ \boldsymbol{C}^{(1)} \ni e^{it} \longmapsto \exp(tH_0) \in K = I(M)_{x_0}$$

and in the previous notation one has $ad(\exp tH_0)=\exp tJ$. Since $\exp \pi J=\theta$, one has $\exp \pi H_0=s_{x_0}$. Thus we see that all symmetries of M are contained in the connected component G°. We shall see later on (§ 8) that the group of holomorphic automorphisms $\mathrm{Hol}(M)$ is a subgroup of $I(M)$ of index 2^r.

§ 4. The Harish-Chandra embedding of a hermitian symmetric space.

Let $(\mathfrak{g}, H_0)$ be a semi-simple Lie algebra of hermitian type with Cartan decomposition $\mathfrak{g}=\mathfrak{k}+\mathfrak{p}$. Let $\mathfrak{c}$ be the center of $\mathfrak{k}$ and $\mathfrak{h}$ a maximal abelian subalgebra of $\mathfrak{k}$. Then one has $H_0 \in \mathfrak{c} \subset \mathfrak{h}$. Since the centralizer of H_0 in $\mathfrak{g}$ is equal to $\mathfrak{k}$, it follows that $\mathfrak{h}$ is a maximal abelian subalgebra in $\mathfrak{g}$; moreover, $ad\,H$ $(H \in \mathfrak{h})$ is semi-simple. Hence (by definition) $\mathfrak{h}$ is a "Cartan subalgebra" of $\mathfrak{g}$, and so $\mathfrak{h}_C$ is a Cartan subalgebra of $\mathfrak{g}_C$. ($\mathfrak{h}_C$ is the Lie algebra of a maximal torus in G_C and hence algebraic.) Let $\mathfrak{r}$ denote the (absolute) root system of $\mathfrak{g}_C$ relative to $\mathfrak{h}_C$. Then, for every $\alpha \in \mathfrak{r}$ and $H \in \mathfrak{h}$, $\alpha(H)$ is purely imaginary. Hence, if we denote by $\mathfrak{h}^*$ the dual space of $\mathfrak{h}$ embedded in the complex dual space $\mathfrak{h}_C^*$ of $\mathfrak{h}_C$, we have $\mathfrak{r} \subset i\mathfrak{h}^*$. We fix a linear order in $i\mathfrak{h}^*$ such that for $\lambda \in i\mathfrak{h}^*$ one has

$$\frac{1}{i}\lambda(H_0) > 0 \Longrightarrow \lambda > 0 \tag{4. 1}$$

and let $\Delta = \{\alpha_1, \cdots, \alpha_l\}$ be the corresponding "fundamental system" (I, § 5). A root $\alpha \in \mathfrak{r}$ is called *compact* (resp. *non-compact*) if

$$\alpha(H_0) = 0 \qquad (\text{resp. } \alpha(H_0) = \pm i),$$

and the set of all compact (resp. non-compact positive or negative) roots are denoted by $\tilde{\mathfrak{r}}_c$ (resp. $\tilde{\mathfrak{r}}_\pm$). Thus one has

(4. 2) $$\tilde{\mathfrak{r}} = \tilde{\mathfrak{r}}_c \cup \tilde{\mathfrak{r}}_+ \cup \tilde{\mathfrak{r}}_-.$$

If we denote by $\mathfrak{g}_\alpha$ the "root space" for $\alpha \in \tilde{\mathfrak{r}}$, i. e.,

$$\mathfrak{g}_\alpha = \{X \in \mathfrak{g}_C \mid [H, X] = \alpha(H)X \text{ for all } H \in \mathfrak{h}_C\},$$

we have

(4. 3) $$\mathfrak{p}_+ = \sum_{\alpha \in \tilde{\mathfrak{r}}_+} \mathfrak{g}_\alpha, \qquad \mathfrak{p}_- = \sum_{\alpha \in \tilde{\mathfrak{r}}_-} \mathfrak{g}_\alpha.$$

We also put $\mathfrak{k}^s = [\mathfrak{k}, \mathfrak{k}]$ (the semi-simple part of $\mathfrak{k}$), $\mathfrak{h}^{(s)} = \mathfrak{h} \cap \mathfrak{k}^s$, and $\tilde{\Delta}_c = \tilde{\Delta} \cap \tilde{\mathfrak{r}}_c$. Then, $\mathfrak{h} = \mathfrak{h}^{(s)} \oplus \mathfrak{c}$, $\tilde{\mathfrak{r}}_c$ is a root system of $\mathfrak{k}^s_C$ relative to $\mathfrak{h}^{(s)}_C$, and it is easy to see that $\tilde{\Delta}_c$ is a fundamental system of $\tilde{\mathfrak{r}}_c$ (with respect to the induced linear order on $i\mathfrak{h}^{(s)*}$).

In the following, we fix once and for all a Zariski connected $\boldsymbol{R}$-group G with Lie algebra $\mathfrak{g}$. Then there exists a natural $\boldsymbol{R}$-homomorphism ("adjoint representation")

(4. 4) $$\alpha (= ad^G) : G \longrightarrow \operatorname{Aut} \mathfrak{g}$$

defined by $\alpha(g) = d_e \nu_g$, ν_g denoting the inner automorphism of G defined by $g \in G$. The image $\alpha(G)$ is contained in the "algebraic adjoint group" $\operatorname{Ad} \mathfrak{g} (= (\operatorname{Aut} \mathfrak{g})^z)$, which may be identified with $I(M)^z$ by Theorem 2. 5. We define the "isometric action" of G on M by the homomorphism α, i. e., we put $g \cdot x = \alpha(g) \cdot x$ for $g \in G$, $x \in M$. For $x \in M$, we denote by G_x the isotropy subgroup of G at x, i. e., $G_x = \alpha^{-1}(I(M)_x)$. Then, since the kernel of α is finite, it is clear that G_x is a maximal compact subgroup of G and every maximal compact subgroup of G is obtained in this way. In particular, we put $K = G_{x_0}$. Then one has Lie $K = \mathfrak{k}$ and

(4. 5) $$G = (\exp \mathfrak{p}) \cdot K \quad (\text{topological direct product}).$$

By I, § 3, K is an $\boldsymbol{R}$-closed subgroup of G, and one has $K^z = K^\circ$. By (4. 5) one has $G^\circ = (\exp \mathfrak{p}) \cdot K^\circ$. Let K_C (resp. $(K^\circ)_C$) be the complexification of K (resp. K°), viewed as $\boldsymbol{C}$-closed subgroups of G_C. Then one has $(K_C)^z = (K_C)^\circ = (K^\circ)_C$, which we denote by K°_C. We note that for $k \in K^\circ_C$ one has $(ad\, k)H_0 = H_0$ and so $(ad\, k)\mathfrak{p}_\pm = \mathfrak{p}_\pm$.

We notice that the subalgebras $\mathfrak{c}$, $\mathfrak{h}$, $\mathfrak{p}_\pm$ are all algebraic. Let T denote the (Zariski connected) $\boldsymbol{R}$-subgroup of G corresponding to $\mathfrak{h}$, and let $P_\pm$ be the $\boldsymbol{C}$-subgroups of G_C corresponding to $\mathfrak{p}_\pm$. Then T is a compact $\boldsymbol{R}$-torus and the $P_\pm$ are unipotent. As K°_C is reductive and normalizes P_-, $K^\circ_C P_-$ is a semi-direct product. In the terminology of I, § 5, $K^\circ_C P_-$ is a "parabolic subgroup" of G_C relative to $(\mathfrak{h}_C, \tilde{\Delta})$ corresponding to the subset $\tilde{\Delta}_c$ of $\tilde{\Delta}$.

Lemma 4. 1. *One has*

(4. 6) $$K^\circ_C = \{k \in K_C \mid (ad\, k)H_0 = H_0\}.$$

In fact, if we denote the right-hand side of (4. 6) by K^h_C, then clearly $K^\circ_C \subset K^h_C$ and K^h_C normalizes $K^\circ_C P_-$. Since a parabolic subgroup coincides with its own normalizer (I, § 5), one has $K_c^h \subset K_c^\circ P_-$ and, since $K_c \cap P_- = \{1\}$, one has $K_c^h = K_c^\circ$. It follows from (4. 6) that, if G is simple, $(K_c : K_c^\circ)(=(K : K^\circ)=(G : G^\circ))$ is at most two.

The following lemma is well-known (e. g., Helgason [1], VIII, § 7) :

Lemma 4. 2. *One has the following relations :*

(4. 7 a) $$P_+ \cap K_c^\circ P_- = \{1\},$$

(4. 7 b) $$G^\circ \subset P_+ K_c^\circ P_-,$$

(4. 7 c) $$G^\circ \cap K_c^\circ P_- = K^\circ.$$

Since our setting is slightly different from that in Helgason [1], we will reproduce the proofs of (4. 7 a, b). The proof of (4. 7 c) is left for an exercise of the reader.

Proof of (4. 7 a). Let $p_+ \in P_+ \cap K_c^\circ P_-$ and write $p_+ = \exp X$ with $X \in \mathfrak{p}_+$. Then, since $p_+ \in K_c^\circ P_-$, one has $ad(p_+)\mathfrak{p}_- \subset \mathfrak{p}_-$. But for $Y \in \mathfrak{p}_-$ one has

$$ad(p_+)Y = (\exp(ad\, X))Y = \sum_{\nu=0}^{\infty} \frac{1}{\nu!}(ad\, X)^\nu Y,$$

where $(ad\, X)Y \in \mathfrak{k}_c$, $(ad\, X)^2 Y \in \mathfrak{p}_+$, and $(ad\, X)^\nu Y = 0$ for $\nu \geq 3$. Hence one has $[X, Y]=0$ for all $Y \in \mathfrak{p}_-$, which implies $X=0$, i. e., $p_+ = 1$.

Proof of (4. 7 b). Let $g \in G^\circ$ and write $g = pk$ with $k \in K^\circ$, $p = \exp X$, $X \in \mathfrak{p}$. Clearly it is enough to show that $p \in P_+ K_c^\circ P_-$. We use the Iwasawa decomposition $G_c = U \cdot \exp(i\mathfrak{h}) \cdot N_+$, where U is the compact real form of G_c corresponding to $\mathfrak{k} + i\mathfrak{p}$ and N_+ is the maximal unipotent subgroup corresponding to $\sum_{\alpha>0} \mathfrak{g}_\alpha$ (cf. Helgason [1], VI). Then one can write

$$p^{\frac{1}{2}} = \exp\left(\frac{1}{2}X\right) = uhn$$

with $u \in U$, $h \in \exp(i\mathfrak{h})$, $n \in N_+$. Denoting by σ the complex conjugation of G_c with respect to U, one has

$$(p^{\frac{1}{2}})^\sigma = p^{-\frac{1}{2}} = uh^{-1}n^\sigma.$$

Hence

$$p = n^{-\sigma} h^2 n \in N_- \cdot \exp(i\mathfrak{h}) \cdot N_+,$$

where $N_- = N_+^\sigma$ is the maximal unipotent subgroup corresponding to $\sum_{\alpha<0} \mathfrak{g}_\alpha$. Since $K_c^\circ P_\pm$ contain the Borel subgroups $\exp(i\mathfrak{h})N_\pm$, one has $N_- \cdot \exp(i\mathfrak{h}) \cdot N_+ \subset P_- K_c^\circ P_+$, which proves our assertion, q. e. d.

It follows from this Lemma by a simple dimension argument that both $G^\circ K_c^\circ P_-$ and $P_+ K_c^\circ P_-$ are open in G_c. Hence by (4. 7 b, c) one obtains the following natural (open) embeddings :

(4. 8) $$M \approx G^\circ/K^\circ \approx G^\circ K_c^\circ P_-/K_c^\circ P_- \hookrightarrow P_+ K_c^\circ P_-/K_c^\circ P_- \hookrightarrow G_c/K_c^\circ P_-.$$

Since by (4. 7 a)

$$P_+K_C^\circ P_-/K_C^\circ P_- \approx P_+ \approx \mathfrak{p}_+,$$

the first map in (4. 8) gives rise to a homeomorphism of M onto a domain $\mathscr{D}$ in $\mathfrak{p}_+$. The embedding $\psi: M \hookrightarrow \mathfrak{p}_+$ thus obtained is called the *Harish-Chandra embedding* of M (at x_0). By definition ψ is given by the following rule:

$$(4.9) \qquad M \ni x = g_1 x_0 \stackrel{\psi}{\longmapsto} z \in \mathscr{D} \subset \mathfrak{p}_+ \qquad (g_1 \in G^\circ)$$
$$\Longleftrightarrow (\exp z)^{-1} g_1 \in K_C^\circ P_-;$$

or in other words, $\exp z$ is the "P_+-part" of g_1 in the direct product decomposition $P_+K_C^\circ P_-$. It is clear that the Harish-Chandra embedding ψ preserves the almost complex structure, so that we may identify (M, J) with the domain $\mathscr{D} \subset \mathfrak{p}_+$ as (almost) complex manifold. After this identification, $M = \mathscr{D}$ is often called a "symmetric domain" (or "symmetric bounded domain", since $\mathscr{D}$ is shown to be bounded). The (isometric) action of $g \in G^\circ$ on $\mathscr{D}$ is given by the rule

$$(4.10) \qquad \exp(g(z)) = (g \cdot \exp z)_+,$$

$(\ )_+$ denoting the P_+-part. For a symmetric domain $\mathscr{D}$, the origin x_0 will be denoted by o. We note that the holomorphic tangent space $T'_o(\mathscr{D})$ at o is naturally identified with $\mathfrak{p}_+$, and this identification is compatible with the one $T_o(\mathscr{D}) = \mathfrak{p}$ given in § 1.

Remark. It can also be shown that one has $UK_C^\circ P_- = G_C$ and $U \cap K_C^\circ P_- = K^\circ$. Hence, by a mapping similar to the Harish-Chandra embedding, the "compact dual" $M^* = U/K^\circ$ of M may be identified with the compact complex manifold $G_C/K_C^\circ P_-$. Thus the composed map in (4. 8), called the "Borel embedding", gives a (holomorphic) embedding of $M = \mathscr{D}$ into its compact dual.

Example. Let $G = SU(1, 1)$ be the special unitary group for the standard hermitian form of signature (1, 1) and let $\mathfrak{g} = \mathfrak{su}(1, 1)$ be its Lie algebra. Then one has $\mathfrak{su}(1, 1) = \{ih, h', ih''\}_R$, where

$$(4.11) \qquad h = \begin{pmatrix} 1 & 0 \\ 0 & -1 \end{pmatrix}, \quad h' = \begin{pmatrix} 0 & 1 \\ 1 & 0 \end{pmatrix}, \quad h'' = \begin{pmatrix} 0 & 1 \\ -1 & 0 \end{pmatrix}.$$

Then $(\mathfrak{su}(1, 1), \frac{i}{2}h)$ is a semi-simple Lie algebra of hermitian type, for which the canonical decomposition $\mathfrak{g}_C = \mathfrak{p}_+ + \mathfrak{k}_C + \mathfrak{p}_-$ is given by

$$\mathfrak{k}_C = \{h\}_C, \quad \mathfrak{p}_\pm = \{e_\pm\}_C, \qquad \text{where } e_+ = \begin{pmatrix} 0 & 1 \\ 0 & 0 \end{pmatrix}, \quad e_- = \begin{pmatrix} 0 & 0 \\ 1 & 0 \end{pmatrix}.$$

In this case, one has $\mathfrak{k} = \mathfrak{h} = \mathfrak{c}$, $\Delta = \{\alpha_1\}$, $\Delta_c = \phi$, where α_1 is a unique root relative to $\mathfrak{h}_C$ determined by $\alpha_1(h) = 2$, and $\mathfrak{p}_\pm = \mathfrak{g}_{\pm\alpha_1}$. For simplicity, we identify $\mathfrak{p}_+$ with C by the correspondence $\zeta e_+ \leftrightarrow \zeta (\in C)$. Every element $g \in G = SU(1, 1)$ can be written in the form:

$$(4.12) \qquad g = \begin{pmatrix} \alpha & \beta \\ \bar\beta & \bar\alpha \end{pmatrix} = \begin{pmatrix} 1 & \beta\bar\alpha^{-1} \\ 0 & 1 \end{pmatrix} \begin{pmatrix} \bar\alpha^{-1} & 0 \\ 0 & \bar\alpha \end{pmatrix} \begin{pmatrix} 1 & 0 \\ \bar\alpha^{-1}\bar\beta & 1 \end{pmatrix}$$

where $\alpha, \beta \in C$ and $|\alpha|^2 - |\beta|^2 = 1$. (In this case, one has $K \cong C^{(1)}$ and so G is connected.) Hence the Harish-Chandra embedding is given by

$$\psi(gx_0) = \zeta = \beta\bar{\alpha}^{-1}. \tag{4.13}$$

From $|\alpha|^2-|\beta|^2=1$, one has $|\zeta|<1$. Conversely, for any $|\zeta|<1$, taking $\alpha\in\boldsymbol{C}$ such that $|\alpha|=(1-|\zeta|^2)^{-1/2}$ and putting $\beta=\zeta\bar{\alpha}$, one obtains $g\in G$ for which $\psi(gx_0)=\zeta$. Thus the domain $\mathcal{D}=\psi(M)$ is the "unit disc" $\mathcal{D}^1=\{\zeta\in\boldsymbol{C}\,|\,|\zeta|<1\}$.

A simple computation shows that the action of $g=\begin{pmatrix}\alpha & \beta\\ \bar{\beta} & \bar{\alpha}\end{pmatrix}\in G$ on $\mathcal{D}$ is given by a linear fractional transformation:

$$\zeta \longmapsto (\alpha\zeta+\beta)(\bar{\beta}\zeta+\bar{\alpha})^{-1}.$$

Also, as a special case of the decomposition (4. 12) one has

$$\exp\lambda h' = \begin{pmatrix}\mathrm{ch}\,\lambda & \mathrm{sh}\,\lambda\\ \mathrm{sh}\,\lambda & \mathrm{ch}\,\lambda\end{pmatrix} = \begin{pmatrix}1 & \mathrm{th}\,\lambda\\ 0 & 1\end{pmatrix}\begin{pmatrix}(\mathrm{ch}\,\lambda)^{-1} & 0\\ 0 & \mathrm{ch}\,\lambda\end{pmatrix}\begin{pmatrix}1 & 0\\ \mathrm{th}\,\lambda & 1\end{pmatrix} \qquad (\lambda\in\boldsymbol{R}). \tag{4.14}$$

Thus by the Harish-Chandra map the geodesic $\{(\exp\lambda h')x_0\ (\lambda\in\boldsymbol{R})\}$ in M is identified with a line segment $\{\mathrm{th}\,\lambda\ (\lambda\in\boldsymbol{R})\}$ in the unit disc.

In the study of symmetric domains the following property of roots is fundamental:

Lemma 4. 3. *Let $\mathfrak{g}$ be a semi-simple Lie algebra of hermitian type with Cartan decomposition $\mathfrak{g}=\mathfrak{k}+\mathfrak{p}$, and $\mathfrak{h}$ a Cartan subalgebra of $\mathfrak{g}$ contained in $\mathfrak{k}$. Let r be the real rank of $\mathfrak{g}$. Then there exist r linearly independent positive non-compact roots $\beta_1,\cdots,\beta_r$ (relative to $\mathfrak{h}_c$) such that $\beta_j\pm\beta_k$ is not a root for any $j\neq k$.*

For a proof, see Harish-Chandra [2], VI or Helgason [1], VIII, § 7.

Now we take a generator o_k of $\mathfrak{g}_{\beta_k}$ normalized by the condition:

$$[[o_k, \bar{o}_k], o_k] = 2o_k \tag{4.15}$$

and put

$$\begin{cases} H_k = i[o_k, \bar{o}_k],\\ X_k = o_k+\bar{o}_k,\\ Y_k = -i(o_k-\bar{o}_k).\end{cases} \tag{4.16}$$

Then H_k, X_k, Y_k $(1\leq k\leq r)$ are all real and by (4. 15) the following relations are satisfied:

$$[H_k, X_k] = -2Y_k, \qquad [H_k, Y_k] = 2X_k, \qquad [X_k, Y_k] = 2H_k.$$

Therefore one has injective homomorphisms $\tilde{\kappa}_k : \mathfrak{su}(1,1)\rightarrow\mathfrak{g}$ defined by

$$\tilde{\kappa}_k(ih) = H_k, \qquad \tilde{\kappa}_k(h') = X_k, \qquad \tilde{\kappa}_k(-ih'') = Y_k. \tag{4.17}$$

These r homomorphisms are mutually "commutative" or "orthogonal" in the sense that

$$[\tilde{\kappa}_k(x), \tilde{\kappa}_{k'}(y)] = 0 \qquad \text{for all } x, y\in\mathfrak{su}(1,1) \text{ and } k\neq k'.$$

In particular, if we put $\mathfrak{a}=\{X_1,\cdots,X_r\}_{\boldsymbol{R}}$, then $\mathfrak{a}$ is an abelian subalgebra of $\mathfrak{g}$ contained in $\mathfrak{p}$. Since $r=\boldsymbol{R}$-rank $\mathfrak{g}$, $\mathfrak{a}$ is actually maximal among the subalgebras having this property. (An alternate proof, without using Lemma 4. 3, of the existence of r orthogonal (H_1)-homomorphisms κ_k will be given in V, § 6.)

Now let $z \in \mathfrak{p}_+$. Then, since $z+\bar{z} \in \mathfrak{p}=(ad\, K^\circ)\mathfrak{a}$ (I, § 5), there exist $k \in K^\circ$ and $\xi_i \in \boldsymbol{R}$ $(1 \leq i \leq r)$ such that one has

$$z+\bar{z} = (ad\, k)\Big(\sum_{j=1}^{r} \xi_j X_j\Big).$$

Taking the $\mathfrak{p}_+$-part, one obtains

$$z = (ad\, k)\Big(\sum_{j=1}^{r} \xi_j o_j\Big). \tag{4. 18}$$

Though irrelevant to our discussion, we note that in (4. 18) the r-tuple (ξ_j) is uniquely determined by z up to the order and the signs. This follows from the fact that $\sum \xi_j X_j$ is uniquely determined by $z+\bar{z}$ up to an operation of the Weyl group relative to $\mathfrak{a}$ (cf. e. g., "Sophus Lie" [1], Exp. 23), and from the Remark 2 after Proposition 4. 4.

We shall now prove the following

Proposition 4. 4. *An element z in $\mathfrak{p}_+$ belongs to $\mathcal{D}=\psi(M)$ if and only if z has an expression of the form* (4. 18) *in which $|\xi_i|<1$ for all $1 \leq i \leq r$. In particular, $\mathcal{D}$ is bounded.*

Proof. First suppose $z \in \mathcal{D}$. Then one has $z=go$ with $g \in G^\circ$. Write $g=k_1ak_2$ with $k_1, k_2 \in K^\circ$, $a=\exp X$, $X \in \mathfrak{a}$. Let further $X=\sum_{i=1}^{r} \lambda_i X_i$ $(\lambda_i \in \boldsymbol{R})$ and $a=p'_+k'p'_-$ with $k' \in K^\circ_C$, $p'_\pm \in P_\pm$. Then by (4. 14) one has

$$p'_+ = \exp\Big(\sum_{i=1}^{r} (\mathrm{th}\, \lambda_i) o_i\Big)$$

and hence

$$z = (ad\, k_1)\Big(\sum_{i=1}^{r} (\mathrm{th}\, \lambda_i) o_i\Big).$$

This gives an expression of the form (4. 18) with $|\xi_i|=|\mathrm{th}\, \lambda_i|<1$, proving the "only if" part. Conversely, given $|\xi_i|<1$ $(1 \leq i \leq r)$, one can find $\lambda_i \in \boldsymbol{R}$ such that $\xi_i = \mathrm{th}\, \lambda_i$. Hence one has $\sum \xi_i o_i \in \mathcal{D}$. Since $\mathcal{D}$ is stable under the adjoint action of K°, this proves the "if" part. Since K is compact, the boundedness of $\mathcal{D}$ follows from the "only if" part, q. e. d.

Remark 1. Since $\exp(tH_0)z=e^{it}z$ $(t \in \boldsymbol{R})$, the domain $\mathcal{D}=\psi(M)$ is "circular" (i. e., stable under the scalar multiplication by $\boldsymbol{C}^{(1)}$). If $\psi' : M \hookrightarrow \boldsymbol{C}^N$ is another realization of M such that $\psi'(M)$ is a circular bounded domain and $\psi'(x_0)=0$, then by "uniqueness theorem" of H. Cartan (cf. e. g., Narasimhan [1]) there exists a linear isomorphism $\varphi : \mathfrak{p}_+ \to \boldsymbol{C}^N$ such that $\psi'=\varphi\circ\psi$. In this sense, the Harish-Chandra realization ψ is unique.

Remark 2. We have shown above that any hermitian symmetric space of non-compact type can be realized as a bounded domain. In É. Cartan [5], this was verified by a case-by-case argument; our (intrinsic) proof is due to Harish-Chandra [2], VI. Conversely, if one starts from a bounded domain $\mathcal{D}$ in $\boldsymbol{C}^N$, one can first show that $\mathcal{D}$ endowed with the Bergman metric becomes a hermitian manifold and $\mathrm{Hol}(\mathcal{D})$, the group of holomorphic automorphisms of $\mathcal{D}$, is a closed subgroup of $I(\mathcal{D})$ (cf. § 6). $\mathcal{D}$ is called (holomorphically) *symmetric*, if for each point $x \in \mathcal{D}$ there exists a symmetry $s_x \in \mathrm{Hol}(\mathcal{D})$. It was shown by É. Cartan [5] that a symmetric bounded domain (with Bergman metric) is a hermitian symmetric space, so that $\mathrm{Hol}(\mathcal{D})$ is semi-simple (K-N, XI, § 9 ; Helgason [1], VIII, §7). It is known, furthermore, that a bounded domain which is a homogeneous space of a semi-simple Lie group is symmetric (Borel [3], Koszul [2]), and the condition of semi-simplicity can be weakened to the unimodularity (Hano [1]). The first examples of non-symmetric homogeneous bounded domains were given by Piatetskii-Shapiro [1], [2] (cf. Ch. V).

Remark 3. Let $(\xi_1, \cdots, \xi_r)$ be the basis of $\mathfrak{a}^*$ dual to $(X_1, \cdots, X_r)$, i. e., such that $\xi_i(X_j)=\delta_{ij}$. It will be shown in III, § 4 that, when $\mathfrak{g}$ is simple, the $\boldsymbol{R}$-root system $\mathfrak{r}$ of $\mathfrak{g}$ relative to $\mathfrak{a}$ is given by

$$\mathfrak{r} = \begin{cases} \{\pm\xi_j\pm\xi_k\,(j\neq k),\ \pm 2\xi_j\} & \text{(type } (C_r)\text{), or} \\ \{\pm\xi_j\pm\xi_k\,(j\neq k),\ \pm 2\xi_j,\ \pm\xi_j\} & \text{(type } (BC_r)\text{).} \end{cases} \tag{4. 19}$$

(For a direct proof, see Moore [1], Baily-Borel [1], Ash et al. [1].) Moreover (III, § 8), $\mathfrak{r}$ is of type (C) if and only if $\mathcal{D}$ is "of tube type" (i. e., holomorphically equivalent to a tube domain). The multiplicity of roots can be found in Loos [1], Vol. II, p. 162, Table 1.

Now, assuming the result in the above Remark 3, we shall prove the convexity of $\mathcal{D}$. Recall first that $-B(X, \theta\bar{Y})$ $(X, Y \in \mathfrak{g}_C)$ is a positive definite hermitian form on $\mathfrak{g}_C$. We define a norm on $\mathfrak{g}_C$ by

$$\|X\| = (-B(X, \theta\bar{X}))^{\frac{1}{2}} \qquad (X \in \mathfrak{g}_C). \tag{4. 20}$$

We also define the "operator norm" of $ad\, X$ $(X \in \mathfrak{g}_C)$ by

$$\|ad\, X\| = \mathrm{Sup}\,\{\|(ad\, X)Y\| \mid Y \in \mathfrak{g}_C,\ \|Y\| \leq 1\}. \tag{4. 21}$$

Clearly these norms are invariant under the adjoint action of K.

Lemma 4. 5. *For* $z \in \mathfrak{p}_+$, *written in the form* (4. 18), *one has*

$$\|ad(\mathrm{Re}\, z)\| = \max_{1\leq i\leq r} |\xi_i|. \tag{4. 22}$$

Proof. Let $\mathfrak{r}$ be the $\boldsymbol{R}$-root system relative to $\mathfrak{a}$ and

$$\mathfrak{g} = \mathfrak{g}_0 + \sum_{\beta\in\mathfrak{r}} \mathfrak{g}_\beta$$

the corresponding root space decomposition of $\mathfrak{g}$, where $\mathfrak{g}_0 = \mathfrak{c}(\mathfrak{a})$. It is easy to see that $\mathfrak{g}_0$ and the $\mathfrak{g}_\beta$'s are mutually orthogonal with respect to the inner product $-B(X, \theta Y)$. Let $Y \in \mathfrak{g}$ and write

$$Y = Y_0 + \sum_{\beta\in\mathfrak{r}} Y_\beta$$

with $Y_0 \in \mathfrak{c}(\mathfrak{a})$, $Y_\beta \in \mathfrak{g}_\beta$. Then, for $X \in \mathfrak{a}$, one has

$$(ad\, X)Y = \sum_{\beta\in\mathfrak{r}} \beta(X) Y_\beta.$$

Hence

$$\begin{aligned}\|ad\,X\|^2 &= \operatorname*{Sup}_{Y\neq 0}\|(ad\,X)\,Y\|^2/\|Y\|^2 \\ &= \operatorname*{Sup}_{Y\neq 0}\,(\Sigma|\beta(X)|^2\|Y_\beta\|^2)/(\|Y_0\|^2+\Sigma\|Y_\beta\|^2) \\ &= \operatorname*{Max}_{\beta\in\mathfrak{r}}|\beta(X)|^2,\end{aligned}$$

i. e.,

$$\|ad\,X\| = \operatorname*{Max}_{\beta\in\mathfrak{r}}|\beta(X)| \qquad (X\in\mathfrak{a}). \tag{4.23}$$

Now, since $\|ad(\mathrm{Re}\,z)\|$ is invariant under the transformation $z\mapsto(ad\,k')z$ $(k'\in K^\circ)$, it is enough to prove (4. 22) for $z=\sum_{i=1}^{r}\xi_i o_i$ $(\xi_i\in\boldsymbol{R})$. One then has

$$\mathrm{Re}\,z = \frac{1}{2}\sum_{i=1}^{r}\xi_i X_i\in\mathfrak{a}$$

and so $\beta(\mathrm{Re}\,z)$ is given by one of the following expressions:

$$\pm\xi_i,\quad \pm\frac{1}{2}\xi_i,\quad \pm\frac{1}{2}\xi_i\pm\frac{1}{2}\xi_j\quad (i\neq j),$$

in which the $\pm\xi_i$ occurs always. Hence (4. 22) follows from (4. 23), q. e. d.

Combining Lemma 4. 5 with Proposition 4. 4, we obtain

Proposition 4. 6 (*Hermann*). *One has*

$$\mathcal{D} = \{z\in\mathfrak{p}_+ \mid \|ad(\mathrm{Re}\,z)\|<1\}. \tag{4.24}$$

In particular, $\mathcal{D}$ is convex.

Exercises

1. The notation being as in the text, we identify $\mathfrak{h}_C$ with its own dual by the restriction of the Killing form B on $\mathfrak{h}_C$, which we denote by $\langle\ \rangle$. Prove the formula

$$H_0 = 2\sqrt{-1}\sum_{\alpha\in\tilde{\mathfrak{r}}_+}\alpha. \tag{4.25}$$

Hint. Compute $\langle H_0, H\rangle=\mathrm{tr}(ad\,H_0\cdot ad\,H)$ for $H\in\mathfrak{h}_C$.

2. Let $\tilde{\Delta}=\{\alpha_1,\cdots,\alpha_l\}$ be the fundamental system of $\tilde{\mathfrak{r}}$ defined in the text. When $\mathfrak{g}$ is simple, there exists a unique element, say α_1, in $\tilde{\Delta}$ such that $\alpha_1(H_0)=\sqrt{-1}$ and $\alpha_j(H_0)=0$ for $2\leq j\leq l$. Define the "coroots" $\hat{\alpha}_j=2\alpha_j/\langle\alpha_j,\alpha_j\rangle$ and write

$$H_0 = \sqrt{-1}\sum_{j=1}^{l}a_j\hat{\alpha}_j \tag{4.26}$$

with $a_j\in\boldsymbol{R}$. Prove the following relations:

$$\langle\alpha_1,\alpha_1\rangle^{-1} = \sum_{\alpha\in\tilde{\mathfrak{r}}_+}\langle\alpha,\hat{\alpha}_1\rangle, \tag{4.27}$$

$$a_1 = \frac{1}{2}\dim\mathfrak{p}\cdot\langle\alpha_1,\alpha_1\rangle. \tag{4.28}$$

(4. 27) implies that $\langle\alpha_1,\alpha_1\rangle^{-1}$ is an integer.

Hint. (4. 27) follows from (4. 25). To obtain (4. 28), compute $\langle H_0, H_0\rangle$ in two different manners.

Remark. One has $a_1\leq r/2$, and the equality sign holds if and only if $\mathcal{D}$ is of tube type (Satake [8]).

§ 5. The canonical automorphy factors and kernel functions.

Let $\mathfrak{g}$ be a semi-simple Lie algebra of hermitian type. We keep the notation of § 4, fixing a (Zariski connected) $\boldsymbol{R}$-group G with Lie algebra $\mathfrak{g}$, acting (isometrically) on the symmetric domain $\mathcal{D}$ in $\mathfrak{p}_+$. For $g \in P_+ K_C^\circ P_- \subset G_C$, we denote by $(g)_0$ (resp. $(g)_\pm$) the K_C°-part (resp. $P_\pm$-parts) of g, i. e.,

$$g = (g)_+ (g)_0 (g)_-, \qquad (g)_0 \in K_C^\circ, \ (g)_\pm \in P_\pm.$$

For $(g, z) \in G_C \times \mathfrak{p}_+$, such that $g \cdot \exp z \in P_+ K_C^\circ P_-$, we define $g(z) \in \mathfrak{p}_+$ and $J(g, z) \in K_C^\circ$ by

$$\exp g(z) = (g \cdot \exp z)_+, \tag{5. 1}$$

$$J(g, z) = (g \cdot \exp z)_0. \tag{5. 2}$$

When $g \cdot \exp z \in P_+ K_C^\circ P_-$, we say that the *holomorphic action* $g(z)$ is defined. This action of G_C on $\mathfrak{p}_+$ extends the "isometric" action of G° on $\mathcal{D}$ (defined in § 4) as well as the linear action of K_C° on $\mathfrak{p}_+$ defined by the restriction of the adjoint representation. (In general, the holomorphic and isometric actions of $g \in G$ on $\mathcal{D}$ coincide if and only if one has $\alpha(g) \in \mathrm{Hol}(\mathcal{D})$ in the notation of § 4.) The K_C°-valued function J is called the *canonical automorphy factor* of G_C (relative to o). Both $g(z)$ and $J(g, z)$ are defined on an open subset of $G_C \times \mathfrak{p}_+$ (containing $G^\circ \times \mathcal{D}$ and $K_C^\circ \times \mathfrak{p}_+$) and holomorphic in $g \in G_C$ and $z \in \mathfrak{p}_+$ (as long as they are defined).

Lemma 5. 1. *The canonical automorphy factor J satisfies the following relations:*

$$J(g, o) = (g)_0 \quad \textit{for all} \ \ g \in P_+ K_C^\circ P_-, \tag{5. 3}$$

$$J(k, z) = k \quad \textit{for all} \ \ k \in K_C^\circ \ \textit{and} \ z \in \mathfrak{p}_+. \tag{5. 4}$$

For $g, g' \in G_C$ and $z \in \mathfrak{p}_+$, if $g'(z)$ and $g(g'(z))$ are defined, then $(gg')(z)$ is also defined and

$$J(gg', z) = J(g, g'(z)) J(g', z). \tag{5. 5}$$

Proof. (5. 3) and (5. 4) are obvious. To prove (5. 5), suppose $g'(z)$ and $g(g'(z))$ are defined. Then, by the definition, one has

$$g' \cdot \exp z = \exp g'(z) \cdot J(g', z) \cdot p_-,$$
$$g \cdot \exp g'(z) = \exp g(g'(z)) \cdot J(g, g'(z)) \cdot p'_-$$

with $p_-, p'_- \in P_-$. Hence

$$gg' \cdot \exp z = \exp g(g'(z)) \cdot J(g, g'(z)) \cdot J(g', z) \cdot p''_-,$$

where $p''_- = J(g', z)^{-1} p'_- J(g', z) \cdot p_- \in P_-$. Hence $(gg')(z)$ is defined, $(gg')(z) = g(g'(z))$, and one has (5. 5), q. e. d.

Next, for $z, w \in \mathfrak{p}_+$ such that $(\exp \bar{w})^{-1} \exp z \in P_+ K_C^\circ P_-$ one defines $K(z, w) \in K_C^\circ$ by

$$K(z, w) = J((\exp \bar{w})^{-1}, z)^{-1} = (((\exp \bar{w})^{-1} \exp z)_0)^{-1}. \tag{5. 6}$$

For $z, w \in \mathcal{D}$, $K(z, w)$ is always defined, since one has

$$(\exp \bar{w})^{-1} \exp z \in (\overline{G^\circ K_c^\circ P_-})^{-1}(G^\circ K_c^\circ P_-) = P_+ K_c^\circ G^\circ K_c^\circ P_- = P_+ K_c^\circ P_-.$$

The K_c°-valued function K defined on an open subset of $\mathfrak{p}_+ \times \mathfrak{p}_+$ (containing $\mathcal{D} \times \mathcal{D}$) is called the *canonical kernel function* of G_c (relative to o). Clearly $K(z, w)$ is holomorphic in z and antiholomorphic in w (as long as it is defined).

Lemma 5. 2. *The canonical kernel function K satisfies the following relations:*

(5. 7) $$K(w, z) = \overline{K(z, w)}^{-1} \qquad \text{if } K(z, w) \text{ is defined,}$$

(5. 8) $$K(o, z) = K(z, o) = 1 \qquad \text{for all } z \in \mathfrak{p}_+.$$

For $g \in G_c$ and $z, w \in \mathfrak{p}_+$ such that $g(z)$, $\bar{g}(w)$ and $K(z, w)$ are defined, $K(g(z), \bar{g}(w))$ is also defined and one has

(5. 9) $$K(g(z), \bar{g}(w)) = J(g, z)\, K(z, w)\, \overline{J(\bar{g}, w)}^{-1}.$$

Proof. (5. 7) and (5. 8) are obvious. To prove (5. 9), suppose $g(z)$, $\bar{g}(w)$ and $K(z, w)$ are defined. Then

$$g \cdot \exp z = \exp g(z) \cdot J(g, z) \cdot p_-,$$
$$\bar{g} \cdot \exp w = \exp \bar{g}(w) \cdot J(\bar{g}, w) \cdot p'_-$$

with $p_-, p'_- \in P_-$. Hence

(*) $$(\exp \bar{w})^{-1} \exp z = \bar{p}_-'^{-1} \cdot \overline{J(\bar{g}, w)}^{-1} \cdot \exp \overline{\bar{g}(w)}^{-1} \cdot \exp g(z) \cdot J(g, z) \cdot p_-$$

where $\bar{p}_-'^{-1} \in P_+$ and $(\exp \bar{w})^{-1} \exp z \in P_+ K_c^\circ P_-$. Hence one has

$$(\exp \overline{\bar{g}(w)})^{-1} \exp g(z) \in P_+ K_c^\circ P_-$$

and, comparing the K_c°-parts of both sides of (*), one has

$$K(z, w)^{-1} = \overline{J(\bar{g}, w)}^{-1}\, K(g(z), \bar{g}(w))^{-1}\, J(g, z),$$

i. e., (5. 9), q. e. d.

Lemma 5. 3. *For $g \in G_c$, the Jacobian (linear map) of the holomorphic map $z \mapsto g(z)$, where it is defined, is given by*

(5. 10) $$\mathrm{Jac}(z \mapsto g(z)) = ad_{\mathfrak{p}_+}(J(g, z)),$$

$ad_{\mathfrak{p}_+}(\ )$ denoting the restriction of $ad(\ )$ on $\mathfrak{p}_+$.

Proof. Put $(g \cdot \exp z)_- = \exp \bar{w}$ with $w \in \mathfrak{p}_+$. Then, for $t \in \boldsymbol{R}$, $z_1 \in \mathfrak{p}_+$, one has

$$\begin{aligned} g \cdot \exp(z + t z_1) &= \exp g(z) \cdot J(g, z) \cdot \exp \bar{w} \cdot \exp(t z_1) \\ &= \exp g(z) \cdot J(g, z) \cdot \exp\Big(t(z_1 + [\bar{w}, z_1] + \frac{1}{2}[\bar{w}, [\bar{w}, z_1]])\Big) \cdot \exp \bar{w}. \end{aligned}$$

Hence for a sufficiently small t one has $g \cdot \exp(z + t z_1) \in P_+ K_c^\circ P_-$ and

$$(g \cdot \exp(z + t z_1))_+ = \exp(g(z) + t(ad_{\mathfrak{p}_+} J(g, z)) z_1 + O(t^2)),$$

whence follows (5. 10), q. e. d.

Corollary 5. 4 (*Baily-Borel* [1]). *When $K(z, w)$ is defined, put*

$$(\exp \bar{w})^{-1} \exp z = \exp z' \cdot K(z, w)^{-1} \cdot (\exp \bar{w}')^{-1}$$

with $z', w' \in \mathfrak{p}_+$. *Then*

$$\text{Jac}(z \mapsto z') = ad_{\mathfrak{p}_+} K(z, w)^{-1}. \tag{5. 11}$$

This is a special case of Lemma 5. 3, since by the definitions one has $z' = (\exp \bar{w})^{-1}(z)$ and $K(z, w)^{-1} = J((\exp \bar{w})^{-1}, z)$. The relation (5. 10) implies that the holomorphic tangent bundle of the symmetric domain $\mathcal{D}$ can naturally be identified with $\mathcal{D} \times \mathfrak{p}_+$ on which the action of G° is defined by

$$G^\circ \ni g: \ (z, z_1) \longmapsto (g(z), (ad_{\mathfrak{p}_+} J(g, z)) z_1).$$

For a (holomorphic) character $\chi : K_c^\circ \to \boldsymbol{C}^\times$ we define the *canonical automorphy factor* and *kernel function of type* χ by

$$\begin{aligned} j_\chi(g, z) &= \chi(J(g, z)), \\ k_\chi(z, w) &= \chi(K(z, w)). \end{aligned} \tag{5. 12}$$

Since $\chi(\bar{k}) = \overline{\chi(k)}^{-1}$, one has

$$k_\chi(w, z) = \overline{k_\chi(z, w)}, \tag{5. 7'}$$

$$k_\chi(g(z), \bar{g}(w)) = j_\chi(g, z) k_\chi(z, w) \overline{j_\chi(\bar{g}, w)} \tag{5. 9'}$$

in place of (5. 7) and (5. 9). The character defined by

$$\chi_1(k) = \det(ad_{\mathfrak{p}_+}(k)) \qquad \text{for} \quad k \in K_c^\circ \tag{5. 13}$$

is of particular importance, since by Lemma 5. 3 the corresponding automorphy factor $j = j_{\chi_1}$ is the "jacobian" of the holomorphic transformation $z \mapsto g(z)$ in the usual sense.

Example. The case $\mathfrak{g} = \mathfrak{su}(1, 1)$. For $g = \begin{pmatrix} \alpha & \beta \\ \bar{\beta} & \bar{\alpha} \end{pmatrix} \in G = SU(1, 1)$ one has

$$J(g, z) = \begin{pmatrix} (\bar{\beta} z + \bar{\alpha})^{-1} & 0 \\ 0 & \bar{\beta} z + \bar{\alpha} \end{pmatrix},$$

$$K(z, w) = \begin{pmatrix} 1 - z\bar{w} & 0 \\ 0 & (1 - z\bar{w})^{-1} \end{pmatrix},$$

and

$$\chi_1\left(\begin{pmatrix} \alpha & 0 \\ 0 & \alpha^{-1} \end{pmatrix}\right) = \alpha^2 \qquad (\alpha \in \boldsymbol{C}^\times).$$

Hence

$$j_{\chi_1}(g, z) = (\bar{\beta} z + \bar{\alpha})^{-2}, \qquad k_{\chi_1}(z, w) = (1 - z\bar{w})^2.$$

In general, a map $j' : G^\circ \times \mathcal{D} \to \boldsymbol{C}^\times$ satisfying the following two conditions (J 1), (J 2) is called a *holomorphic automorphy factor* for $(G^\circ, \mathcal{D})$:

(J 1) $j'(g, z)$ *is* C^∞ *in* $g \in G^\circ$ *and holomorphic in* $z \in \mathcal{D}$;

(J 2) $j'(gg', z) = j'(g, g'(z)) j'(g', z)$ *for all* $g, g' \in G^\circ$, $z \in \mathcal{D}$.

Two (holomorphic) automorphy factors j' and j'' are called (holomorphically) *equivalent* if there exists a (holomorphic) function $\varphi : \mathcal{D} \to \boldsymbol{C}^\times$ such that

$$j''(g, z) = \varphi(g(z))j'(g, z)\varphi(z)^{-1}. \tag{5.14}$$

This is clearly an equivalence relation. From (J 2) we see that, for an automorphy factor j, if we put $\chi(k)=j(k, o)$, χ is a character of K°; j is then called *of type* χ. We recall that any (continuous) character χ of K° is uniquely extendible to a holomorphic character of K°_C, which we also denote by the same letter χ.

Proposition 5. 5. *Let j' be a holomorphic automorphy factor of type χ for $(G^\circ, \mathcal{D})$. Then j' is (holomorphically) equivalent to the canonical automorphy factor j_χ.*

Proof (Gunning [3], Murakami [4]). Put $\varphi(g)=j'(g, o)j_\chi(g, o)^{-1}(g\in G^\circ)$. Then by (J 2) one has $\varphi(gk)=\varphi(g)$ for $k\in K^\circ$, i. e., $\varphi(g)$ depends only on $z=go$. Hence we write $\varphi(g)=\varphi(z)$. Then again by (J 2) one has

$$\varphi(g(z)) = j'(g, z)\varphi(z)j_\chi(g, z)^{-1}, \tag{*}$$

which implies the relation similar to (5. 14). Hence it remains only to show that φ is holomorphic. Put $\omega=\varphi^{-1}d''\varphi$. Then ω is a $\boldsymbol{C}$-valued C^∞-form on $\mathcal{D}$ and by (J 1) and (*) satisfies the relation

$$(\omega\circ g)_z = {}^t(d\alpha(g))\omega_{g(z)} = \omega_z \qquad \text{for all } g\in G^\circ,\ z\in\mathcal{D}. \tag{$\sharp$}$$

In particular, one has

$$(\omega\circ k)_o = {}^t(d\alpha(k))\omega_o = \omega_o \qquad \text{for all } k\in K^\circ. \tag{$\sharp\sharp$}$$

But, if one identifies $T_o(\mathcal{D})$ with $\mathfrak{p}_+$ in the natural manner, the differential $d\alpha(k)$ (at o) is identified with $ad_{\mathfrak{p}_+} k$. Since there exists an element k in K° such that $ad_{\mathfrak{p}_+}k$ does not have eigenvalue 1 (e. g., $k=\exp \pi H_0$), it follows from ($\sharp\sharp$) that $\omega_o=0$, and hence by ($\sharp$) $\omega_z=0$ for all $z\in\mathcal{D}$. Hence φ is holomorphic, q. e. d.

In the rest of this section, we shall determine the explicit form of the "standard" kernel function k_{χ_1} and show that the domain $\mathcal{D}$ coincides with one of the connected components of the (real) Zariski open set $\{z\in\mathfrak{p}_+ \mid k_{\chi_1}(z, z)\neq 0\}$. First we prove some identities.

Lemma 5. 6. *For z, $w\in\mathfrak{p}_+$ we have the following identities :*

$$[[z, \bar{w}], [z, [\bar{w}, [z, \bar{w}]]]] = 0, \tag{5.15}$$

$$(ad\, z)^2(ad\, \bar{w})^2 = 2(ad[z, \bar{w}])^2 - ad[z, [\bar{w}, [z, \bar{w}]]] \qquad \text{on } \mathfrak{p}_+. \tag{5.16}$$

Proof of (5. 15). Put $A=ad[z, \bar{w}]$. Then the left-hand side of (5. 15) can be transformed as follows :

$$
\begin{aligned}
&= [[z,[z,[\bar{w},[z,\bar{w}]]]],\bar{w}]+[z,[\bar{w},[z,[\bar{w},[z,\bar{w}]]]]] \\
&= [[z,[\bar{w},[z,[z,\bar{w}]]]],\bar{w}]+[z,[[\bar{w},z],[\bar{w},[z,\bar{w}]]]] \\
&= [[[z,\bar{w}],[z,[z,\bar{w}]]],\bar{w}]+[z,[[z,\bar{w}],[[z,\bar{w}],\bar{w}]]] \\
&= -[A^2z,\bar{w}]+[z,A^2\bar{w}] \\
&= -A[Az,\bar{w}]+A[z,A\bar{w}] \\
&= -A^2[z,\bar{w}]+2A[z,A\bar{w}].
\end{aligned}
$$

But, since the left-hand side of (5. 15) is also equal to $-A[z, A\bar{w}]$, and $A^2[z,\bar{w}]=0$, one has $A[z, A\bar{w}]=0$, i. e., (5. 15).

Proof of (5. 16). For $X \in \mathfrak{p}_+$, one has

$$
\begin{aligned}
(ad\, z)^2(ad\, \bar{w})^2X &= [z,[z,[\bar{w},[\bar{w},X]]]] \\
&= [z,[[z,\bar{w}],[\bar{w},X]]+[\bar{w},[z,[\bar{w},X]]]] \\
&= [z,[[[z,\bar{w}],\bar{w}],X]+2[\bar{w},[[z,\bar{w}],X]]] \\
&= [[z,[[z,\bar{w}],\bar{w}]],X]+2[[z,\bar{w}],[[z,\bar{w}],X]], \qquad \text{q. e. d.}
\end{aligned}
$$

From (5. 15), (5. 16) (and the above proof) we also obtain

(5. 17) $$[(ad\, z)^2(ad\, \bar{w})^2, ad[z,\bar{w}]] = 0 \quad \text{on } \mathfrak{p}_+,$$

(5. 18) $$(ad\, z)^2(ad\, \bar{w})^2 z = (ad[z,\bar{w}])^2 z.$$

For $z \in \mathfrak{p}_+$, we denote by φ_z the $\boldsymbol{C}$-linear map $\mathfrak{k}_C \to \mathfrak{p}_+$ defined by the restriction of $ad\, z$, i. e., $\varphi_z = (ad\, z)|\mathfrak{k}_C$. Then one has

$$ad\, z|\mathfrak{p}_- = -{}^t\varphi_z, \qquad ad\, \bar{z}|\mathfrak{k}_C = \bar{\varphi}_z, \qquad ad\, \bar{z}|\mathfrak{p}_+ = -{}^t\bar{\varphi}_z,$$

where t denotes the transpose with respect to the Killing form B. Hence, in the matrix expression according to the canonical decomposition $\mathfrak{g}_C = \mathfrak{p}_+ + \mathfrak{k}_C + \mathfrak{p}_-$, one has

$$ad(\exp z) = \begin{pmatrix} 1 & \varphi_z & -\frac{1}{2}\varphi_z{}^t\varphi_z \\ 0 & 1 & -{}^t\varphi_z \\ 0 & 0 & 1 \end{pmatrix}.$$

Similarly, for $w \in \mathfrak{p}_+$ one has

$$ad(\exp \bar{w})^{-1} = \begin{pmatrix} 1 & 0 & 0 \\ {}^t\bar{\varphi}_w & 1 & 0 \\ -\frac{1}{2}\bar{\varphi}_w{}^t\bar{\varphi}_w & -\bar{\varphi}_w & 1 \end{pmatrix}.$$

Hence

(5. 19) $$ad((\exp \bar{w})^{-1}(\exp z))$$

$$= \begin{pmatrix} 1 & \varphi_z & -\frac{1}{2}\varphi_z{}^t\varphi_z \\ {}^t\bar{\varphi}_w & 1+{}^t\bar{\varphi}_w\varphi_z & -{}^t\varphi_z-\frac{1}{2}{}^t\bar{\varphi}_w\varphi_z{}^t\varphi_z \\ -\frac{1}{2}\bar{\varphi}_w{}^t\bar{\varphi}_w & -\bar{\varphi}_w-\frac{1}{2}\bar{\varphi}_w{}^t\bar{\varphi}_w\varphi_z & 1+\bar{\varphi}_w{}^t\varphi_z+\frac{1}{4}\bar{\varphi}_w{}^t\bar{\varphi}_w\varphi_z{}^t\varphi_z \end{pmatrix}.$$

If $(\exp \bar{w})^{-1} \exp z \in P_+ K_c^\circ P_-$, one has in the notation of Corollary 5.4

$$ad((\exp \bar{w})^{-1}(\exp z)) = ad((\exp z')K(z, w)^{-1}(\exp \bar{w}')^{-1}) \tag{5.20}$$

$$= \begin{pmatrix} 1 & \varphi_{z'} & -\frac{1}{2}\varphi_{z'}{}^t\varphi_{z'} \\ 0 & 1 & -{}^t\varphi_{z'} \\ 0 & 0 & 1 \end{pmatrix} \begin{pmatrix} k_+ & 0 & 0 \\ 0 & k_0 & 0 \\ 0 & 0 & k_- \end{pmatrix} \begin{pmatrix} 1 & 0 & 0 \\ {}^t\bar{\varphi}_{w'} & 1 & 0 \\ -\frac{1}{2}\bar{\varphi}_{w'}{}^t\bar{\varphi}_{w'} & -\bar{\varphi}_{w'} & 1 \end{pmatrix},$$

where $z', w' \in \mathfrak{p}_+$ and $k_0 = ad_{\mathfrak{k}_c}K(z, w)^{-1}$, $k_\pm = ad_{\mathfrak{p}_\pm}K(z, w)^{-1}$. Comparing the (3, 3) and (2, 3) components of the matrices in (5.19) and (5.20), one obtains

$$k_- = 1 + \bar{\varphi}_w{}^t\varphi_z + \frac{1}{4}\bar{\varphi}_w{}^t\bar{\varphi}_w\varphi_z{}^t\varphi_z,$$

$$-{}^t\varphi_{z'}k_- = -{}^t\varphi_z - \frac{1}{2}{}^t\bar{\varphi}_w\varphi_z{}^t\varphi_z.$$

It follows that

$$(5.21)\qquad \begin{aligned} k_+ &= {}^tk_-^{-1} = \left(1 + \varphi_z{}^t\bar{\varphi}_w + \frac{1}{4}\varphi_z{}^t\varphi_z\bar{\varphi}_w{}^t\bar{\varphi}_w\right)^{-1}, \\ \varphi_{z'} &= k_+\left(\varphi_z + \frac{1}{2}\varphi_z{}^t\varphi_z\bar{\varphi}_w\right). \end{aligned}$$

Proposition 5.7. *Let $z, w \in \mathfrak{p}_+$ and suppose that $K(z, w)$ is defined. Then one has*

$$ad_{\mathfrak{p}_+}K(z, w) = 1 - ad[z, \bar{w}] + \frac{1}{4}(ad\, z)^2(ad\, \bar{w})^2 \qquad (on\ \mathfrak{p}_+). \tag{5.22}$$

If, moreover, $1 - \frac{1}{2}ad_{\mathfrak{p}_+}[z, \bar{w}]$ is non-singular, one has

$$(\exp \bar{w})^{-1}(z) = \left(1 - \frac{1}{2}ad_{\mathfrak{p}_+}[z, \bar{w}]\right)^{-1} z. \tag{5.23}$$

Proof. In the above notation, one has $k_+ = ad_{\mathfrak{p}_+}K(z, w)^{-1}$ and $z' = (\exp \bar{w})^{-1}(z)$. Hence (5.22) follows from (5.21). On the other hand, one has

$$\begin{aligned} z' &= i\varphi_{z'}(H_0) \\ &= k_+\left(i\varphi_z(H_0) + \frac{i}{2}\varphi_z{}^t\varphi_z\bar{\varphi}_w(H_0)\right) \\ &= k_+\left(z + \frac{1}{2}[z, [z, \bar{w}]]\right) \\ &= k_+\left(1 - \frac{1}{2}ad_{\mathfrak{p}_+}[z, \bar{w}]\right)z. \end{aligned}$$

In view of (5.22), (5.17), and (5.18), this is equal to $(1 - \frac{1}{2}ad_{\mathfrak{p}_+}[z, \bar{w}])^{-1}z$, if $1 - \frac{1}{2}ad_{\mathfrak{p}_+}[z, \bar{w}]$ is non-singular, q. e. d.

Remark. In the notation of JTS, (5.22) and (5.23) can be expressed as

(5. 22′) $$ad_{\mathfrak{p}_+}K(z, w) = 1-2\, z\Box w+P(z)P(w),$$

(5. 23′) $$(\exp \bar{w})^{-1}(z) = (1-z\Box w)^{-1} z$$

(cf. I, § 6, Exerc. 5, 6).

Lemma 5. 8. *For $z\in\mathfrak{p}_+$, one has $1-z\Box z\gg 0$ (with respect to the hermitian inner product on $\mathfrak{p}_+$ defined by the Killing form) if and only if z has an expression of the form (4. 18) in which one has $|\xi_j|<1$ for all j. Also one has $ad_{\mathfrak{p}_+}K(z, z)\gg 0$ if $|\xi_j|<1$ for all j.*

Proof. Put $z_1=\sum \xi_j o_j$. Then

$$1-z\Box z = (ad_{\mathfrak{p}_+}k)(1-z_1\Box z_1)(ad_{\mathfrak{p}_+}k)^{-1},$$
$$ad_{\mathfrak{p}_+}K(z, z) = (ad_{\mathfrak{p}_+}k)(ad_{\mathfrak{p}_+}K(z_1, z_1))(ad_{\mathfrak{p}_+}k)^{-1}$$

and $(ad_{\mathfrak{p}_+}k)^{-1}={}^t\overline{(ad_{\mathfrak{p}_+}k)}$. Hence, for the proof of the Lemma, we may assume $z=z_1$. Then in the notation of § 4

$$[z, \bar{z}] = \sum_{j=1}^{r} |\xi_j|^2 [o_j, \bar{o}_j]$$
$$= -i\sum_{j=1}^{r} |\xi_j|^2 H_j.$$

For simplicity, we identify $\mathfrak{h}_c$ with its dual $\mathfrak{h}_c^*$ by the restriction of Killing form B on $\mathfrak{h}_c$, which we denote by $\langle\ \rangle$. Then from our normalization (4. 15), it is easy to see that

(5. 24) $$-iH_j = \frac{2\beta_j}{\langle\beta_j, \beta_j\rangle}.$$

Hence one has

$$ad[z, \bar{z}] = \sum_{j=1}^{r}\frac{2\langle\alpha, \beta_j\rangle}{\langle\beta_j, \beta_j\rangle}|\xi_j|^2 \qquad \text{on } \mathfrak{g}_\alpha.$$

Therefore one has $1-z\Box z\gg 0$ if and only if

(5. 25) $$1-\sum_{j=1}^{r}\frac{\langle\alpha, \beta_j\rangle}{\langle\beta_j, \beta_j\rangle}|\xi_j|^2 > 0$$

for all $\alpha\in\tilde{\mathfrak{r}}_+$. As is well-known, there exists an (inner) automorphism of $\mathfrak{g}_c$ (called a "Cayley transformation") which maps iH_j to X_j for all $1\leq j\leq r$ (III, § 1). Clearly we may assume that $\mathfrak{g}$ is simple. Then, by the Remark 3 following Proposition 4. 4, we see that the left-hand side of (5. 25) is equal to one of the following expressions, unless it is trivially positive:

$$1-|\xi_j|^2, \quad 1-\frac{1}{2}(|\xi_j|^2\pm|\xi_k|^2) \quad (j\neq k),$$
$$\left(1-\frac{1}{2}|\xi_j|^2 \quad \text{if } \mathfrak{r} \text{ is of type } (BC)\right).$$

Note that, for each j, the first expression $1-|\xi_j|^2$ actually occurs. Hence it is clear that (5. 25) holds for all $\alpha\in\tilde{\mathfrak{r}}_+$ if and only if $|\xi_j|<1$ for all j. Similarly, from (5. 22) and (5. 16) one obtains

$$ad\,K(z, z) = 1-\sum_j \frac{2\langle\alpha, \beta_j\rangle}{\langle\beta_j, \beta_j\rangle}|\xi_j|^2+2\Big(\sum_j \frac{\langle\alpha, \beta_j\rangle}{\langle\beta_j, \beta_j\rangle}|\xi_j|^2\Big)^2$$
$$-\sum_j \frac{\langle\alpha, \beta_j\rangle}{\langle\beta_j, \beta_j\rangle}|\xi_j|^4 \quad \text{on } \mathfrak{g}_\alpha.$$

This is equal to one of the following expressions, unless it is trivially positive:

$$(1-|\xi_j|^2)^2, \quad (1-|\xi_j|^2)(1-|\xi_k|^2),$$
$$(1-|\xi_j|^2)(1+|\xi_k|^2)+|\xi_k|^4 \quad (j\neq k),$$
$$(1-|\xi_j|^2 \text{ if } \mathfrak{r} \text{ is of type } (BC)),$$

which are clearly positive if $|\xi_j|<1$ for all j, q. e. d.

Theorem 5. 9 (*Koecher* [9]). *The symmetric domain $\mathcal{D}$ (embedded in $\mathfrak{p}_+$ by the Harish-Chandra map) is given by*

(5. 26) $$\mathcal{D} = \{z\,|\,z\in\mathfrak{p}_+,\ 1-z\square\bar{z}\gg 0\}.$$

$\mathcal{D}$ is also equal to the connected component (containing 0) of

(5. 27) $$\{z\,|\,z\in\mathfrak{p}_+,\ ad_{\mathfrak{p}_+}K(z, z)\gg 0\}.$$

Proof. By Proposition 4. 4, z belongs to $\mathcal{D}$ if and only if z has an expression of the form (4. 18) with $|\xi_j|<1$ for all j. Hence, by Lemma 5. 8, we obtain (5. 26). It also follows from Lemma 5. 8 that $\mathcal{D}$ is contained in the open set defined by (5. 27). From the proof of Lemma 5. 8, we see that, if z is in the connected component of this open set containing 0, then one has $|\xi_j|<1$ for all j, and hence $z\in\mathcal{D}$. This proves the second assertion, q. e. d.

Exercises

1. Prove (5. 24).

2. For $z\in\mathcal{D}$, put

(5. 28) $$g_z = \exp z\cdot K(z, z)^{\frac{1}{2}}\cdot\exp\bar{z}.$$

[Note that $K(z, z)$ is in the P-part of K_C° in its Cartan decomposition, so that $K(z, z)^{\frac{1}{2}}=\exp\big(\frac{1}{2}\log K(z, z)\big)$ is defined.] Show that g_z is the unique element in $\exp\mathfrak{p}$ such that $g_z(o)=z$. (From this one obtains another proof of the second assertion in Theorem 5. 9.)

§ 6. The Bergman metric of a symmetric domain.

First we recall the notion of Bergman kernel function. Let $\mathcal{D}$ be a (not necessarily symmetric) domain in a complex Euclidean space C^N with fixed coordinate functions $z^\alpha=x^\alpha+iy^\alpha$ $(1\leq\alpha\leq N)$. The Euclidean volume element of C^N is then given by

$$d\mu(z) = \Big(\frac{i}{2}\Big)^N\prod_{\alpha=1}^N dz^\alpha\wedge d\bar{z}^\alpha.$$

(A positive $2N$-form with respect to the natural orientation is always identified with the corresponding volume element.) For a $\boldsymbol{C}$-valued measurable function f on $\mathcal{D}$, we define the L^2-norm by

$$\|f\| = \left(\int_{\mathcal{D}} |f(z)|^2 d\mu(z)\right)^{\frac{1}{2}}.$$

Then, as is well-known (e. g., Weil [2]), the space $\mathcal{H}^2(\mathcal{D})$ of all square-integrable holomorphic functions on $\mathcal{D}$ becomes a (separable) Hilbert space with respect to the norm $\|\ \|$.

By definition, the *Bergman kernel function* of $\mathcal{D}$ is a function $k_1(z, w)$ on $\mathcal{D}\times\mathcal{D}$ satisfying the following conditions:

(B 1) *For each* $w \in \mathcal{D}$, $k_1(z, w)$ *(viewed as a function of* z*) belongs to* $\mathcal{H}^2(\mathcal{D})$, *and*

(B 2) $$k_1(w, z) = \overline{k_1(z, w)},$$

(B 3) $$\int_{\mathcal{D}} k_1(z, w) f(w) d\mu(w) = f(z) \qquad \textit{for all } f \in \mathcal{H}^2(\mathcal{D}).$$

Clearly the function k_1 (or rather the differential form $k_1(z, w)d\mu(w)$) is uniquely characterized by these properties. The Bergman kernel function of $\mathcal{D}$ (when it exists) will be denoted by $k_{\mathcal{D}}$.

It is known (e. g., Weil, loc. cit.; Helgason [1], VIII, § 3) that when $\mathcal{D}$ is bounded or more generally, holomorphically equivalent to a bounded domain, the Bergman kernel function $k_{\mathcal{D}}$ exists. Moreover, if $\{\varphi_\nu (\nu=1, 2, \cdots)\}$ is any orthonormal basis of $\mathcal{H}^2(\mathcal{D})$, then one has

(6. 1) $$k_{\mathcal{D}}(z, w) = \sum_{\nu=1}^{\infty} \varphi_\nu(z)\overline{\varphi_\nu(w)},$$

where the infinite series on the right-hand side converges absolutely and uniformly on any compact set in $\mathcal{D}\times\mathcal{D}$.

Let $\mathcal{D}'$ be another domain holomorphically equivalent to $\mathcal{D}$ by a biholomorphic map $\varphi: \mathcal{D}\to\mathcal{D}'$. Then it is clear that the correspondence $f'(\in \mathcal{H}^2(\mathcal{D}'))\mapsto f(\in \mathcal{H}^2(\mathcal{D}))$ defined by

(6. 2) $$f(z) = f'(\varphi(z))\mathrm{jac}(\varphi, z),$$

where jac denotes the (usual) jacobian, gives an isomorphism of Hilbert space: $\mathcal{H}^2(\mathcal{D})\cong\mathcal{H}^2(\mathcal{D}')$. Hence one has

(6. 3) $$k_{\mathcal{D}}(z, w) = \mathrm{jac}(\varphi, z)\, k_{\mathcal{D}'}(\varphi(z), \varphi(w))\,\overline{\mathrm{jac}(\varphi, w)} \qquad (z, w \in \mathcal{D}).$$

Lemma 6. 1. *Suppose the domain* $\mathcal{D}$ *is holomorphically equivalent to a bounded domain and is homogeneous, and let* G_1 *be a subgroup of* Hol$(\mathcal{D})$ *which acts transitively on* $\mathcal{D}$. *Then a* $\boldsymbol{C}$*-valued function* k *on* $\mathcal{D}\times\mathcal{D}$ *is a constant multiple of the Bergman kernel function* $k_{\mathcal{D}}$ *if and only if the following two conditions are satisfied:*

(K 1) $k(z, w)$ *is holomorphic in* z *and satisfies* (B 2).

(K 2) *For any* $g \in G_1$ *one has*

(6. 4) $$k(z, w) = \mathrm{jac}(g, z)\, k(g(z), g(w))\,\overline{\mathrm{jac}(g, w)} \qquad (z, w \in \mathcal{D}).$$

Proof. By the definition and (6. 3), it is clear that $k=k_{\mathcal{D}}$ satisfies the conditions (K 1), (K 2). To prove the converse, put $\phi(z, w)=k(z, w)k_{\mathcal{D}}(z, w)^{-1}$. Then, by (K 2) one has

$$\phi(g(z), g(w)) = \phi(z, w)$$

for all $g \in G_1$. Since G_1 is transitive, it follows that

$$\phi(z, z) = \phi(o, o),$$

where o is a fixed element in $\mathcal{D}$. But, since by (K 1) $\phi(z, w)$ is holomorphic in z and antiholomorphic in w, this implies that $\phi(z, w)$ is identically equal to the constant $c=\phi(o, o)$, i. e., $k=ck_{\mathcal{D}}$, q. e. d.

It follows from this Lemma and (5. 7′), (5. 9′) that, for a symmetric domain $\mathcal{D}$, the inverse of the Bergman kernel function coincides with the canonical kernel function of type χ_1 up to a positive constant, i. e.,

$$k_{\mathcal{D}}(z, w)^{-1} = ck_{\chi_1}(z, w). \tag{6. 5}$$

In particular, by (5. 8) one has $k_{\mathcal{D}}(o, w)=c^{-1}$, so that by (B 3) applied to $f=1$ one sees that c is equal to the Euclidean volume of the symmetric domain $\mathcal{D}$.

In general, it can be shown by (6. 1) that for a domain $\mathcal{D}$ in C^N, holomorphically equivalent to a bounded domain, one has $k_{\mathcal{D}}(z, z)>0$ for all $z \in \mathcal{D}$ and the "Hessian" $(h_{\alpha\beta})$ defined by

$$h_{\alpha\beta} = 2\frac{\partial^2}{\partial z^\alpha \partial \bar{z}^\beta} \log k_{\mathcal{D}}(z, z) \tag{6. 6}$$

is a positive definite hermitian matrix, where as usual one sets

$$\frac{\partial}{\partial z^\alpha} = p_+\left(\frac{\partial}{\partial x^\alpha}\right) = \frac{1}{2}\left(\frac{\partial}{\partial x^\alpha} - i\frac{\partial}{\partial y^\alpha}\right),$$
$$\frac{\partial}{\partial \bar{z}^\alpha} = p_-\left(\frac{\partial}{\partial x^\alpha}\right) = \frac{1}{2}\left(\frac{\partial}{\partial x^\alpha} + i\frac{\partial}{\partial y^\alpha}\right).$$

When these conditions are satisfied, one can define a hermitian metric on $\mathcal{D}$ by

$$\begin{aligned} q\left(\frac{\partial}{\partial x^\alpha}, \frac{\partial}{\partial x^\beta}\right) &= q\left(\frac{\partial}{\partial y^\alpha}, \frac{\partial}{\partial y^\beta}\right) = \operatorname{Re} h_{\alpha\beta}, \\ q\left(\frac{\partial}{\partial y^\alpha}, \frac{\partial}{\partial x^\beta}\right) &= \operatorname{Im} h_{\alpha\beta}, \end{aligned} \tag{6. 7}$$

or symbolically, by

$$ds^2 = \sum_{\alpha,\beta=1}^{N} h_{\alpha\beta} dz^\alpha d\bar{z}^\beta. \tag{6. 8}$$

The corresponding hermitian form h, defined by (3. 3), is then determined by the relation

$$h\left(\frac{\partial}{\partial x^\alpha}, \frac{\partial}{\partial x^\beta}\right) = 2q\left(\frac{\partial}{\partial z^\alpha}, \frac{\partial}{\partial \bar{z}^\beta}\right) = h_{\alpha\beta}. \tag{6. 7′}$$

[Note that $q(\partial/\partial z^\alpha, \partial/\partial z^\beta)=q(\partial/\partial \bar{z}^\alpha, \partial/\partial \bar{z}^\beta)=0$.] This hermitian metric h (or the corresponding Riemannian metric q), which is clearly independent of the choice of coordinates, is called the *Bergman metric* of $\mathcal{D}$, and denoted by $h_{\mathcal{D}}$ (or $q_{\mathcal{D}}$).

The "fundamental 2-form" ω associated with the Bergman metric $h_{\mathcal{D}}$ is given by

$$\begin{aligned}(6.9)\qquad \omega &= \frac{i}{2}\sum_{\alpha,\beta=1}^{N} h_{\alpha\beta}dz^{\beta}\wedge d\bar{z}^{\alpha}\\ &= id'd''(\log k_{\mathcal{D}}(z,z)).\end{aligned}$$

Clearly one has $d\omega=0$, which shows that the Bergman metric is "Kählerian".

Following the convention in I, § 7, we put

$$\partial_{z_1}=\sum_{\alpha=1}^{N} z_1^{\alpha}\frac{\partial}{\partial z^{\alpha}},\qquad \bar{\partial}_{z_1}=\sum_{\alpha=1}^{N} \bar{z}_1^{\alpha}\frac{\partial}{\partial \bar{z}^{\alpha}}$$

for $z_1=(z_1^{\alpha})\in \boldsymbol{C}^N$ and identify the holomorphic (resp. antiholomorphic) tangent vector ∂_{z_1} (resp. $\bar{\partial}_{z_1}$) at any point $z\in\mathcal{D}$ with z_1 (resp. $\bar{z}_1$). Then, by (6. 6), (6. 7') one has

$$(6.10)\qquad (q_{\mathcal{D}})_z(\bar{z}_1, z_2)=\bar{\partial}_{z_1}\partial_{z_2}\log k_{\mathcal{D}}(z,z)\qquad (z\in\mathcal{D},\ z_1,z_2\in\boldsymbol{C}^N).$$

It is clear from the definition that the Bergman metric is invariant under holomorphic automorphisms of $\mathcal{D}$. It follows that $\mathrm{Hol}(\mathcal{D})$ is a closed subgroup of the group of isometries $I(\mathcal{D})$ (with respect to the Bergman metric). Thus $\mathrm{Hol}(\mathcal{D})$ has a structure of Lie group (with respect to the compact-open topology) and, for any $z\in\mathcal{D}$, the isotropy subgroup $\mathrm{Hol}(\mathcal{D})_z$ is compact (H. Cartan [1]).

In the case of a symmetric domain $\mathcal{D}$, we identify $\mathfrak{p}_+$ with $\boldsymbol{C}^N$ for a fixed basis (though the symbols like ∂_{z_1}, $\bar{\partial}_{z_1}$ have intrinsic meanings independent of the coordinates). Then an explicit form of the Bergman metric can be given as follows.

Proposition 6. 2. *For a symmetric domain $\mathcal{D}\subset\mathfrak{p}_+$, the Bergman metric is given by*

$$(6.11)\qquad (q_{\mathcal{D}})_z(\bar{z}_1, z_2)=\frac{1}{2}B(\bar{z}_1, ad(K(z,z)^{-1})z_2)\qquad (z\in\mathcal{D},\ z_1,z_2\in\mathfrak{p}_+),$$

where K is the canonical kernel function of $G_{\boldsymbol{C}}$. (As before, G is a fixed Zariski connected $\boldsymbol{R}$-group with Lie algebra $\mathfrak{g}$.)

Proof. Since by (5. 9) and Lemma 5. 3 both sides of (6. 11) are invariant under the G°-action on $\mathcal{D}\times\mathfrak{p}_+\times\mathfrak{p}_+$:

$$G^{\circ}\ni g:\ (z,z_1,z_2)\longmapsto (g(z), ad(J(g,z))z_1, ad(J(g,z))z_2),$$

it is enough to prove (6. 11) for $z=o$, i. e.,

$$(6.12)\qquad q_o(\bar{z}_1,z_2)=\frac{1}{2}B(\bar{z}_1,z_2)\qquad (z_1,z_2\in\mathfrak{p}_+),$$

or equivalently,

$$(6.12')\qquad q_o=\frac{1}{2}B\quad \text{on}\ \ \mathfrak{p},$$

where q_o stands for $(q_{\mathcal{D}})_o$. From (6. 10) one has

$$q_o(z_1, z_2) = \bar{\partial}_{z_1}\partial_{z_2} \log k_{\mathscr{D}}(z, z)\Big|_{z=0},$$

where by Proposition 5. 7 and (6. 5)

(6. 13) $$k_{\mathscr{D}}(z, z) = c^{-1} \det\Big(1_{\mathfrak{p}_+} - ad_{\mathfrak{p}_+}[z, \bar{z}] + \frac{1}{4}((ad\, z)^2(ad\, \bar{z})^2)|\mathfrak{p}_+\Big)^{-1}.$$

Hence

(6. 14) $$q_o(z_1, z_2) = \mathrm{tr}(ad_{\mathfrak{p}_+}[z_2, \bar{z}_1]).$$

Thus the relation (6. 12) is reduced to

(6. 15) $$\mathrm{tr}(ad_{\mathfrak{p}_+}[z_2, \bar{z}_1]) = \frac{1}{2}B(\bar{z}_1, z_2).$$

But, since $z_2 \square z_1 = \frac{1}{2} ad_{\mathfrak{p}_+}[z_2, \bar{z}_1]$, this follows immediately from (I. 7. 9 a), or from Lemma 6. 4 below. [In the notation of § 3, (6. 15) means $\tau'(z_1, z_2) = \frac{1}{4}B(\bar{z}_1, z_2)$.] q. e. d.

Proposition 6. 2 shows that, for a symmetric domain $\mathscr{D}$, the Bergman metric $q_{\mathscr{D}}$ coincides with the "normalized" invariant metric (§ 1). In the following, unless otherwise specified, symmetric domains will always be endowed with the Bergman metric.

Corollary 6. 3. *In the case of a symmetric domain, one has*

(6. 16) $$\bar{\partial}_{z_1}\partial_{z_2}\bar{\partial}_{z_3}\partial_{z_4} \log k_{\mathscr{D}}(z, z)\Big|_{z=0} = B(\bar{z}_1, \{z_2, \bar{z}_3, z_4\})$$

for $z_1, z_2, z_3, z_4 \in \mathfrak{p}_+$.

This follows from (6. 10), (6. 11), since by Proposition 5. 7 one has

$$\frac{1}{2}\bar{\partial}_{z_3}\partial_{z_4} ad(K(z, z)^{-1})z_2\Big|_{z=0} = \frac{1}{2}ad([z_4, \bar{z}_3])z_2$$
$$= \{z_4, \bar{z}_3, z_2\} = \{z_2, \bar{z}_3, z_4\}.$$

We shall add here the relation between the Bergman metric and the Ricci curvature*). Let $\mathscr{D}$ be a symmetric domain and let $X, Y, Z \in T_o(\mathscr{D}) = \mathfrak{p}$. Then by definition (and (1. 5)) the "Ricci curvature" of $q_{\mathscr{D}}$ at the origin o is given by

(6. 17) $$r_o(X, Y) = -\mathrm{tr}(Z \mapsto R_o(Y, Z)X)$$
$$= -\mathrm{tr}((ad\, X)(ad\, Y)|\mathfrak{p}).$$

Lemma 6. 4. *For a symmetric domain one has*

(6. 18) $$r_o = -\frac{1}{2}B \quad \text{on } \mathfrak{p}.$$

*) Here we followed the definition of Ricci curvature given in Kobayashi-Nomizu [1]. The definition in Helgason [1] differs from ours in the sign.

(This Lemma is valid for any Riemannian symmetric space.)

Proof. In view of (6. 17) (since $r_0(X, Y)$ is symmetric in X and Y) it is enough to show that

$$(*) \qquad \mathrm{tr}((ad\,X)^2|\mathfrak{p}) = \frac{1}{2}B(X, X)$$

for all $X \in \mathfrak{p}$. Let $ad\,X = \begin{pmatrix} 0 & A_1 \\ A_2 & 0 \end{pmatrix}$ be the matrix expression according to the decomposition $\mathfrak{g} = \mathfrak{k} + \mathfrak{p}$. Then one has

$$\begin{aligned} B(X, X) &= \mathrm{tr}(ad\,X)^2 = \mathrm{tr}(A_1A_2 + A_2A_1) \\ &= 2\,\mathrm{tr}(A_2A_1) = 2\,\mathrm{tr}((ad\,X)^2|\mathfrak{p}), \end{aligned}$$

which proves (*), q. e. d.

Combining this with Proposition 6. 2, we have $r = -q_{\mathcal{D}}$. Thus the Bergman metric of a symmetric domain is "Einstein" (i. e., proportional to its Ricci curvature).

Remark. The relation $r = -q_{\mathcal{D}}$ is true for any homogeneous bounded domain. In general, let M be a complex manifold with a Kählerian metric h and, in a neighbourhood with coordinates $\{z^\alpha\}$, write the volume element in the form $k(z)d\mu(z)$ where $k(z) = |\det(h_{\alpha\beta})|^2$. Then it is known that the Ricci curvature r of h is given by

$$(6.\ 19) \qquad \begin{aligned} &r\left(\frac{\partial}{\partial \bar{z}^\alpha}, \frac{\partial}{\partial z^\beta}\right) = -\frac{\partial^2}{\partial \bar{z}^\alpha \partial z^\beta}\log k(z), \\ &r\left(\frac{\partial}{\partial z^\alpha}, \frac{\partial}{\partial z^\beta}\right) = r\left(\frac{\partial}{\partial \bar{z}^\alpha}, \frac{\partial}{\partial \bar{z}^\beta}\right) = 0 \qquad (1 \le \alpha, \beta \le N) \end{aligned}$$

(cf. K-N, IX, § 5 ; Helgason [1], VIII, Prop. 2. 5). In the case of homogeneous bounded domain $\mathcal{D}$, Lemma 6. 1 and the uniqueness of the invariant volume element yield $k(z) = c'k_{\mathcal{D}}(z, z)$. Hence, comparing (6. 19) with (6. 6), (6. 7), one obtains $r = -q_{\mathcal{D}}$. This, together with (6. 18), gives another proof of Proposition 6. 2.

For a more explicit determination of the Bergman kernels in terms of the orthonormal basis of $\mathcal{H}^2(\mathcal{D})$, cf. Hua [1], Takeuchi [7].

Exercise

1. For the unit disc $\mathcal{D}^1 = \{|z| < 1\}$ in $\mathfrak{p}_+ = \boldsymbol{C}$, prove the following.

1. 1) The Bergman kernel function is given by

$$k_{\mathcal{D}^1}(z, w) = \pi^{-1}(1 - z\bar{w})^{-1}.$$

1. 2) The Bergman metric is given by

$$ds^2 = 4(1 - |z|^2)^{-1}\,dz d\bar{z}.$$

1. 3) The Gaussian curvature (=sectional curvature) of $\mathcal{D}^1$ is constant -1.

§ 7. The Siegel space.

In this section, we introduce the notion of Siegel space which is the most important example of symmetric domains.

A *symplectic vector space* (V, A) *over* F (a field of characteristic zero) is a pair formed of a (finite-dimensional) vector space V over F and a non-degenerate alternating bilinear form $A: V\times V\to F$. The dimension of V is then necessarily even, say $2n$, and there exists a basis $(e_1, \cdots, e_{2n})$ of V over F such that

$$A(e_i, e_j) = \begin{cases} 1 & \text{if } i=j+n, \\ -1 & \text{if } i=j-n, \\ 0 & \text{otherwise.} \end{cases}$$

Such a basis will be called a *symplectic basis* of (V, A). The group of automorphisms of (V, A), i. e.,

$$\{g\in GL(V) | A(gv, gv')=A(v, v') \text{ for all } v, v'\in V\}$$

is called the *symplectic group* and denoted by $Sp(V, A)$, or by $Sp_{2n}(F)$ when a symplectic basis is fixed. We note that $g\in Sp(V, A)$ implies $\det(g)=1$.

Let (V, A) be a symplectic vector space over $\boldsymbol{R}$. We consider the set $\mathfrak{S}=\mathfrak{S}(V, A)$ of all complex structures I on V such that the bilinear form $A(v, Iv')$ $(v, v'\in V)$ is symmetric and positive definite, or in notation, $AI\gg 0$. $\mathfrak{S}$ is non-empty, because for any symplectic basis (e_i) the complex structure I on V defined by

$$Ie_i = \begin{cases} -e_{i+n} & \text{for } 1\leq i\leq n, \\ e_{i-n} & \text{for } n+1\leq i\leq 2n \end{cases}$$

clearly satisfies the condition $AI\gg 0$. Let V_C be the complexification of V and extend A to an (alternating) $\boldsymbol{C}$-bilinear form on $V_C\times V_C$. For $I\in\mathfrak{S}$ we put

$$V_\pm = V_\pm(I) = V_C(I; \pm\sqrt{-1}).$$

Then we have a direct sum decomposition

$$(7.1) \qquad V_C = V_+\oplus V_-, \qquad \bar{V}_+ = V_-.$$

Since on V_+ the bilinear form $A(v, Iv')=\sqrt{-1}\,A(v, v')$ $(v, v'\in V_+)$ is at the same time symmetric and alternating, it must be identically equal to zero, i. e., $A|V_+\times V_+=0$. Similarly, we have $A|V_-\times V_-=0$. Hence $\sqrt{-1}\,A|V_-\times V_+$ is non-degenerate and gives a canonical pairing $\langle\ \rangle$, by which we identify V_- with the dual space V_+^* of V_+. Thus we have

$$(7.2) \qquad \begin{cases} A|V_+\times V_+ = 0, \quad A|V_-\times V_- = 0, \\ \sqrt{-1}\,A(v, v') = \langle v, v'\rangle \quad \text{for } v\in V_-(=V_+^*),\ v'\in V_+. \end{cases}$$

We define a hermitian form h_A on V_C by

$$(7.3) \qquad h_A(v, v') = \sqrt{-1}\,A(\bar{v}, v') \qquad (v, v'\in V_C).$$

Then from (7.2) and the condition $AI\gg 0$ it is clear that

$$(7.4) \qquad \begin{cases} h_A|V_+\times V_+ \gg 0, \quad h_A|V_-\times V_- \ll 0, \\ h_A|V_+\times V_- = 0. \end{cases}$$

Thus h_A is of signature (n, n). We have proved a part of the following

Lemma 7.1. *The correspondence $I\mapsto V_-(I)$ gives a one-to-one correspondence between $\mathfrak{S}$ and the set of all n-dimensional complex subspaces V_- of V_C satisfying the conditions*

$$(7.5) \qquad A|V_-\times V_- = 0 \quad \text{and} \quad h_A|V_-\times V_- \ll 0.$$

Proof. To complete the proof, let V_- be an n-dimensional complex subspace of V_C satisfying (7. 5) and put $V_+ = \bar{V}_-$. Then one has $h_A|V_+ \times V_- = 0$ and hence $V_+ \cap V_- = \{0\}$. Thus one has a direct sum decomposition of V_C of the form (7. 1). Hence there exists a uniquely determined complex structure I on V such that $V_\pm = V_C(I; \pm i)$. Let $v \in V$, $v \neq 0$ and write $v = v_+ + v_-$ with $v_\pm \in V_\pm$. Then $\bar{v}_+ = v_-$ and

$$\begin{aligned} A(v, Iv) &= A(v_+ + v_-, iv_+ - iv_-) \\ &= -2i\, A(v_+, v_-) \\ &= -2h_A(v_-, v_-), \end{aligned}$$

which is positive by (7. 5). Hence one has $I \in \mathfrak{S}$, q. e. d.

We denote by $\mathcal{G}_n(V_C)$ the Grassmannian manifold of n-dimensional complex subspaces of V_C and put

$$M^* = \{W \in \mathcal{G}_n(V_C) \mid A|W \times W = 0\}. \tag{7. 6}$$

Then M^* is an algebraic submanifold of $\mathcal{G}_n(V_C)$ and, by Lemma 7. 1, $\mathfrak{S}$ can be identified with an open subset of M^* defined by the inequality $h_A|W \times W \ll 0$. The space $\mathfrak{S} = \mathfrak{S}(V, A)$ with this induced structure of complex manifold is called the *Siegel space* associated with the symplectic space (V, A).

Let $G = Sp(V, A)$ be the symplectic group. Then its complexification $G_C = Sp(V_C, A)$ acts on M^* by

$$(g, W) \longmapsto gW \qquad (g \in G,\ W \in M^*).$$

Restricting this to $G \times \mathfrak{S}$, one obtains an action of G on $\mathfrak{S}$:

$$(g, I) \longmapsto gIg^{-1} \qquad (g \in G,\ I \in \mathfrak{S}).$$

By a theorem of Witt (cf. e. g., Bourbaki [1]), G_C is transitive on M^*. We shall see below that the action of G on $\mathfrak{S}$ is also transitive.

Now, to obtain matrix expressions, we fix $o = I_0 \in \mathfrak{S}$ and $V_\pm = V_\pm(I_0)$. Let $(e_1, \cdots, e_n)$ be an orthonormal basis of V_+ with respect to $h_A|V_+ \times V_+$, and let $e_{j+n} = \bar{e}_j\ (1 \leq j \leq n)$. Then $(e_{n+1}, \cdots, e_{2n})$ is the dual basis for $V_- = V_+^*$ with respect to $\langle\ \rangle$, i. e., $(e_1, \cdots, e_{2n})$ is a symplectic basis for $(V_C, \sqrt{-1}\,A)$. In the following, we will identify various linear maps and bilinear (or hermitian) forms with the corresponding matrices with respect to this basis. For instance, we have

$$\begin{aligned} &A = \begin{pmatrix} 0 & i1_n \\ -i1_n & 0 \end{pmatrix}, \quad h_A = \begin{pmatrix} 1_n & 0 \\ 0 & -1_n \end{pmatrix}, \\ &I_0 = \begin{pmatrix} i1_n & 0 \\ 0 & -i1_n \end{pmatrix}. \end{aligned} \tag{7. 7}$$

It should be noted that, since the basis (e_i) is *not* real, this identification is *not* compatible with the operation of taking complex conjugates. In order to avoid possible confusions, we will reserve the bar to denote the usual complex conjugation of matrices, while the symbol σ_0 is used to denote the complex conjugation relative to the real form V of V_C. Thus for $v = \begin{pmatrix} v_+ \\ v_- \end{pmatrix} \in V_C (= C^{2n})$ one has

$$v^{\sigma_0} = \begin{pmatrix} \bar{v}_- \\ \bar{v}_+ \end{pmatrix} = \begin{pmatrix} 0 & 1_n \\ 1_n & 0 \end{pmatrix} \bar{v}.$$

Similarly, for $X \in \mathrm{End}(V_C)$ $(=\mathcal{M}_{2n}(\boldsymbol{C}))$, one has

$$(7.8) \qquad X^{\sigma_0} = \begin{pmatrix} 0 & 1_n \\ 1_n & 0 \end{pmatrix} \bar{X} \begin{pmatrix} 0 & 1_n \\ 1_n & 0 \end{pmatrix}.$$

Now, from the definition, one has for $g = \begin{pmatrix} a & b \\ c & d \end{pmatrix} \in \mathrm{End}(V_C)$,

$$g \in G_C \Longleftrightarrow {}^t g A g = A$$
$$\Longleftrightarrow g^{-1} = A^{-1}\,{}^t g A = \begin{pmatrix} {}^t d & -{}^t d \\ -{}^t c & {}^t a \end{pmatrix}.$$

Hence we see that either one of the following conditions is necessary and sufficient for $g \in G_C$:

$$(7.9) \qquad \begin{cases} a\,{}^t d - b\,{}^t c = 1_n, \\ a\,{}^t b = b\,{}^t a, \quad c\,{}^t d = d\,{}^t c. \end{cases}$$

$$(7.9') \qquad \begin{cases} {}^t a d - {}^t c b = 1_n, \\ {}^t a c = {}^t c a, \quad {}^t b d = {}^t d b. \end{cases}$$

It follows that

$$(7.10) \qquad g \in G \Longleftrightarrow {}^t g A g = A \quad \text{and} \quad g^{\sigma_0} = g$$
$$\Longleftrightarrow g = \begin{pmatrix} a & b \\ \bar{b} & \bar{a} \end{pmatrix} = \begin{pmatrix} {}^t\bar{a} & -{}^t b \\ -{}^t\bar{b} & {}^t a \end{pmatrix}^{-1}.$$

We denote by $K = G_o$ the isotropy subgroup of G at o. Then

$$K = \left\{ g = \begin{pmatrix} a & 0 \\ 0 & \bar{a} \end{pmatrix} \middle| a = {}^t a^{-1} \right\}.$$

Hence the Lie algebras $\mathfrak{g} = \mathrm{Lie}\, G$ and $\mathfrak{k} = \mathrm{Lie}\, K$ are given by

$$(7.11) \qquad \begin{aligned} \mathfrak{g} &= \left\{ X = \begin{pmatrix} X_1 & X_{12} \\ \bar{X}_{12} & \bar{X}_1 \end{pmatrix} \middle| X_1 = -{}^t\bar{X}_1,\ X_{12} = {}^t X_{12} \right\}, \\ \mathfrak{k} &= \left\{ X = \begin{pmatrix} X_1 & 0 \\ 0 & \bar{X}_1 \end{pmatrix} \middle| X_1 = -{}^t\bar{X}_1 \right\}. \end{aligned}$$

It is clear that (for the fixed basis) the standard Cartan involution $\sigma : g \mapsto {}^t\bar{g}^{-1}$ of $GL(V_C)$ leaves G_C invariant and commutes with σ_0. Hence, by I, § 4, $\theta = \sigma | G$ is a (global) Cartan involution of G. Clearly, for $g \in G$, one has

$$g^\theta = {}^t\bar{g}^{-1} = A\bar{g}A^{-1} = \begin{pmatrix} 1_n & 0 \\ 0 & -1_n \end{pmatrix} g \begin{pmatrix} 1_n & 0 \\ 0 & -1_n \end{pmatrix}.$$

Hence one has $K = \{g \in G | g^\theta = g\}$, which shows that K is the maximal compact subgroup of G corresponding to θ. [It follows that, since $K \cong U(V_+)$ is connected, so is G, which implies $G \subset SL_{2n}(\boldsymbol{R})$.] If we denote the corresponding Cartan involution of $\mathfrak{g}$ by the same letter θ, one has $\mathfrak{k} = \mathfrak{g}(\theta\,;1)$. It is then easy to see that $H_0 = \frac{1}{2} I_0$ is in the center of $\mathfrak{k}$ and $J_0 = ad_{\mathfrak{p}} H_0$ is a complex structure on $\mathfrak{p} = \mathfrak{g}(\theta\,;-1)$. Thus $(\mathfrak{g}, H_0)$ is a (simple) semi-simple Lie algebra of hermitian type and the homogeneous space G/K is the associated hermitian symmetric space.

As in the preceding sections, we put $\mathfrak{p}_\pm = \mathfrak{p}_C(J_0; \pm i)$ and $P_\pm = \exp \mathfrak{p}_\pm$. Then

$$(7.12)\qquad \begin{aligned} \mathfrak{p}_+ &= \left\{ \begin{pmatrix} 0 & z \\ 0 & 0 \end{pmatrix} \middle| z \in \mathrm{Sym}_n(\boldsymbol{C}) \right\}, \\ \mathfrak{p}_- &= \left\{ \begin{pmatrix} 0 & 0 \\ \bar{z} & 0 \end{pmatrix} \middle| z \in \mathrm{Sym}_n(\boldsymbol{C}) \right\}, \end{aligned}$$

where $\mathrm{Sym}_n(\boldsymbol{C})$ denotes the space of $n\times n$ symmetric complex matrices. In this notation, the isotropy subgroup $(G_C)_o$ of G_C at $o = V_-$ is given by $K_C P_-$. Hence one has a natural biholomorphic map

$$M^* \approx G_C/(G_C)_o = G_C/K_C P_-.$$

Thus M^* is the compact dual of G/K. In the following, we often identify $\mathfrak{p}_+$ with $\mathrm{Sym}_n(\boldsymbol{C})$ (or more intrinsically, with the space of symmetric linear maps $V_- \to V_+ = V_-^*$ with respect to $\langle\ \rangle$) by the correspondence $\begin{pmatrix} 0 & z \\ 0 & 0 \end{pmatrix} \leftrightarrow z$. $\Bigl($However, for $z \in \mathfrak{p}_+$, the notation $\exp z$ will always mean $\exp\begin{pmatrix} 0 & z \\ 0 & 0 \end{pmatrix} = \begin{pmatrix} 1 & z \\ 0 & 1 \end{pmatrix}$.$\Bigr)$

Lemma 7.2. *Let W be an n-dimensional complex subspace of V_C. Then W satisfies the condition* (7.5) *if and only if W can be written in the form*

$$(7.13)\qquad W = \begin{pmatrix} z \\ 1_n \end{pmatrix} V_- = \left\{ \begin{pmatrix} zv_- \\ v_- \end{pmatrix} \middle| v_- \in V_- \right\}$$

with $z \in \mathrm{Sym}_n(\boldsymbol{C})$ satisfying the condition $1_n - z\bar{z} \gg 0$.

Proof. Let $p_\pm$ denote the canonical projections: $V_C \to V_\pm$. If W satisfies the condition (7.5), then $p_-|W$ is non-singular, since $\mathrm{Ker}(p_-|W) = V_+ \cap W = \{0\}$. Let z be a linear map $V_- \to V_+$ (or the corresponding $n\times n$ matrix) defined by $z = p_+ \circ (p_-|W)^{-1}$. Then for any $w, w' \in W$ one has $w = \begin{pmatrix} zv_- \\ v_- \end{pmatrix}$, $w' = \begin{pmatrix} zv'_- \\ v'_- \end{pmatrix}$ with $v_-, v'_- \in V_-$ and hence

$$\begin{aligned} A(w, w') &= ({}^t v_- {}^t z, {}^t v_-) \begin{pmatrix} 0 & i1_n \\ -i1_n & 0 \end{pmatrix} \begin{pmatrix} zv'_- \\ v'_- \end{pmatrix} \\ &= i\,{}^t v_- ({}^t z - z) v'_- = 0, \\ h_A(w, w') &= ({}^t \bar{v}_- {}^t \bar{z}, {}^t \bar{v}_-) \begin{pmatrix} 1_n & 0 \\ 0 & -1_n \end{pmatrix} \begin{pmatrix} zv'_- \\ v'_- \end{pmatrix} \\ &= {}^t \bar{v}_- ({}^t \bar{z} z - 1_n) v'_- \ll 0. \end{aligned}$$

It follows that $z = {}^t z$ and $1_n - z\bar{z} \gg 0$. The converse is also clear from the above argument, q. e. d.

Lemma 7.3. *Let $g = \begin{pmatrix} a & b \\ c & d \end{pmatrix} \in G_C$. Then one has $g \in P_+ K_C P_-$ if and only if the matrix d is non-singular, and when this is the case the canonical decomposition of g is given as follows:*

(7. 14) $$g=\begin{pmatrix}1_n & bd^{-1}\\ 0 & 1_n\end{pmatrix}\begin{pmatrix}{}^t d^{-1} & 0\\ 0 & d\end{pmatrix}\begin{pmatrix}1_n & 0\\ d^{-1}c & 1_n\end{pmatrix}.$$

This follows immediately by solving the equation

$$\begin{pmatrix}a & b\\ c & d\end{pmatrix}=\begin{pmatrix}1 & z_1\\ 0 & 1\end{pmatrix}\begin{pmatrix}a_1 & 0\\ 0 & {}^t a_1^{-1}\end{pmatrix}\begin{pmatrix}1 & 0\\ \varpi_1 & 1\end{pmatrix}$$
$$=\begin{pmatrix}a_1+z_1{}^t a_1^{-1}\varpi_1 & z_1{}^t a_1^{-1}\\ {}^t a_1^{-1}\varpi_1 & {}^t a_1^{-1}\end{pmatrix}.$$

We note that one has $gV_-=\begin{pmatrix}z_1\\ 1\end{pmatrix}V_-$ if and only if $(g)_+=\exp z_1$, i. e., $g(o)=z_1$.

Corollary 7. 4. *Let* $g=\begin{pmatrix}a & b\\ c & d\end{pmatrix}\in G_C$ *and* $z\in\mathfrak{p}_+=\mathrm{Sym}_n(\boldsymbol{C})$. *Then the holomorphic action* $g(z)$ *is defined if and only if* $cz+d$ *is non-singular and, when that is so, one has*

(7. 15) $$g(z)=(az+b)(cz+d)^{-1},$$

(7. 16) $$J_+(g,z)={}^t(cz+d)^{-1}=-g(z)c+a,$$

where $J_+(g,z)$ *stands for* $J(g,z)|V_+$.

This follows from Lemma 7. 3 applied to

$$\begin{pmatrix}a & b\\ c & d\end{pmatrix}\begin{pmatrix}1_n & z\\ 0 & 1_n\end{pmatrix}=\begin{pmatrix}a & az+b\\ c & cz+d\end{pmatrix}.$$

Now, for $g=\begin{pmatrix}a & b\\ \bar b & \bar a\end{pmatrix}\in G$, one has by (7. 9′) or (7. 10)

(7. 10 a) $${}^t a\bar a-{}^t\bar b b=1_n,\qquad {}^t\bar a b={}^t b\bar a.$$

Hence ${}^t a\bar a=1_n+{}^t\bar b b\gg 0$, which shows that $\bar a$ is non-singular. Therefore, by Lemma 7. 3, one has $g\in P_+K_CP_-$ (cf. Lem. 4. 2, (4. 7 b)). If we put $z_1=b\bar a^{-1}$, then $gV_-=\begin{pmatrix}z_1\\ 1\end{pmatrix}V_-$ and from (7. 10 a) (or from Lem. 7. 2) one has ${}^t z_1=z_1$ and $1_n-z_1\bar z_1\gg 0$. Conversely, suppose there is given an $n\times n$ complex matrix z_1 satisfying these conditions. Take $a\in GL(V_+)$ in such a way that $a{}^t\bar a=(1-z_1\bar z_1)^{-1}$ and put $b=z_1\bar a$. Then one has $g=\begin{pmatrix}a & b\\ \bar b & \bar a\end{pmatrix}\in G$ and $gV_-=\begin{pmatrix}z_1\\ 1\end{pmatrix}V_-$, i. e., $g(o)=z_1$. Therefore, by Lemmas 7. 1, 7. 2, we can conclude that G is transitive on $\mathfrak{S}$. Also we see that the embedding of $\mathfrak{S}$ into $\mathfrak{p}_+=\mathrm{Sym}_n(\boldsymbol{C})$ defined by the relation $V_-(I)=\begin{pmatrix}z\\ 1\end{pmatrix}V_-$ coincides with the Harish-Chandra embedding of $\mathfrak{S}=G/K$. Thus we have proved the first part of the following theorem.

Theorem 7. 5. *The Siegel space* $\mathfrak{S}=\mathfrak{S}(V,A)$ *is a hermitian symmetric space and by Harish-Chandra embedding* $\mathfrak{S}$ *is identified with a symmetric bounded domain (of type* (III_n)*)*

(7. 17) $$\{z\mid z\in\mathrm{Sym}_n(\boldsymbol{C}),\ 1_n-z\bar z\gg 0\}.$$

The symplectic group $G=Sp(V,A)$ *acts on* $\mathfrak{S}$ *by*

$$g=\begin{pmatrix} a & b \\ \bar{b} & \bar{a} \end{pmatrix}: z \longmapsto g(z)=(az+b)(\bar{b}z+\bar{a})^{-1},$$

and the group of holomorphic automorphisms $\mathrm{Hol}(\mathfrak{S})$ *can be identified with* $G/\{\pm 1_{2n}\}$.

To prove the last assertion, we first note that $g \in G$ acts trivially on $\mathfrak{S}$ if and only if $g=\pm 1_{2n}$. Since G is connected, one has a natural identification $G/\{\pm 1_{2n}\}=I(\mathfrak{S})^\circ$ (§ 1). On the other hand, since $\mathfrak{g}_C$ is simple and of type (C), one has $\mathrm{Aut}\,\mathfrak{g}_C=\mathrm{Ad}\,\mathfrak{g}_C$. Hence $I(\mathfrak{S})\cong \mathrm{Aut}\,\mathfrak{g}$ is Zariski connected. It follows from the results to be proved in the next section (Prop. 8. 5) that $(I(\mathfrak{S}):\mathrm{Hol}(\mathfrak{S}))=2$ and $\mathrm{Hol}(\mathfrak{S})=I(\mathfrak{S})^\circ=G/\{\pm 1_{2n}\}$.

Remark. If we define the group of "symplectic similarities" by

$$\tilde{G}=GSp(V,A)=\{g\in GL(V)\,|\,{}^t gAg=\lambda A \text{ with } \lambda\in \boldsymbol{R}^\times\},$$

then $I(\mathfrak{S})$ can be identified with $\tilde{G}/\{\lambda 1_{2n}\,|\,\lambda\in\boldsymbol{R}^\times\}$. The antiholomorphic automorphism of $\mathfrak{S}$: $z\mapsto \bar{z}$ corresponds to $g=\begin{pmatrix} 0 & 1_n \\ 1_n & 0\end{pmatrix}\in\tilde{G}$ (cf. Exerc. 2).

As a special case of Corollary 7. 4, we have that $(\exp w^{\sigma_0})^{-1}(z)$ is defined if and only if $1_n-z\bar{w}$ is non-singular and in that case, putting $K_+(z,w)=K(z,w)|V_+$, one has

$$K_+(z,w)=1_n-z\bar{w}. \tag{7. 18}$$

In general, for $k\in K_C$, we put $k_\pm=k|V_\pm$. Then clearly

$$ad_{\mathfrak{p}_+}k: z\longmapsto k_+z{}^tk_+.$$

Hence one has

$$\chi_1(k)=\det(ad_{\mathfrak{p}_+}k)=\det(k_+)^{n+1}. \tag{7. 19}$$

It follows that the jacobian of the linear fractional transformation $z\mapsto g(z)=(az+b)(cz+d)^{-1}$ is given by

$$j_{\chi_1}(g,z)=\det(cz+d)^{-(n+1)}. \tag{7. 20}$$

On the other hand, the Bergman kernel function of $\mathfrak{S}$ is given by

$$k_{\mathfrak{S}}(z,w)=c_0^{-1}\det(1_n-z\bar{w})^{-(n+1)} \qquad (c_0=\mathrm{vol}(\mathfrak{S})). \tag{7. 21}$$

It follows that the Bergman metric $h_{\mathfrak{S}}$ and the associated fundamental 2-form ω are given by

$$\begin{aligned} h_{\alpha\beta}&=2\frac{\partial^2}{\partial\bar{z}^\alpha\partial z^\beta}\log k_1(z,z)\\ &=2(n+1)\mathrm{tr}\Big((1_n-\bar{z}z)^{-1}\frac{\partial z}{\partial\bar{z}^\alpha}(1_n-z\bar{z})^{-1}\frac{\partial z}{\partial z^\beta}\Big),\\ \omega&=id'd''\log k_1(z,z)\\ &=i(n+1)\mathrm{tr}((1_n-z\bar{z})^{-1}dz\wedge(1_n-\bar{z}z)^{-1}d\bar{z}). \end{aligned} \tag{7. 22}$$

Finally we remark that, if we use a real symplectic basis (e_i') instead of the basis (e_i) used above, then as an analogue of Lemma 7. 2 we obtain

Lemma 7. 2′. *An n-dimensional complex subspace W of V_C satisfies the condition* (7. 5) *if and only if W has a basis of the form* $(e'_1, \cdots, e'_{2n})\begin{pmatrix} z \\ 1 \end{pmatrix}$ *with* $z \in \mathrm{Sym}_n(\boldsymbol{C})$ *satisfying the condition* $\mathrm{Im}\, z \gg 0$.

From this we see that the Siegel space $\mathfrak{S}$ is holomorphically equivalent to a tube domain

$$\{z \mid z \in \mathrm{Sym}_n(\boldsymbol{C}), \mathrm{Im}\, z \gg 0\}, \tag{7. 23}$$

called "the generalized upper half-plane of Siegel". For more on the geometry of the Siegel space, see Siegel [4]. (Cf. also Siegel [1], H. Cartan [4], Maass [7].)

Exercises

1. For $v \in V$ and $z \in \mathfrak{S}(V, A)$, set

$$v_z = v_+ - z v_- \in V_+, \tag{7. 24}$$

where $v_\pm$ denote the $V_\pm$-components of v $(v_- = \bar{v}_+)$. Show that the correspondence $v \mapsto v_z$ gives a $\boldsymbol{C}$-linear isomorphism $(V, I_z) \cong V_+$. Show also that, for $w \in V_+$, one has

$$w = v_z \Longleftrightarrow v_+ = (1 - z\bar{z})^{-1}(w + z\bar{w}). \tag{7. 25}$$

2. We identify $I(\mathfrak{S})$ with $GSp(V, A)/\boldsymbol{R}^\times$. Show that, for $g = \begin{pmatrix} a & b \\ \bar{b} & \bar{a} \end{pmatrix} \in GSp(V, A)$ with $\lambda < 0$, the "holomorphic" and "isometric" (antiholomorphic) actions of g on $\mathfrak{S}$ are given, respectively, by

$$z \longmapsto (az + b)(\bar{b}z + \bar{a})^{-1}, \tag{7. 26 a}$$

$$z \longmapsto (b\bar{z} + a)(\bar{a}\bar{z} + \bar{b})^{-1}. \tag{7. 26 b}$$

3. Prove Lemma 7. 2′.

4. Find formulas analogous to (7. 22) in terms of the tube domain expression of $\mathfrak{S}$.

§ 8. Equivariant holomorphic maps of symmetric domains.

Let $\mathscr{D}$ be a symmetric domain with origin o and $(\mathfrak{g}, H_0)$ the corresponding semi-simple Lie algebra of hermitian type. We then have the canonical decomposition

$$\mathfrak{g}_C = \mathfrak{p}_+ + \mathfrak{k}_C + \mathfrak{p}_- \tag{8. 1}$$

which, together with the Cartan involution $\sigma : X \mapsto \theta\bar{X}$, defines a structure of symmetric Lie algebra on $\mathfrak{g}_C$ (viewed as a real Lie algebra). $\mathfrak{p}_+$ endowed with a triple product

$$\{z_1, z_2, z_3\} = \frac{1}{2}[[z_1, \bar{z}_2], z_3] \qquad (z_1, z_2, z_3 \in \mathfrak{p}_+) \tag{8. 2}$$

is a positive definite hermitian JTS (§ 3). Let $\mathscr{D}'$ be another symmetric domain with origin o'. The objects relative to $(\mathscr{D}', o')$ will be denoted by the corresponding symbols with a prime; e. g., $(\mathfrak{g}', H'_0)$ is the corresponding semi-simple Lie algebra of hermitian type.

A (C^∞-)map $\rho_{\mathscr{D}} : \mathscr{D} \to \mathscr{D}'$ with $\rho_{\mathscr{D}}(o) = o'$, or in abbreviation, $\rho_{\mathscr{D}} : (\mathscr{D}, o) \to (\mathscr{D}', o')$, is called *(strongly) equivariant* if there exists a Lie algebra homomorphism $\rho(=\rho_{\mathfrak{g}}):$ $\mathfrak{g}\to\mathfrak{g}'$ such that

(8. 3) $$\rho\circ\theta=\theta'\circ\rho, \quad \text{and}$$

(8. 4) $$\rho_{\mathcal{D}}((\exp X)z)=(\exp\rho(X))\rho_{\mathcal{D}}(z) \qquad \text{for all } X\in\mathfrak{g},\ z\in\mathcal{D}.$$

Thus $(\rho, \rho_{\mathcal{D}})$ is a "(strongly) equivariant pair" (at o) in the sense of § 2. Recall that under these conditions, $\rho_{\mathcal{D}}$ and ρ determine each other uniquely.

Let $\rho_{\mathcal{D}}:(\mathcal{D}, o)\to(\mathcal{D}', o')$ be an equivariant holomorphic map and ρ the associated Lie algebra homomorphism. Then, by the condition (8. 3), one has $\rho(\mathfrak{k})\subset\mathfrak{k}'$, $\rho(\mathfrak{p})\subset\mathfrak{p}'$, and if one identifies the tangent spaces $T_o(\mathcal{D})$ and $T_{o'}(\mathcal{D}')$ with $\mathfrak{p}$ and $\mathfrak{p}'$, respectively, the condition (8. 4) implies that the tangential map $d\rho_{\mathcal{D}}: T_o(\mathcal{D})\to T_{o'}(\mathcal{D}')$ is identified with $\rho|\mathfrak{p}:\mathfrak{p}\to\mathfrak{p}'$. Since $\rho_{\mathcal{D}}$ is holomorphic, $\rho|\mathfrak{p}$ is $\boldsymbol{C}$-linear with respect to the complex structures $J=ad_{\mathfrak{p}}H_0$ and $J'=ad_{\mathfrak{p}'}H_0'$. Therefore ρ satisfies the condition

(H_1) $$\rho([H_0, X])=[H_0', \rho(X)] \qquad \text{for all } X\in\mathfrak{g}.$$

If we extend ρ to a complex Lie algebra homomorphism $\mathfrak{g}_C\to\mathfrak{g}_C'$ (denoted sometimes by ρ_C, but more often simply by ρ), it is clear that the condition (H_1) is equivalent to saying that ρ_C preserves the canonical decomposition, i. e.,

(8. 5) $$\rho_C(\mathfrak{k}_C)\subset\mathfrak{k}_C', \qquad \rho_C(\mathfrak{p}_\pm)\subset\mathfrak{p}_\pm'.$$

Clearly (8. 5) implies (8. 3).

It will also be useful to consider the condition

(H_2) $$\rho(H_0)=H_0',$$

which clearly implies (H_1). For brevity, a homomorphism of Lie algebra of hermitian type satisfying the condition (H_i) ($i=1, 2$) will be called an (H_i)*-homomorphism*. When a homomorphism $\rho:\mathfrak{g}\to\mathfrak{g}'$ satisfies the condition (H_1) with respect to the H-elements H_0 and H_0', we write $\rho:(\mathfrak{g}, H_0)\to(\mathfrak{g}', H_0')$.

Now, suppose there is given an (H_1)-homomorphism $\rho:(\mathfrak{g}, H_0)\to(\mathfrak{g}', H_0')$. We extend ρ to a complex Lie algebra homomorphism $\rho_C:\mathfrak{g}_C\to\mathfrak{g}_C'$, satisfying (8. 5), and set $\rho_+=\rho_C|\mathfrak{p}_+$. As usual, we suppose that $\mathcal{D}$ and $\mathcal{D}'$ are realized as symmetric bounded domains in $\mathfrak{p}_+$ and $\mathfrak{p}_+'$ through Harish-Chandra embeddings. Then ρ_+ induces on $\mathcal{D}$ an equivariant holomorphic map of $\mathcal{D}$ into $\mathcal{D}'$ associated with ρ. In fact, let G and G' be Zariski connected $\boldsymbol{R}$-groups with Lie algebras $\mathfrak{g}$ and $\mathfrak{g}'$. Replacing G by a (finite) covering group if necessary, we may assume that the Lie algebra homomorphism ρ can be lifted to an $\boldsymbol{R}$-homomorphism $\rho_G: G\to G'$. Let $z\in\mathcal{D}$, $z=g\cdot o$ with $g\in G^\circ$. Then one has $(g)_+=\exp z$. Since we have

$$\rho_G(K_C^\circ)\subset K_C'^\circ, \qquad \rho_G(P_\pm)\subset P_\pm'$$

by (8. 5), this implies $(\rho_G(g))_+=\exp\rho_+(z)$, i. e.,

$$\rho_+(z)=\rho_G(g)\cdot o'.$$

It follows that one has

(*) $$\rho_+(g_1 z)=\rho_G(g_1)\cdot\rho_+(z) \qquad \text{for all } g_1\in G^\circ,\ z\in\mathcal{D};$$

in particular, $\rho_+(\mathcal{D}) \subset \mathcal{D}'$. Thus the restriction $\rho_{\mathcal{D}} = \rho_+ | \mathcal{D}$, which is clearly holomorphic, is an equivariant map associated with ρ. [Note that the relation (*) remains true for any $g_1 \in G_C$ as long as the holomorphic action $g_1(z)$ is defined.]

Summing up, we obtain the following

Proposition 8. 1. *Let $\rho_{\mathcal{D}} : (\mathcal{D}, o) \to (\mathcal{D}', o')$ be a strongly equivariant holomorphic map and $\rho : \mathfrak{g} \to \mathfrak{g}'$ the associated Lie algebra homomorphism. Then ρ satisfies the condition* (H_1) *and, if $\mathcal{D}$ and $\mathcal{D}'$ are viewed as symmetric bounded domains in $\mathfrak{p}_+$ and $\mathfrak{p}'_+$ through Harish-Chandra embeddings, then $\rho_{\mathcal{D}}$ coincides with the restriction to $\mathcal{D}$ of a $\boldsymbol{C}$-linear map $\rho_+ : \mathfrak{p}_+ \to \mathfrak{p}'_+$ induced by the $\boldsymbol{C}$-linear extension of ρ to $\mathfrak{g}_C$. Conversely, if ρ is an* (H_1)*-homomorphism $(\mathfrak{g}, H_0) \to (\mathfrak{g}', H'_0)$, then $\rho_{\mathcal{D}} = \rho_+ | \mathcal{D}$ is a (strongly) equivariant holomorphic map of $(\mathcal{D}, o)$ into $(\mathcal{D}', o')$ associated with ρ.*

Now the condition (8. 5) means that $\rho : \mathfrak{g}_C \to \mathfrak{g}'_C$ preserves the gradation given by the canonical decomposition. Moreover, since ρ is real (i. e., $\bar{\rho} = \rho$), the condition (8. 3) is equivalent to

$$\rho \circ \sigma = \sigma' \circ \rho, \tag{8. 6}$$

σ' denoting the Cartan involution of $\mathfrak{g}'_C$ (viewed as a real Lie algebra) given by $\sigma' X' = \theta' \bar{X}'$ $(X' \in \mathfrak{g}'_C)$. Thus, $\rho : (\mathfrak{g}, H_0) \to (\mathfrak{g}', H'_0)$ gives rise to a symmetric Lie algebra homomorphism $(\mathfrak{g}_C, \sigma) \to (\mathfrak{g}'_C, \sigma')$, and hence by Proposition I. 9. 1 to a JTS homomorphism $\rho_+ : \mathfrak{p}_+ \to \mathfrak{p}'_+$, which is $\boldsymbol{C}$-linear. Conversely, when there is given a $\boldsymbol{C}$-linear JTS homomorphism $\rho_+ : \mathfrak{p}_+ \to \mathfrak{p}'_+$, Proposition I. 9. 1 assures that ρ_+ can uniquely be extended to a symmetric Lie algebra homomorphism $\rho : (\mathfrak{g}_C, \sigma) \to (\mathfrak{g}'_C, \sigma')$ which is also $\boldsymbol{C}$-linear. Since this ρ satisfies (8. 3) and (8. 6), one has $\bar{\rho} = \rho$, i. e., ρ comes from a real Lie algebra homomorphism $\mathfrak{g} \to \mathfrak{g}'$ satisfying (H_1). Thus one obtains the following

Proposition 8. 2. *A $\boldsymbol{C}$-linear map $\rho_+ : \mathfrak{p}_+ \to \mathfrak{p}'_+$ is a restriction to $\mathfrak{p}_+$ of a $\boldsymbol{C}$-linear extension to $\mathfrak{g}_C$ of an* (H_1)*-homomorphism $\rho : (\mathfrak{g}, H_0) \to (\mathfrak{g}', H'_0)$ if and only if ρ_+ is a JTS homomorphism.*

From these Propositions we can conclude that the following three categories are equivalent to one another:

($\mathcal{SD}$) The category of symmetric domains $\mathcal{D}$ with origin o, a morphism $\rho_{\mathcal{D}} : (\mathcal{D}, o) \to (\mathcal{D}', o')$ being a (strongly) equivariant holomorphic map $\mathcal{D} \to \mathcal{D}'$ with $\rho_{\mathcal{D}}(o) = o'$.

($\mathcal{HL}$) The category of semi-simple Lie algebras of hermitian type $(\mathfrak{g}, H_0)$ (without compact factors), a morphism $\rho : (\mathfrak{g}, H_0) \to (\mathfrak{g}', H'_0)$ being an (H_1)-homomorphism.

($\mathcal{HT}$) The category of positive definite hermitian JTS's $(\mathfrak{p}_+, \{\ \})$, a morphism $\rho_+ : (\mathfrak{p}_+, \{\ \}) \to (\mathfrak{p}'_+, \{\ \})$ being a $\boldsymbol{C}$-linear JTS homomorphism.

The functors giving these equivalences are given as follows:

$$(\mathcal{SD})\rightarrow(\mathcal{HL}):\ \mathfrak{g}=\mathrm{Lie}\,I(\mathcal{D}),\quad H_0\in\mathrm{Cent}(\mathrm{Lie}\,I(\mathcal{D})_o),\quad ad_{\mathfrak{p}}H_0=J;$$
$$\rho|\mathfrak{p}=d_o\rho_{\mathcal{D}}.$$
$$(\mathcal{HL})\rightarrow(\mathcal{HT}):\ \mathfrak{p}_+=\mathfrak{g}_C(J;i),\quad \{z_1,z_2,z_3\}=\frac{1}{2}[[z_1,\bar{z}_2],z_3]\quad(z_1,z_2,z_3\in\mathfrak{p}_+);$$
$$\rho_+=\rho_C|\mathfrak{p}_+.$$
$$(\mathcal{HT})\rightarrow(\mathcal{SD}):\ \mathcal{D}=\{z\in\mathfrak{p}_+|1-z\square z\gg0\};$$
$$\rho_{\mathcal{D}}=\rho_+|\mathcal{D}.$$

Remark. Strongly equivariant maps of symmetric domains (or symmetric spaces) have been used implicitly in many works on automorphic forms (before 1960, e. g., Siegel [9], Eichler [3]). Various types of equivariant holomorphic embeddings have been studied by Klingen [9], Hammond [2], Freitag-Schneider [1], K. Iyanaga [1]. Two important special cases, $(III_1)\rightarrow\mathcal{D}$ and $\mathcal{D}\rightarrow(III_n)$, will be discussed in Chapters III and IV. (For the latter, cf. Kuga [1], Satake [4], [6].) Equivariant holomorphic embeddings of symmetric domains were classified by S. Ihara [2]. (For equivariant embeddings of compact symmetric spaces, see e. g., Takeuchi-Kobayashi [1], Chen-Nagano [1].)

As an immediate application, we consider the relation between $I(\mathcal{D})$ (the group of isometries of $\mathcal{D}$ with respect to the Bergman metric) and $\mathrm{Hol}(\mathcal{D})$ (the group of holomorphic automorphisms of $\mathcal{D}$). We start with

Lemma 8. 3. *Let $\mathcal{D}$ and $\mathcal{D}'$ be irreducible symmetric domains and let $\varphi:\mathcal{D}\rightarrow\mathcal{D}'$ be a diffeomorphism of $\mathcal{D}$ onto $\mathcal{D}'$. Then φ is an isometry (with respect to the Bergman metrics) if and only if φ is holomorphic or antiholomorphic.*

Proof. Let $o\in\mathcal{D}$, $o'=\varphi(o)$, and let $(\mathfrak{g}, H_0)$ and $(\mathfrak{g}', H_0')$ be Lie algebras of hermitian type corresponding to $(\mathcal{D}, o)$ and $(\mathcal{D}', o')$. If φ is isometric, then by Theorem 2. 4, there exists a Lie algebra isomorphism $\rho:\mathfrak{g}\rightarrow\mathfrak{g}'$ such that (ρ,φ) is an equivariant pair (at o). Since $\mathcal{D}'$ is irreducible, there are only two H-elements (i. e., $\pm H_0'$) of $\mathfrak{g}'$ relative to o'. Hence one has $\rho(H_0)=\pm H_0'$. If $\rho(H_0)=H_0'$, then by Proposition 8. 1 φ is holomorphic. If $\rho(H_0)=-H_0'$, then $\rho:(\mathfrak{g}, H_0)\rightarrow(\mathfrak{g}', -H_0')$ satisfies (H_1), and hence φ is antiholomorphic. Conversely, if φ is holomorphic or antiholomorphic, it is clear from the definitions that φ preserves the Bergman metric, q. e. d.

Lemma 8. 4. *Let $\mathcal{D}$ be a symmetric domain. Then there exists an involutive antiholomorphic automorphism of $\mathcal{D}$.*

Proof. As in § 4, let $\mathfrak{h}$ be a Cartan subalgebra of $\mathfrak{g}$ contained in $\mathfrak{k}$. Then it is known from the theory of complex semi-simple Lie algebras (cf. e. g., Humphreys [1], or Serre [3]) that one can take generators X_α of $\mathfrak{g}_\alpha$ $(\alpha\in\tilde{\mathfrak{r}})$ in such a way that the following conditions are satisfied (the so-called "Chevalley basis"):

(8. 7 a) $$\sigma X_\alpha=-X_{-\alpha},$$

(8. 7 b) $$[[X_\alpha, X_{-\alpha}], X_\alpha]=2X_\alpha,$$

(8. 7 c) $$[X_\alpha, X_\beta] = N_{\alpha,\beta}X_{\alpha+\beta} \quad \text{if} \quad \alpha, \beta, \alpha+\beta \in \mathfrak{r},$$

where $N_{\alpha,\beta} \in \boldsymbol{R}$. Note that these conditions imply that $N_{\alpha,\beta} = -N_{-\alpha,-\beta} (\in \boldsymbol{Z})$. We define a $\boldsymbol{C}$-linear transformation ψ of $\mathfrak{g}_C$ by

(8. 8) $$\begin{cases} \psi H = -H & \text{for all } H \in \mathfrak{h}_C, \\ \psi X_\alpha = -X_{-\alpha} & \text{for all } \alpha \in \mathfrak{r}. \end{cases}$$

Then it is clear that ψ is a complex Lie algebra automorphism of $\mathfrak{g}_C$. It is also clear that

(8. 9) $$\psi^2 = id, \qquad \sigma\psi = \psi\sigma, \qquad \theta\psi = \psi\theta.$$

Since $\sigma X = \theta \bar{X}$ $(X \in \mathfrak{g}_C)$, it follows that $\bar{\psi} = \psi$. Therefore ψ induces a real Lie algebra automorphism of $\mathfrak{g}$, which by (8. 8) satisfies $\psi(H_0) = -H_0$. Hence, by Proposition 8. 1, ψ gives rise to an involutive antiholomorphic automorphism $\psi_{\mathcal{D}}$ of $\mathcal{D}$, q. e. d.

Remark. The involutive antiholomorphic automorphisms of irreducible symmetric domains were determined by Jaffee [1], [2]. In view of the equivalence of three categories mentioned above, this is essentially equivalent to the determination of positive definite (real) JTS given (from Lie theoretic view point) by Kobayashi-Nagano [1] (cf. I, § 7).

Proposition 8. 5. *Let $\mathcal{D}$ be a symmetric domain. Then* $\mathrm{Hol}(\mathcal{D})$ *is a subgroup of $I(\mathcal{D})$ (with respect to the Bergman metric) of index 2^s, where s is the number of irreducible components of $\mathcal{D}$. Moreover, one has*

(8. 10) $$I(\mathcal{D})^z \cap \mathrm{Hol}(\mathcal{D}) = I(\mathcal{D})^\circ.$$

Proof. Since any holomorphic automorphism of $\mathcal{D}$ leaves the Bergman metric invariant, it is clear that $\mathrm{Hol}(\mathcal{D}) \subset I(\mathcal{D})$. When $\mathcal{D}$ is irreducible, it follows from Lemmas 8. 3, 8. 4 that $(I(\mathcal{D}) : \mathrm{Hol}(\mathcal{D})) = 2$. In the general case, let

$$\mathcal{D} = \mathcal{D}_1 \times \cdots \times \mathcal{D}_s,$$

where the $\mathcal{D}_i$ $(1 \leq i \leq s)$ are irreducible. We denote by π_i the canonical projection $\mathcal{D} \to \mathcal{D}_i$. Then for every $\varphi \in I(\mathcal{D})$ there corresponds a permutation of $\{1, \cdots, s\}$, which we also denote by φ, and isometries $\varphi_i : \mathcal{D}_i \to \mathcal{D}_{\varphi(i)}$ $(1 \leq i \leq s)$ such that $\pi_{\varphi(i)} \circ \varphi = \varphi_i \circ \pi_i$. By Lemma 8. 3 each φ_i is holomorphic or antiholomorphic. Put

$$I(\mathcal{D})^{\sharp} = \{\varphi \in I(\mathcal{D}) \mid \varphi(i) = i \ (1 \leq i \leq s)\},$$
$$\mathrm{Hol}(\mathcal{D})^{\sharp} = \mathrm{Hol}(\mathcal{D}) \cap I(\mathcal{D})^{\sharp}.$$

Then $I(\mathcal{D})^{\sharp}$ is a normal subgroup of $I(\mathcal{D})$ and by Lemma 8. 4 one has $I(\mathcal{D}) = \mathrm{Hol}(\mathcal{D}) \cdot I(\mathcal{D})^{\sharp}$. Hence one has

$$\begin{aligned}(I(\mathcal{D}) : \mathrm{Hol}(\mathcal{D})) &= (I(\mathcal{D})^{\sharp} : \mathrm{Hol}(\mathcal{D})^{\sharp}) \\ &= \prod_{i=1}^{s} (I(\mathcal{D}_i) : \mathrm{Hol}(\mathcal{D}_i)) = 2^s.\end{aligned}$$

To prove (8. 10) it is enough to show

$$(I(\mathcal{D})^z)_o \cap \mathrm{Hol}(\mathcal{D}) = I(\mathcal{D})^\circ_o.$$

This follows from Lemma 4. 1, applied to $G = I(\mathcal{D})^z$, q. e. d.

We note that by the isomorphism $I(\mathcal{D}) \cong \mathrm{Aut}\,\mathfrak{g}$ the Zariski connected component $I(\mathcal{D})^z$ corresponds to the "algebraic adjoint group" $\mathrm{Ad}\,\mathfrak{g} = (\mathrm{Aut}\,\mathfrak{g}) \cap (\mathrm{Ad}\,\mathfrak{g}_C)$. It is known (Takeuchi [1], Matsumoto [2]) that, in general, the factor group $\mathrm{Aut}\,\mathfrak{g}/\mathrm{Ad}\,\mathfrak{g}$ is isomorphic to the automorphism group of the "marked diagram" for $\tilde{\Delta}$ of $\mathfrak{g}$ (cf. I, § 5). Hence, for a simple Lie algebra of hermitian type $\mathfrak{g}$, one has $(\mathrm{Aut}\,\mathfrak{g} : \mathrm{Ad}\,\mathfrak{g}) \leq 2$ (III, § 4). Combining this with Proposition 8. 5, we see that, for an irreducible symmetric domain, if $\mathrm{Hol}(\mathcal{D}) \neq I(\mathcal{D})^\circ$, one has $I(\mathcal{D}) = \mathrm{Hol}(\mathcal{D}) \cdot I(\mathcal{D})^z$ and the factor group $I(\mathcal{D})/I(\mathcal{D})^\circ$ is an abelian group of type (2, 2). Otherwise, one has $\mathrm{Hol}(\mathcal{D}) = I(\mathcal{D})^\circ$ and $(I(\mathcal{D}) : I(\mathcal{D})^\circ) = 2$.

Thus we have the following three possibilities*):

1. $(I(\mathcal{D}) : I(\mathcal{D})^\circ) = 2, \quad \mathrm{Hol}(\mathcal{D}) = I(\mathcal{D})^\circ, \quad I(\mathcal{D})^z = I(\mathcal{D})$:
   $(I_{1,1})$, (II_n) (n even, ≥ 6), (III_n), (IV_p) (p odd, ≥ 3), (VI).
2. $(I(\mathcal{D}) : I(\mathcal{D})^\circ) = 2, \quad \mathrm{Hol}(\mathcal{D}) = I(\mathcal{D})^z = I(\mathcal{D})^\circ$:
   $(I_{p,q})$ $(p > q)$, (II_n) (n odd, ≥ 3), (V).
3. $(I(\mathcal{D}) : I(\mathcal{D})^\circ) = 4, \quad (\mathrm{Hol}(\mathcal{D}) : I(\mathcal{D})^\circ) = (I(\mathcal{D})^z : I(\mathcal{D})^\circ) = 2$,
   $\mathrm{Hol}(\mathcal{D}) \cap I(\mathcal{D})^z = I(\mathcal{D})^\circ$:
   $(I_{p,p})$ $(p \geq 2)$, (II_4), (IV_p) (p even, ≥ 4).

The case (III_n) (the Siegel space) was discussed in § 7. For the treatment of other classical domains, see Appendix, §§ 3 and 6. Notice that, in the cases 1 and 3, $\mathcal{D}$ is of tube type (cf. III, § 4).

Exercises

1. The notation being as in the text, let $\rho : (\mathfrak{g}, H_0) \to (\mathfrak{g}', H_0')$ be an (H_1)-homomorphism and let $g' \in G'^\circ$. Show that ρ is an (H_1)-homomorphism with respect to the H-elements H_0 and $(ad\, g')H_0'$ if and only if $g' \in C_{G'}(\rho(G)) \cdot K'^\circ$. (Use § 2, Exerc. 1 and Lem. 4. 1.)

2. 2. 1) Let $\mathcal{D}$ be a symmetric domain. Show that $\mathrm{Hol}(\mathcal{D})^\circ$ is generated (as analytic group) by the symmetries s_z $(z \in \mathcal{D})$.

2. 2) Let $\mathcal{D}'$ be another symmetric domain. Show that a holomorphic map $\varphi : \mathcal{D} \to \mathcal{D}'$ is equivariant (i. e., totally geodesic) if and only if one has $\varphi \circ s_z = s_{\varphi(z)} \circ \varphi$ for all $z \in \mathcal{D}$.

*) For the numbering of irreducible symmetric domains, we follow Siegel [8], see III, § 4.

Chapter III

Unbounded Realizations of Symmetric Domains (Theory of Wolf-Korányi)

§ 1. The (H_1)-homomorphisms κ of $\mathfrak{sl}_2(\boldsymbol{R})$ into $\mathfrak{g}$.

Let $(\mathfrak{g}, H_0)$ and $(\mathfrak{g}', H_0')$ be two semi-simple Lie algebras of hermitian type. We recall that a homomorphism $\rho: \mathfrak{g} \to \mathfrak{g}'$ is said to satisfy the condition (H_1) (with respect to H_0 and H_0'), if $\rho \circ ad\, H_0 = ad\, H_0' \circ \rho$, and the condition (H_2), if $\rho(H_0) = H_0'$ (II, § 8). For brevity, such a homomorphism is also called an (H_i)-homomorphism $(i=1, 2)$. In this chapter, we will study (non-trivial) (H_1)-homomorphisms κ: $\mathfrak{sl}_2(\boldsymbol{R}) \to \mathfrak{g}$ in their connection with the "boundary components" of the symmetric domain $\mathcal{D}$ associated with $\mathfrak{g}$. We shall see that such a homomorphism gives rise to a Siegel domain expression of $\mathcal{D}$ relative to the corresponding boundary component.

The following notation will be fixed throughout the chapter.

$$\begin{aligned}
&\mathfrak{g}^1 = \mathfrak{sl}_2(\boldsymbol{R}) = \{h, e_+, e_-\}_{\boldsymbol{R}} = \{h, h', h''\}_{\boldsymbol{R}},\\
&h = \begin{pmatrix} 1 & 0 \\ 0 & -1 \end{pmatrix}, \quad e_+ = \begin{pmatrix} 0 & 1 \\ 0 & 0 \end{pmatrix}, \quad e_- = \begin{pmatrix} 0 & 0 \\ 1 & 0 \end{pmatrix},\\
&h' = e_+ + e_-, \qquad h'' = e_+ - e_-.
\end{aligned} \tag{1.1}$$

Clearly one has

$$\begin{aligned}
&[h, e_+] = 2e_+, \quad [h, e_-] = -2e_-, \quad [e_+, e_-] = h,\\
&[h, h'] = 2h'', \quad [h, h''] = 2h', \quad [h', h''] = -2h.
\end{aligned} \tag{1.2}$$

A Cartan decomposition of $\mathfrak{g}^1$ is given by

$$\mathfrak{g}^1 = \mathfrak{k}^1 + \mathfrak{p}^1, \qquad \mathfrak{k}^1 = \{h''\}_{\boldsymbol{R}}, \qquad \mathfrak{p}^1 = \{h, h'\}_{\boldsymbol{R}},$$

and $H_0^1 = \frac{1}{2} h''$ is an H-element of $\mathfrak{g}^1$.

First we consider the representations of $\mathfrak{g}^1 = \mathfrak{sl}_2(\boldsymbol{R})$ in general. Let u, v be two (real) variables, and write

$$\begin{pmatrix} u \\ v \end{pmatrix}^{(\nu)} = \begin{pmatrix} u^\nu \\ u^{\nu-1}v \\ \vdots \\ v^\nu \end{pmatrix} \qquad (\nu = 0, 1, 2, \cdots).$$

For each $g \in SL_2(\boldsymbol{R})$, we define $\rho_\nu(g) \in SL_{\nu+1}(\boldsymbol{R})$ by the relation

$$\left(g \begin{pmatrix} u \\ v \end{pmatrix}\right)^{(\nu)} = \rho_\nu(g) \begin{pmatrix} u \\ v \end{pmatrix}^{(\nu)}. \tag{1.3}$$

Then, clearly, ρ_ν is a representation of $SL_2(\boldsymbol{R})$ of dimension $\nu+1$ (the so-called symmetric tensor representation). It is known that the ρ_ν ($\nu=0, 1, 2, \cdots$) are absolutely irreducible and constitute a full set of representatives for equivalence classes of (finite-dimensional) irreducible representations of $SL_2(\boldsymbol{R})$. Denoting the corresponding representation of the Lie algebra $\mathfrak{sl}_2(\boldsymbol{R})$ by the same letter ρ_ν, one has

$$\rho_\nu(h) = \begin{pmatrix} \nu & & & & 0 \\ & \nu-2 & & & \\ & & \ddots & & \\ & & & -\nu+2 & \\ 0 & & & & -\nu \end{pmatrix}, \qquad \rho_\nu(e_+) = \begin{pmatrix} 0 & \nu & & & 0 \\ & 0 & \nu-1 & & \\ & & \ddots & \ddots & \\ & & & \ddots & 1 \\ 0 & & & & 0 \end{pmatrix},$$

$$\rho_\nu(e_-) = \begin{pmatrix} 0 & & & & 0 \\ 1 & 0 & & & \\ & 2 & \ddots & & \\ & & \ddots & \ddots & \\ 0 & & & \nu & 0 \end{pmatrix}.$$

Now let ρ be an arbitrary representation of $\mathfrak{sl}_2(\boldsymbol{R})$ on a finite-dimensional real vector space V. Then ρ is fully reducible and V is decomposed uniquely into the direct sum

$$V = V^{[0]} \oplus V^{[1]} \oplus V^{[2]} \oplus \cdots,$$

where $V^{[\nu]}$ is the "primary component" belonging to ρ_ν, i. e., the sum of all irreducible invariant subspaces of V transmitting representations equivalent to ρ_ν (IV, § 1).

Lemma 1. 1. *In the above notation, if* $\rho(e_+)^n=0$, *then one has* $V^{[\nu]}=\{0\}$ *for all* $\nu\geq n$.

This is clear from the relations $\rho_\nu(e_+)^\nu\neq 0$, $\rho_\nu(e_+)^{\nu+1}=0$.

Now, let $(\mathfrak{g}, H_0)$ be a semi-simple Lie algebra of hermitian type and suppose there is given a non-trivial (H_1)-homomorphism $\kappa: \mathfrak{g}^1\rightarrow\mathfrak{g}$ with respect to H_0^1 $(=\frac{1}{2}h'')$ and H_0. (Since these H-elements are kept fixed throughout the following discussion, the reference to H_0^1 and H_0 will often be omitted.) Then $\rho=ad^{\mathfrak{g}}\circ\kappa$ is a representation of $\mathfrak{g}^1$ on the underlying vector space of $\mathfrak{g}$. By the general theory, we obtain the primary decomposition $\mathfrak{g}=\bigoplus_\nu\mathfrak{g}^{[\nu]}$ of $\mathfrak{g}$ viewed as a representation-space of ρ; in particular, $\mathfrak{g}^{[0]}=\mathfrak{c}(\kappa(\mathfrak{g}^1))$.

Lemma 1. 2. *For* $\rho=ad^{\mathfrak{g}}\circ\kappa$ *with* κ *satisfying* (H_1), *one has* $\mathfrak{g}^{[\nu]}=\{0\}$ *for* $\nu\geq 3$, *i. e.*,

$$\mathfrak{g} = \mathfrak{g}^{[0]}+\mathfrak{g}^{[1]}+\mathfrak{g}^{[2]}. \tag{1.4}$$

Proof. Let

$$\mathfrak{g}_c^1 = \mathfrak{p}_+^1+\mathfrak{k}_c^1+\mathfrak{p}_-^1 \quad \text{and} \quad \mathfrak{g}_c = \mathfrak{p}_++\mathfrak{k}_c+\mathfrak{p}_-$$

be the canonical decompositions of $\mathfrak{g}_C^1$ and $\mathfrak{g}_C$ with respect to H_0^1 and H_0 (II, § 3). Then, the condition (H_1) implies that the subspaces $\mathfrak{p}_+^1$, $\mathfrak{k}_C^1$, $\mathfrak{p}_-^1$ are mapped into $\mathfrak{p}_+$, $\mathfrak{k}_C$, $\mathfrak{p}_-$, respectively, by κ. On the other hand, $ad\,\mathfrak{p}_+$ maps $\mathfrak{p}_+$, $\mathfrak{k}_C$, $\mathfrak{p}_-$ into $\{0\}$, $\mathfrak{p}_+$, $\mathfrak{k}_C$, respectively, so that $(ad\,\mathfrak{p}_+)^3=0$. Therefore, one has $(ad^{\mathfrak{g}}\circ\kappa(\mathfrak{p}_+^1))^3=0$. Since all non-zero nilpotent elements in $\mathfrak{g}_C^1$ are conjugate, it follows that $\rho(e_+)^3=(ad^{\mathfrak{g}}\circ\kappa(e_+))^3=0$. Hence, by Lemma 1. 1, one has $\mathfrak{g}^{[\nu]}=0$ for $\nu\geq 3$, q. e. d.

The condition (H_1) implies that the subspaces $\mathfrak{g}^{[\nu]}$ $(\nu=0, 1, 2)$ are stable under the Cartan involution $\theta=\exp(\pi\, ad\, H_0)$. Hence, if $\mathfrak{g}=\mathfrak{k}+\mathfrak{p}$ is the corresponding Cartan decomposition, one has

$$\mathfrak{g}^{[\nu]} = \mathfrak{k}^{[\nu]}+\mathfrak{p}^{[\nu]} \qquad (\nu=0, 1, 2) \tag{1. 5}$$

with $\mathfrak{k}^{[\nu]}=\mathfrak{k}\cap\mathfrak{g}^{[\nu]}$, $\mathfrak{p}^{[\nu]}=\mathfrak{p}\cap\mathfrak{g}^{[\nu]}$. We also write $\mathfrak{g}^{[\text{even}]}=\mathfrak{g}^{[0]}+\mathfrak{g}^{[2]}$, $\mathfrak{k}^{[\text{even}]}=\mathfrak{k}\cap\mathfrak{g}^{[\text{even}]}$, $\mathfrak{p}^{[\text{even}]}=\mathfrak{p}\cap\mathfrak{g}^{[\text{even}]}$.

For the given (non-trivial) (H_1)-homomorphism κ, we put

$$X_\kappa = \kappa(h), \qquad Y_\kappa = \kappa(h'), \qquad H_\kappa = \kappa(h''). \tag{1. 6}$$

Then $\{H_\kappa, X_\kappa, Y_\kappa\}_{\boldsymbol{R}}$ is a subalgebra of $\mathfrak{g}$ isomorphic to $\mathfrak{sl}_2(\boldsymbol{R})$ and one has

$$\begin{aligned} [H_\kappa, X_\kappa] &= -2Y_\kappa, \\ [H_\kappa, Y_\kappa] &= 2X_\kappa, \\ [X_\kappa, Y_\kappa] &= 2H_\kappa. \end{aligned} \tag{1. 7}$$

Clearly one has $\kappa(H_0^1)=\frac{1}{2}H_\kappa\in\mathfrak{k}$ and $X_\kappa, Y_\kappa\in\mathfrak{p}$. The condition (H_1) implies that $H_0-\frac{1}{2}H_\kappa=H_0-\kappa(H_0^1)\in\mathfrak{g}^{[0]}$. On the other hand, since the adjoint representation $ad^{\mathfrak{g}^1}$ is equivalent to ρ_2, one has $\kappa(\mathfrak{g}^1)\subset\mathfrak{g}^{[2]}$. Hence one has

$$H_0-\frac{1}{2}H_\kappa\in\mathfrak{k}^{[0]}, \qquad H_\kappa\in\mathfrak{k}^{[2]}, \qquad X_\kappa, Y_\kappa\in\mathfrak{p}^{[2]}. \tag{1. 8}$$

We define the *partial Cayley transformation* c_κ of $\mathfrak{g}_C$ associated with κ by

$$c_\kappa = \exp\left(\frac{\pi i}{4} ad\, Y_\kappa\right)\left(=\kappa\left(\frac{1}{\sqrt{2}}\begin{pmatrix}1 & i\\ i & 1\end{pmatrix}\right)\right), \tag{1. 9}$$

where exp denotes the exponential map $\mathfrak{g}_C\to G_C=\mathrm{Ad}(\mathfrak{g}_C)$. Then by (1. 7) one has

$$c_\kappa : \begin{cases} H_\kappa \longrightarrow -iX_\kappa, \\ X_\kappa \longrightarrow -iH_\kappa, \\ Y_\kappa \longrightarrow Y_\kappa. \end{cases} \tag{1. 10}$$

Put $\gamma=c_\kappa^2$. Then, since $ad\,Y_\kappa$ is semi-simple and has eigenvalues 0 on $\mathfrak{g}^{[0]}$, ± 1 on $\mathfrak{g}^{[1]}$, and 0, ± 2 on $\mathfrak{g}^{[2]}$, the transformation γ is real on $\mathfrak{g}^{[\text{even}]}$ and has eigenvalues 1 on $\mathfrak{g}^{[0]}$, $\pm i$ on $\mathfrak{g}_C^{[1]}$, and ± 1 on $\mathfrak{g}^{[2]}$. Thus one has

$$\gamma^4 = c_\kappa^8 = 1 \tag{1. 11}$$

and

$$\begin{cases} \mathfrak{g}^{[0]} \subset \mathfrak{g}(ad\, Y_\kappa\,; 0) = \mathfrak{g}^{[\text{even}]}(\gamma\,; 1), \\ \mathfrak{g}^{[\text{even}]} = \mathfrak{g}(ad\, Y_\kappa\,; 0, \pm 2) = \mathfrak{g}(\gamma^2\,; 1), \\ \mathfrak{g}^{[1]} = \mathfrak{g}(ad\, Y_\kappa\,; \pm 1) = \mathfrak{g}(\gamma^2\,; -1). \end{cases} \tag{1. 12}$$

For the Cartan involution θ, one has $\theta Y_\kappa = -Y_\kappa$ and hence

$$\theta c_\kappa \theta = c_\kappa^{-1} = \bar{c}_\kappa, \qquad \theta\gamma\theta = \gamma^{-1} = \bar{\gamma}. \tag{1. 13}$$

In particular, one has $\theta\gamma = \gamma\theta$ on $\mathfrak{g}^{[\mathrm{even}]}$.

For convenience, we extend the definition of "hermitian type" as follows. A reductive $\boldsymbol{R}$-group G_1 (or its Lie algebra $\mathfrak{g}_1 = \mathrm{Lie}\ G_1$) is called *of hermitian type*, if the abelian part G_1^a is compact and if the semi-simple part $\mathfrak{g}_1^s$ is of hermitian type in the previous sense. This condition is equivalent to saying that, for any Cartan decomposition $\mathfrak{g}_1 = \mathfrak{k}_1 + \mathfrak{p}_1$ (relative to G_1), there exists an element H_1 in the center of $\mathfrak{k}_1$ such that $(ad\, H_1)^2 = -1$ on $\mathfrak{p}_1$ [e. g., an H-element in $\mathfrak{g}_1^s$ belonging to the Cartan decomposition $\mathfrak{g}_1^s = (\mathfrak{k}_1 \cap \mathfrak{g}_1^s) + \mathfrak{p}_1$]. Such an element H_1 (determined modulo $\mathfrak{g}_1^a$) will be called an "H-element" of G_1 (or of $\mathfrak{g}_1$). For brevity, the pair $(\mathfrak{g}_1, H_1)$ is also called of hermitian type. The notion of (H_i)-homomorphisms can also be extended in a natural manner.

Proposition 1. 3. *$\mathfrak{g}^{[0]}$ and $\mathfrak{g}^{[\mathrm{even}]} = \mathfrak{g}^{[0]} + \mathfrak{g}^{[2]}$ are reductive algebraic Lie subalgebras of $\mathfrak{g}$ and the corresponding (Zariski connected) $\boldsymbol{R}$-subgroups of* $\mathrm{Ad}\,\mathfrak{g}$ *are of hermitian type with H-elements $H_0 - \frac{1}{2}H_\kappa$ and H_0, respectively.*

Proof. In virtue of the relations $\mathfrak{g}^{[0]} = \mathfrak{c}(\kappa(\mathfrak{g}^1))$ and (1. 12), $\mathfrak{g}^{[0]}$ and $\mathfrak{g}^{[\mathrm{even}]}$ are algebraic subalgebras of $\mathfrak{g}$. The corresponding $\boldsymbol{R}$-subgroups of $\mathrm{Ad}\,\mathfrak{g}$ are (algebraically) reductive, as they are stable under the Cartan involution θ (I, § 4). Since $ad\, H_\kappa = 0$ on $\mathfrak{g}^{[0]}$, one has

$$ad\left(H_0 - \frac{1}{2}H_\kappa\right) = ad\, H_0 \qquad \text{on} \quad \mathfrak{g}^{[0]}.$$

Hence, by (1. 8), $H_0 - \frac{1}{2}H_\kappa$ is in the center of $\mathfrak{k}^{[0]}$ and defines a complex structure on $\mathfrak{p}^{[0]}$ (which coincides with the restriction to $\mathfrak{p}$ of the one defined by $ad\, H_0$). Thus $(\mathfrak{g}^{[0]}, H_0 - \frac{1}{2}H_\kappa)$ is of hermitian type. Next, from $H_0 - \frac{1}{2}H_\kappa \in \mathfrak{k}^{[0]}$ and $H_\kappa \in \mathfrak{k}^{[2]}$, one has $H_0 \in \mathfrak{k}^{[\mathrm{even}]}$. Therefore, H_0 is in the center of $\mathfrak{k}^{[\mathrm{even}]}$ and defines a complex structure on $\mathfrak{p}^{[\mathrm{even}]}$, i. e., $(\mathfrak{g}^{[\mathrm{even}]}, H_0)$ is also of hermitian type, q. e. d.

Though $\mathfrak{g}^{[0]}$ is a subalgebra of $\mathfrak{g}$, $\mathfrak{g}^{[2]}$ is not so in general. Hence our next step is to adjust the decomposition $\mathfrak{g}^{[\mathrm{even}]} = \mathfrak{g}^{[0]} + \mathfrak{g}^{[2]}$ to obtain a direct sum of ideals.

Lemma 1. 4. *The following conditions are equivalent:*

(a) $\gamma^2 = 1$, (a′) $\mathfrak{g}^{[1]} = 0$,

(b) $\theta\gamma = \gamma\theta$, (b′) $\gamma\mathfrak{k} = \mathfrak{k}$.

When $\mathfrak{g}$ is simple, these conditions are also equivalent to

(c) $\gamma H_0 = -H_0$,

(d) $H_0 = \frac{1}{2}H_\kappa$, *i. e., κ satisfies the condition* (H_2).

Proof. The equivalences (a)$\Leftrightarrow$(b), (a)$\Leftrightarrow$(a′), (b)$\Leftrightarrow$(b′) follow from the relations $\theta\gamma\theta=\gamma^{-1}$, $\mathfrak{g}^{[1]}=\mathfrak{g}(\gamma^2;-1)$, $\mathfrak{k}=\mathfrak{g}(\theta;1)$, and $\mathfrak{p}=\mathfrak{k}^{\perp}$. Next, assume that $\mathfrak{g}$ is simple. Since we are assuming that κ is non-trivial, $\mathfrak{g}$ is then necessarily non-compact, and the center of $\mathfrak{k}$ is of dimension 1. Hence one has

$$(\mathrm{b}') \Longleftrightarrow \gamma(H_0)=\varepsilon H_0 \qquad \text{where} \quad \varepsilon=\pm 1.$$

But, in general, since $H_0-\frac{1}{2}H_\kappa\in\mathfrak{k}^{[0]}$, one has by (1. 10)

$$H_0-\frac{1}{2}H_\kappa=\gamma\Big(H_0-\frac{1}{2}H_\kappa\Big)=\gamma(H_0)+\frac{1}{2}H_\kappa.$$

Thus $H_\kappa=H_0-\gamma(H_0)$, and so $\gamma(H_0)$ cannot be equal to H_0. Combining these, we obtain (b′)$\Leftrightarrow$(c)$\Leftrightarrow$(d), q. e. d.

Now we consider the decomposition

$$\mathfrak{g}^{[\mathrm{even}]}=\bigoplus_{i=0}^{s}\mathfrak{g}_i,$$

where $\mathfrak{g}_0$ is the largest compact ideal of $\mathfrak{g}^{[\mathrm{even}]}$ (where "compact" means that the corresponding subgroup of $\mathrm{Ad}\,\mathfrak{g}$ is compact) and the $\mathfrak{g}_i\,(i\geq 1)$ are simple and non-compact. Then, all $\mathfrak{g}_i$'s are of hermitian type, and the abelian part of $\mathfrak{g}^{[\mathrm{even}]}$ is contained in $\mathfrak{g}_0$. We denote by π_i the canonical projections $\mathfrak{g}\to\mathfrak{g}_i$. Then one has

$$(1.\ 14)\qquad \begin{aligned}\mathfrak{g}_i\subset\mathfrak{g}^{[0]} &\Longleftrightarrow [\mathfrak{g}_i,\kappa(\mathfrak{g}^1)]=0\\ &\Longleftrightarrow \pi_i(\kappa(\mathfrak{g}^1))=0.\end{aligned}$$

[The second $\Rightarrow$ follows from the fact that, if $\pi_i(\kappa(\mathfrak{g}^1))\neq 0$, then $\pi_i(\kappa(\mathfrak{g}^1))\cong\mathfrak{g}^1$ is non-abelian.] For $i=0$, one has $\pi_0(\kappa(\mathfrak{g}^1))=0$ and so $\mathfrak{g}_0\subset\mathfrak{g}^{[0]}$. For $i\geq 1$, $\pi_i\circ\kappa:\mathfrak{g}^1\to\mathfrak{g}_i$ is an (H_1)-homomorphism with respect to H_0^1 and $\pi_i(H_0)\,(\neq 0)$ and, if $\mathfrak{g}_i\not\subset\mathfrak{g}^{[0]}$, then $\pi_i\circ\kappa$ is non-trivial and satisfies the condition (a′) in Lemma 1. 4. Hence, applying Lemma 1. 4 to $\pi_i\circ\kappa$, one has

$$(1.\ 15)\qquad \begin{aligned}\mathfrak{g}_i\not\subset\mathfrak{g}^{[0]} &\Longleftrightarrow \pi_i(H_0)=\frac{1}{2}\pi_i(H_\kappa)\\ &\Longleftrightarrow \Big[\mathfrak{g}_i,\,H_0-\frac{1}{2}H_\kappa\Big]=0.\end{aligned}$$

[To obtain the second $\Leftarrow$, note that the center of $\mathfrak{g}_i\,(i\geq 1)$ is trivial.]

Now define

$$(1.\ 16)\qquad \begin{cases}\mathfrak{g}_\kappa^{(1)}=\displaystyle\bigoplus_{\mathfrak{g}_i\subset\mathfrak{g}^{[0]}}\mathfrak{g}_i,\\ \mathfrak{l}_2=\mathfrak{g}_0,\qquad \mathfrak{g}_\kappa=\displaystyle\bigoplus_{\substack{\mathfrak{g}_i\subset\mathfrak{g}^{[0]}\\ i\geq 1}}\mathfrak{g}_i,\\ \mathfrak{g}_\kappa^*=\displaystyle\bigoplus_{\mathfrak{g}_i\not\subset\mathfrak{g}^{[0]}}\mathfrak{g}_i.\end{cases}$$

Clearly, $\mathfrak{g}_\kappa^{(1)}\subset\mathfrak{g}^{[0]}$; as $\mathfrak{g}^{[2]}=\bigoplus_i(\mathfrak{g}_i\cap\mathfrak{g}^{[2]})$, it follows that $\mathfrak{g}_\kappa^*\supset\mathfrak{g}^{[2]}$.

Proposition 1. 5. *One has*

(1. 17) $$\mathfrak{g}^{[\text{even}]} = \mathfrak{g}_\kappa^{(1)} \oplus \mathfrak{g}_\kappa^*, \qquad \mathfrak{g}_\kappa^{(1)} = \mathfrak{l}_2 \oplus \mathfrak{g}_\kappa,$$

which are direct sums of ideals. The ideal $\mathfrak{l}_2$ is compact, $\mathfrak{g}_\kappa$ and $\mathfrak{g}_\kappa^*$ are semi-simple, of hermitian type (without compact factors), and one has $\kappa(\mathfrak{g}^1) \subset \mathfrak{g}_\kappa^*$. $H_0 - \frac{1}{2}H_\kappa$ and $\frac{1}{2}H_\kappa$ are H-elements of $\mathfrak{g}_\kappa^{(1)}$ and $\mathfrak{g}_\kappa^*$, respectively.

Proof. From the definitions, all statements are obvious, except for the last one concerning H-elements. From (1. 15) and the definition of $\mathfrak{g}_\kappa^{(1)}$, one has $H_0 - \frac{1}{2}H_\kappa \in \mathfrak{g}_\kappa^{(1)}$. On the other hand, one has $H_\kappa \in \mathfrak{g}^{[2]} \subset \mathfrak{g}_\kappa^*$. Since H_0 is an H-element of $\mathfrak{g}^{[\text{even}]}$, it follows from (1. 17) that $H_0 - \frac{1}{2}H_\kappa$ and $\frac{1}{2}H_\kappa$ are H-elements of $\mathfrak{g}_\kappa^{(1)}$ and $\mathfrak{g}_\kappa^*$, respectively, q. e. d.

It follows that the (restricted) homomorphism $\kappa : \mathfrak{g}^1 \to \mathfrak{g}_\kappa^*$ satisfies the stronger condition (H_2). Putting further $\mathfrak{l}_1 = \mathfrak{g}^{[0]} \cap \mathfrak{g}_\kappa^*$, one has

(1. 18) $$\mathfrak{g}^{[0]} = \mathfrak{l}_1 \oplus \mathfrak{l}_2 \oplus \mathfrak{g}_\kappa, \qquad \mathfrak{g}_\kappa^* = \mathfrak{l}_1 + \mathfrak{g}^{[2]}.$$

Since one has $\mathfrak{l}_1 \subset \mathfrak{g}^{[0]} \cap \mathfrak{c}(H_0 - \frac{1}{2}H_\kappa) = \mathfrak{k}^{[0]}$, $\mathfrak{l}_1$ is compact. In the following, we write $H_0^{(1)}$ for $H_0 - \frac{1}{2}H_\kappa$.

Corollary 1. 6. *For an (H_1)-homomorphism $\kappa : \mathfrak{g}^1 \to \mathfrak{g}$, the following conditions are equivalent:*

(i) *κ satisfies (H_2), i. e., $H_0 = \frac{1}{2}H_\kappa$.*

(ii) *$\mathfrak{g}^{[1]} = 0$ and $\mathfrak{g}^{[0]}$ (or $\mathfrak{g}_\kappa^{(1)}$) is compact.*

Proof. Clearly the proof can be reduced to the case where $\mathfrak{g}$ is simple (but including the case κ is trivial). Hence we have to check the following three cases.

1. $\mathfrak{g}$ compact and κ trivial. In this case, (H_2) is always satisfied and one has $\mathfrak{g} = \mathfrak{g}^{[0]} = \mathfrak{g}_\kappa^{(1)} = \mathfrak{l}_2$ (compact), $\mathfrak{g}_\kappa^* = 0$, and $\mathfrak{g}^{[1]} = 0$.

2. $\mathfrak{g}$ non-compact and κ trivial. In this case, (H_2) is *not* satisfied and one has $\mathfrak{g} = \mathfrak{g}^{[0]} = \mathfrak{g}_\kappa^{(1)} = \mathfrak{g}_\kappa$ (non-compact), $\mathfrak{g}_\kappa^* = 0$, and $\mathfrak{g}^{[1]} = 0$.

3. $\mathfrak{g}$ simple, non-compact and κ non-trivial. By Lemma 1. 4, κ satisfies (H_2) if and only if $\mathfrak{g}^{[1]} = 0$. When this condition is satisfied, one has $\mathfrak{g}_\kappa^{(1)} = 0$, $\mathfrak{g} = \mathfrak{g}_\kappa^*$, and $\mathfrak{g}^{[0]} = \mathfrak{l}_1$ (compact).

From these we see that the conditions (i), (ii) are always equivalent, q. e. d.

We shall see later on (§ 8, Rem. 1) that the conditions (i), (ii) above are also equivalent to

(iii) $\mathcal{D}$ is "of tube type", i. e., holomorphically equivalent to a tube domain $U + i\Omega$ and the boundary component of $\mathcal{D}$ corresponding to κ reduces to a point.

Remark. Let F be a subfield of $\boldsymbol{R}$, and suppose that $\mathfrak{g}$ is given an F-structure and $\kappa : \mathfrak{g}^1 \to \mathfrak{g}$ is defined over F. [$\mathfrak{g}^1 = \mathfrak{sl}_2(\boldsymbol{R})$ is always given an F-structure defined by the basis $\{h, h', h''\}$.]

Then the elements H_κ, X_κ, Y_κ are F-rational, and the subspaces $\mathfrak{g}^{[\nu]}(\nu=0,1,2)$ are defined over F. The subalgebra $\mathfrak{g}_\kappa^{(1)}$ is also defined over F, since it is the largest ideal of $\mathfrak{g}^{[\mathrm{even}]}$ contained in $\mathfrak{g}^{[0]}$. It follows that $\mathfrak{g}_\kappa^*$, which is the unique complement of $\mathfrak{g}_\kappa^{(1)}$ in $\mathfrak{g}^{[\mathrm{even}]}$, and $\mathfrak{l}_1=\mathfrak{g}^{[0]}\cap\mathfrak{g}_\kappa^*$ are both defined over F. However, the subalgebras $\mathfrak{l}_2$ and $\mathfrak{g}_\kappa$, which are the compact and non-compact parts of $\mathfrak{g}_\kappa^{(1)}$, may not be defined over F, since in general the compactness does not make sense over F. The Cayley transformation c_κ, which is imaginary, is of course not defined over F.

Exercise

1. Let $\rho:(\mathfrak{g},H_0)\to(\mathfrak{g}',H_0')$ be an (H_1)-homomorphism of reductive Lie algebras of hermitian type, and set

(1.19) $$\mathfrak{c}(\rho)=\mathfrak{c}_{\mathfrak{g}'}(\rho(\mathfrak{g})),$$

(1.20) $$\mathfrak{c}^*(\rho)=\mathfrak{c}_{\mathfrak{g}'}(H_0'-\rho(H_0)),$$

$\mathfrak{c}_{\mathfrak{g}'}$ denoting the centralizer in $\mathfrak{g}'$. Prove the following.

1.1) $(\mathfrak{c}(\rho), H_0'-\rho(H_0))$ is reductive, of hermitian type and the inclusion map $\mathfrak{c}(\rho)\to\mathfrak{g}'$ satisfies (H_1).

1.2) $(\mathfrak{c}^*(\rho), \rho(H_0))$ is reductive, of hermitian type and the inclusion map $\mathfrak{c}^*(\rho)\to\mathfrak{g}'$ satisfies (H_1). Moreover, $\rho(\mathfrak{g})$ is contained in $\mathfrak{c}^*(\rho)$ and the (restricted) homomorphism $\rho:\mathfrak{g}\to\mathfrak{c}^*(\rho)$ satisfies (H_2). *$\mathfrak{c}^*(\rho)$ is the largest subalgebra of $\mathfrak{g}'$ having these properties.*

1.3) For the (H_1)-homomorphism $\kappa:(\mathfrak{g}^1,H_0^1)\to(\mathfrak{g},H_0)$ in the text, one has

$$\mathfrak{c}(\kappa)=\mathfrak{g}^{[0]}=\mathfrak{l}_1\oplus\mathfrak{l}_2\oplus\mathfrak{g}_\kappa,$$
$$\mathfrak{c}^*(\kappa)=\mathfrak{k}^{[0]}+\mathfrak{g}^{[2]}=\mathfrak{l}_2\oplus\mathfrak{k}_\kappa\oplus\mathfrak{g}_\kappa^*\qquad(\mathfrak{k}_\kappa=\mathfrak{k}\cap\mathfrak{g}_\kappa).$$

§ 2. The parabolic subalgebra attached to a homomorphism κ.

Using the decompositions obtained in § 1, we shall now study the parabolic subalgebra

(2.1) $$\mathfrak{b}_\kappa=\mathfrak{c}(X_\kappa)+V_\kappa+U_\kappa,$$

where

$$V_\kappa=\mathfrak{g}(ad\,X_\kappa\,;1),\qquad U_\kappa=\mathfrak{g}(ad\,X_\kappa\,;2).$$

When κ is kept fixed, we often omit the subscript κ and write V, U for V_κ, U_κ. The following lemma will be useful.

Lemma 2.1. 1) *For $X\in\mathfrak{g}_C$, one has*

(2.2) $$X\in c_\kappa^{-1}\mathfrak{k}_C\iff\theta X=\gamma X,\qquad X\in c_\kappa^{-1}\mathfrak{p}_C\iff\theta X=-\gamma X.$$

2) *The subspaces $\mathfrak{k}_C(\gamma\,;1)$ and $\mathfrak{p}_C(\gamma\,;1)$ are stable under c_κ, while $\mathfrak{k}_C(\gamma\,;-1)\,(\subset\mathfrak{k}_C^{[2]})$ and $\mathfrak{p}_C(\gamma\,;-1)\,(\subset\mathfrak{p}_C^{[2]})$ are interchanged by c_κ:*

$$c_\kappa:\ \mathfrak{k}_C^{[2]}(\gamma\,;-1)\rightleftarrows\mathfrak{p}_C^{[2]}(\gamma\,;-1).$$

3) *One has*

(2.3) $$\mathfrak{k}^{[0]}=\mathfrak{k}^{[\mathrm{even}]}(\gamma\,;1),\qquad\mathfrak{k}^{[2]}=\mathfrak{k}^{[\mathrm{even}]}(\gamma\,;-1).$$

Proof. 1) follows from (1. 13) immediately, and 2) follows from 1). For instance, for $X \in \mathfrak{g}_C(\gamma; -1)$, one has by (2. 2)

$$c_\kappa X \in \mathfrak{p}_C(\gamma; -1) \Longleftrightarrow \theta X = -\gamma X = X$$
$$\Longleftrightarrow X \in \mathfrak{k}_C^{[2]}(\gamma; -1),$$

which proves $c_\kappa(\mathfrak{k}_C(\gamma; -1)) = \mathfrak{p}_C(\gamma; -1)$.

3) We know that the subspaces $\mathfrak{k}^{[\nu]}$ $(\nu=0, 2)$ are γ-stable and

$$\mathfrak{k}^{[0]} + \mathfrak{k}^{[2]} = \mathfrak{k}^{[\mathrm{even}]} = \mathfrak{k}^{[\mathrm{even}]}(\gamma; 1) + \mathfrak{k}^{[\mathrm{even}]}(\gamma; -1), \qquad \mathfrak{k}^{[0]} \subset \mathfrak{k}^{[\mathrm{even}]}(\gamma; 1).$$

Hence, to prove (2. 3), it is enough to show that $\mathfrak{k}^{[2]}(\gamma; 1) = 0$. Clearly by (1. 12) $\mathfrak{k}^{[2]}(\gamma; 1) \subset \mathfrak{c}(Y_\kappa)$ and, since $\frac{1}{2}H_\kappa$ is an H-element of $\mathfrak{g}_\kappa^*$, one has $\mathfrak{k}^{[2]} \subset \mathfrak{c}(H_\kappa)$. Therefore one has $\mathfrak{k}^{[2]}(\gamma; 1) \subset \mathfrak{k}^{[2]} \cap \mathfrak{c}(\kappa(\mathfrak{g}^1)) = \mathfrak{k}^{[2]} \cap \mathfrak{g}^{[0]} = 0$, as desired, q. e. d.

Now, from the decompositions (1. 17) and (1. 18), putting $\mathfrak{k}_\kappa^* = \mathfrak{k} \cap \mathfrak{g}_\kappa^*$, one has

$$\text{(2. 4)} \qquad \begin{cases} \mathfrak{c}(H_\kappa) = \mathfrak{g}_\kappa^{(1)} \oplus \mathfrak{k}_\kappa^*, \\ \mathfrak{k}_\kappa^* = \mathfrak{l}_1 + \mathfrak{k}^{[2]}. \end{cases}$$

Applying c_κ^{-1} to (2. 4) and noting that $c_\kappa^{-1}(H_\kappa) = iX_\kappa$ and $c_\kappa = id$ on $(\mathfrak{g}_\kappa^{(1)})_C$ and $\mathfrak{l}_{1C} (\subset \mathfrak{g}_C^{[0]})$, one has

$$\begin{cases} \mathfrak{c}(X_\kappa)_C = (\mathfrak{g}_\kappa^{(1)})_C \oplus c_\kappa^{-1}(\mathfrak{k}_\kappa^*)_C, \\ c_\kappa^{-1}(\mathfrak{k}_\kappa^*)_C = \mathfrak{l}_{1C} + c_\kappa^{-1}(\mathfrak{k}_C^{[2]}). \end{cases}$$

Since $c_\kappa^{-1}(\mathfrak{k}_\kappa^*)_C$ is the unique complement of $(\mathfrak{g}_\kappa^{(1)})_C$ in the reductive Lie algebra $\mathfrak{c}(X_\kappa)_C$, it is stable under the complex conjugation and θ. Hence we define

$$\mathfrak{g}_\kappa^{(2)} = \mathrm{Re}(c_\kappa^{-1}(\mathfrak{k}_\kappa^*)_C) = c_\kappa^{-1}(\mathfrak{k}_\kappa^*)_C \cap \mathfrak{g}.$$

Then one has

$$\text{(2. 5)} \qquad \mathfrak{c}(X_\kappa) = \mathfrak{g}_\kappa^{(1)} \oplus \mathfrak{g}_\kappa^{(2)}.$$

We also put $\mathfrak{k}_\kappa^{(2)} = \mathfrak{k} \cap \mathfrak{g}_\kappa^{(2)}$, $\mathfrak{p}_\kappa^{(2)} = \mathfrak{p} \cap \mathfrak{g}_\kappa^{(2)}$.

Lemma 2. 2. *One has*

$$\text{(2. 6)} \qquad \mathfrak{k}_\kappa^{(2)} = \mathfrak{l}_1, \qquad \mathfrak{p}_\kappa^{(2)} = ic_\kappa^{-1}(\mathfrak{k}^{[2]}).$$

In other words, the Cartan decomposition of $\mathfrak{g}_\kappa^{(2)}$ *is given by* $\mathfrak{g}_\kappa^{(2)} = \mathfrak{l}_1 + ic_\kappa^{-1}(\mathfrak{k}^{[2]})$. *The center of* $\mathfrak{g}_\kappa^{(2)}$ *is contained in the* $\mathfrak{p}$*-part* $ic_\kappa^{-1}(\mathfrak{k}^{[2]})$.

Proof. We know that $\mathfrak{l}_1 \subset \mathfrak{k}_\kappa^{(2)}$. Since $(\mathfrak{g}_\kappa^{(2)})_C = c_\kappa^{-1}(\mathfrak{k}_\kappa^*)_C$, one has from (2. 2), (2. 3)

$$(\mathfrak{p}_\kappa^{(2)})_C = (\mathfrak{g}_\kappa^{(2)})_C(\gamma; -1) = c_\kappa^{-1}(\mathfrak{k}_{\kappa C}^*(\gamma; -1)) = c_\kappa^{-1}(\mathfrak{k}_C^{[2]}).$$

Moreover, for $X \in \mathfrak{k}_C^{[2]}$, one has

$$\overline{c_\kappa^{-1} X} = c_\kappa^{-1} X \Longleftrightarrow \gamma \bar{X} = X \Longleftrightarrow \bar{X} = -X.$$

Hence one has

$$\mathfrak{p}_\kappa^{(2)} = \mathrm{Re}(c_\kappa^{-1}(\mathfrak{k}_C^{[2]})) = ic_\kappa^{-1}(\mathfrak{k}^{[2]}),$$

and so $\mathfrak{k}_\kappa^{(2)} = \mathfrak{l}_1$. In view of (1. 15), for each i such that $\mathfrak{g}_i \not\subset \mathfrak{g}^{[0]}$, the center of $\mathfrak{g}_i \cap \mathfrak{k}$

is generated by $\pi_i(H_\kappa) \in \mathfrak{g}_i \cap \mathfrak{k}^{[2]}$. Hence the center of $\mathfrak{k}_\kappa^*$ is contained in $\mathfrak{k}^{[2]}$, and so the center of $\mathfrak{g}_\kappa^{(2)}$ is contained in $\mathfrak{p}_\kappa^{(2)} = ic_\kappa^{-1}(\mathfrak{k}^{[2]})$, q. e. d.

We denote by $G_\kappa^{(i)}$ $(i=1, 2)$ the Zariski connected $\boldsymbol{R}$-subgroups of $G = \mathrm{Ad}\, \mathfrak{g}$ corresponding to $\mathfrak{g}_\kappa^{(i)}$. Since $C_G(X_\kappa)$ (the centralizer of X_κ in G) is Zariski connected, $C_G(X_\kappa)$ is the almost direct product of $G_\kappa^{(1)}$ and $G_\kappa^{(2)}$, and so $C_G(X_\kappa)^\circ = G_\kappa^{(1)\circ} \cdot G_\kappa^{(2)\circ}$. By (1. 17), the abelian part $(G_\kappa^{(1)})^a$ (which corresponds to $\mathfrak{l}_2^a$) is a compact torus and, by Lemma 2. 2, $(G_\kappa^{(2)})^a$ is an $\boldsymbol{R}$-split torus [i. e., isomorphic to $(\boldsymbol{R}^\times)^r$]. We note that one has $H_\kappa \in (\mathfrak{k}_\kappa^*)^a \subset \mathfrak{k}^{[2]}$ and $X_\kappa \in (\mathfrak{g}_\kappa^{(2)})^a \subset \mathfrak{p}_\kappa^{(2)}$.

Next, we consider the representations of $G_\kappa^{(i)}$ $(i=1, 2)$ on $U = U_\kappa$ obtained by restricting the adjoint representation, which we denote by ad_U. First, since $\frac{1}{2}H_\kappa$ is an H-element of $\mathfrak{g}_\kappa^* = \mathfrak{k}_\kappa^* + \mathfrak{p}^{[2]}$, we have

$$U_C = \mathfrak{g}_C(ad\, X_\kappa\,; 2) = c_\kappa^{-1}(\mathfrak{g}_C(ad\, H_\kappa\,; 2i)) \\ = c_\kappa^{-1}(\mathfrak{p}_+^{[2]}), \tag{2. 7}$$

where $\mathfrak{p}_+^{[2]} = \mathfrak{p}_+ \cap \mathfrak{p}_C^{[2]}$. In view of (2. 2) one has $\theta = -\gamma$ on U_C. Hence, from $\theta X_\kappa = -X_\kappa$ one has

$$\gamma U_\kappa = \theta U_\kappa = \mathfrak{g}(ad\, X_\kappa\,; -2). \tag{2. 8}$$

We put $e = e_\kappa = \kappa(e_+)$. Then, from $e_+ = \frac{1}{2}(h' + h'')$ and $[h, e_+] = 2e_+$, one has

$$e_\kappa = \frac{1}{2}(Y_\kappa + H_\kappa) \in U_\kappa. \tag{2. 9}$$

Theorem 2. 3. *The representation $ad_U|\mathfrak{g}_\kappa^{(1)}$ is trivial and $ad_U|\mathfrak{g}_\kappa^{(2)}$ is faithful. The orbit $\Omega_\kappa = (G_\kappa^{(2)\circ})e_\kappa$ is a self-dual homogeneous cone in U_κ, and $ad_U(G_\kappa^{(2)\circ})$ coincides with the connected component of the automorphism group $G(\Omega_\kappa)$.*

Proof. $ad_U|\mathfrak{g}_\kappa^{(1)}$ is trivial, because $U \subset \mathfrak{g}_\kappa^*$ and $[\mathfrak{g}_\kappa^{(1)}, \mathfrak{g}_\kappa^*] = 0$. On the other hand, $ad_U|\mathfrak{g}_\kappa^{(2)}$ is faithful, because $\mathfrak{g}_\kappa^*$ is semi-simple, of hermitian type and so $(\mathfrak{k}_\kappa^*)_C (= c_\kappa(\mathfrak{g}_\kappa^{(2)})_C)$ acts faithfully on $\mathfrak{p}_+^{[2]} (= c_\kappa(U_C))$. To prove the last statement, we apply Lemmas I. 8. 3 and I. 8. 6 to $G_1 = ad_U(G_\kappa^{(2)\circ})$. We define a (positive definite) inner product $\langle\ \rangle$ on U by

$$\langle u, u' \rangle = -B(u, \theta u') \qquad (u, u' \in U_\kappa), \tag{2. 10}$$

where B is the Killing form of $\mathfrak{g}$. [$\langle\ \rangle$ is positive definite, because θ is a Cartan involution. We note that, in view of (2. 2), one can rewrite (2. 10) as $\langle u, u' \rangle = B(u, \gamma u')$, which shows that, if κ is defined over $F \subset \boldsymbol{R}$, so is $\langle\ \rangle$.] For $g \in G_\kappa^{(2)}$, one has

$$\langle gu, u' \rangle = -B(gu, \theta u') = -B(u, g^{-1}\theta u') \\ = \langle u, \theta g^{-1} \theta u' \rangle,$$

i. e.,

$$ad_U(g^\theta) = {}^t(ad_U\, g)^{-1}. \tag{2. 11}$$

Thus G_1 is self-adjoint. Hence it remains to check the conditions (ii), (iii) of Lemma I. 8. 6. We show that, for $X \in \mathfrak{g}_\kappa^{(2)}$, one has

$$(ad\, X)e_\kappa = 0 \Longleftrightarrow \theta X = X.$$

In fact, if $\theta X = X$, then by (2. 6) one has $X \in \mathfrak{l}_1 \subset \mathfrak{g}^{[0]} = \mathfrak{c}(\kappa(\mathfrak{g}^1))$ and so $[X, e_\kappa] = 0$. To prove the converse, it is enough to show that $[X, e_\kappa] = 0$ and $\theta X = -X$ imply $X = 0$. Applying θ to the first equation, which says $[X, Y_\kappa + H_\kappa] = 0$, one obtains $[-X, -Y_\kappa + H_\kappa] = 0$. These two equations imply $[X, Y_\kappa] = [X, H_\kappa] = 0$ and so $X \in \mathfrak{g}^{[0]}$. But, since $X \in \mathfrak{p}_\kappa^{(2)} \subset \mathfrak{g}^{[2]}$, one obtains $X = 0$. This proves the condition (ii). To prove the condition (iii), it suffices to show that $\dim \mathfrak{p}_\kappa^{(2)} = \dim U$. By (2. 6), (2. 3) and Lemma 2. 1, 2) one has

$$\begin{aligned}\dim \mathfrak{p}_\kappa^{(2)} &= \dim \mathfrak{k}^{[2]} = \dim \mathfrak{k}^{[\text{even}]}(\gamma; -1) \\ &= \frac{1}{2} \dim \mathfrak{g}^{[\text{even}]}(\gamma; -1) \\ &= \frac{1}{2} \dim \mathfrak{g}(ad\, Y_\kappa; \pm 2) \\ &= \dim \mathfrak{g}(ad\, Y_\kappa; 2) = \dim U_\kappa,\end{aligned}$$

which completes the proof, q. e. d.

We note that, since $ad\, X_\kappa = 2$ on U, $\exp(\lambda X_\kappa)$ $(\lambda \in \boldsymbol{R})$ gives a dilation $u \mapsto e^{2\lambda} u$ of the cone Ω_κ. From this, without using Lemma I. 8. 6, it is possible to give another proof of the last statement of Theorem 2. 3, by showing directly the convexity of Ω_κ (cf. Exerc. 2).

Exercises

1. For the irreducible symmetric domain of type $(I_{p,q})$ $(p \geq q)$, verify the following relations.

$$\mathfrak{g} = \mathfrak{su}(p, q) = \left\{ X = \begin{pmatrix} X_1 & X_{12} \\ {}^t\bar{X}_{12} & X_2 \end{pmatrix} \middle| X_i = -{}^t\bar{X}_i\, (i=1, 2),\ \mathrm{tr}(X_1 + X_2) = 0 \right\}.$$

For the standard Cartan involution $\theta : X \mapsto -{}^t\bar{X}$, one has

$$\mathfrak{k} = \left\{ \begin{pmatrix} X_1 & 0 \\ 0 & X_2 \end{pmatrix} \right\}, \quad \mathfrak{p} = \left\{ \begin{pmatrix} 0 & X_{12} \\ {}^t\bar{X}_{12} & 0 \end{pmatrix} \right\},$$

$$H_0 = \begin{pmatrix} \dfrac{iq}{p+q} 1_p & 0 \\ 0 & -\dfrac{ip}{p+q} 1_q \end{pmatrix} = \frac{i}{2} \begin{pmatrix} 1_p & 0 \\ 0 & -1_q \end{pmatrix} - \frac{i}{2} \frac{p-q}{p+q} 1_{p+q},$$

$$\mathfrak{p}_+ = \left\{ \begin{pmatrix} 0 & z \\ 0 & 0 \end{pmatrix} \middle| z \in \mathscr{M}_{p,q}(\boldsymbol{C}) \right\},$$

$$\mathscr{D} = \{ z \in \mathscr{M}_{p,q}(\boldsymbol{C}) \mid 1_q - {}^t\bar{z}z \gg 0 \}.$$

For $1 \leq k \leq q (= \mathrm{rank}\, \mathfrak{g})$, define an (H_1)-homomorphism $\kappa (= \kappa^{(k)}) : \mathfrak{g}^1 \to \mathfrak{g}$ by

(2.12)

$$X_\kappa = \begin{bmatrix} \overbrace{0}^{p-k} & \overbrace{0}^{k} & \overbrace{0}^{q-k} & \overbrace{0}^{k} \\ 0 & 0 & 0 & 1_k \\ 0 & 0 & 0 & 0 \\ 0 & 1_k & 0 & 0 \end{bmatrix}, \quad Y_\kappa = \begin{bmatrix} 0 & 0 & 0 & 0 \\ 0 & 0 & 0 & -i1_k \\ 0 & 0 & 0 & 0 \\ 0 & i1_k & 0 & 0 \end{bmatrix}, \quad H_\kappa = \begin{bmatrix} 0 & 0 & 0 & 0 \\ 0 & i1_k & 0 & 0 \\ 0 & 0 & 0 & 0 \\ 0 & 0 & 0 & -i1_k \end{bmatrix}.$$

Then one has

$$e_\kappa = \begin{bmatrix} 0 & 0 & 0 & 0 \\ 0 & \frac{i}{2}1_k & 0 & -\frac{i}{2}1_k \\ 0 & 0 & 0 & 0 \\ 0 & \frac{i}{2}1_k & 0 & -\frac{i}{2}1_k \end{bmatrix}, \quad o_\kappa = \begin{bmatrix} 0 & 0 & 0 & 0 \\ 0 & 0 & 0 & 1_k \\ 0 & 0 & 0 & 0 \\ 0 & 0 & 0 & 0 \end{bmatrix} \quad \text{(see § 4)},$$

$$c_\kappa\,(\text{in } SL(p+q, \boldsymbol{C})) = \begin{bmatrix} 1_{p-k} & 0 & 0 & 0 \\ 0 & \frac{1}{\sqrt{2}}1_k & 0 & \frac{1}{\sqrt{2}}1_k \\ 0 & 0 & 1_{q-k} & 0 \\ 0 & -\frac{1}{\sqrt{2}}1_k & 0 & \frac{1}{\sqrt{2}}1_k \end{bmatrix},$$

$$\mathfrak{g}(X_\kappa; 2) = \left\{ \begin{bmatrix} 0 & 0 & 0 & 0 \\ 0 & \frac{i}{2}u & 0 & -\frac{i}{2}u \\ 0 & 0 & 0 & 0 \\ 0 & \frac{i}{2}u & 0 & -\frac{i}{2}u \end{bmatrix} \middle| u \in \mathcal{H}_k(\boldsymbol{C}) \right\},$$

$$\mathfrak{g}(X_\kappa; 1) = \left\{ \begin{bmatrix} 0 & v_1 & 0 & -v_1 \\ -{}^t\bar{v}_1 & 0 & {}^t\bar{v}_2 & 0 \\ 0 & v_2 & 0 & -v_2 \\ -{}^t\bar{v}_1 & 0 & {}^t\bar{v}_2 & 0 \end{bmatrix} \middle| v_1 \in \mathcal{M}_{p-k,k}(\boldsymbol{C}),\ v_2 \in \mathcal{M}_{q-k,k}(\boldsymbol{C}) \right\},$$

$$\mathfrak{c}(X_\kappa) = \mathfrak{g}_\kappa^{(1)} \oplus \mathfrak{g}_\kappa^{(2)},$$

where

$$\mathfrak{g}_\kappa^{(1)} = \left\{ \begin{bmatrix} b_1 & 0 & b_{12} & 0 \\ 0 & 0 & 0 & 0 \\ {}^t\bar{b}_{12} & 0 & b_2 & 0 \\ 0 & 0 & 0 & 0 \end{bmatrix} - \frac{1}{p+q}\operatorname{tr}(b_1+b_2)1_{p+q} \middle| b_i = -{}^t\bar{b}_i\ (i=1,2) \right\},$$

$$\mathfrak{l}_2 = \begin{cases} \left\{ \operatorname{diag}(\lambda 1_{p-k}, \mu 1_k, \lambda 1_{q-k}, \mu 1_k) \middle| \lambda \in i\boldsymbol{R},\ \mu = \left(1 - \frac{p+q}{2k}\right)\lambda \right\} & (1 \leq k < q), \\ \mathfrak{g}_\kappa^{(1)} & (k = q < p), \\ \{0\} & (k = q = p), \end{cases}$$

$$H_0^{(1)} = H_0 - \frac{1}{2}H_\kappa = \operatorname{diag}\left(\frac{i}{2}1_{p-k}, 0, -\frac{i}{2}1_{q-k}, 0\right) - \frac{i}{2}\frac{p-q}{p+q}1_{p+q},$$

$$\mathfrak{g}_\kappa^{(2)} = \left\{ \begin{bmatrix} 0 & 0 & 0 & 0 \\ 0 & a_1 & 0 & a_2 \\ 0 & 0 & 0 & 0 \\ 0 & a_2 & 0 & a_1 \end{bmatrix} \middle| a_1 = -{}^t\bar{a}_1,\ a_2 = {}^t\bar{a}_2,\ \operatorname{tr} a_1 = 0 \right\}.$$

For simplicity, we make the following identifications :

$$U = \mathcal{H}_k(\boldsymbol{C}), \qquad V = \mathcal{M}_{p+q-2k,k}(\boldsymbol{C}),$$

$$\mathfrak{g}_\kappa^{(1)} = \left\{b=\begin{pmatrix} b_1 & b_{12} \\ {}^t\bar{b}_{12} & b_2 \end{pmatrix}\right\} = \mathfrak{u}(p-k, q-k), \tag{2.13}$$
$$\mathfrak{g}_\kappa^{(2)} = \{a=a_1+a_2\} = \mathfrak{gl}^0(k, \boldsymbol{C}).$$

Then the representations of $\mathfrak{g}_\kappa^{(i)}$ $(i=1, 2)$ on U and V are given as follows:

$$\mathfrak{g}_\kappa^{(2)} \ni a: \begin{cases} u \longmapsto au+u{}^t\bar{a}, \\ \begin{pmatrix} v_1 \\ v_2 \end{pmatrix} \longmapsto \begin{pmatrix} v_1 \\ v_2 \end{pmatrix}{}^t\bar{a}, \end{cases} \tag{2.14}$$

$$\mathfrak{g}_\kappa^{(1)} \ni b: \begin{cases} u \longmapsto 0, \\ \begin{pmatrix} v_1 \\ v_2 \end{pmatrix} \longmapsto \begin{pmatrix} b_1 & b_{12} \\ {}^t\bar{b}_{12} & b_2 \end{pmatrix}\begin{pmatrix} v_1 \\ v_2 \end{pmatrix}. \end{cases} \tag{2.15}$$

Thus one has $\Omega_\kappa = G_\kappa^{(2)\circ} e_\kappa = \mathcal{P}_\kappa(\boldsymbol{C})$.

2. In the notation of the text, show first that $B|\mathfrak{g}_\kappa^{(2)} \times \mathfrak{g}_\kappa^{(2)}$ is non-degenerate. Then, denote by $\mathfrak{g}_\kappa^{(2)\prime}$ the orthogonal complement of $\{X_\kappa\}_{\boldsymbol{R}}$ in $\mathfrak{g}_\kappa^{(2)}$. Let $X \in \mathfrak{g}_\kappa^{(2)\prime}$, $X \neq 0$, and put $\phi(\lambda) = \langle e, \exp(\lambda\, ad\, X)e\rangle$ $(\lambda \in \boldsymbol{R})$. Show that $\phi'(0)=0$ and $\phi''(0)>0$.

§ 3. The representations of $G_\kappa^{(i)}$ $(i=1, 2)$ on V_κ.

In this section, we study the representations of $G_\kappa^{(i)}$ on $V=V_\kappa$. We shall show that there exist a complex structure I_0 on V and a non-degenerate alternating bilinear form A_e on $V\times V$ such that $A_e I_0 \gg 0$ and

$$ad_V(G_\kappa^{(1)}) \subset Sp(V, A_e), \qquad ad_V(G_\kappa^{(2)}) \subset GL(V, I_0).$$

First we recall that $V_\kappa = \mathfrak{g}(ad\, X_\kappa; 1) \subset \mathfrak{g}^{[1]}$ and

$$\mathfrak{g}^{[1]} = \mathfrak{g}(ad\, X_\kappa; \pm 1) = \mathfrak{g}(ad\, Y_\kappa; \pm 1) = \mathfrak{g}(\gamma^2; -1).$$

By (1. 13) one has $\theta\gamma\theta = \bar{\gamma} = -\gamma$ on $\mathfrak{g}_C^{[1]}$ so that γ interchanges $\mathfrak{k}_C^{[1]}$ and $\mathfrak{p}_C^{[1]}$:

$$\gamma:\ \mathfrak{k}_C^{[1]} \rightleftarrows \mathfrak{p}_C^{[1]}. \tag{3.1}$$

Since $\theta X_\kappa = \gamma X_\kappa = -X_\kappa$, one has

$$\theta V_\kappa = i\gamma V_\kappa = \mathfrak{g}(ad\, X_\kappa; -1). \tag{3.2}$$

We define

$$I_0 = i\gamma\theta = -i\theta\gamma \tag{3.3}$$
$$(= -(ad\, Y_\kappa)\theta = \theta(ad\, Y_\kappa)) \qquad \text{on} \quad V_\kappa.$$

Then, by (3. 2), I_0 is a real endomorphism of V and one has $I_0^2 = -(\gamma\theta)^2 = -1_V$, i. e., I_0 is a complex structure of V. (Actually, the same definition gives a complex structure on $\mathfrak{g}^{[1]}$.) If we put $V_\pm = V_C(I_0; \pm i)$, then from (2. 2) we obtain

$$\begin{cases} V_+ = V_C \cap c_\kappa^{-1}\mathfrak{p}_C, \\ V_- = V_C \cap c_\kappa^{-1}\mathfrak{k}_C. \end{cases} \tag{3.4}$$

We also recall that

$$H_0^{(1)} = H_0 - \frac{1}{2} H_\kappa \tag{3.5}$$

is an H-element of $\mathfrak{g}_\kappa^{(1)} = \mathfrak{k}_\kappa^{(1)} + \mathfrak{p}_\kappa^{(1)}$.

Lemma 3.1. *One has*

$$ad_V(H_0^{(1)}) = \frac{1}{2} I_0. \tag{3.6}$$

In other words, one has

$$ad(H_0^{(1)}) = \begin{cases} \dfrac{i}{2} & \text{on } V_+, \\ -\dfrac{i}{2} & \text{on } V_-. \end{cases} \tag{3.6'}$$

Proof. Let $v \in V_-$ and put $X = c_\kappa v$. Then by (3.4) one has $X \in \mathfrak{k}_C^{[1]}$ and so $[H_0, X] = 0$. On the other hand, from $c_\kappa X_\kappa = -iH_\kappa$, one obtains $X \in c_\kappa V_C = \mathfrak{g}_C(ad\, H_\kappa; i)$, i. e., $[H_\kappa, X] = iX$. Hence by (3.5) one has $[H_0^{(1)}, X] = -\frac{i}{2} X$. Since $c_\kappa H_0^{(1)} = H_0^{(1)}$, this gives the second equation of (3.6'). The first one is then obtained by taking the complex conjugate of the second, q. e. d.

From this Lemma we obtain the following relations:

$$\begin{aligned} V_+ &= c_\kappa^{-1} \mathfrak{p}_+^{[1]}, & V_- &= c_\kappa \mathfrak{p}_-^{[1]}, \\ \theta V_+ &= c_\kappa \mathfrak{p}_+^{[1]}, & \theta V_- &= c_\kappa^{-1} \mathfrak{p}_-^{[1]}, \end{aligned} \tag{3.7}$$

where $\mathfrak{p}_\pm^{[1]} = \mathfrak{p}_C^{[1]} \cap \mathfrak{p}_\pm$. In fact, let $v \in V_+$ and put $X = c_\kappa v$. Then one has $[H_\kappa, X] = iX$ and, from the first equation of (3.6'), $[H_0^{(1)}, X] = \frac{i}{2} X$. Hence $[H_0, X] = iX$, i. e., $X \in \mathfrak{p}_+^{[1]}$. Thus one has $V_+ \subset c_\kappa^{-1} \mathfrak{p}_+^{[1]}$. But, by (3.1), one has

$$\dim \mathfrak{p}_+^{[1]} = \frac{1}{2} \dim \mathfrak{p}_C^{[1]} = \frac{1}{4} \dim \mathfrak{g}_C^{[1]} = \dim V_+.$$

Therefore $V_+ = c_\kappa^{-1} \mathfrak{p}_+^{[1]}$. Taking the complex conjugate, one obtains $V_- = c_\kappa \mathfrak{p}_-^{[1]}$. Finally, applying θ to these equations and using (1.13), one obtains the last two equations.

Summarizing the above result, we obtain the following list:

	$ad\, H_0$	$\frac{1}{2} ad\, H_\kappa$	$ad\, H_0^{(1)}$
$c_\kappa V_+ = \mathfrak{p}_+^{[1]}$	i	$\frac{i}{2}$	$\frac{i}{2}$
$c_\kappa V_- = \gamma \mathfrak{p}_-^{[1]}$	0	$\frac{i}{2}$	$-\frac{i}{2}$
$c_\kappa \theta V_+ = \gamma \mathfrak{p}_+^{[1]}$	0	$-\frac{i}{2}$	$\frac{i}{2}$
$c_\kappa \theta V_- = \mathfrak{p}_-^{[1]}$	$-i$	$-\frac{i}{2}$	$-\frac{i}{2}$

Next, for $u \in U$, $v, v' \in V$, we put

$$A_u(v, v') = -\frac{1}{4} \langle u, [v, v'] \rangle. \tag{3.8}$$

(See (2. 10). We also put $A(v, v') = -\frac{1}{4}[v, v']$.) For each u, A_u is an alternating bilinear form on $V \times V$. For $g \in C_G(X_\kappa)$, we obtain by (2. 11)

$$\begin{aligned} A_{g^\theta u}(v, v') &= -\frac{1}{4}\langle g^\theta u, [v, v']\rangle \\ &= -\frac{1}{4}\langle u, g^{-1}[v, v']\rangle \\ &= A_u(g^{-1}v, g^{-1}v'), \end{aligned}$$

i. e.,

$$A_{(ad_U g^\theta)u} = {}^t(ad_V g)^{-1} A_u (ad_V g)^{-1}. \tag{3. 9}$$

Lemma 3. 2. *The bilinear form $A_u(v, I_0 v')$ $(v, v' \in V)$ is symmetric. It is positive definite for $u \in \Omega$.*

Proof. By (3. 9), since Ω is open (in U) and homogeneous, it is enough to prove these statements for $u = e$. We first prove the relation

$$A_e(v, v') = -\frac{i}{4} B(v, \gamma v'). \tag{3. 10}$$

In fact, by the definitions

$$\begin{aligned} 4A_e(v, v') &= -\langle e, [v, v']\rangle = B(e, \theta[v, v']) \\ &= B(\theta e, [v, v']) = -B(v, [\theta e, v']) \\ &= B(v, [Y_\kappa, v']). \end{aligned}$$

The last equality follows from the equations $e = \frac{1}{2}(Y_\kappa + H_\kappa)$, $\theta e = \frac{1}{2}(-Y_\kappa + H_\kappa)$ and $[e, v'] \in [\mathfrak{g}(ad\, X_\kappa; 2), \mathfrak{g}(ad\, X_\kappa; 1)] = 0$. Since one has $\gamma = i\, ad\, Y_\kappa$ on $\mathfrak{g}_C^{[1]}$, this gives (3. 10). Combining (3. 10) with (3. 3), we obtain

$$A_e(v, I_0 v') = -\frac{1}{4} B(v, \theta v'), \tag{3. 11}$$

which proves that $A_e I_0$ is symmetric and positive definite, q. e. d.

We put

$$S_e = A_e I_0 \quad \text{and} \quad h = S_e + i A_e. \tag{3. 12}$$

Then h is a positive definite hermitian form on the complex vector space (V, I_0). We denote by $*$ the adjoint of $\boldsymbol{C}$-linear transformations of V with respect to h (which is the same thing as the transpose with respect to A_e or S_e). $\mathscr{H}(V, I_0, h)$ (or $\mathscr{H}(V, I_0, A_e)$) will denote the space of all hermitian (i. e., self-adjoint) $\boldsymbol{C}$-linear transformations of (V, I_0) with respect to h, and $\mathcal{P}(V, I_0, h)$ (or $\mathcal{P}(V, I_0, A_e)$) will denote the cone of positive definite transformations in $\mathscr{H}(V, I_0, h)$. For each $u \in U$, we define a linear transformation R_u (written sometimes as $R(u)$) of V by

$$A_u(v, v') = 2A_e(v, R_u v') \ (= 2A_e(R_u v, v')), \tag{3. 13}$$

i. e., $R_u=\frac{1}{2}A_e^{-1}A_u$. Then, from Lemma 3. 2, one has $R_u\in\mathcal{H}(V, I_0, h)$ for all $u\in U$ and $R_u\in\mathcal{P}(V, I_0, h)$ for all $u\in\Omega$. In particular, one has $R_e=\frac{1}{2}1_V$.

Theorem 3. 3. *The representation $ad_V|G_\kappa^{(1)}$ gives a homomorphism*

$$G_\kappa^{(1)}\longrightarrow Sp(V, A_e) \tag{3. 14}$$

satisfying the condition (H_2) *with respect to the H-elements $H_0^{(1)}(=H_0-\frac{1}{2}H_\kappa)$ and $\frac{1}{2}I_0$, i. e., satisfying the relation* (3. 6). *The representation $ad_V|G_\kappa^{(2)}$ gives a homomorphism*

$$G_\kappa^{(2)}\longrightarrow GL(V, I_0) \tag{3. 15}$$

with which the linear map

$$\Omega_\kappa\ni u\longmapsto R_u\in\mathcal{P}(V, I_0, h) \tag{3. 16}$$

is (strongly) equivariant, i. e., the following conditions are satisfied:

$$ad_V(g^\theta)=(ad_V g)^{*-1}, \tag{3. 17}$$

$$R_{(ad_U g)u}=(ad_V g)R_u(ad_V g)^* \qquad (g\in G_\kappa^{(2)}). \tag{3. 18}$$

Proof. Since $ad_U g=1$ for $g\in G_\kappa^{(1)}$, the relation (3. 9) for $u=e$ implies that $ad_V(G_\kappa^{(1)})\subset Sp(V, A_e)$. [Actually, in the notation of § 6, one has $ad_V(G_\kappa^{(1)})\subset Sp(V, A)$.] The relation (3. 6) was already proved in Lemma 3. 1. Next, from (3. 6) and $[H_0^{(1)}, \mathfrak{g}_\kappa^{(2)}]=0$, one obtains $[I_0, ad_V(\mathfrak{g}_\kappa^{(2)})]=0$. Hence one has $ad_V(G_\kappa^{(2)})\subset GL(V, I_0)$. To prove the equivariance of the maps (3. 15), (3. 16), we first note that (3. 17) follows from (3. 11). Then from (3. 9) one has

$$A_{(ad_U g)u}={}^t(ad_V g)^*A_u(ad_V g)^* \tag{3. 19}$$

which implies (3. 18), q. e. d.

By Theorem 2. 3 (and its proof) there exists, for each $u\in U$, a unique element X_u in $\mathfrak{p}_\kappa^{(2)}$ such that $ad_U X_u=T_u$, T_u denoting the multiplication $u'\mapsto uu'$ in the Jordan algebra (U, e) (I, § 8). Then, by (3. 17) and (3. 18) one has

$$R_{T_u u'}=(ad_V X_u)R_{u'}+R_{u'}(ad_V X_u).$$

Hence, putting $u'=e$, one has $R_u=ad_V X_u$ and so

$$R_{uu'}=R_uR_{u'}+R_{u'}R_u \qquad (u, u'\in U). \tag{3. 20}$$

This shows that the map $u\mapsto 2R_u$ is a (unital) Jordan algebra homomorphism $(U, e)\to(\mathcal{H}(V, I_0, h), 1_V)$.

Remark. As we shall show later on (§ 4), the representations ad_U, ad_V of $\mathfrak{c}(X_\kappa)$ are irreducible over $\boldsymbol{R}$, when $\mathfrak{g}$ is simple. More generally, if $\mathfrak{g}$ is defined and simple over a field $F\subset\boldsymbol{R}$, these representations are also defined and irreducible over F. The symplectic representations satisfying the condition (H_2) will be discussed in detail (full reducibility, classification, etc.) in Ch. IV. A similar result for equivariant linear maps of self-dual homogeneous cones will be given in V, § 5.

Now, by (2. 7) and (3. 7) one has

$$c_\kappa^{-1}(\mathfrak{p}_+) = U_{\kappa,C} + V_{\kappa,+} + \mathfrak{p}_{\kappa,+}, \tag{3.21}$$

where $\mathfrak{p}_{\kappa,+} = \mathfrak{p}_+^{[0]} = \mathfrak{p}_+ \cap \mathfrak{g}_{\kappa,C}$.

Proposition 3. 4. *The space $c_\kappa^{-1}(\mathfrak{p}_+)$ is closed under the triple product*

$$\{X, Y, Z\} = -\frac{1}{2}[[X, \theta\bar{Y}], Z], \tag{3.22}$$

and c_κ gives a JTS isomorphism $c_\kappa^{-1}(\mathfrak{p}_+) \to \mathfrak{p}_+$. Moreover, $e_\kappa = \kappa(e_+)$ is an "idempotent" of $c_\kappa^{-1}(\mathfrak{p}_+)$, i. e.,

$$\{e_\kappa, e_\kappa, e_\kappa\} = e_\kappa. \tag{3.23}$$

Proof. We know that $\mathfrak{p}_+$ is a hermitian JTS with the triple product (II. 3. 12) (which is essentially identical with (3. 22)). Let $X, Y, Z \in c_\kappa^{-1}(\mathfrak{p}_+)$ and define $\{X, Y, Z\}$ by (3. 22). Then one has by (1. 13)

$$\begin{aligned} c_\kappa\{X, Y, Z\} &= -\frac{1}{2}[[c_\kappa X, c_\kappa\theta\bar{Y}], c_\kappa Z] \\ &= -\frac{1}{2}[[c_\kappa X, \theta\overline{c_\kappa Y}], c_\kappa Z] \\ &= \frac{1}{2}[[c_\kappa X, \overline{c_\kappa Y}], c_\kappa Z] = \{c_\kappa X, c_\kappa Y, c_\kappa Z\}, \end{aligned}$$

which proves the first half of the Proposition. Next, from (2. 9) one obtains

$$[e_\kappa, \theta e_\kappa] = -X_\kappa, \tag{3.24}$$

and so

$$e_\kappa \square e_\kappa = \frac{1}{2} ad\, X_\kappa|_{c_\kappa^{-1}(\mathfrak{p}_+)}. \tag{3.24'}$$

Since $e_\kappa \in U_\kappa$, this implies (3. 23), q. e. d.

Remark. (3. 24′) shows that the decomposition (3. 21) is nothing but the "Peirce decomposition" of $c_\kappa^{-1}(\mathfrak{p}_+)$ with respect to the idempotent e_κ (cf. V, § 6). A large part of the results in this chapter can be reproduced from this view point. For such an approach to Wolf-Korányi theory, see Loos [2], [9].

Proposition 3. 5. *One has*

$$c_\kappa^{-1}(\mathfrak{k}_C) = V_{\kappa,-} + (\mathfrak{k}_{\kappa,C}^{(1)} \oplus \mathfrak{g}_{\kappa,C}^{(2)}) + \theta V_{\kappa,+}, \tag{3.25}$$

$$c_\kappa^{-1}(\mathfrak{p}_-) = \theta U_{\kappa,C} + \theta V_{\kappa,-} + \mathfrak{p}_{\kappa,-}. \tag{3.21'}$$

Proof. (3. 21′) is obtained from (3. 21) by applying the Cartan involution $x \mapsto \theta\bar{x}$ and using (1. 13). Next, since the action of c_κ on $\mathfrak{g}_{\kappa,C}^{(1)}$ is trivial and $c_\kappa(\mathfrak{g}_{\kappa,C}^{(2)}) = \mathfrak{k}_{\kappa,C}^*$ by the definition (§ 2), one has by (1. 17)

$$\mathfrak{k}_{\kappa,C}^{(1)} \oplus \mathfrak{g}_{\kappa,C}^{(2)} = c_\kappa^{-1}(\mathfrak{k}_C^{[\mathrm{even}]}).$$

On the other hand, by (3. 7) and (3. 1) one has

$$V_{\kappa,-}+\theta V_{\kappa,+} = c_\kappa(\mathfrak{p}_C^{[1]}) = c_\kappa^{-1}(\mathfrak{k}_C^{[1]}).$$

From these one obtains (3. 25), q. e. d.

Corollary 3. 6. *One has*

(3. 26) $$[U, \theta U]\subset \mathfrak{g}^{(2)},$$

(3. 27) $$[U_C, \theta V_\pm]\subset V_\pm,$$

(3. 28) $$[V_+, \theta V_+]\subset \mathfrak{p}_+^{(1)},$$

(3. 29) $$[V_+, \theta V_-]\subset \mathfrak{k}_C^{(1)}\oplus \mathfrak{g}_C^{(2)},$$

(3. 30) $$[\mathfrak{k}_C^{(1)}\oplus \mathfrak{g}_C^{(2)}, V_+]\subset V_+,$$

(3. 31) $$[\mathfrak{p}_+^{(1)}, V_+] = \{0\}, \quad [\mathfrak{p}_-^{(1)}, V_+]\subset V_-.$$

Proof. From $U_C, \theta U_C \subset c_\kappa^{-1}(\mathfrak{p}_C^{[2]})$, one has

$$[U_C, \theta U_C]\subset c_\kappa^{-1}(\mathfrak{k}_{\kappa,C}^{*}) = \mathfrak{g}_C^{(2)},$$

whence follows (3. 26). All other relations are immediate consequences of the comparison of the following two gradations of $\mathfrak{g}_C$:

(3. 32) $$\begin{aligned}\mathfrak{g}_C &= U_C+(V_++V_-)+(\mathfrak{g}_C^{(1)}+\mathfrak{g}_C^{(2)})+(\theta V_++\theta V_-)+\theta U_C\\ &= (U_C+V_++\mathfrak{p}_+^{(1)})+(V_-+\mathfrak{k}_C^{(1)}+\mathfrak{g}_C^{(2)}+\theta V_+)+(\mathfrak{p}_-^{(1)}+\theta V_-+\theta U_C).\end{aligned}$$

For instance, V_+ (resp. θV_+) is in the second and first (resp. the fourth and second) terms in these gradations, so that $[V_+, \theta V_+]$ is contained in the third and first terms, which proves (3. 28), q. e. d.

Exercises

1. In the notation of § 2, Exerc. 1, show that

(3. 33) $$I_0\begin{pmatrix} v_1 \\ v_2\end{pmatrix} = \begin{pmatrix} iv_1 \\ -iv_2\end{pmatrix},$$

(3. 34) $$H\left(\begin{pmatrix} v_1 \\ v_2\end{pmatrix}, \begin{pmatrix} v_1' \\ v_2'\end{pmatrix}\right) = {}^t\bar{v}_1' v_1' + {}^t\bar{v}_2' v_2,$$

(3. 35) $$R_u\begin{pmatrix} v_1 \\ v_2\end{pmatrix} = \frac{1}{2}\begin{pmatrix} v_1 u \\ v_2 {}^t\bar{u}\end{pmatrix} \qquad (u\in U_C=\mathfrak{gl}_k(\boldsymbol{C})),$$

where we set $H(v, v')=A(v, I_0 v')+iA(v, v')$, $A(v, v')=-\frac{1}{4}[v, v']$ for $v, v'\in V$.

2. *The case of* (III_n). Let (V_1, A_1) be a real symplectic space of dimension $2n$ and fix a symplectic basis $(e_1, \cdots, e_{2n})$ of V. For an integer k, $1\leq k\leq n$, let $\kappa=\kappa^{(k)}$ be a homomorphism $\mathfrak{g}^1=\mathfrak{sl}(2, \boldsymbol{R})\to \mathfrak{sp}(2n, \boldsymbol{R})$ defined by

$$X_\kappa = \begin{bmatrix} 0 & 0 & 0 & 0 \\ 0 & 1_k & 0 & 0 \\ 0 & 0 & 0 & 0 \\ 0 & 0 & 0 & -1_k \end{bmatrix} \begin{matrix} \}n-k \\ \}k \\ \}n-k \\ \}k \end{matrix}, \qquad Y_\kappa = \begin{bmatrix} 0 & 0 & 0 & 0 \\ 0 & 0 & 0 & 1_k \\ 0 & 0 & 0 & 0 \\ 0 & 1_k & 0 & 0 \end{bmatrix}, \tag{3.36}$$

$$H_\kappa = \begin{bmatrix} 0 & 0 & 0 & 0 \\ 0 & 0 & 0 & 1_k \\ 0 & 0 & 0 & 0 \\ 0 & -1_k & 0 & 0 \end{bmatrix}.$$

Then κ satisfies the condition (H_1) with respect to

$$H_0 = \frac{1}{2}\begin{pmatrix} 0 & 1_n \\ -1_n & 0 \end{pmatrix}.$$

Show that the corresponding parabolic subalgebra $\mathfrak{b}_\kappa$ consists of the matrices of the form

$$X = \begin{bmatrix} x_1^{(1)} & 0 & x_{12}^{(1)} & v_1 \\ -{}^t v_2 & x^{(2)} & {}^t v_1 & u \\ x_{21}^{(1)} & 0 & x_2^{(1)} & v_2 \\ 0 & 0 & 0 & -{}^t x^{(2)} \end{bmatrix} \begin{matrix} \}n-k \\ \}k \\ \}n-k \\ \}k \end{matrix}, \tag{3.37}$$

where

$$x^{(1)} = \begin{pmatrix} x_1^{(1)} & x_{12}^{(1)} \\ x_{21}^{(1)} & x_2^{(1)} \end{pmatrix} \in \mathfrak{sp}(2(n-k), \boldsymbol{R}), \qquad x^{(2)} \in \mathfrak{gl}(k, \boldsymbol{R}),$$

$$v_1, v_2 \in \mathcal{M}_{n-k,k}(\boldsymbol{R}), \qquad u \in \mathrm{Sym}_k(\boldsymbol{R}).$$

Check Theorems 2. 3, 3. 3 for this case. In particular, show that

$$I_0\begin{pmatrix} v_1 \\ v_2 \end{pmatrix} = \begin{pmatrix} v_2 \\ -v_1 \end{pmatrix}, \tag{3.38}$$

$$A\left(\begin{pmatrix} v_1 \\ v_2 \end{pmatrix}, \begin{pmatrix} v_1' \\ v_2' \end{pmatrix}\right) = \frac{1}{4}({}^t v_2 v_1' - {}^t v_1 v_2' + {}^t v_1' v_2 - {}^t v_2' v_1), \tag{3.39}$$

$$R_u\begin{pmatrix} v_1 \\ v_2 \end{pmatrix} = \frac{1}{2}\begin{pmatrix} v_1 u \\ v_2 u \end{pmatrix}. \tag{3.40}$$

3. In the notation of §§ 2, 3, prove the following relations :

$$[u, \theta u'] = -2(X_{uu'} + [X_u, X_{u'}]), \tag{3.41}$$

$$[u, \theta v] = -2I_0 R_u v, \tag{3.42}$$

$$\begin{cases} [v, \theta v']^{\mathfrak{k}} = \dfrac{1}{2}([v, \theta v'] + [\theta v, v']), \\ [v, \theta v']^{\mathfrak{p}^{(1)}} = \dfrac{1}{2}([v, \theta v'] - [I_0 v, \theta I_0 v']), \\ [v, \theta v']^{\mathfrak{p}^{(2)}} = X_{[v, I_0 v']} \qquad (u, u' \in U,\ v, v' \in V), \end{cases} \tag{3.43}$$

where X_u is the unique element in $\mathfrak{p}^{(2)}$ such that $ad_U X_u = T_u$, $ad_V X_u = R_u$, and $[\cdots]^{\mathfrak{k}}$ (resp. $[\cdots]^{\mathfrak{p}^{(i)}}$) denotes the $\mathfrak{k}$- (resp. $\mathfrak{p}^{(i)}$-) part of $[\cdots]$.

Hint. To prove (3. 41) and (3. 42), first establish the special cases where u or u' is equal to e. To prove (3. 43), observe that $\gamma\theta = \theta\gamma = 1$ on $\mathfrak{k}^{(1)} \oplus \mathfrak{g}^{(2)}$ and $= -1$ on $\mathfrak{p}^{(1)}$, so that one has

$$[v, \theta v']^{\mathfrak{p}^{(1)}} = \frac{1}{2}([v, \theta v'] - \gamma\theta[v, \theta v']).$$

4. Prove the following relations :

$$[[v, \theta v'], u] = -8 \operatorname{Re} H(v, R_u v'), \tag{3.44}$$

(3.45)
$$[[v,\theta v'],v''] = P(v,v')v''+P(v',v'')v-P(v,v'')v' \\ -4(R_{H(v',v)}v''+R_{H(v',v'')}v-R_{H(v,v'')}v'),$$

where we set

$$P(v,v') = ad_V([v,\theta v']^{\mathfrak{p}^{(1)}}) = \frac{1}{2}ad_V([v,\theta v']-[\theta v,v'])-R_{[v,I_0v']}.$$

Remark. In view of (3.28)~(3.31), one has

$$(P(v,v')v'')_+ = [[v_+,\theta v'_+],v''_-],$$

which shows that $P(v,v')v''$ is $\boldsymbol{C}$-linear in v, v' and $\boldsymbol{C}$-antilinear in v'' (with respect to the complex structure I_0). From (3.45) and (3.43) one has

$$\begin{cases} ad_V([v,\theta v']^{\mathfrak{k}}) = P^{(1)}(v,v')-P^{(1)}(v',v)+4I_0R_{A(v,v')}, \\ ad_V([v,\theta v']^{\mathfrak{p}}) = P(v,v')-4R_{A(v,I_0v')}, \end{cases}$$

where we set

$$P^{(1)}(v,v')v'' = P(v',v'')v+4R_{H(v,v'')}v'.$$

$P^{(1)}(v,v')v''$ is $\boldsymbol{C}$-linear in v', v'' and $\boldsymbol{C}$-antilinear in v.

§ 4. The root structure of $\mathfrak{g}$ and the determination of $ad_V|G_\kappa^{(i)}$.

We shall first determine the $\boldsymbol{R}$-roots of $\mathfrak{g}$ by using the orthogonal decomposition of the homomorphisms κ and then give an explicit determination of the representations ad_V of $G_\kappa^{(i)}$ $(i=1,2)$. We start with some supplementary observations on the Cayley transformations c_κ.

We denote by c_1 the Cayley transformation in $\mathrm{Ad}(\mathfrak{sl}_2(\boldsymbol{C}))$ corresponding to the identical isomorphism $\mathfrak{g}^1\to\mathfrak{g}^1$, i. e., the inner automorphism of $\mathfrak{g}^1_{\boldsymbol{C}}$ given by

(4.1)
$$c_1 = ad\left(\exp\left(\frac{\pi i}{4}h'\right)\right) = ad\left(\frac{1}{\sqrt{2}}\begin{pmatrix}1 & i\\ i & 1\end{pmatrix}\right).$$

Then one has

(4.2)
$$c_1 : \begin{cases} ih \longrightarrow h'' \\ h' \longrightarrow h' \\ ih'' \longrightarrow h. \end{cases}$$

Thus c_1 induces an isomorphism $\mathfrak{su}(1,1)\to\mathfrak{g}^1=\mathfrak{sl}_2(\boldsymbol{R})$. We also consider another inner automorphism of $\mathfrak{g}^1_{\boldsymbol{C}}$:

(4.3)
$$ad\left(\exp\left(\frac{\pi i}{4}h\right)\right) = ad\left(\begin{pmatrix}i & 0\\ 0 & -i\end{pmatrix}\right) : \begin{cases} ih \longrightarrow ih \\ h' \longrightarrow ih'' \\ ih'' \longrightarrow -h', \end{cases}$$

which leaves $\mathfrak{su}(1,1)$ stable. Then it is clear that $c_1\circ ad\left(\exp\left(\frac{\pi i}{4}h\right)\right)$ gives an isomorphism $\mathfrak{su}(1,1)\to\mathfrak{sl}_2(\boldsymbol{R})$ satisfying (H_2) for H-elements $\frac{i}{2}h$ and $\frac{1}{2}h''$. Therefore, to each (non-trivial) (H_1)-homomorphism $\kappa : (\mathfrak{g}^1, \frac{1}{2}h'')\to(\mathfrak{g}, H_0)$, we can associate an (H_1)-homomorphism $\tilde{\kappa} : (\mathfrak{su}(1,1), \frac{i}{2}h)\to(\mathfrak{g}, H_0)$ by the relation

$$\tilde{\kappa} = \kappa \circ c_1 \circ ad\left(\exp\left(\frac{\pi i}{4} h\right)\right) = c_\kappa \circ \exp\left(\frac{\pi i}{4} X_\kappa\right) \circ \kappa. \tag{4.4}$$

The situation is illustrated by the following commutative diagram:

$$\begin{array}{ccc}
\mathfrak{su}(1,1) = \{h', -ih'', ih\} & \xrightarrow{\ \kappa\ } & \{Y_\kappa, -iH_\kappa, iX_\kappa\} \subset c_\kappa^{-1}\mathfrak{g} \\
\Big\downarrow {\scriptstyle ad\left(\exp\left(\frac{\pi i}{4} h\right)\right)} & \searrow^{\tilde{\kappa}} & \Big\downarrow {\scriptstyle \exp\left(\frac{\pi i}{4} X_\kappa\right)} \\
\mathfrak{su}(1,1) = \{ih'', h', ih\} & & \{iH_\kappa, Y_\kappa, iX_\kappa\} \subset c_\kappa^{-1}\mathfrak{g} \\
\Big\downarrow {\scriptstyle c_1 = ad\left(\exp\left(\frac{\pi i}{4} h'\right)\right)} & & \Big\downarrow {\scriptstyle c_\kappa = \exp\left(\frac{\pi i}{4} Y_\kappa\right)} \\
\mathfrak{sl}_2(\boldsymbol{R}) = \{h, h', h''\} & \xrightarrow{\ \kappa\ } & \{X_\kappa, Y_\kappa, H_\kappa\} \subset \mathfrak{g},
\end{array} \tag{4.5}$$

where the horizontal arrows are induced by $\kappa : \mathfrak{sl}_2(\boldsymbol{C}) \to \mathfrak{g}_C$.

Let $\mathfrak{g}_C^1 = \mathfrak{p}_+^1 + \mathfrak{k}_C^1 + \mathfrak{p}_-^1$ be the canonical decomposition of $\mathfrak{g}_C^1 = \mathfrak{sl}_2(\boldsymbol{C})$ corresponding to the H-element $H_0^1 = \frac{1}{2}h''$. Then the one corresponding to $\frac{i}{2}h$ is given by

$$\begin{aligned} \mathfrak{g}_C^1 &= c_1^{-1}\mathfrak{p}_+^1 + c_1^{-1}\mathfrak{k}_C^1 + c_1^{-1}\mathfrak{p}_-^1, \\ c_1^{-1}\mathfrak{k}_C^1 &= \{h\}_C, \qquad c_1^{-1}\mathfrak{p}_\pm^1 = \{e_\pm\}_C. \end{aligned} \tag{4.6}$$

We note that $ad\left(\exp\left(\frac{\pi i}{4} h\right)\right)\Big|_{c_1^{-1}\mathfrak{p}^1_+} = i$ $\left(\text{resp. } \exp\left(\frac{\pi i}{4} X_\kappa\right)\Big|_{c_\kappa^{-1}\mathfrak{p}_+}\right)$ is a JTS automorphism of $c_1^{-1}\mathfrak{p}_+^1$ (resp. $c_\kappa^{-1}\mathfrak{p}_+$).

Proposition 4.1. *Let $(\mathfrak{g}, H_0)$ be a semi-simple Lie algebra of hermitian type. Then there exists a one-to-one correspondence between* (H_1)*-homomorphisms $\kappa : (\mathfrak{g}^1, H_0^1) \to (\mathfrak{g}, H_0)$ and the idempotents z in $\mathfrak{p}_+$ by the relation $z = \tilde{\kappa}(e_+)$.*

Clearly e_+ is an idempotent in $c_1^{-1}\mathfrak{p}_+^1$. Hence to give a $\boldsymbol{C}$-linear JTS homomorphism $c_1^{-1}\mathfrak{p}_+^1 = \{e_+\}_C \to \mathfrak{p}_+$ is equivalent to giving an idempotent z in $\mathfrak{p}_+$. Hence the assertion of our Proposition follows from Proposition II. 8. 2 applied to the homomorphism $\tilde{\kappa} : c_1^{-1}\mathfrak{g}^1 \to \mathfrak{g}$.

The idempotent in $\mathfrak{p}_+$ corresponding to κ will be denoted by o_κ, i. e.,

$$o_\kappa = \tilde{\kappa}(e_+). \tag{4.7}$$

Then by the ($\boldsymbol{C}$-linear extensions of the) homomorphisms in the diagram (4. 5) one has the following correspondence:

$$\begin{array}{lcl}
e_+ \in c_1^{-1}\mathfrak{p}_+^1 \cap \mathfrak{g}^1 & \longrightarrow & e_\kappa = \kappa(e_+) \in U_\kappa \\
\downarrow & & \downarrow \\
ie_+ \in c_1^{-1}\mathfrak{p}_+^1 & \longrightarrow & ie_\kappa \in U_{\kappa, C} = c_\kappa^{-1}\mathfrak{p}_+^{[2]} \\
\downarrow & & \downarrow \\
\frac{1}{2}(h + ih') \in \mathfrak{p}_+^1 & \longrightarrow & o_\kappa = \frac{1}{2}(X_\kappa + iY_\kappa) \in \mathfrak{p}_+^{[2]}.
\end{array}$$

In particular, one has the relation

(4. 8) $$o_\kappa + \bar{o}_\kappa = X_\kappa.$$

Thus o_κ and X_κ determine each other uniquely. We also notice that, if we denote by θ^1 the Cartan involution of $\mathfrak{g}^1$ corresponding to H_0^1, the composite $\kappa\circ\theta^1$ is also an (H_1)-homomorphism $(\mathfrak{g}^1, H_0^1)\to(\mathfrak{g}, H_0)$ for which one has

(4. 9) $$o_{\kappa\circ\theta^1} = -o_\kappa.$$

We insert here a general definition concerning Lie algebra homomorphisms. Let ρ and ρ' be two Lie algebra homomorphisms $\mathfrak{g}\to\mathfrak{g}'$. ρ and ρ' are said to be *commutative* (or orthogonal) if one has

$$[\rho(X), \rho'(Y)] = 0 \qquad \text{for all } X, Y\in\mathfrak{g}.$$

Under this condition, the sum $\rho+\rho'$ defined by

$$(\rho+\rho')(X) = \rho(X)+\rho'(X) \qquad (X\in\mathfrak{g})$$

is also a Lie algebra homomorphism $\mathfrak{g}\to\mathfrak{g}'$, called the *commutative sum* of ρ and ρ'. When $(\mathfrak{g}, H_0)$ and $(\mathfrak{g}', H_0')$ are of hermitian type and ρ and ρ' are (H_1)-homomorphisms into the same Lie algebra of hermitian type, it is clear that the commutative sum $\rho+\rho'$ is also an (H_1)-homomorphism with respect to the same H-elements.

Now, in II, § 4, using a system of strongly orthogonal non-compact roots β_j ($1\leq j\leq r$, $r=\boldsymbol{R}$-rank $\mathfrak{g}$), we constructed a system of mutually commutative (H_1)-homomorphisms $\tilde{\kappa}_j : (\mathfrak{su}(1, 1), \frac{i}{2}h)\to(\mathfrak{g}, H_0)$; the $\tilde{\kappa}_j$ are defined by

(4. 10) $$\tilde{\kappa}_j(ih) = H_j, \qquad \tilde{\kappa}_j(h') = X_j, \qquad \tilde{\kappa}_j(-ih'') = Y_j,$$

where H_j, X_j, Y_j are given by (II. 4. 16). (Cf. also Rem. following V, § 6, Exerc. 9.) In view of the diagram (4. 5) we see that $\tilde{\kappa}_j$ corresponds to the homomorphism $\kappa_j : \mathfrak{g}^1\to\mathfrak{g}$ defined by

(4. 10′) $$\kappa_j(h) = X_j, \qquad \kappa_j(h') = Y_j, \qquad \kappa_j(h'') = H_j,$$

i. e., in the present notation, X_j, Y_j, H_j coincide with $X_{\kappa_j}, Y_{\kappa_j}, H_{\kappa_j}$. In what follows, we fix such a system $(\kappa_1, \cdots, \kappa_r)$ once and for all, and set $\mathfrak{a}=\{X_1, \cdots, X_r\}_{\boldsymbol{R}}$. Then $\mathfrak{a}$ is a maximal abelian subalgebra of $\mathfrak{g}$ contained in $\mathfrak{p}$. We say that κ *is belonging to* the system $(\kappa_1, \cdots, \kappa_r)$, or to $\mathfrak{a}$, if $X_\kappa\in\mathfrak{a}$.

Proposition 4. 2. *Let $\kappa : (\mathfrak{g}^1, H_0^1)\to(\mathfrak{g}, H_0)$ be an* (H_1)*-homomorphism belonging to the system $(\kappa_1, \cdots, \kappa_r)$. Then one has*

(4. 11) $$X_\kappa = \sum_{j=1}^{r}\lambda_j X_j$$

with $\lambda_j=0, \pm 1$.

Proof. Comparing the $\mathfrak{p}_+$-parts in (II. 4. 16) and (4. 8), we see that the relation (4. 11) is equivalent to

$$o_\kappa = \sum\lambda_j o_j.$$

Since the roots β_j are strongly orthogonal, one has

$$\{o_i, o_j, o_k\} = \begin{cases} o_i & \text{if } i = j = k, \\ 0 & \text{otherwise.} \end{cases}$$

Hence, from $\{o_\kappa, o_\kappa, o_\kappa\} = o_\kappa$, one has $\lambda_j^3 = \lambda_j \ (1 \leq j \leq r)$, i. e., $\lambda_j = 0, \pm 1$, q. e. d.

From Propositions 4. 1, 4. 2, we see that there are exactly 3^r (H_1)-homomorphisms $\kappa : \mathfrak{g}^1 \to \mathfrak{g}$ belonging to $\mathfrak{a}$, including the trivial one for which $o_\kappa = 0$. When the relation (4. 11) holds, we define the *multiplicity* $m(\kappa)$ by

$$m(\kappa) = \sum_{j=1}^{r} |\lambda_j|. \tag{4. 12}$$

Clearly one has $0 \leq m(\kappa) \leq r$.

Now, let $\mathfrak{a}^*$ be the dual space of $\mathfrak{a}$ and $\{\xi_1, \cdots, \xi_r\}$ the basis of $\mathfrak{a}^*$ dual to $\{X_1, \cdots, X_r\}$. Let $\mathfrak{r}$ be the set of "$\boldsymbol{R}$-roots" relative to $\mathfrak{a}$ (I, § 5). Then, since

$$[X_i, H_j + Y_j] = 2\delta_{ij}(H_j + Y_j),$$
$$[X_i, H_j - Y_j] = -2\delta_{ij}(H_j - Y_j),$$

$\pm 2\xi_j \ (1 \leq j \leq r)$ are $\boldsymbol{R}$-roots. In general, let $\alpha \in \mathfrak{r}$, $\alpha = \sum m_j \xi_j$ and $Y \in \mathfrak{g}_\alpha$. Then, for $X_\kappa = \sum_j \lambda_j X_j$, one has

$$[X_\kappa, Y] = \alpha(X_\kappa) Y = (\sum_j \lambda_j m_j) Y.$$

Hence

$$\mathfrak{g}_\alpha \subset \mathfrak{g}(X_\kappa, \sum_j \lambda_j m_j)\ ; \tag{4. 13}$$

in particular, one has

$$\sum_j \lambda_j m_j \in \{0, \pm 1, \pm 2\}.$$

Since this condition holds for all κ belonging to $\mathfrak{a}$, it follows that

$$m_j \in \boldsymbol{Z} \quad \text{and} \quad \sum |m_j| \leq 2.$$

Therefore α is of the form $\pm \xi_i$, or $\pm 2\xi_i$, or $\pm \xi_i \pm \xi_j \ (i \neq j)$. Thus one has

$$\{\pm 2\xi_i \ (1 \leq i \leq r)\} \subset \mathfrak{r} \subset \{\pm \xi_i, \pm 2\xi_i, \pm \xi_i \pm \xi_j \ (1 \leq i, j \leq r, \ i \neq j)\}.$$

From the list of irreducible $\boldsymbol{R}$-root systems, it can be seen at once that, if $\mathfrak{g}$ is simple (i. e., if $\mathfrak{r}$ is irreducible), then $\mathfrak{r}$ is either of type (BC_r) or (C_r). This proves the statement in II, § 4, Remark 3.

Let $\Delta = \{\gamma_1, \cdots, \gamma_r\}$ be the fundamental system of $\mathfrak{r}$ defined by the "lexicographical" linear order of $\mathfrak{a}^*$ with respect to the (ordered) basis $\{\xi_1, \cdots, \xi_r\}$. [This means that, for $\alpha, \beta \in \mathfrak{a}^*$, we have $\alpha < \beta$ if and only if $\alpha(X_i) = \beta(X_i) \ (1 \leq i < i_0)$ and $\alpha(X_{i_0}) < \beta(X_{i_0})$ for some $1 \leq i_0 \leq r$.] Then, when $\mathfrak{g}$ is simple, one has

$$\begin{cases} \gamma_i = \xi_i - \xi_{i+1} & (1 \leq i \leq r-1), \\ \gamma_r = \xi_r \ \text{or} \ 2\xi_r & (\text{according as } \mathfrak{r} \text{ is of type } (BC_r) \text{ or } (C_r)). \end{cases} \tag{4. 14}$$

Thus Δ is of type (B_r) or (C_r).

The "Weyl chamber" corresponding to Δ is given by

$$\mathfrak{a}^+ = \{X \in \mathfrak{a} \,|\, \gamma_i(X) \geq 0 \ \text{for all} \ \gamma_i \in \Delta\}. \tag{4. 15}$$

The homomorphism κ is said to *be belonging to* $(\mathfrak{a}, \Delta)$ if $X_\kappa \in \mathfrak{a}^+$. Clearly this condition implies that in the expression (4. 11) all λ_j are ≥ 0. In particular, when $\mathfrak{g}$ is simple, κ belongs to $(\mathfrak{a}, \Delta)$ if and only if $\lambda_1 \geq \lambda_2 \geq \cdots \geq \lambda_r \geq 0$, i. e., X_κ is of the form

$$X_\kappa = X_1 + X_2 + \cdots + X_b, \tag{4. 16}$$

where $b = m(\kappa)$. Thus there are exactly $r+1$ (H_1)-homomorphisms κ belonging to $(\mathfrak{a}, \Delta)$ (including the trivial one). The homomorphism κ for which (4. 16) holds, i. e., the commutative sum $\kappa_1 + \cdots + \kappa_b$, will be called the *b-th canonical homomorphism* belonging to $(\mathfrak{a}, \Delta)$ (the trivial one being counted as the 0-th). As explained in I, § 5, every element in $\mathfrak{p}$ is conjugate (under $ad\, K^\circ$) to an element in $\mathfrak{a}^+$. Hence, for a given (H_1)-homomorphism $\kappa : (\mathfrak{g}^1, H_0^1) \to (\mathfrak{g}, H_0)$, we can always find a system $(\kappa_1, \cdots, \kappa_r)$ such that $X_\kappa \in \mathfrak{a}^+$. Thus, whenever it is convenient, we may assume without loss of generality that κ is a canonical homomorphism belonging to $(\mathfrak{a}, \Delta)$.

Proposition 4. 3. *Let* $\kappa : \mathfrak{sl}_2(\boldsymbol{R}) \to \mathfrak{g}$ *be an* (H_1)*-homomorphism. Then the multiplicity* $m(\kappa)$ *is uniquely determined by* κ *independently of the choice of the system* $(\kappa_1, \cdots, \kappa_r)$ *to which it belongs. Moreover, when* $\mathfrak{g}$ *is simple, if* κ' *is another* (H_1)*-homomorphism* $\mathfrak{sl}_2(\boldsymbol{R}) \to \mathfrak{g}$, *then one has* $m(\kappa) = m(\kappa')$ *if and only if there exists* $\varphi \in (\mathrm{Ad}\,\mathfrak{g})^\circ$ $(= \mathrm{Inn}\,\mathfrak{g})$ *such that* $\kappa' = \varphi \circ \kappa$.

Proof. To prove the first assertion, let $(\kappa_1', \cdots, \kappa_r')$ be another (maximal) system of mutually commutative (H_1)-homomorphisms $\mathfrak{g}^1 \to \mathfrak{g}$ such that $X_\kappa \in \mathfrak{a}' = \{X_1', \cdots, X_r'\}_{\boldsymbol{R}}$, where $X_j' = X_{\kappa'_j}$, and let $X_\kappa = \sum \lambda_j' X_j'$. We want to show that $\sum |\lambda_j'| = \sum |\lambda_j|$. Suppose that κ and $\kappa_1', \cdots, \kappa_r'$ satisfy (H_1) with respect to H-elements H_0^1 and H_0'. Then there exists $\varphi_1 \in C(\kappa)^\circ$ (the analytic subgroup of $G = \mathrm{Ad}\,\mathfrak{g}$ corresponding to $\mathfrak{g}^{[0]}$) such that $\varphi_1(H_0) = H_0'$ (II, § 8, Exerc. 1). Next, since $\mathfrak{c}(X_\kappa)$ is a θ-stable reductive subalgebra of $\mathfrak{g}$ and since both $\mathfrak{a}$ and $\varphi_1^{-1}(\mathfrak{a}')$ are maximal abelian subalgebras of $\mathfrak{c}(X_\kappa)$ contained in $\mathfrak{c}(X_\kappa) \cap \mathfrak{p}$, there exists $\varphi_2 \in (C_G(X_\kappa) \cap K)^\circ$ such that $\varphi_1\varphi_2(\mathfrak{a}) = \mathfrak{a}'$. Then from (4. 11) one has

$$X_\kappa = \sum_{i=1}^{r} \lambda_i \varphi_1\varphi_2(X_i)$$

and, as the homomorphisms $\varphi_1\varphi_2\kappa_i$ $(1 \leq i \leq r)$ also belong to $(\kappa_1', \cdots, \kappa_r')$,

$$\varphi_1\varphi_2(X_i) = X_{\varphi_1\varphi_2\kappa_i} = \sum \lambda_{ij} X_j'$$

with $\lambda_{ij} \in \{0, \pm 1\}$. Since $\varphi_1\varphi_2\kappa_i$ $(1 \leq i \leq r)$ are mutually commutative, the sets of indices $I_i = \{j \mid \lambda_{ij} \neq 0\}$ $(1 \leq i \leq r)$ are mutually disjoint. It follows that each I_i consists of a single element. Hence, after a suitable rearrangement of indices, we may assume that $\varphi_1\varphi_2(X_i) = \pm X_i'$. Then we obtain $\lambda_i' = \pm \lambda_i$ $(1 \leq i \leq r)$ and so $\sum |\lambda_i'| = \sum |\lambda_i| = m(\kappa)$, as desired. In the second assertion, the "if" part is obvious (without the assumption of simplicity of $\mathfrak{g}$). To prove the "only if" part, we may, by the conjugacy of the pair $(\mathfrak{a}, \Delta)$ (I, § 5) and the invariance of $m(\kappa)$ (i. e., the "if" part), assume that both κ and κ' belong to the same H-element H_0 and $(\mathfrak{a}, \Delta)$. Then, as we mentioned above, $m(\kappa) = m(\kappa')$ implies $\kappa = \kappa'$, q. e. d.

By a similar argument, one can show that, if two (H_1)-homomorphisms κ, κ': $\mathfrak{g}^1 \to \mathfrak{g}$ (with respect to the same H-elements) are commutative, then

$$m(\kappa+\kappa') = m(\kappa)+m(\kappa'). \tag{4.17}$$

It follows, in particular, that $m(\kappa)$ is equal to the maximal number m such that κ can be written as a commutative sum of m (H_1)-homomorphisms $\mathfrak{g}^1 \to \mathfrak{g}$ (with respect to common H-elements).

In the rest of this section, we assume that $\mathfrak{g}$ is simple. Let $\kappa = \kappa^{(b)}$ be the b-th canonical homomorphism belonging to $(\mathfrak{a}, \Delta)$ and set $\Gamma = \Delta - \{\gamma_b\}$ (set-theoretical difference). Then, in the notation of I, § 5, one has $\mathfrak{a}_\Gamma = \{X_{\kappa^{(b)}}\}$ and hence, comparing (I. 5. 3) and (2. 1), one has $\mathfrak{b}_\Gamma = \mathfrak{b}_{\kappa^{(b)}}$. Thus, by (I. 5. 4), $\mathfrak{b}_\kappa$ is a maximal parabolic subalgebra containing $\mathfrak{b}_\phi$, and hence is a maximal (proper) Lie subalgebra of $\mathfrak{g}$. [Conversely, it can be shown that if $\mathfrak{b}_\kappa \supset \mathfrak{b}_\phi$ then κ belongs to $(\mathfrak{a}, \Delta)$ and so $\kappa = \kappa^{(b)}$ for some b.] We also note that, in view of (4. 13), $V_\kappa = \mathfrak{g}(ad\, X_\kappa; 1)$ reduces to $\{0\}$, if and only if $\sum \lambda_i m_i$ is even for all $\alpha \in \mathfrak{r}$. Clearly this is the case if and only if $\mathfrak{r}$ is of type (C_r) and κ is the r-th (i. e., the last) canonical homomorphism (cf. Cor. 1. 6).

Proposition 4. 4. *When $\mathfrak{g}$ is simple, the representations ad_V and ad_U of $\mathfrak{c}(X_\kappa)$ are irreducible over $\boldsymbol{R}$ (but not necessarily absolutely irreducible).*

Proof *). For a moment, we set $\mathfrak{g}(ad\, X_\kappa; \nu) = \mathfrak{g}_{-\nu/2}$. Then

$$\mathfrak{b}_\kappa = \mathfrak{c}(X_\kappa) + V_\kappa + U_\kappa = \mathfrak{g}_0 + \mathfrak{g}_{-1/2} + \mathfrak{g}_{-1}.$$

Suppose there is a non-zero proper $\mathfrak{g}_0$-invariant subspace $\mathfrak{m}$ in $\mathfrak{g}_{-1/2}$, and let $\mathfrak{m}' = \mathfrak{m}^\perp$ be the annihilator of $\mathfrak{m}$ in $\mathfrak{g}_{1/2}$ with respect to the Killing form of $\mathfrak{g}$. Then $\mathfrak{m}'$ is also a non-zero, proper, $\mathfrak{g}_0$-invariant subspace in $\mathfrak{g}_{1/2}$. Hence

$$\mathfrak{b}' = \mathfrak{g}_{-1} + \mathfrak{g}_{-1/2} + \mathfrak{g}_0 + \mathfrak{m}' + [\mathfrak{m}', \mathfrak{m}']$$

is a proper subalgebra of $\mathfrak{g}$, properly containing $\mathfrak{b}_\kappa$, which contradicts the maximality of $\mathfrak{b}_\kappa$. This proves the irreducibility of the representation ad_V of $\mathfrak{c}(X_\kappa)$. The irreducibility of ad_U can be proved similarly, q. e. d.

Corollary 4. 5. *Suppose $\mathfrak{g}$ is simple and $V_\kappa \neq \{0\}$. Then one has*

$$U_\kappa = [V_\kappa, V_\kappa], \tag{4.18}$$

and the representation ad_V of $\mathfrak{c}(X_\kappa)$ is faithful.

Proof. Clearly $[V_\kappa, V_\kappa]$ is a $\mathfrak{g}_0$-invariant subspace of U_κ. In the notation of § 3, the symmetric bilinear form

$$A_e(v, I_0 v') = -\frac{1}{4}\langle e, [v, I_0 v']\rangle$$

is positive definite. Hence $[V, V] \neq \{0\}$ and so, by the irreducibility of ad_U, one has

*) This proof is due to M. Takeuchi.

$U=[V, V]$. Next let $\mathfrak{n}$ be the kernel of ad_V in $\mathfrak{c}(X_\kappa)$. Then from (4. 18) one has $[\mathfrak{n}, U+V]=0$. Since $\mathfrak{n}$ is an ideal of $\mathfrak{c}(X_\kappa)$, it is θ-stable and so one has also $[\mathfrak{n}, \theta U+\theta V]=0$. Thus $\mathfrak{n}$ is an ideal of $\mathfrak{g}$ and hence reduces to $\{0\}$, q. e. d.

We note that the relation (4. 18) implies that the map $u\mapsto R_u$ defined in § 3 is injective.

Now in order to describe the irreducible representations ad_V and ad_U of $\mathfrak{c}(X_\kappa)$ explicitly, we use the following decompositions given in § 2:

$$\mathfrak{c}(X_\kappa) = \mathfrak{g}_\kappa^{(1)} \oplus \mathfrak{g}_\kappa^{(2)},$$
$$\mathfrak{g}_\kappa^{(1)} = \mathfrak{l}_2 \oplus \mathfrak{g}_\kappa, \qquad \mathfrak{g}_\kappa^{(2)} = \{X_\kappa\}_{\boldsymbol{R}} \oplus \mathfrak{g}_\kappa^{(2)\prime},$$

where $\mathfrak{l}_2$ is compact, $\mathfrak{g}_\kappa$ is semi-simple, non-compact, of hermitian type, and $\mathfrak{g}_\kappa^{(2)\prime}$ is the orthogonal complement of $\{X_\kappa\}_{\boldsymbol{R}}$ in $\mathfrak{g}_\kappa^{(2)}$ with respect to $B^{\mathfrak{g}}$. Since the representation $ad_U|\mathfrak{g}_\kappa^{(2)}$ is faithful and irreducible (Th. 2. 3, Prop. 4. 4), the self-dual cone Ω_κ is irreducible and so $\mathfrak{g}_\kappa^{(2)\prime}$ is simple (or reduces to $\{0\}$). Thus $\mathfrak{g}_\kappa^{(2)\prime}=\mathfrak{g}_\kappa^{(2)s}$. In the Dynkin diagram of Δ (which is linear), if one deletes one vertex corresponding to γ_b, then the diagram splits into two parts, which are the diagrams corresponding to $\mathfrak{g}_\kappa$ and $\mathfrak{g}_\kappa^{(2)\prime}$, respectively. Thus $\mathfrak{g}_\kappa$ is also simple, and of $\boldsymbol{R}$-rank $r-b$. If $\mathfrak{r}^{(1)}$ and $\mathfrak{r}^{(2)}$ denote the $\boldsymbol{R}$-root systems for $\mathfrak{g}_\kappa$ and $\mathfrak{g}_\kappa^{(2)\prime}$, respectively, then $\mathfrak{r}^{(1)}$ $(b\leq r-1)$ is of type (BC_{r-b}) or (C_{r-b}), according as $\mathfrak{r}$ is of type (BC_r) or (C_r), while $\mathfrak{r}^{(2)}$ $(b\geq 2)$ is always of type (A_{b-1}).

By Theorem 3. 3, the irreducible representation ad_V of $C(X_\kappa)$ satisfies the condition (H_2) for $G_\kappa^{(1)}$ and a certain linearity condition for $G_\kappa^{(2)}$. As we shall explain in Chapters IV and V, such a representation is obtained in the following manner. There are three different cases.

($\boldsymbol{R}$-type) There are absolutely irreducible representations $(V^{(i)}, \rho^{(i)})$ $(i=1, 2)$ of $G_\kappa^{(i)}$ over $\boldsymbol{R}$ such that

$$V = V^{(1)} \otimes_{\boldsymbol{R}} V^{(2)}, \qquad ad_V = \rho^{(1)} \otimes \rho^{(2)},$$
$$A_e = A^{(1)} \otimes S^{(2)}, \qquad I_0 = I^{(1)} \otimes 1_{V^{(2)}},$$

where $A^{(1)}$ (resp. $S^{(2)}$) is a $\rho^{(1)}$-invariant, alternating (resp. symmetric) bilinear form on $V^{(1)}$ (resp. $V^{(2)}$), and $I^{(1)}$ is a complex structure on $V^{(1)}$ such that $A^{(1)}I^{(1)}\gg 0, S^{(2)}\gg 0$. Moreover

$$\rho^{(1)}: G_\kappa^{(1)} \longrightarrow Sp(V^{(1)}, A^{(1)})$$

satisfies (H_2), i. e., $\rho^{(1)}(H_0^{(1)})=\frac{1}{2}I^{(1)}$, and $\rho^{(2)}: G_\kappa^{(2)}\to GL(V^{(2)})$ is equivariant with a linear embedding $\Omega_\kappa\to\mathcal{P}(V^{(2)})$ which maps e_κ to $S^{(2)}$.

($\boldsymbol{C}$-type) There are irreducible representations $(V^{(i)}, \rho^{(i)})$ $(i=1, 2)$ of $G_\kappa^{(i)}$ over $\boldsymbol{C}$ such that

$$V = R_{\boldsymbol{C}/\boldsymbol{R}}(V^{(1)} \otimes_{\boldsymbol{C}} V^{(2)}), \qquad ad_V = \rho^{(1)} \otimes \rho^{(2)},$$
$$A_e = \mathrm{Im}(h^{(1)}\otimes h^{(2)}), \qquad I_0 = I^{(1)} \otimes 1_{V^{(2)}},$$

where $h^{(i)}$ $(i=1, 2)$ are hermitian forms on $V^{(i)}$, $h^{(1)}$ is $\rho^{(1)}$-invariant, and $I^{(1)}$ is a

($\boldsymbol{C}$-linear) complex structure on $V^{(1)}$ (viewed as an $\boldsymbol{R}$-space) such that $-ih^{(1)}I^{(1)}\gg 0$, $h^{(2)}\gg 0$. Moreover

$$\rho^{(1)}: G_{\kappa}^{(1)} \longrightarrow SU(V^{(1)}, h^{(1)})$$

satisfies (H_2), i. e., $\rho^{(1)}(H_0^{(1)})=H_0^{(1)\prime}$, where $H_0^{(1)\prime}=\frac{1}{2}I^{(1)}-\frac{i}{2}\frac{p_1-q_1}{p_1+q_1}1_{V^{(1)}}$ is an H-element for $SU(V^{(1)}, h^{(1)})$, (p_1, q_1) denoting the signature of $h^{(1)}$, and $\rho^{(2)}: G_{\kappa}^{(2)}\to GL(V^{(2)})$ is equivariant with a linear embedding $\Omega_{\kappa}\to\mathcal{P}(V^{(2)})$ which maps e_{κ} to $h^{(2)}$.

($\boldsymbol{H}$-type) There are (absolutely) irreducible representations $(V^{(i)}, \rho^{(i)})$ $(i=1,2)$ of $G_{\kappa}^{(i)}$ over $\boldsymbol{H}$, $V^{(1)}$ (resp. $V^{(2)}$) being a right (resp. left) vector space over $\boldsymbol{H}$, such that

$$\begin{aligned} V &= V^{(1)} \otimes_{\boldsymbol{H}} V^{(2)} \quad \text{(viewed as an } \boldsymbol{R}\text{-space)},\\ ad_V &= \rho^{(1)} \otimes \rho^{(2)},\\ A_e(x_1\otimes x_2, y_1\otimes y_2) &= \mathrm{tr}(h^{(1)}(x_1, y_1)h^{(2)}(y_2, x_2)) \qquad (x_1, y_1\in V^{(1)},\ x_2, y_2\in V^{(2)}),\\ I_0 &= I^{(1)} \otimes 1_{V^{(2)}}, \end{aligned}$$

where $h^{(1)}$ (resp. $h^{(2)}$) is a quaternion skew-hermitian (resp. hermitian) form on $V^{(1)}$ (resp. $V^{(2)}$), $h^{(1)}$ is $\rho^{(1)}$-invariant, and $I^{(1)}$ is an ($\boldsymbol{H}$-linear) complex structure on $V^{(1)}$ such that $h^{(1)}I^{(1)}\gg 0$, $h^{(2)}\gg 0$. Moreover

$$\rho^{(1)}: G_{\kappa}^{(1)} \longrightarrow SU(V^{(1)}, h^{(1)})$$

satisfies (H_2), i. e., $\rho^{(1)}(H_0^{(1)})=\frac{1}{2}I^{(1)}$, and $\rho^{(2)}: G_{\kappa}^{(2)}\to GL(V^{(2)})$ is equivariant with a linear embedding $\Omega_{\kappa}\to\mathcal{P}(V^{(2)})$ which maps e_{κ} to $h^{(2)}$. [$\mathcal{P}(V^{(2)})$ denotes the cone of positive definite hermitian forms on $V^{(2)}$ over $\boldsymbol{R}$, $\boldsymbol{C}$, or $\boldsymbol{H}$, according to the type.]

Consulting the lists of possible $\rho^{(1)}$ and $\rho^{(2)}$ (IV, § 5, V, § 5), it is not difficult to determine $\rho^{(1)}$ and $\rho^{(2)}$ case by case for each simple group G of hermitian type. So we give only the result. In the following list, the group G for the classical cases $(I)\sim(IV)$ is taken in the usual matrix expression (not the adjoint expression Ad $\mathfrak{g}$). The dimensions of G, V, U are to be understood as the real one, while dim $\mathcal{D}$ is always the complex one. The symbol $\sim$ stands for an isogeny. We fix a θ-stable Cartan subalgebra $\mathfrak{h}'$ of $\mathfrak{g}$ containing $\mathfrak{a}$ and let $\tilde{\Delta}$ denote a fundamental system relative to $\mathfrak{h}'_c$ as defined in I, § 5. The meaning of the marked diagram for $\tilde{\Delta}$ was also explained there. The letter n on the diagrams indicates the place where the (unique) non-compact root appears in the corresponding fundamental system relative to a compact Cartan subalgebra $\mathfrak{h}$. In the parentheses following the diagrams of Δ, we indicate the type of $\mathfrak{r}$ (not of Δ). In most cases the representations $\rho^{(i)}$ are given by the identical representation. A special attention will be paid for the cases where the compact factor $\mathfrak{l}_2$ appears.

$(I_{p,q})$ $G=SU(p,q)$ $(p+q=n\geq 2,\ p\geq q)$ (cf. § 2, Exerc. 1).
$r=\boldsymbol{R}\text{-rank}\, G=q,\quad \dim G=n^2-1,\quad \dim\mathcal{D}=pq.$

$\tilde{\Delta}$ 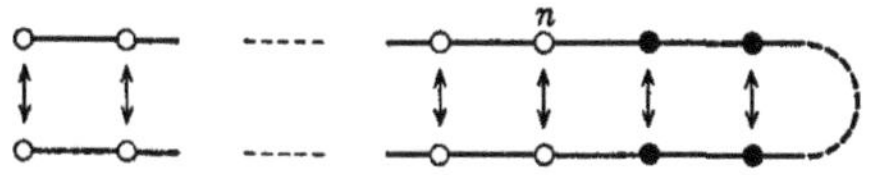

$(p>q)$

Δ  (type (BC_q))

$\tilde{\Delta}$ 

$(p=q)$

Δ  (type (C_p))

For $1<b\leq q$, one has

$$G_\kappa^{(1)} = SU(p-b, q-b)\times \boldsymbol{C}^{(1)}, \qquad V^{(1)} = \boldsymbol{C}^{n-2b}, \qquad \rho^{(1)} = id\times sc,$$
$$G_\kappa^{(2)} = GL^0(b, \boldsymbol{C}), \qquad V^{(2)} = \boldsymbol{C}^b, \qquad \rho^{(2)} = id,$$

where $\boldsymbol{C}^{(1)} = \{\zeta\in\boldsymbol{C}^\times \mid |\zeta|=1\}$, $GL^0(b, \boldsymbol{C}) = \{g\in GL(b, \boldsymbol{C}) \mid \det(g)\in\boldsymbol{R}^\times\}$ and sc denotes the scalar representation.

$$\begin{cases} \dim V = 2b(n-2b) & (\boldsymbol{C}\text{-type}), \\ \dim U = b^2 & (U\cong\mathscr{H}_b(\boldsymbol{C})). \end{cases}$$

In particular, for $b=q$, $G_\kappa^{(1)}$ is compact; $SU(p-q, 0)$ is $=\{1\}$ if $p-q=0$ or 1, and simple if $p-q\geq 2$.

For $b=1$, one has

$$G_\kappa^{(1)} = SU(p-1, q-1)\times\boldsymbol{C}^{(1)}, \qquad V^{(1)} = \boldsymbol{C}^{n-2} \text{ (viewed as an } \boldsymbol{R}\text{-space)},$$
$$G_\kappa^{(2)} = \boldsymbol{R}^\times, \qquad V^{(2)} = \boldsymbol{R}.$$

This is of $\boldsymbol{R}$-type, but the above formula for dim V remains valid.

(II_n) $G = SU^-(n, \boldsymbol{H})$ $(n\geq 3)$.

$$r = \left[\frac{n}{2}\right], \quad \dim G = 2n^2-n, \quad \dim\mathscr{D} = \frac{1}{2}n(n-1).$$

$\tilde{\Delta}$ 

(n even)

Δ  (type (C_r))

$\tilde{\Delta}$ 

(n odd)

Δ  (type (BC_r))

For $1<b\le[n/2]$, one has

$$\begin{aligned} G_\kappa^{(1)} &= SU^-(n-2b, \boldsymbol{H}), & V^{(1)} &= \boldsymbol{H}^{n-2b}, & \rho^{(1)} &= id, \\ G_\kappa^{(2)} &= GL(b, \boldsymbol{H}), & V^{(2)} &= \boldsymbol{H}^b, & \rho^{(2)} &= id, \end{aligned}$$

$$\begin{cases} \dim V = 4b(n-2b) & (\boldsymbol{H}\text{-type}), \\ \dim U = 2b^2-b & (U\cong\mathscr{H}_b(\boldsymbol{H})). \end{cases}$$

Note that

$$G_\kappa^{(1)} = \begin{cases} SU^-(2,\boldsymbol{H}) \sim SL(2,\boldsymbol{R})\times SL(1,\boldsymbol{H}) & \text{for } b=\dfrac{n}{2}-1 \ (n \text{ even}) \\ SU^-(1,\boldsymbol{H}) \cong \boldsymbol{C}^{(1)} & \text{for } b=\dfrac{n-1}{2} \ (n \text{ odd}), \end{cases}$$

where $SL(1,\boldsymbol{H})$ and $SU^-(1,\boldsymbol{H})$ are compact.

For $b=1$, the compact factor $SL(1,\boldsymbol{H})$ should be shifted from $G_\kappa^{(2)}$ to $G_\kappa^{(1)}$, so that one has

$$\begin{aligned} G_\kappa^{(1)} &= SU^-(n-2,\boldsymbol{H})\times SL(1,\boldsymbol{H}), & V^{(1)} &= \boldsymbol{H}^{n-2} \text{ (viewed as an } \boldsymbol{R}\text{-space)}; \\ G_\kappa^{(2)} &= \boldsymbol{R}^\times, & V^{(2)} &= \boldsymbol{R}, \end{aligned}$$

where the two factors of $G_\kappa^{(1)}$ act on $V^{(1)}$ from the left and right. Thus this is of $\boldsymbol{R}$-type, but the above dimension formula remains valid.

(III_n) $G=Sp(2n,\boldsymbol{R})$ (cf. § 3, Exerc. 2).

$$r=n, \quad \dim G = 2n^2+n, \quad \dim\mathscr{D}=\frac{1}{2}n(n+1).$$

$\tilde{\Delta}=\Delta$ 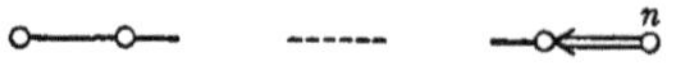 (type (C_n))

$$\begin{aligned} G_\kappa^{(1)} &= Sp(2(n-b),\boldsymbol{R}), & V^{(1)} &= \boldsymbol{R}^{2(n-b)}, & \rho^{(1)} &= id, \\ G_\kappa^{(2)} &= GL(b,\boldsymbol{R}), & V^{(2)} &= \boldsymbol{R}^b, & \rho^{(2)} &= id, \end{aligned}$$

$$\begin{cases} \dim V = 2b(n-b) & (\boldsymbol{R}\text{-type}), \\ \dim U = \dfrac{1}{2}b(b+1) & (U\cong\mathrm{Sym}_b(\boldsymbol{R})). \end{cases}$$

(IV_p) $G=SO(p,2)$ $(p\ge3)$.

$$r=2, \quad \dim G=\frac{1}{2}(p+1)(p+2), \quad \dim\mathscr{D}=p.$$

$\tilde{\Delta}$ 
(p even, ↕ if $p\equiv0$ (4))
(p odd)

Δ 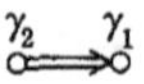 (type (C_2))

For $b=1$, one has

$$G_K^{(1)} = SL(2, \boldsymbol{R}) \times SO(p-2), \qquad V^{(1)} = \boldsymbol{R}^{2(p-2)}, \qquad \rho^{(1)} = id \otimes id,$$
$$G_K^{(2)} = \boldsymbol{R}^{\times}, \qquad V^{(2)} = \boldsymbol{R}, \qquad \rho^{(2)} = id,$$
$$\begin{cases} \dim V = 2(p-2) & (\boldsymbol{R}\text{-type}), \\ \dim U = 1. \end{cases}$$

For $b=2$, one has

$$G_K^{(1)} = \{1\}, \qquad G_K^{(2)} = \boldsymbol{R}^{\times} \times SO(p-1, 1),$$
$$\begin{cases} \dim V = 0, \\ \dim U = p. \end{cases}$$

(*V*) $G = ({}^2E_{6,2})$ $(K \sim Spin(10) \times SO(2))$.
$r = 2, \quad \dim G = 78, \quad \dim \mathcal{D} = 16.$

$\tilde{\Delta}$ 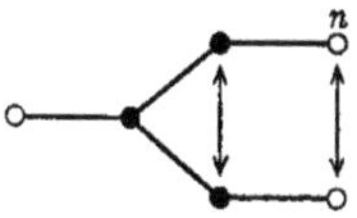

Δ 

(type (BC_2))

For $b=1$, one has

$$G_K^{(1)} \sim SU(5, 1), \qquad V^{(1)} = \boldsymbol{R}^{20} \,(= {}_{\boldsymbol{R}}\wedge^3(\boldsymbol{C}^6)), \qquad \rho^{(1)} = {}_{\boldsymbol{R}}\wedge^3 \text{ (see pp. 186, 278)},$$
$$G_K^{(2)} = \boldsymbol{R}^{\times}, \qquad V^{(2)} = \boldsymbol{R}, \qquad \rho^{(2)} = id,$$
$$\begin{cases} \dim V = 20 & (\boldsymbol{R}\text{-type}), \\ \dim U = 1. \end{cases}$$

For $b=2$, one has

$$G_K^{(1)} = \boldsymbol{C}^{(1)}, \qquad V^{(1)} = \boldsymbol{C}, \qquad \rho^{(1)} = id,$$
$$G_K^{(2)} \sim \boldsymbol{R}^{\times} \times Spin(7, 1), \qquad V^{(2)} = \boldsymbol{C}^8, \qquad \rho^{(2)} = spin,$$
$$\begin{cases} \dim V = 16 & (\boldsymbol{C}\text{-type}), \\ \dim U = 8. \end{cases}$$

(*VI*) $G = (E_{7,3})$ $(K \sim (E_6) \times SO(2))$.
$r = 3, \quad \dim G = 133, \quad \dim \mathcal{D} = 27.$

$\tilde{\Delta}$ 

Δ 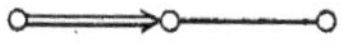
(type (C_3))

For $b=1$, one has

$$G_\kappa^{(1)} = Spin(10,2), \quad V^{(1)} = \boldsymbol{R}^{32}, \quad \rho^{(1)} = spin,$$
$$G_\kappa^{(2)} = \boldsymbol{R}^\times, \quad V^{(2)} = \boldsymbol{R}, \quad \rho^{(2)} = id,$$
$$\begin{cases} \dim V = 32 & (\boldsymbol{R}\text{-type}), \\ \dim U = 1. \end{cases}$$

For $b=2$, one has

$$G_\kappa^{(1)} = SL(2,\boldsymbol{R}), \quad V^{(1)} = \boldsymbol{R}^2, \quad \rho^{(1)} = id,$$
$$G_\kappa^{(2)} \sim \boldsymbol{R}^\times \times Spin(9,1), \quad V^{(2)} = \boldsymbol{R}^{16}, \quad \rho^{(2)} = spin,$$
$$\begin{cases} \dim V = 32 & (\boldsymbol{R}\text{-type}), \\ \dim U = 10. \end{cases}$$

For $b=3$, one has

$$G_\kappa^{(1)} = \{1\}, \quad G_\kappa^{(2)} \sim \boldsymbol{R}^\times \times (E_{6,2}),$$
$$\begin{cases} \dim V = 0, \\ \dim U = 27 & (U \cong \mathcal{H}_3(\boldsymbol{O})). \end{cases}$$

Remark. The absolute root system $\tilde{\mathfrak{r}}^{(1)}$ of $\mathfrak{g}_\kappa$ $(b>0)$ is always of classical type, reflecting the fact that the exceptional simple Lie algebras of hermitian type have no non-trivial symplectic representation satisfying (H_1) (IV, § 5).

Exercise

1. Let $\mathcal{E}(\mathfrak{p}_+)$ denote the set of all idempotents in $\mathfrak{p}_+$. We define a partial order in $\mathcal{E}(\mathfrak{p}_+)$ as follows. For $z_1, z_2 \in \mathcal{E}(\mathfrak{p}_+)$ we write $z_2 \prec z_1$ if and only if there exists $z_3 \in \mathcal{E}(\mathfrak{p}_+)$ such that $z_1 = z_2 + z_3$ and $z_2 \square z_3 = 0$. Show that, if κ_1 and κ_2 are (H_1)-homomorphisms of $\mathfrak{g}^1$ into $\mathfrak{g}$ corresponding to z_1 and z_2, then $z_2 \prec z_1$ implies $m(\kappa_2) \leq m(\kappa_1)$, and that z_1 is minimal (resp. maximal) if and only if $m(\kappa_1)=1$ (resp. $=r$). A minimal (resp. maximal) idempotent is called a "primitive" (resp. "principal") idempotent.

§ 5. Groups of Harish-Chandra type.

For our discussion of Siegel domains, it will be useful to generalize the notions of Harish-Chandra embeddings and the canonical automorphy factors given in II, §§ 4, 5 to a wider class of groups to be called of H-C type (cf. Satake [12]).

Let G be a Zariski connected $\boldsymbol{R}$-group with Lie algebra $\mathfrak{g}$. Then the complexification G_C (defined independently of the matrix realization of G) is a connected complex Lie group. Suppose there are given a Zariski connected $\boldsymbol{R}$-subgroup K of G with Lie algebra $\mathfrak{k}$ and (connected) unipotent $\boldsymbol{C}$-subgroups $P_\pm$ of G_C with Lie algebras $\mathfrak{p}_\pm$. We call G *of Harish-Chandra type,* or *of H-C type* for short, if the following two conditions are satisfied.

(HC 1) (Infinitesimal condition) One has

$$\begin{aligned} &\mathfrak{g}_C = \mathfrak{p}_+ + \mathfrak{k}_C + \mathfrak{p}_- \quad \text{(direct sum of vector spaces)}, \\ &[\mathfrak{k}_C, \mathfrak{p}_\pm] \subset \mathfrak{p}_\pm, \quad \bar{\mathfrak{p}}_+ = \mathfrak{p}_-. \end{aligned} \tag{5.1}$$

[It follows that $\mathfrak{k}_C + \mathfrak{p}_\pm$ is an (algebraic) Lie subalgebra of $\mathfrak{g}_C$.] The decomposition

(5. 1) will be called a "canonical decomposition" of $\mathfrak{g}_C$.

(HC 2) (Global condition) One has a holomorphic injection

$$P_+ \times K_C \times P_- \longrightarrow P_+K_CP_- \subset G_C. \tag{5. 2}$$

Namely, the natural mapping $(p_+, k, p_-) \mapsto p_+kp_-$ gives a biholomorphic homeomorphism of $P_+ \times K_C \times P_-$ onto its image $P_+K_CP_-$ in G_C. (It follows that $P_+K_CP_-$ is open in G_C and the $K_CP_\pm$ are semi-direct products.) Moreover one has

$$G^\circ \subset P_+K_CP_- \quad \text{and} \quad G^\circ \cap K_CP_- = K^\circ. \tag{5. 3}$$

Thus we are in a situation quite similar to that for groups of hermitian type except that here we do not assume K to be compact, nor impose any condition on $[\mathfrak{p}_+, \mathfrak{p}_-]$.

From the condition (HC 2) we have the following natural inclusion mappings

$$\mathcal{D} = G^\circ/K^\circ \longrightarrow \underset{\underset{\textstyle P_+}{\wr\wr}}{P_+K_CP_-/K_CP_-} \subset G_C/K_CP_-. \tag{5. 4}$$

As G_C/K_CP_- has an invariant complex structure, we may endow the homogeneous space $\mathcal{D}$ with a G°-invariant complex structure. As the exponential map gives an analytic isomorphism $\mathfrak{p}_+ \to P_+$, we may consider $\mathcal{D}$ as embedded in $\mathfrak{p}_+$, where a coset $gK^\circ (g \in G^\circ)$ is identified with $z \in \mathfrak{p}_+$ if and only if

$$\exp z = (g)_+ \quad (\text{the } P_+\text{-part of } g). \tag{5. 5}$$

This is a generalization of the Harish-Chandra embedding.

In this setting, we can also define the (holomorphic) action of G° on $\mathcal{D}$, the canonical automorphy factor $J: G^\circ \times \mathcal{D} \to K_C$, and the canonical kernel function K: $\mathcal{D} \times \mathcal{D} \to K_C$ exactly in the same manner as in II, § 5, i. e., by the formulas

$$\exp g(z) = (g \cdot \exp z)_+, \qquad J(g, z) = (g \cdot \exp z)_0, \tag{5. 6}$$

$$K(z, z') = (((\exp \bar{z}')^{-1} \exp z)_0)^{-1}. \tag{5. 7}$$

The subscript 0 denoting always the K_C-component in the decomposition (5. 2). Actually these definitions can be extended for any $g \in G_C$, $z, z' \in \mathfrak{p}_+$ as long as the right-hand sides of (5. 6) or (5. 7) are defined. It is then clear that Lemmas II. 5. 1, II. 5. 2 remain valid.

Example 1 (*Generalized Heisenberg groups*). Suppose there are given two (finite-dimensional) real vector spaces U and V together with an alternating bilinear map $A: V \times V \to U$ and a complex structure I_0 on V such that AI_0 is symmetric. (Here we do not assume any positivity condition on AI_0.) Let $\tilde{V} = U \times V$ and define a "product" in $\tilde{V}$ by

$$(u, v)(u', v') = (u+u'-2A(v, v'), v+v'). \tag{5. 8}$$

Then clearly $\tilde{V}$ becomes a unipotent $\boldsymbol{R}$-group. From (5. 8) we obtain

$$\frac{d}{d\lambda}(\lambda u, \lambda v)(u', v')(\lambda u, \lambda v)^{-1}\Big|_{\lambda=0} = (-4A(v, v'), 0).$$

Hence the Lie algebra $\tilde{\mathfrak{v}}$ of $\tilde{V}$ is canonically identified with $U\oplus V$ with bracket product

$$(5.9) \qquad [u+v, u'+v'] = -4A(v, v').$$

Then U is a Lie subalgebra contained in the center of $\tilde{\mathfrak{v}}$.

For instance, when $U=\boldsymbol{R}$, $V=\boldsymbol{R}^{2n}$ and

$$A(v, v') = -\frac{1}{4}({}^t v_1 v_2' - {}^t v_1' v_2)$$

for $v=\begin{pmatrix} v_1 \\ v_2 \end{pmatrix}$, $v'=\begin{pmatrix} v_1' \\ v_2' \end{pmatrix} \in \boldsymbol{R}^{2n}$, we obtain the classical "Heisenberg group" $\tilde{V}$. A matrix realization of $\tilde{\mathfrak{v}}$ is given by

$$\left\{ u+v = \begin{pmatrix} 0 & {}^t v_1 & u \\ 0 & 0 & v_2 \\ 0 & 0 & 0 \end{pmatrix} \begin{matrix} \}1 \\ \}n \\ \}1 \end{matrix} \right\}.$$

In the general case, let

$$(5.10) \qquad V_c = V_+ \oplus V_-, \qquad \text{where} \quad V_\pm = V_c(I_0; \pm i).$$

Then $A|V_+\times V_+$ is identically equal to zero, since $A(v, I_0 v') = iA(v, v')$ $(v, v' \in V_+)$ is at the same time symmetric and alternating. Hence V_+ (and so $V_- = \bar{V}_+$, too) is an abelian subalgebra of $\tilde{\mathfrak{v}}_c$. If we set

$$K = \exp U = \{(u, 0) \,|\, u \in U\},$$
$$P_\pm = \exp V_\pm = \{(0, w) \,|\, w \in V_\pm\},$$

then the condition (HC 1) is clearly satisfied for $\mathfrak{k}=U$, $\mathfrak{p}_\pm = V_\pm$. To check the condition (HC 2), we first show that

$$(5.11) \qquad \tilde{V}_c = (\exp V_+)(\exp U_c)(\exp V_-).$$

In fact, we will show more precisely that every element $(u, v) \in \tilde{V}_c$ can uniquely be expressed in the form

$$(u, v) = (0, w)(u', 0)(0, \bar{w}')$$

with $u' \in U_c$, $w, w' \in V_+$. In view of (5.8), this equation is equivalent to

$$\begin{cases} u = u' - 2A(w, \bar{w}'), \\ v = w + \bar{w}', \end{cases}$$

which has a unique solution

$$\begin{cases} w = w' = v_+ \quad (\text{the } V_+\text{-part of } v), \\ u' = u + 2A(v_+, v_-) \quad (v_- = \bar{v}_+). \end{cases}$$

Thus

$$(5.12) \qquad \begin{cases} (u, v)_\pm = \exp v_\pm, \\ (u, v)_0 = \exp(u + 2A(v_+, v_-)), \end{cases}$$

which proves that one has the direct product decomposition (5.11). (5.12) also implies that

$$\tilde{V} \cap (\exp U_c)(\exp V_-) = \exp U.$$

Thus the condition (HC 2) is satisfied. In this case, the embedding $\tilde{V}/\exp U \to V_+$ is bijective, so that $\tilde{V}/\exp U$ can be identified with V_+ itself. The (holomorphic) action of $\tilde{V}$ on V_+ is then given by the translation

$$(a, b) : w \longmapsto w + b_+.$$

Also from (5. 12) we obtain

$$J((a, b), w) = \exp\left(a - 4A\left(b, w + \frac{1}{2} b_+\right)\right),$$
$$K(w, w') = \exp(4A(w, \bar{w}')).$$

These are special cases of the formulas to be given in the next example.

Example 2. Let G_1 be a group of H-C type and let $\tilde{V}$ be as in Example 1. Suppose there are given representations ρ_U and ρ_V of G_1 (or of $\mathfrak{g}_1 = \operatorname{Lie} G_1$) on U and V satisfying the following conditions:

(5. 13) $$A(\rho_V(g_1)v, \rho_V(g_1)v') = \rho_U(g_1)A(v, v') \qquad (g_1 \in G_1,\ v, v' \in V),$$

(5. 14) $$\begin{aligned} &\rho_V(\mathfrak{p}_{1+})V_- \subset V_+, \qquad \rho_V(\mathfrak{p}_{1+})V_+ = \{0\},\\ &\rho_U(\mathfrak{p}_{1+})U_C = \{0\}, \qquad \rho_V(\mathfrak{k}_{1C})V_+ \subset V_+, \end{aligned}$$

where $\mathfrak{g}_{1C} = \mathfrak{p}_{1+} + \mathfrak{k}_{1C} + \mathfrak{p}_{1-}$ is a canonical decomposition of $\mathfrak{g}_{1C}$. The condition (5. 14) (which is an analogue of (H_1)) implies that there exists a $\boldsymbol{C}$-linear map $\varphi : \mathfrak{p}_{1+} \to \operatorname{Hom}(V_-, V_+)$ such that in the matrix expression according to the decomposition (5. 10) one has

(5. 15) $$\rho_V(z_1) = \begin{pmatrix} 0 & \varphi(z_1) \\ 0 & 0 \end{pmatrix} \qquad \text{for} \quad z_1 \in \mathfrak{p}_{1+}.$$

By (5. 13) one has

(5. 16) $$A(\varphi(z_1)v, v') + A(v, \varphi(z_1)v') = 0 \qquad (z_1 \in \mathfrak{p}_{1+},\ v, v' \in V_-),$$

i. e., $\varphi(z_1)$ is symmetric with respect to the bilinear map $A|V_- \times V_+$.

Under these conditions, G_1 acts on $\tilde{V}$ by

$$g_1 \cdot (u, v) = (\rho_U(g_1)u, \rho_V(g_1)v).$$

For brevity, we will often write $g_1 u$, $g_1 v$ for $\rho_U(g_1)u$, $\rho_V(g_1)v$. Then we obtain a semi-direct product $G = \tilde{V} \cdot G_1$, in which the multiplication is given by

(5. 17) $$\begin{aligned} (u, v, g_1) \cdot (u', v', g_1') &= ((u, v) \cdot (g_1 u', g_1 v')) \cdot (g_1 g_1') \\ &= (u + g_1 u' - 2A(v, g_1 v'), v + g_1 v', g_1 g_1'). \end{aligned}$$

Clearly G is an $\boldsymbol{R}$-group and the underlying vector space of $\mathfrak{g} = \operatorname{Lie} G$ may be identified with $U \oplus V \oplus \mathfrak{g}_1$. As K_1 normalizes U, the $\boldsymbol{R}$-subgroup K of G generated by K_1 and $\exp U$ is the product $(\exp U)K_1$ and $\operatorname{Lie} K = \mathfrak{k} = U + \mathfrak{k}_1$. On the other hand, as $[\mathfrak{p}_{1+}, V_+] = 0$, we have a (connected) unipotent $\boldsymbol{C}$-subgroup $P_+ = (\exp V_+)P_{1+} (\cong (\exp V_+) \times P_{1+})$ of G_C and $\operatorname{Lie} P_+ = \mathfrak{p}_+ = V_+ \oplus \mathfrak{p}_{1+}$. Similarly, we have $P_- = (\exp V_-)P_{1-}$ and $\operatorname{Lie} P_- = \mathfrak{p}_- = V_- \oplus \mathfrak{p}_{1-}$. It is then clear that

$$\mathfrak{g}_C = \mathfrak{p}_+ + \mathfrak{k}_C + \mathfrak{p}_- \qquad \text{(a direct sum of vector spaces)}$$

and

$$[\mathfrak{k}_{\mathbf{C}}, \mathfrak{p}_+] = [\mathfrak{k}_{1\mathbf{C}}, \mathfrak{p}_+] \subset \mathfrak{p}_+, \qquad \bar{\mathfrak{p}}_+ = \mathfrak{p}_-.$$

Thus the condition (HC 1) is satisfied. Moreover one has

(5. 18)
$$\begin{aligned} P_+ K_{\mathbf{C}} P_- &= (\exp V_+) P_{1+} (\exp U_{\mathbf{C}}) K_{1\mathbf{C}} (\exp V_-) P_{1-} \\ &= P_{1+} (\exp V_+)(\exp U_{\mathbf{C}})(\exp V_-) K_{1\mathbf{C}} P_{1-} \\ &= P_{1+} \cdot \tilde{V}_{\mathbf{C}} \cdot (K_{1\mathbf{C}} P_{1-}) = \tilde{V}_{\mathbf{C}} \cdot (P_{1+} K_{1\mathbf{C}} P_{1-}). \end{aligned}$$

From this and the condition (HC 2) for G_1, one sees that the condition (HC 2) is also satisfied for $G = \tilde{V} G_1$. Thus G is of H-C type. We consider $\mathcal{D} = G^\circ / K^\circ$ as embedded in $\mathfrak{p}_+ = V_+ \oplus \mathfrak{p}_{1+}$ by the generalized Harish-Chandra map.

Now we want to determine the explicit form of the (holomorphic) action of G° on $\mathcal{D}$ and the canonical automorphy factor J. Let

$$g = (a, b, g_1) \in G, \qquad z = (w, z_1) \in \mathcal{D}$$

and put

$$\begin{cases} g(z) = (w', z_1') \ (\in \mathcal{D}), \\ J(g, z) = (\mathcal{J}_U, 0, J_1) \ (\in K_{\mathbf{C}}). \end{cases}$$

Then, by the definition, one has

(5. 19)
$$g \cdot \exp z = (0, w', \exp z_1')(\mathcal{J}_U, 0, J_1)(0, w_-, p_{1-})$$

with $(0, w_-, p_{1-}) \in P_-$. Comparing the corresponding components on both sides, we see that (5. 19) is equivalent to the following three equations:

(5. 20 a)
$$a - 2A(b, g_1 w) = \mathcal{J}_U - 2A(w', (\exp z_1') J_1 w_-),$$

(5. 20 b)
$$b + g_1 w = w' + (\exp z_1') J_1 w_-,$$

(5. 20 c)
$$g_1 (\exp z_1) = (\exp z_1') J_1 p_{1-}.$$

First, (5. 20 c) implies

(5. 21)
$$z_1' = g_1(z_1), \qquad J_1 = J_1(g_1, z_1),$$

where $J_1(g_1, z_1)$ is the canonical automorphy factor for G_1. By (5. 15) one has

$$\rho_V(\exp z_1) = \begin{pmatrix} 1 & \varphi(z_1) \\ 0 & 1 \end{pmatrix} \qquad \text{for} \quad z_1 \in \mathfrak{p}_{1+}.$$

We also write

$$\rho_V(g_1) = \begin{pmatrix} \alpha & \beta \\ \gamma & \delta \end{pmatrix}, \qquad \rho_V(J_1) = \begin{pmatrix} J_+ & 0 \\ 0 & J_- \end{pmatrix}$$

according to the decomposition (5. 10). Then, applying ρ_V to (5. 20 c), we obtain

$$\begin{pmatrix} \alpha & \beta \\ \gamma & \delta \end{pmatrix} \begin{pmatrix} 1 & \varphi(z_1) \\ 0 & 1 \end{pmatrix} = \begin{pmatrix} 1 & \varphi(z_1') \\ 0 & 1 \end{pmatrix} \begin{pmatrix} J_+ & 0 \\ 0 & J_- \end{pmatrix} \begin{pmatrix} 1 & 0 \\ * & 1 \end{pmatrix},$$

whence

(5. 22)
$$\begin{cases} \varphi(z_1') = (\alpha \varphi(z_1) + \beta)(\gamma \varphi(z_1) + \delta)^{-1}, \\ J_+ = \alpha - \varphi(z_1') \gamma, \qquad J_- = \gamma \varphi(z_1) + \delta. \end{cases}$$

Putting $\rho_V(g_1)^{-1} = \begin{pmatrix} \alpha' & \beta' \\ \gamma' & \delta' \end{pmatrix}$, we also obtain

$$(5.22')\qquad \begin{cases} \varphi(z_1') = (\alpha'-\varphi(z_1)\gamma')^{-1}(-\beta'+\varphi(z_1)\delta'), \\ J_+ = (\alpha'-\varphi(z_1)\gamma')^{-1}, \quad J_- = (\gamma'\varphi(z_1')+\delta')^{-1}. \end{cases}$$

In these notations, we can rewrite (5. 20 b) as

$$(5.20\,b')\qquad \begin{cases} b_+ + \alpha w = w' + \varphi(z_1')J_- w_-, \\ b_- + \gamma w = J_- w_-, \end{cases}$$

where $b_\pm$ denote the $V_\pm$-components of b $(b_- = \bar{b}_+)$. Hence

$$\begin{aligned} w' &= (b_+ + \alpha w) - \varphi(z_1')(b_- + \gamma w) \\ &= (b_+ - \varphi(z_1')b_-) + (\alpha - \varphi(z_1')\gamma)w. \end{aligned}$$

Putting $b_{z_1'} = b_+ - \varphi(z_1')b_-$ (II, § 7, Exerc. 1), we obtain

$$(5.23)\qquad w' = b_{z_1'} + J_+ w.$$

In particular, letting $z=(0, 0)$, one obtains

$$g(0, 0) = (b_{g_1(0)}, g_1(0)),$$

which implies that $\mathcal{D} = V_+ \times \mathcal{D}_1$, where $\mathcal{D}_1$ is the Harish-Chandra realization of G_1°/K_1° in $\mathfrak{p}_{1+}$.

Finally, from (5. 20 a), (5. 20 b'), and (5. 23), we have

$$\begin{aligned} \mathcal{J}_U &= a - 2A(b, g_1 w) + 2A(w', (\exp z_1')J_1 w_-) \\ &= a - 2A(b, g_1 w) + 2A(b_{z_1'} + J_1 w, b_- + \gamma w). \end{aligned}$$

By a simple computation, one has

$$-A(b, g_1 w) + A(b_{z_1'}, \gamma w) = -A(b, J_1 w).$$

Hence

$$(5.24)\qquad \mathcal{J}_U = a - 2A(b, b_{g_1(z_1)}) - 4A(b, J_1 w) - 2A(g_1 w, J_1 w),$$

where $J_1 = J_1(g_1, z_1)$. This is nothing but a generalization of the well-known automorphy factor for theta-functions (cf. Igusa [8]).

Summing up, we obtain

Proposition 5. 1. *The homogeneous space* $\mathcal{D} = G^\circ/K^\circ$ *may be identified with* $V_+ \times \mathcal{D}_1$ *by the generalized Harish-Chandra map. For* $g=(a, b, g_1) \in G^\circ$ *and* $z=(w, z_1) \in \mathcal{D}$, *one has*

$$(5.25)\qquad \begin{cases} g(z) = (b_{g_1(z_1)} + J_1(g_1, z_1)w, g_1(z_1)), \\ J(g, z) = (\mathcal{J}_U, 0, J_1(g_1, z_1)), \end{cases}$$

where $\mathcal{J}_U$ *is given by* (5. 24).

The above formulas remain valid for any $g \in G_c$ and $z \in \mathfrak{p}_+$ as long as $g(z)$ is defined. In particular, applying them to $g = (\exp \bar{z}')^{-1}$, we obtain formulas for $K(z, z') = J((\exp \bar{z}')^{-1}, z)^{-1}$ $(z, z' \in \mathcal{D})$. In this case, setting $z' = (w', z_1')$, one has

$$\begin{aligned} (\exp \bar{z}')^{-1} &= (0, -\bar{w}', (\exp \bar{z}_1')^{-1}), \\ \rho_V((\exp \bar{z}_1')^{-1}) &= \begin{pmatrix} 1 & 0 \\ -\overline{\varphi(z_1')} & 1 \end{pmatrix}. \end{aligned}$$

Hence by (5. 22), denoting by K_1 the canonical kernel function for G_1, one has

$$\rho_V(K_1(z_1, z_1')) = \rho_V(J_1((\exp \bar{z}_1')^{-1}, z_1))^{-1} = \begin{pmatrix} 1-\varphi(z_1)\overline{\varphi(z_1')} & 0 \\ 0 & (1-\overline{\varphi(z_1')}\varphi(z_1))^{-1} \end{pmatrix}. \tag{5. 26}$$

We note that, since $K_1(z_1, z_1')$ is in the subgroup of G generated by P_{1+} and P_{1-}, the action of $K_1(z_1, z_1')$ on U is trivial, so that by (5. 13) one has

$$\begin{aligned} A(K_1(z_1, z_1')v, v') &= A(v, K_1(z_1, z_1')^{-1}v') \\ &= A(v, \overline{K_1(z_1', z_1)}v') \\ &\qquad (z_1, z_1' \in \mathcal{D}_1,\ v, v' \in V_C). \end{aligned} \tag{5. 27}$$

Hence by the definition we have

$$\begin{aligned} K(z, z') &= J((\exp \bar{z}')^{-1}, z)^{-1} \\ &= (\mathcal{J}_U, 0, J_1((\exp \bar{z}_1')^{-1}, z_1))^{-1} \\ &= (-\mathcal{J}_U, 0, J_1((\exp \bar{z}_1')^{-1}, z_1)^{-1}), \end{aligned}$$

where $\mathcal{J}_U$ stands for $\mathcal{J}_U((\exp \bar{z}')^{-1}, z)$. Hence by Proposition 5. 1 we obtain

Proposition 5. 2. *The canonical kernel function K for $\mathcal{D}$ is given by*

$$\begin{aligned} K(z, z') &= (\mathcal{K}_U, 0, K_1(z_1, z_1')) \\ &\qquad (z=(w, z_1),\ z'=(w', z_1') \in \mathcal{D}), \\ \mathcal{K}_U(z, z') &= -2A(\bar{w}', K_1(z_1, z_1')^{-1}\varphi(z_1)\bar{w}') \\ &\quad -4A(\bar{w}', K_1(z_1, z_1')^{-1}w) \\ &\quad -2A(\overline{\varphi(z_1')}w, K_1(z_1, z_1')^{-1}w). \end{aligned} \tag{5. 28}$$

By Lemma II. 5. 2, and noting that $J_1(g_1, z_1)u = g_1u$ by (5. 14), we see that $\mathcal{K}_U$ has the following properties:

$$\mathcal{K}_U(z', z) = -\overline{\mathcal{K}_U(z, z')}, \tag{5. 29}$$

$$\mathcal{K}_U(g(z), g(z')) = \mathcal{J}_U(g, z) + g_1\cdot\mathcal{K}_U(z, z') - \overline{\mathcal{J}_U(g, z')}. \tag{5. 30}$$

In particular, putting $\mathcal{K}_U[z] = \mathcal{K}_U(z, z)$, one has

$$\mathcal{K}_U[g(z)] - g_1\cdot\mathcal{K}_U[z] = 2i \operatorname{Im} \mathcal{J}_U(g, z). \tag{5. 31}$$

The results in Proposition 5. 1 can be rephrased as follows. As we have seen above, the natural homomorphism $G \to G_1$ induces the projection

$$G^\circ/K^\circ = \mathcal{D} = V_+ \times \mathcal{D}_1 \longrightarrow G_1^\circ/K_1^\circ = \mathcal{D}_1.$$

On the other hand, since $\exp U_C$ and P_+ commute elementwise, the exponential map gives rise to a natural embedding

$$U_C \oplus \mathfrak{p}_+ \longrightarrow G_C/K_{1C}P_-,$$

which induces the natural embedding $\mathfrak{p}_+ \to G_C/K_CP_-$. Therefore we obtain the following double fibering

$$\begin{array}{ccc} U_c \times V_+ \times \mathcal{D}_1 & \longrightarrow & G_c/K_{1c}P_- \\ \downarrow & & \downarrow \\ V_+ \times \mathcal{D}_1 & \longrightarrow & G_c/K_c P_- \\ \downarrow & & \downarrow \\ \mathcal{D}_1 & \longrightarrow & G_{1c}/K_{1c}P_{1-}, \end{array} \tag{5. 32}$$

where the vertical arrows are natural projections and the horizontal ones (the generalized Borel embeddings) are all defined by exponential mappings. By means of the embedding on the top floor, we can define a (holomorphic) action of G° on $U_c \times V_+ \times \mathcal{D}_1$ by

$$\begin{aligned} & g(u, w, z_1) = (u', w', z_1') \\ \Longleftrightarrow\ & g\cdot \exp(u+w+z_1) \in \exp(u'+w'+z_1')K_{1c}P_-, \end{aligned} \tag{5. 33}$$

where $g \in G^\circ$, $u, u' \in U_c$, $w, w' \in V_+$, $z_1, z_1' \in \mathcal{D}_1$. Let

$$g\cdot \exp(u+w+z_1) = \exp(u'+w'+z_1')J_1 p_-$$

with $J_1 \in K_{1c}, p_- \in P_-$. Then, going down to the second floor, we have

$$g\cdot \exp(w+z_1) = \exp(w'+z_1')\cdot \exp(u'-g_1 u)J_1 p_-.$$

Hence, by Proposition 5. 1, one has

$$u' = \mathcal{J}_U + g_1 u$$

and z_1', w_1' are given by (5. 21), (5. 23). Thus one has

$$\left\{\begin{aligned} u' &= a - 2A(b, b_{g_1(z_1)}) - 4A(b, J_1 w) - 2A(g_1 w, J_1 w) + g_1 u, \\ w' &= b_{g_1(z_1)} + J_1 w, \\ z_1' &= g_1(z_1), \end{aligned}\right. \tag{5. 34}$$

where $J_1 = J_1(g_1, z_1)$. The projection $U_c \times V_+ \times \mathcal{D}_1 \to \mathcal{D}_1$ is equivariant with respect to the homomorphism $G^\circ \to G_1^\circ$. (5. 34) is a so-called "quasi-linear transformation" (Piatetskii-Shapiro [2]). In particular, when $g=1$, one obtains a "parallel translation":

$$\left\{\begin{aligned} u' &= a - 2A(b, b_{z_1}) - 4A(b, w) + u, \\ w' &= b_{z_1} + w, \\ z_1' &= z_1. \end{aligned}\right. \tag{5. 35}$$

Now, though G° is transitive on $\mathcal{D} = V_+ \times \mathcal{D}_1$, it is not so in general on $U_c \times V_+ \times \mathcal{D}_1$. In order to find G°-orbits, consider a mapping $\Phi : U_c \times V_+ \times \mathcal{D}_1 \to U$ defined by

$$\Phi(u, w, z_1) = \operatorname{Im} u - \frac{1}{2i}\mathcal{K}_U[(w, z_1)], \tag{5. 36}$$

where in virtue of (5. 29) $\mathcal{K}_U[(w, z_1)]$ is purely imaginary.

Proposition 5. 3. *The mapping Φ defined by* (5. 36) *is G°-equivariant, i. e., one has*

$$\Phi(g\cdot(u, w, z_1)) = g_1 \cdot \Phi(u, w, z_1) \tag{5. 37}$$

for all $g = (a, b, g_1) \in G^\circ$ and $(u, w, z_1) \in U_c \times V_+ \times \mathcal{D}_1$. If (Ker $\rho_U)^\circ$ *is transitive on $\mathcal{D}_1$, then Φ gives rise to a one-to-one correspondence between G°-orbits in $U_c \times V_+ \times \mathcal{D}_1$ and G_1°-orbits in U.*

Proof. By (5. 34) and (5. 31) one has

$$\begin{aligned}&\Phi(g\cdot(u, w, z_1))-g_1\cdot\Phi(u, w, z_1)\\ &= \mathrm{Im}(\mathscr{J}_U+g_1u)-\frac{1}{2i}\mathscr{K}_U[g(z)]-g_1\Big(\mathrm{Im}\, u-\frac{1}{2i}\mathscr{K}_U[z]\Big)\\ &= 0 \qquad (z=(w, z_1)),\end{aligned}$$

which proves (5. 37). To prove the second assertion, it is enough to show that $\tilde{V}(\mathrm{Ker}\,\rho_U)^\circ$ is transitive on each fiber of Φ. Let $(u, w, z_1), (u', w', z_1') \in U_c\times V_+\times\mathscr{D}_1$ and suppose

$$\Phi(u, w, z_1) = \Phi(u', w', z_1').$$

We will show that there exists $g\in\tilde{V}(\mathrm{Ker}\,\rho_U)^\circ$ such that $g(u, w, z_1)=(u', w', z_1')$. Since, by the assumption, $\tilde{V}(\mathrm{Ker}\,\rho_U)^\circ$ is transitive on $V_+\times\mathscr{D}_1$, one may assume that $(w, z_1)=(w', z_1')$. Then one has $\mathrm{Im}\, u=\mathrm{Im}\, u'$, i. e., $u-u'\in U$. Hence, putting $a=u'-u$, one has $\exp a\in\tilde{V}$ and $(\exp a)(u, w, z_1)=(u', w, z_1)$, as was to be shown, q. e. d.

When there exists an open G_1°-orbit Ω in U, the inverse image $\mathscr{S}=\Phi^{-1}(\Omega)$ is an open G°-orbit, which is a generalized form of the (homogeneous) "Siegel domain of the third kind", to be defined in the next section. It is important to observe that $\mathscr{S}$ has the following two (equivariant) projections:

$$\begin{array}{ccc} & \mathscr{S}\subset U_c\times V_+\times\mathfrak{p}_{1+} & \\ {}^{\Phi}\swarrow & & \searrow \\ \Omega\subset U & & \mathscr{D}_1\subset\mathfrak{p}_{1+}\end{array}$$

Exercises

In these Exercises, we study the "change of variables" in the domain $\mathscr{D}_1$ and $\mathscr{D}$. Let G_1 be a group of H-C type and fix an element σ of G_{1C} such that $\sigma(z_1)$ is defined for all $z_1\in\mathscr{D}_1$, i. e., such that $\sigma G_1^\circ\subset P_{1+}K_{1C}P_{1-}$. Define a new variable $z_1^\wedge$ by

(5. 38) $$z_1^\wedge = (ad\,\sigma)^{-1}\sigma(z_1)$$

for $z_1\in\mathscr{D}_1\subset\mathfrak{p}_{1+}$. Then

(5. 39) $$z_1^\wedge\in\mathscr{D}_1^\wedge = (ad\,\sigma)^{-1}\sigma(\mathscr{D}_1)\subset(ad\,\sigma)^{-1}\mathfrak{p}_{1+}.$$

Hence one has $\sigma(\exp z_1^\wedge)\sigma^{-1}\in P_{1+}$. For $g_1\in G_1^\circ$ and $z_1^\wedge, z_1'^\wedge\in\mathscr{D}_1^\wedge$, we define

(5. 40) $$\begin{cases} g_1(z_1^\wedge) = (ad\,\sigma)^{-1}\log((\sigma g_1(\exp z_1^\wedge)\sigma^{-1})_+),\\ J_1^\wedge(g_1, z_1^\wedge) = \sigma^{-1}(\sigma g_1(\exp z_1^\wedge)\sigma^{-1})_0\sigma,\\ K_1^\wedge(z_1^\wedge, z_1'^\wedge) = \sigma^{-1}((\bar{\sigma}(\exp\overline{z_1'^\wedge})^{-1}\bar{\sigma}^{-1}\sigma(\exp z_1^\wedge)\sigma^{-1})_0)^{-1}\bar{\sigma}.\end{cases}$$

1. Prove the following formulas:

(5. 41) $$\begin{cases} g_1(z_1^\wedge) = (ad\,\sigma)^{-1}\sigma g_1(z_1) = g_1(z_1)^\wedge,\\ J_1^\wedge(g_1, z_1^\wedge) = \sigma^{-1}J_1(\sigma, g_1(z_1))J_1(g_1, z_1)J_1(\sigma, z_1)^{-1}\sigma,\\ K_1^\wedge(z_1^\wedge, z_1'^\wedge) = \sigma^{-1}J_1(\sigma, z_1)K_1(z_1, z_1')\overline{J_1(\sigma, z_1')}^{-1}\bar{\sigma}.\end{cases}$$

Also check that the formulas similar to (II. 5. 5), (II. 5. 9) remain valid.

2. Let $G=\tilde{V}G_1$ be the semi-direct product defined in Example 2. Then from $\sigma G_1^\circ\subset P_{1+}K_{1C}P_{1-}$

we obtain $\sigma G^\circ \subset P_+ K_C P_-$. Hence we can define a new variable

$$z^\wedge = (ad\,\sigma)^{-1}\sigma(z) = (w^\wedge, z_1{}^\wedge)$$

for $z=(w, z_1) \in V_+ \times \mathcal{D}_1$. Then $z_1{}^\wedge$ is given by (5. 38) and

$$w^\wedge = (ad\,\sigma)^{-1} J(\sigma, z_1) w \in (ad\,\sigma)^{-1} V_+. \tag{5. 42}$$

(Note that $w^\wedge$ depends not only on w but also on z_1.) Show that, for $g=(a, b, g_1) \in G^\circ$, one has

$$g(z^\wedge) = g(z)^\wedge = (b_{g_1(z_1)^\wedge} + J_1{}^\wedge(g_1, z_1{}^\wedge) w^\wedge, g_1(z_1{}^\wedge)), \tag{5. 43}$$

where $b_{g_1(z_1)^\wedge} = \sigma^{-1}(\sigma(b)_{\sigma g_1(z_1)})$. [In general, setting $v_{z_1{}^\wedge} = \sigma^{-1}(\sigma(v)_{\sigma(z_1)})$, prove the relation $(v_{z_1}, z_1)^\wedge = (v_{z_1{}^\wedge}, z_1{}^\wedge)$.]

3. Suppose further that $\rho_U(\sigma)=1$. [Then one has also $\rho_U(J_1(\sigma, z_1))=1$.] Defining $J^\wedge$, $K^\wedge$ by formulas similar to (5. 40) or (5. 41) (which coincide), show that

$$J^\wedge(g, z^\wedge) = (\mathcal{J}_U{}^\wedge, 0, J_1{}^\wedge(g_1, z_1{}^\wedge)),$$
$$K^\wedge(z^\wedge, z'^\wedge) = (\mathcal{K}_U{}^\wedge, 0, K_1{}^\wedge(z_1{}^\wedge, z_1'{}^\wedge)),$$

where

$$\mathcal{J}_U{}^\wedge = \mathcal{J}_U - 2A(b_{g_1(z_1)} + J_{g_1} w, b_{g_1(z_1)^\wedge} + J_{g_1}{}^\wedge w^\wedge) + 2A(w, w^\wedge), \tag{5. 44}$$

$$\mathcal{K}_U{}^\wedge = \mathcal{K}_U - 2A(w, w^\wedge) + 2A(\overline{w'}, \overline{w'^\wedge}), \tag{5. 45}$$

where $J_{g_1} = J_1(g_1, z_1)$, $J_{g_1}{}^\wedge = J_1{}^\wedge(g_1, z_1{}^\wedge)$. [Note that $J(\sigma, z) = (-2A(w, w^\wedge), 0, J_1(\sigma, z_1))$.]

4. Assume further that $(ad\,\sigma)^{-1} V_\pm$ are real. This means that there exist a pair of totally isotropic subspaces V_1, V_2 of V such that $V_1 \oplus V_2 = V$ and

$$(ad\,\sigma)^{-1} V_+ = V_{1C}, \qquad (ad\,\sigma)^{-1} V_- = V_{2C}. \tag{5. 46}$$

Let $\varphi^\wedge(z_1{}^\wedge) = \rho_V(z_1{}^\wedge)|V_{2C}$. Then, in the matrix expression according to the direct decomposition $V_C = V_{1C} \oplus V_{2C}$, one has

$$\rho_V(K_1{}^\wedge(z_1{}^\wedge, z_1'{}^\wedge)) = \begin{pmatrix} 0 & -(\varphi^\wedge(z_1{}^\wedge) - \overline{\varphi^\wedge(z_1'{}^\wedge)}) \\ (\varphi^\wedge(z_1{}^\wedge) - \overline{\varphi^\wedge(z_1'{}^\wedge)})^{-1} & 0 \end{pmatrix}, \tag{5. 47}$$

$$\mathcal{K}_U{}^\wedge = -2A(w^\wedge - \overline{w'^\wedge}, (\varphi^\wedge(z_1{}^\wedge) - \overline{\varphi^\wedge(z_1'{}^\wedge)})^{-1}(w^\wedge - \overline{w'^\wedge})). \tag{5. 48}$$

Hint. In the matrix expression relative to (V_+, V_-), let $\rho_V(\sigma) = \begin{pmatrix} \alpha & \beta \\ \gamma & \delta \end{pmatrix}$, $\rho_V(\sigma)^{-1} = \begin{pmatrix} \alpha' & \beta' \\ \gamma' & \delta' \end{pmatrix}$. Then $\rho_V(\bar\sigma) = \begin{pmatrix} \bar\delta & \bar\gamma \\ \bar\beta & \bar\alpha \end{pmatrix}$ and by the assumption $\rho_V(\bar\sigma\sigma^{-1})$ is of the form $\begin{pmatrix} 0 & \mu \\ \lambda & 0 \end{pmatrix}$, whence $\bar\alpha = \lambda\beta$, $\bar\beta = \lambda\alpha$, $\bar\gamma = \mu\delta$, $\bar\delta = \mu\gamma$ and so $\bar\lambda = \lambda^{-1}$, $\bar\mu = \mu^{-1}$. Also, putting $\sigma_i = \rho_V(\sigma)|V_{iC}$ $(i=1, 2)$, one has $\bar\sigma_1 = \lambda\sigma_1$, $\bar\sigma_2 = \mu\sigma_2$. Then the equation (5. 47) follows from (5. 26), (5. 41) and the relation

$$\begin{aligned} &\sigma_1^{-1}(\alpha' - \varphi(z_1)\gamma')^{-1}(1 - \varphi(z_1)\overline{\varphi(z_1')})\overline{(\gamma\varphi(z_1') + \delta)}^{-1}\mu\sigma_2 \\ &= -\varphi^\wedge(z_1{}^\wedge) + \overline{\varphi^\wedge(z_1'{}^\wedge)}. \end{aligned} \tag{*}$$

Similarly (5. 48) follows from (5. 28), (5. 45) and the relation

$$\begin{aligned} &\overline{\varphi(z_1')}(1 - \varphi(z_1)\overline{\varphi(z_1')})^{-1} - \gamma'(\alpha' - \varphi(z_1)\gamma')^{-1} \\ &= -(1 - \overline{\varphi(z_1')}\varphi(z_1))^{-1}\overline{(\alpha' - \varphi(z_1')\gamma')}\lambda(\alpha' - \varphi(z_1)\gamma')^{-1}. \end{aligned} \tag{**}$$

§ 6. The notion of Siegel domains.

We start with the definition of a Siegel domain of the second kind, which in this book will simply be called a "Siegel domain".

Suppose there are given data (U, V, Ω, H), where U is a vector space over $\boldsymbol{R}$ of dimension $m(>0)$ and V is a vector space over $\boldsymbol{C}$ of dimension n, which is also viewed as a $2n$-dimensional real vector space with a fixed complex structure I_0;

Ω is a (non-degenerate) open convex cone in U (with vertex at the origin) and H is a hermitian map $V \times V \to U_C$ (C-linear in the second variable and C-antilinear in the first). We assume that H is *Ω-positive,* which means that one has

(6. 1) $$H(v, v) \in \bar{\Omega} - \{0\} \qquad \text{for all } v \in V,\ v \neq 0.$$

Then by definition the *Siegel domain* $\mathcal{S} = \mathcal{S}(U, V, \Omega, H)$ is given by

(6. 2) $$\mathcal{S} = \{(u, v) \in U_C \times V \mid \operatorname{Im} u - H(v, v) \in \Omega\}.$$

A Siegel domain $\mathcal{S}$ is convex, and hence is a domain of holomorphy. In fact, for $(u, v), (u', v') \in \mathcal{S}$ and $0 < \lambda < 1$, one has

$$\begin{aligned}&\operatorname{Im}(\lambda u + (1-\lambda)u') - H(\lambda v + (1-\lambda)v', \lambda v + (1-\lambda)v')\\ &= \lambda(\operatorname{Im} u - H(v, v)) + (1-\lambda)(\operatorname{Im} u' - H(v', v'))\\ &\quad + \lambda(1-\lambda)H(v-v', v-v')\\ &\in \Omega,\end{aligned}$$

which shows that $(\lambda u + (1-\lambda)u', \lambda v + (1-\lambda)v') \in \mathcal{S}$.

Example 1. When $V = \{0\}$, we obtain a "tube domain"

(6. 3) $$\mathcal{S} = U + i\Omega = \{u \in U_C \mid \operatorname{Im} u \in \Omega\},$$

which is also called a *Siegel domain of the first kind.* In general, any Siegel domain (of the second kind) $\mathcal{S}(U, V, \Omega, H)$ contains a tube domain $U + i\Omega$ as the zero section $\{v=0\}$.

Example 2. Let $U = \boldsymbol{R}$, $V = \boldsymbol{C}^n$, and let H be the standard hermitian form

$$H(v, v) = \sum_{\alpha=1}^{n} |v_\alpha|^2 \qquad (v = (v_\alpha) \in \boldsymbol{C}^n).$$

Then the corresponding Siegel domain

(6. 4) $$\mathcal{S} = \left\{(u, v) \in \boldsymbol{C}^{n+1} \middle| \operatorname{Im} u - \sum_{\alpha=1}^{n} |v_\alpha|^2 > 0\right\}$$

is holomorphically equivalent to a "unit ball":

$$\left\{z = (z_k) \in \boldsymbol{C}^{n+1} \middle| \sum_{k=1}^{n+1} |z_k|^2 < 1\right\}.$$

In fact, in the new variables

$$z_1 = \frac{u-i}{u+i}, \qquad z_k = \frac{2v_{k-1}}{u+i} \quad (2 \leq k \leq n+1),$$

one has $(u, v) \in \mathcal{S}$ if and only if

$$1 - \sum_{k=1}^{n+1} |z_k|^2 = \frac{4}{|u+i|^2}\left(\operatorname{Im} u - \sum_{k=2}^{n+1} |v_{k-1}|^2\right) > 0.$$

Proposition 6. 1. *Any Siegel domain is holomorphically equivalent to a bounded domain.*

Proof. Since Ω is non-degenerate, we can take the coordinate system in U in such a way that for $u=(u_k)\in\Omega$ one has $u_1>0, \cdots, u_m>0$. Let $H(v, v')=(H_k(v, v'))$. Then by (6. 1) each H_k is a positive semi-definite hermitian form on V, so that we can write

$$H_k(v, v) = \sum_{j=1}^{s_k} |f_{kj}(v)|^2,$$

where f_{kj}'s are $\boldsymbol{C}$-linear forms on V. Also by (6. 1) one has $\sum H_k \gg 0$, so that the set $\{f_{kj}(1\leq k\leq n, 1\leq j\leq s_k)\}$ contains n linearly independent forms. After a suitable rearrangement of indices, we may assume that the linear forms $f_{kj}(1\leq k\leq m, 1\leq j \leq n_k$, where $\sum_{k=1}^{m} n_k=n)$ are linearly independent. Then, defining new coordinates in V by $v_j^{(k)}=f_{kj}(v)$ $(1\leq k\leq m, 1\leq j\leq n_k)$, we obtain

$$\begin{aligned} \mathcal{S}(U, V, \Omega, H) &\subset \Big\{(u, v) \in U_c \times V \,\Big|\, \operatorname{Im} u_k - \sum_{j=1}^{n_k} |f_{kj}(v)|^2 > 0 \; (1\leq k\leq m)\Big\} \\ &\approx \mathcal{S}_1 \times \cdots \times \mathcal{S}_m, \end{aligned}$$

where

$$\mathcal{S}_k = \Big\{(u_k, v_1^{(k)}, \cdots, v_{n_k}^{(k)}) \in \boldsymbol{C}^{n_k+1} \,\Big|\, \operatorname{Im} u_k - \sum_{j=1}^{n_k} |v_j^{(k)}|^2 > 0\Big\}.$$

As each $\mathcal{S}_k$ is holomorphically equivalent to an open ball (Example 2), $\mathcal{S}$ is equivalent to a bounded domain, q. e. d.

Remark. The above proof is taken from Piatetskii-Shapiro [2]. It has been shown (Vinberg-Gindikin-Piatetskii-Shapiro [1]) that, conversely, every *homogeneous* bounded domain is holomorphically equivalent to a Siegel domain, uniquely determined up to a linear equivalence, and is in one-to-one correspondence with a so-called "normal j-algebra". For symmetric domains of classical type, Piatetskii-Shapiro gave, in the above book, expressions as Siegel domains of the second and third kinds. Here we are following the method of Korányi-Wolf [1] and Wolf-Korányi [1] to prove this for general symmetric domains.

By virtue of Proposition 6. 1, the group of holomorphic automorphisms $\operatorname{Hol}(\mathcal{S})$ has a structure of Lie group (cf. II, § 6). We denote by $\operatorname{Aff}(\mathcal{S})$ the subgroup of $\operatorname{Hol}(\mathcal{S})$ formed of all (complex) affine transformations of $U_c \oplus V$ leaving $\mathcal{S}$ stable.

Proposition 6. 2. *For a Siegel domain $\mathcal{S}=\mathcal{S}(U, V, \Omega, H)$, the group $\operatorname{Aff}(\mathcal{S})$ consists of all affine transformations of the form*

$$\text{(6. 5)} \qquad \begin{cases} u' = gu+2iH(b, lv)+a+iH(b, b), \\ v' = lv+b, \end{cases}$$

where $g\in G(\Omega)$, $l\in GL(V, I_0)$, $a\in U$, $b\in V$ and g, l satisfy the relation

$$\text{(6. 6)} \qquad gH(v, v') = H(lv, lv') \qquad (v, v'\in V).$$

We leave the proof for an exercise of the reader (cf. Piatetskii-Shapiro [2]).

We denote the imaginary part of $H(v, v')$ by $A(v, v')$. Then $A: V\times V\to U$ is an alternating $\boldsymbol{R}$-bilinear map satisfying the condition that AI_0 is symmetric and "Ω-

positive" [i. e., $A(v, I_0v')$ is symmetric in $v, v' \in V$ and $A(v, I_0v) \in \bar{\Omega} - \{0\}$ for $v \neq 0$], and one has

$$H(v, v') = A(v, I_0v') + iA(v, v'). \tag{6.7}$$

Let $V_\pm = V_C(I_0; \pm i)$. Then, as in § 5, Example 1, one has

$$A|V_+ \times V_+ = 0, \qquad A|V_- \times V_- = 0.$$

Hence, denoting the $V_\pm$-components of v, v' by $v_\pm, v'_\pm$, one obtains

$$H(v, v') = 2iA(v_-, v'_+). \tag{6.8}$$

By the natural isomorphism $(V, I_0) \cong V_+$ given by the projection $v \mapsto v_+$, the Siegel domain (6. 2) is transformed into one in $U_C \times V_+$:

$$\{(u, w) \in U_C \times V_+ \,|\, \mathrm{Im}\, u - 2iA(\bar{w}, w) \in \Omega\}. \tag{6.2'}$$

In this expression, the affine transformation (6. 5) becomes a special case of quasi-linear transformations (5. 34) where one takes as $G^\circ = K^\circ$ the group

$$\{(g, l) \,|\, g \in G(\Omega), l \in GL(V, I_0), gH(v, v') = H(lv, lv') \ (v, v' \in V)\}^\circ.$$

It will be convenient to use the "dual expressions", as we did in § 3. Fix an inner product (i. e., a positive definite symmetric bilinear form) $\langle\ \rangle$ in U and, for $u \in U$, put

$$\begin{aligned} H_u(v, v') &= \langle u, H(v, v')\rangle, \\ A_u(v, v') &= \langle u, A(v, v')\rangle \qquad (v, v' \in V). \end{aligned} \tag{6.9}$$

Then H_u is a hermitian form on V and A_u is the imaginary part of H_u. Let Ω^* be the dual cone of Ω (I, § 8) and $e \in \Omega^*$ a reference point. Then by (6. 1) H_u is positive definite for $u \in \Omega^*$; in particular, H_e, and hence A_e, is non-degenerate. Therefore we can write

$$A_u(v, v') = 2A_e(v, R_uv') \tag{6.10}$$

with $R_u \in \mathrm{End}(V)$. Then, since both A_e and A_u are alternating and both A_eI_0 and A_uI_0 are symmetric, R_u is symmetric with respect to A_e and commutes with I_0. Thus we have $R_u \in \mathcal{H}(V, I_0, A_e)$ (cf. § 3). Moreover, if $u \in \Omega^*$, then $A_uI_0 = 2A_eI_0R_u$ is positive definite, so that one has $R_u \in \mathcal{P}(V, I_0, A_e)$; in particular, $R_e = \frac{1}{2}1_V$. Clearly to give an Ω-positive hermitian map $H: (V, I_0) \times (V, I_0) \to U_C$, or what amounts to the same thing, to give an alternating bilinear map $A: V \times V \to U$ such that AI_0 is symmetric and Ω-positive, is equivalent, by the relations (6. 9), (6. 10), to giving a non-degenerate alternating bilinear form A_e on V along with a linear map R: $U \to \mathcal{H}(V, I_0, A_e)$ satisfying the conditions $R(\Omega^*) \subset \mathcal{P}(V, I_0, A_e)$ and $R_e = \frac{1}{2}1_V$.

Now we want to determine how one can deform the complex structure of V without destroying the structure of Siegel domain $\mathcal{S}$. Such a deformation is parametrized by a point of the following space $\mathfrak{S}$ which generalizes the notion of "Siegel space". Namely, for the given data (U, V, Ω, A) (where V is now viewed as a real vector space), we define the *generalized Siegel space* (or "classifying space") $\mathfrak{S} = \mathfrak{S}(V, A, \Omega)$ to be the set of all complex structures I on V such that AI is symmetric and Ω-positive. In view of the fact that for $u \in U$ one has

$$u \in \bar{\Omega} - \{0\} \Longleftrightarrow \langle u, u' \rangle > 0 \quad \text{for all } u' \in \Omega^*,$$

these conditions on I are equivalent to saying

(6. 11 a) $A_u I$ is symmetric for all $u \in U$, and

(6. 11 b) $A_u I$ is positive definite for all $u \in \Omega^*$.

Hence one has

$$\mathfrak{S} = \bigcap_{u \in \Omega^*} \mathfrak{S}(V, A_u), \tag{6. 12}$$

where $\mathfrak{S}(V, A_u)$ is the ordinary Siegel space defined in II, § 7. In particular, one has $I_0 \in \mathfrak{S} \subset \mathfrak{S}(V, A_e)$.

We now use the realization $\mathfrak{S}(V, A_e) \subset \mathrm{Sym}(V_-, V_+)$ (II, § 7). Recall that $I \in \mathfrak{S}(V, A_e)$ corresponds in one-to-one way to a $\boldsymbol{C}$-linear map $z : V_- \to V_+$ satisfying the conditions

$$^t z = z, \qquad 1 - z\bar{z} \gg 0, \tag{6. 13}$$

where t denotes the transpose with respect to the (non-degenerate) bilinear form $A_e | V_- \times V_+$ (Lem. II. 7. 2). The correspondence $I \leftrightarrow z$ is defined by the relation

$$V_-(I) = (1 + z) V_-, \tag{6. 14}$$

where $V_\pm(I) = V_C(I; \pm i)$; in particular, $I_0 \leftrightarrow 0$. In what follows, we write I_z when $I \leftrightarrow z$ and consider $\mathfrak{S}$ and $\mathfrak{S}(V, A_e)$ as the spaces of z's rather than those of I's. Then, since $V_+(I_z) = (1 + \bar{z}) V_+$, one has in the matrix expression according to the decomposition $V_C = V_+ \oplus V_-$

$$\begin{aligned} I_z &= \begin{pmatrix} 1 & z \\ \bar{z} & 1 \end{pmatrix} \begin{pmatrix} i1 & 0 \\ 0 & -i1 \end{pmatrix} \begin{pmatrix} 1 & z \\ \bar{z} & 1 \end{pmatrix}^{-1} \\ &= i \begin{pmatrix} (1 + z\bar{z})(1 - z\bar{z})^{-1} & -2z(1 - \bar{z}z)^{-1} \\ 2\bar{z}(1 - z\bar{z})^{-1} & -(1 + \bar{z}z)(1 - \bar{z}z)^{-1} \end{pmatrix}. \end{aligned} \tag{6. 15}$$

It follows that the projection map $V_C \to V_+(I_z)$ is given by

$$\begin{aligned} \frac{1}{2}(1 - iI_z) &= \begin{pmatrix} (1 - z\bar{z})^{-1} & -(1 - z\bar{z})^{-1} z \\ \bar{z}(1 - z\bar{z})^{-1} & -\bar{z}(1 - z\bar{z})^{-1} z \end{pmatrix} \\ &= \begin{pmatrix} 1 \\ \bar{z} \end{pmatrix} (1 - z\bar{z})^{-1} (1, -z) \end{aligned} \tag{6. 16}$$

and a $\boldsymbol{C}$-linear isomorphism $(V, I_z) \to (V, I_0)$ is given by

$$\begin{aligned} \frac{1}{2}(1 - I_0 I_z) &= \begin{pmatrix} (1 - z\bar{z})^{-1} & -(1 - z\bar{z})^{-1} z \\ -(1 - \bar{z}z)^{-1} \bar{z} & (1 - \bar{z}z)^{-1} \end{pmatrix} \\ &= \begin{pmatrix} 1 & z \\ \bar{z} & 1 \end{pmatrix}^{-1}. \end{aligned} \tag{6. 17}$$

From (6. 17) we see that the map $v \mapsto v_z = v_+ - z v_-$ gives a $\boldsymbol{C}$-linear isomorphism $(V, I_z) \to V_+$ (II, § 7, Exerc. 1).

Now, let $z \in \mathfrak{S}(V, A_e)$. By the definition, one has $z \in \mathfrak{S}$ if and only if I_z satisfies the conditions (6. 11 a, b). Since $A_u I_z = 2 A_e R_u I_z$ and $R_u \in \mathcal{H}(V, I_0, A_e)$, it is clear that (6. 11 a) is equivalent to

$$I_z R_u = R_u I_z \qquad \text{for all } u \in U. \tag{6. 11'}$$

Moreover, one has $A_u I_z = 2(A_e I_0) R_u (I_0^{-1} I_z)$, where $A_e I_0 R_u \gg 0$ for all $u \in \Omega^*$ and $A_e I_z \gg 0$ for all $z \in \mathfrak{S}(V, A_e)$. Hence, under (6. 11′), the condition (6. 11 b) is automatically satisfied. Thus we see that $\mathfrak{S}$ consists of those elements $z \in \mathfrak{S}(V, A_e)$ satisfying the condition (6. 11′) ; in particular, it depends only on (V, A, e) and not on Ω itself. It is clear that in the matrix expression relative to (V_+, V_-), one has

$$(6.\ 18)\qquad R_u = \begin{pmatrix} R_u^+ & 0 \\ 0 & R_u^- \end{pmatrix},\qquad R_u^- = \bar{R}_u^+ = {}^t R_u^+.$$

Hence, in view of (6. 17), I_z commutes with R_u if and only if one has $R_u^+ z = z R_u^-$. Thus we conclude that

$$(6.\ 19)\qquad \mathfrak{S}(V, A, \Omega) = \{z \in \mathfrak{S}(V, A_e) \mid R_u^+ z = z R_u^- \text{ for all } u \in U\}.$$

We define the *generalized symplectic group* $Sp(V, A)$ by

$$(6.\ 20)\qquad Sp(V, A) = \{g \in GL(V) \mid A(gv, gv') = A(v, v') \text{ for all } v, v' \in V\}.$$

Then, by a similar argument as above, we obtain

$$(6.\ 21)\qquad Sp(V, A) = \{g \in Sp(V, A_e) \mid R_u g = g R_u \text{ for all } u \in U\}.$$

Proposition 6. 3. *$Sp(V, A)$ is a reductive $\boldsymbol{R}$-group of hermitian type and $\mathfrak{S} = \mathfrak{S}(V, A, \Omega)$ is the associated symmetric domain.*

Proof. Let $G = Sp(V, A)$, $G' = Sp(V, A_e)$, $\mathfrak{g} = \text{Lie } G$ and $\mathfrak{g}' = \text{Lie } G'$. Then by (6. 21) it is clear that G is an $\boldsymbol{R}$-subgroup of G'. We know (II, § 7) that a global Cartan involution of G' is given by

$$\theta : g' \longmapsto I_0^{-1} g' I_0$$

and $H_0 = \frac{1}{2} I_0$ is the corresponding H-element in $\mathfrak{g}'$. Let $\mathfrak{g}'_c = \mathfrak{p}'_+ + \mathfrak{k}'_c + \mathfrak{p}'_-$ be the corresponding canonical decomposition. As the $R_u\,(u \in U)$ commute with I_0, G is θ-stable and hence reductive. Since $H_0 \in \mathfrak{g}$, $(\mathfrak{g}, H_0)$ is also of hermitian type, and the inclusion map $\mathfrak{g} \to \mathfrak{g}'$ satisfies the condition (H_2). Hence the canonical decomposition of $\mathfrak{g}_c$ is given by $\mathfrak{g}_c = \mathfrak{p}_+ + \mathfrak{k}_c + \mathfrak{p}_-$ where $\mathfrak{k}_c = \mathfrak{k}'_c \cap \mathfrak{g}_c$, $\mathfrak{p}_\pm = \mathfrak{p}'_\pm \cap \mathfrak{g}_c$. Therefore, by (6. 19), it is clear that the symmetric domain (in $\mathfrak{p}_+$) associated with G is contained in $\mathfrak{S}$. It remains to show that G° is transitive on $\mathfrak{S}$. Let $z \in \mathfrak{S}(V, A_e)$ and put

$$g_z = (\exp z) K(z, z)^{\frac{1}{2}} (\exp \bar{z}).$$

Then we know that $g_z \in G'$ and $z = g_z(o)$ (II, § 5, Exerc. 2). It is clear that, if $z R_u^- = R_u^+ z$, then one has $g_z R_u = R_u g_z$. Hence $z \in \mathfrak{S}$ implies $g_z \in G^\circ$. Thus $\mathfrak{S} \subset G^\circ(o)$, as desired, q. e. d.

We note that $\mathfrak{S}$ is always of classical type (*I, II, III*) as is seen from the result of IV, § 5 (cf. Satake [20]).

Now, to each $z \in \mathfrak{S}$, there corresponds an Ω-positive hermitian map $H_z = A I_z + iA$. By (6. 16) one has

$$\begin{aligned} H_z(v, v') &= A(v, (I_z + i1)v') \\ &= 2iA(v, (1+\bar{z})(1-z\bar{z})^{-1}v'_z) \\ &= 2iA(\bar{v}_z, (1-z\bar{z})^{-1}v'_z). \end{aligned}$$

Hence, if we put

(6. 22) $$H_z^+(w, w') = 2iA(\bar{w}, (1-z\bar{z})^{-1}w')$$

for $w, w' \in V_+$, then we have

(6. 23) $$H_z(v, v') = H_z^+(v_z, v'_z).$$

We put further

(6. 24) $$\mathcal{L}_z(w, w') = 2iA(v, w'),$$

where $w = v_z$. Then, since $v_- = (1-\bar{z}z)^{-1}(\bar{w} + \bar{z}w)$ ((II. 7. 25)), one has

(6. 25) $$\begin{aligned} \mathcal{L}_z(w, w') &= 2iA((1-\bar{z}z)^{-1}(\bar{w}+\bar{z}w), w') \\ &= H_z^+(w + z\bar{w}, w'), \end{aligned}$$

which shows that $\mathcal{L}_z$ is "quasi-hermitian", i. e., a sum of a hermitian map and a symmetric $\boldsymbol{C}$-bilinear map.

In these notations, the *universal Siegel domain of the third kind* is defined by

(6. 26) $$\mathcal{S} = \{(u, w, z) \in U_C \times V_+ \times \mathfrak{S} \mid \mathrm{Im}\, u - \mathrm{Re}\, \mathcal{L}_z(w, w) \in \Omega\}.$$

$\mathcal{S}$ may be viewed as a fiber space over $\mathfrak{S} = \mathfrak{S}(V, A, \Omega)$. Then the fiber over $z \in \mathfrak{S}$ is holomorphically equivalent to the Siegel domain (of the second kind) $\mathcal{S}_z = \mathcal{S}(U, V, I_z, \Omega, H_z)$ by the correspondence

$$\mathcal{S}_z \ni (u, v) \longmapsto (u + iH_z^+(z\bar{v}_z, v_z), v_z, z) \in \mathcal{S}.$$

Note that in general this map does not depend holomorphically on $z \in \mathfrak{S}$.

More generally, given a bounded domain $\mathcal{D}$ and a holomorphic map $\varphi : \mathcal{D} \to \mathfrak{S}$, we define the *associated Siegel domain of the third kind* $\mathcal{S}_\varphi$ by pulling back the universal one, i. e., by

(6. 27) $$\mathcal{S}_\varphi = \{(u, w, t) \in U_C \times V_+ \times \mathcal{D} \mid \mathrm{Im}\, u - \mathrm{Re}\, \mathcal{L}_{\varphi(t)}(w, w) \in \Omega\}.$$

It can be shown that $\mathcal{S}_\varphi$ is also holomorphically equivalent to a bounded domain (Piatetskii-Shapiro [2]).

Exercises

Let $G = Sp(V, A)$ be the generalized symplectic group and $K = Sp(V, A) \cap GL(V, I_0)$. Then by Proposition 6. 3 G is of hermitian type and hence of H-C type. Moreover, letting $\rho_V(g) = g$, $\rho_U(g) = 1_U$ for $g \in G$, we see at once that the conditions in § 5, Example 2 are satisfied. Hence the formula obtained there can be applied to this case. In particular, by (5. 26) one has

$$K(z, z') = \begin{pmatrix} 1 - z\bar{z}' & 0 \\ 0 & (1-\bar{z}'z)^{-1} \end{pmatrix}$$

and hence

(6. 22′) $$H_z^+(w, w') = 2iA(\bar{w}, K(z, z)^{-1}w')$$

for $z \in \mathfrak{S}$, $w, w' \in V_+$.

1. Suppose there is given $\sigma \in G_C$ such that both $\sigma^{-1}V_\pm$ are real. Show that $\sigma(z)$ is then defined for all $z\in\mathfrak{S}$. In the notation of § 5, Exercises, one has by (5. 47)

$$K^{\wedge}(z^{\wedge}, z'^{\wedge}) = \begin{pmatrix} 0 & -(z^{\wedge}-\overline{z'^{\wedge}}) \\ (z^{\wedge}-\overline{z'^{\wedge}})^{-1} & 0 \end{pmatrix}.$$

Show that one has

$$H_z^+(w, w') = -A(\overline{w^{\wedge}}, (\operatorname{Im} z^{\wedge})^{-1}w'^{\wedge}). \tag{6. 28}$$

Moreover, putting

$$\mathcal{L}_{z^{\wedge}}(w^{\wedge}, w'^{\wedge}) = 2iA(\operatorname{Im} w^{\wedge}, (\operatorname{Im} z^{\wedge})^{-1}w'^{\wedge}), \tag{6. 29}$$

(which is a quasi-hermitian form on V_{1C}), show that

$$\mathcal{L}_{z^{\wedge}}(w^{\wedge}, w'^{\wedge}) = \mathcal{L}_z(w, w')-2iA(J(\sigma, z)w, \sigma w'). \tag{6. 30}$$

[This follows from the relation (**) used in the proof of (5. 48).] Note that, in this setting, $V_1=(\sigma^{-1}V_+)\cap V$ is a maximal totally isotropic subspace of V stable under all $R_u\,(u\in U)$.

2. Conversely, suppose there is given a maximal totally isotropic subspace V_1 of V which is stable under all $R_u\,(u\in U)$. Let $R_u^{(1)}=R_u|V_{1C}$ and $\sigma_1=\frac{1}{2}(1-iI_0)|V_{1C}$.

2. 1) Show that σ_1 is a $\boldsymbol{C}$-linear isomorphism $V_{1C}\to V_+$ satisfying the following conditions:

$$\sigma_1 R_u^{(1)} = R_u^+\sigma_1, \qquad A_e(\bar{v}_1, I_0 v_1') = 2iA_e(\bar{\sigma}_1(\bar{v}_1), \sigma_1(v_1'))$$

for $v_1, v_1'\in V_{1C}$. It follows that we have the following commutative diagram

$$\begin{array}{ccc} \Omega & \xrightarrow{R^{(1)}} & \mathcal{P}(V_1, S_1) \\ & {\scriptstyle R^+}\searrow & \downarrow\cap \\ & & \mathcal{P}(V_+, H_0^+) \approx \mathcal{P}(V, I_0, A_e) \end{array} \tag{6. 31}$$

where $S_1=A_eI_0|V_1\times V_1$ and the vertical injection is given by $t\mapsto\sigma_1\circ t\circ(\sigma_1|V_{1C})^{-1}$.

2. 2) Let $V_2=I_0V_1$ and $\sigma_2=2i\bar{\sigma}_1\circ(I_0|V_{2C})$. Show that V_2 is also a maximal totally isotropic subspace of V, stable under all $R_u\,(u\in U)$, and one has $\sigma=\sigma_1\oplus\sigma_2\in Sp(V, A)_C$.

2. 3) For $z\in\mathfrak{S}(V, A_e)$, put

$$z^{(1)} = -z^{\wedge}I_0 = 2i\sigma_1^{-1}\circ\sigma(z)\circ\bar{\sigma}_1 \tag{6. 32}$$

and $\mathfrak{S}^{(1)}=\{z^{(1)}\,|\,z\in\mathfrak{S}(V, A, \Omega)\}$. Show that

$$\mathfrak{S}^{(1)} = \{z^{(1)}\in \operatorname{Sym}(V_{1C}, S_1)\,|\,\operatorname{Im} z^{(1)}\gg 0,\ z^{(1)}R_u^{(1)}=R_u^{(1)}z^{(1)} \text{ for all } u\in U\}. \tag{6. 33}$$

Thus the generalized Siegel space $\mathfrak{S}$ is holomorphically equivalent to a tube domain $\mathfrak{S}^{(1)}$. If one puts

$$T = \{t\in \operatorname{Sym}(V_1, S_1)\,|\,tR_u^{(1)}=R_u^{(1)}t \text{ for all } u\in U\},$$

then $\mathfrak{S}^{(1)}=T+i(T\cap\mathcal{P}(V_1, S_1))$.

2. 4) Show that, by the transformation

$$\begin{cases} u^{\wedge} = u+2A(J(\sigma, z)w, \sigma w)\ (\in U_C), \\ w^{\wedge} = \sigma_1^{-1}J(\sigma, z)w\ (\in V_{1C}), \\ z^{(1)} = 2i\sigma_1^{-1}\circ\sigma(z)\circ\bar{\sigma}_1\ (\in T_C), \end{cases} \tag{6. 34}$$

the universal Siegel domain of the third kind $\mathcal{S}$ is holomorphically equivalent to a tube domain in $U_C\times V_{1C}\times T_C$ defined by the conditions

$$\begin{cases} \operatorname{Im} z^{(1)} \gg 0, \\ \operatorname{Im} u^{\wedge}-A(\operatorname{Im} w^{\wedge}, I_0(\operatorname{Im} z^{(1)})^{-1}\operatorname{Im} w^{\wedge})\in\Omega. \end{cases} \tag{6. 35}$$

3. Let $\mathcal{S}=\mathcal{S}(U, V, \Omega, H)$ be a Siegel domain and put

$$\Phi(z) = \Phi(u, v) = \operatorname{Im} u-H(v, v) \tag{6. 36}$$

for $z=(u, v)\in U_C\times V$. Let $k(z, z')$ be the Bergman kernel function of $\mathcal{S}$ (II, § 6) and put $k[z]=k(z, z)$. Prove the following.

3. 1) There exists a real analytic function λ on Ω such that

(6. 37) $$k[z] = \lambda(\Phi(z)).$$

3. 2) One has

(6. 38) $$\lambda(gu) = \det(g)^{-2}|\det(l)|^{-2}\lambda(u)$$

for any $g \in G(\Omega)$, $l \in GL(V, I_0)$ satisfying (6. 6).

3. 3) When Ω is homogeneous, one has

(6. 39) $$\lambda(u) = c_1\phi(u)^2\det(R(u^*)),$$

where ϕ and u^* are as defined in I, § 8 and c_1 is a positive constant.

3. 4) When Ω is homogeneous, self-dual and $2R$ is a Jordan algebra representation (i. e., when $\mathfrak{S}$ is "quasi-symmetric" in the sense to be defined in V), one has

(6. 40) $$\lambda(u) = c_2 \det(P(u))^{-(1+n/2m)},$$

where $m=\dim U$, $n=\dim V$, and c_2 is a positive constant.

Remark. λ can be extended to a holomorphic function on $\Omega+iU$ such that one has

(6. 41) $$k(z, z') = \lambda\left(\frac{1}{2i}(u-\bar{u}')-H(v, v')\right)$$

for $z=(u, v)$, $z'=(u', v') \in \mathfrak{S}$. In general, λ can be expressed by an integral of the form

$$\lambda(u) = c_3\int_{\Omega^*} e^{-\langle u,u'\rangle}\frac{\det R(u')}{\phi^*(u')}du',$$

where Ω^* is the dual of Ω, ϕ^* the "characteristic" of Ω^*, and c_3 is a positive constant (Rothaus [3], Th. 1).

4. For a given Siegel domain $\mathfrak{S}=\mathfrak{S}(U, V, \Omega, H)$, put

$$B(\mathfrak{S}) = \{z \in U_C\times V \mid \Phi(z)=0\}.$$

Prove the following.

4. 1) For any holomorphic function f defined on $\mathfrak{S}$, there exists a sequence $\{z_\nu\}$ in $\mathfrak{S}$ satisfying the following conditions :

(a) $\{z_\nu\}$ is either unbounded or converges to a point in $B(\mathfrak{S})$.

(b) $\lim_{\nu\to\infty} |f(z_\nu)| = \operatorname{Sup}_{z\in\mathfrak{S}}|f(z)|$.

Hint. For $z \in \mathfrak{S}$, $\zeta \in \boldsymbol{C}$, put

$$z(\zeta) = z+(\zeta-i)(\Phi(z), 0).$$

Then $\Phi(z(\zeta))=\operatorname{Im}\zeta\cdot\Phi(z)$, so that one has $z(\zeta)\in\mathfrak{S}$ if $\operatorname{Im}\zeta>0$, and $z(\zeta)\in B(\mathfrak{S})$ if $\operatorname{Im}\zeta=0$. First take any sequence $\{z_\nu\}$ in $\mathfrak{S}$ satisfying (b). Then replace it by a subsequence of $\{z_\nu(\zeta_\nu)\}$ for a suitably chosen $\{\zeta_\nu\}$ so that (a) is also satisfied.

4. 2) Put $z_0=(0, 0)$. There exists a function f holomorphic on $\bar{\mathfrak{S}}$ (i. e., in a neighbourhood of $\bar{\mathfrak{S}}$) such that $|f(z)|<|f(z_0)|$ for all $z\in\bar{\mathfrak{S}}$, $z\neq z_0$.

4. 3) The group $\operatorname{Aff}(\mathfrak{S})$ acts transitively on $B(\mathfrak{S})$. (In view of these properties, $B(\mathfrak{S})$ may be called the "Bergman-Šilov boundary" of the Siegel domain $\mathfrak{S}$.)

For Exercises 3, 4, cf. Rothaus [3] and Murakami [5].

§ 7. The realization of a symmetric domain as a Siegel domain of the third kind.

We now return to the situation of §§ 1~3. $\mathfrak{g}$ is a semi-simple Lie algebra of hermitian type with a fixed H-element H_0, $\kappa : \mathfrak{g}^1=\mathfrak{sl}_2(\boldsymbol{R})\to\mathfrak{g}$ is an (H_1)-homomorphism, and

$$\mathfrak{b}_\kappa = \mathfrak{c}(X_\kappa) + V_\kappa + U_\kappa, \qquad \mathfrak{c}(X_\kappa) = \mathfrak{g}_\kappa^{(1)} \oplus \mathfrak{g}_\kappa^{(2)}$$

is the corresponding parabolic subalgebra of $\mathfrak{g}$. Until § 4, we made an assumption that $G = \mathrm{Ad}\,\mathfrak{g}$ for the sake of simplicity. But, from now on, we let G be any fixed (Zariski connected) $\boldsymbol{R}$-group with Lie algebra $\mathfrak{g}$. Accordingly, $c_\kappa = \exp\left(\frac{\pi i}{4} Y_\kappa\right)$ will denote an element in G_C (called the "Cayley element"); the corresponding element in $\mathrm{Ad}\,\mathfrak{g}_C$ will then be given by $ad\, c_\kappa = \exp\left(\frac{\pi i}{4} ad\, Y_\kappa\right)$. Let B_κ, $C(X_\kappa)$, $G_\kappa^{(i)}$, and $\tilde{V}_\kappa$ denote the Zariski connected $\boldsymbol{R}$-subgroups of G corresponding to $\mathfrak{b}_\kappa$, $\mathfrak{c}(X_\kappa)$, $\mathfrak{g}_\kappa^{(i)}$, and $U_\kappa + V_\kappa$, respectively. Then $\tilde{V}_\kappa$ is the unipotent radical of B_κ and $C(X_\kappa)$ (which coincides with the centralizer of X_κ in G, cf. e. g., Humphreys [2], p. 140) is an almost direct product of $G_\kappa^{(1)}$ and $G_\kappa^{(2)}$.

Our results in §§ 2, 3 show essentially that $B_\kappa = C(X_\kappa)\tilde{V}_\kappa$ is a group of H-C type considered in § 5, Example 2. In fact, first, $\tilde{V} = \tilde{V}_\kappa$ is a generalized Heisenberg group with the alternating bilinear map $A : V \times V \to U$ given by

$$A(v, v') = -\frac{1}{4}[v, v'] \tag{7. 1}$$

and the complex structure I_0 on V given by (3. 3). Second, since $G_\kappa^{(1)}$ is of hermitian type, $G_1 = C(X_\kappa)$ satisfies (HC 1) with a canonical decomposition

$$\mathfrak{c}(X_\kappa)_C = \mathfrak{p}_{\kappa,+} + (\mathfrak{k}_{\kappa,C}^{(1)} \oplus \mathfrak{g}_{\kappa,C}^{(2)}) + \mathfrak{p}_{\kappa,-}, \tag{7. 2}$$

where $\mathfrak{k}_\kappa^{(1)} = \mathrm{Lie}\, K_\kappa^{(1)}$, $K_\kappa^{(1)} = G_\kappa^{(1)} \cap K$, and the $\mathfrak{p}_{\kappa,\pm} = \mathfrak{p}_\pm^{[0]}$ are the $\mathfrak{p}_\pm$-parts of $\mathfrak{g}_{\kappa,C}^{(1)}$. We note that

$$\mathfrak{k}_\kappa^{(1)} \oplus \mathfrak{g}_\kappa^{(2)} = \mathfrak{c}(X_\kappa, H_0^{(1)}),$$
$$\mathfrak{p}_{\kappa,\pm} = \mathfrak{g}_C(ad\, H_0^{(1)}\,;\, \pm i).$$

The condition (HC 2) for $C(X_\kappa)$ is satisfied by taking $C(X_\kappa, H_0^{(1)})$ (which is Zariski connected) and $P_{\kappa,\pm} = \exp \mathfrak{p}_{\kappa,\pm}$ as K_1 and $P_{1\pm}$. ((5. 3) follows from the corresponding equation for $G_\kappa^{(1)}$.) Let ρ_U and ρ_V be the representations of $C(X_\kappa)$ given by ad_U and ad_V. Then the condition (5. 13) is clearly satisfied. From $\rho_V(H_0^{(1)}) = \frac{1}{2} I_0$ (Lem. 3. 1), one has

$$V_\pm = V_C\left(ad\, H_0^{(1)}\,;\, \pm\frac{i}{2}\right)$$

and, from $G_\kappa^{(1)} \subset \mathrm{Ker}\, \rho_U$ (Th. 2. 3), one has $U \subset \mathfrak{c}(H_0^{(1)})$. It follows that the condition (5. 14) is also satisfied. Thus B_κ is of Harish-Chandra type with a canonical decomposition

$$\mathfrak{b}_{\kappa,C} = \tilde{\mathfrak{p}}_{\kappa,+} + \tilde{\mathfrak{k}}_{\kappa,C} + \tilde{\mathfrak{p}}_{\kappa,-}, \tag{7. 3}$$

where

$$\tilde{\mathfrak{k}}_{\kappa,C} = U_C + (\mathfrak{k}_{\kappa,C}^{(1)} \oplus \mathfrak{g}_{\kappa,C}^{(2)}) = \mathfrak{b}_{\kappa,C}(ad\, H_0^{(1)}\,;\, 0),$$
$$\tilde{\mathfrak{p}}_{\kappa,+} = V_+ + \mathfrak{p}_{\kappa,+} = \mathfrak{b}_{\kappa,C}\left(ad\, H_0^{(1)}\,;\, \frac{i}{2}, i\right),$$

and $\tilde{\mathfrak{p}}_{\kappa,-}$ is the complex conjugate of $\tilde{\mathfrak{p}}_{\kappa,+}$. Notice that $\tilde{\mathfrak{p}}_{\kappa,\pm}$ are abelian. In this case, by Theorem 2. 3, the $G_\kappa^{(2)\circ}$-orbit Ω_κ is a self-dual, homogeneous open convex cone in U and, by Lemma 3. 2, one has $A_u I_0 \gg 0$ for all $u \in \Omega_\kappa$, i. e., AI_0 is symmetric and Ω_κ-positive.

Now, in view of Proposition 6. 3 (and its proof), the homomorphism $\rho_V : G_\kappa^{(1)} \longrightarrow Sp(V, A)$ satisfies the condition (H_2) with respect to $H_0^{(1)}$ and $\frac{1}{2}I_0$. Hence the induced map φ defined by

$$\rho_V(\exp t) = \begin{pmatrix} 1 & \varphi(t) \\ 0 & 1 \end{pmatrix} \qquad (t \in \mathcal{D}_\kappa)$$

is a holomorphic map of $\mathcal{D}_\kappa$ into $\mathfrak{S}(V, A, \Omega)$. Therefore we obtain the associated Siegel domain of the third kind $\mathcal{S}_\varphi$, which we denote also by $\mathcal{S}_\kappa$, i. e.,

$$\mathcal{S}_\kappa = \{(u, w, t) \in U_C \times V_+ \times \mathcal{D}_\kappa \,|\, \mathrm{Im}\, u - \mathrm{Re}\, \mathcal{L}_{\varphi(t)}(w, w) \in \Omega_\kappa\}. \tag{7. 4}$$

Comparing the formulas (5. 28) and (6. 25), we see at once that

$$\mathrm{Re}\, \mathcal{L}_{\varphi(t)}(w, w) = \frac{1}{2i} \mathcal{K}_U[(w, t)]. \tag{7. 5}$$

Hence, in the notation of § 5, one has $\mathcal{S}_\kappa = \Phi^{-1}(\Omega_\kappa)$ and so, by Proposition 5. 3, $\mathcal{S}_\kappa$ is a B_κ°-orbit of ie_κ in $U_C \times V_+ \times \mathcal{D}_\kappa$. Thus $\mathcal{S}_\kappa$ is homogeneous and the action of B_κ° on $\mathcal{S}_\kappa$ is given by quasilinear transformations.

We are now in position to state the main theorem of this chapter.

Theorem 7. 1. *The holomorphic action of c_κ is defined at all $z \in \mathcal{D}$ and the map $(ad\, c_\kappa)^{-1} c_\kappa$ gives an analytic isomorphism of $\mathcal{D}$ onto the Siegel domain of the third kind $\mathcal{S}_\kappa$; in particular, one has*

$$(ad\, c_\kappa)^{-1} c_\kappa(o) = ie_\kappa. \tag{7. 6}$$

Moreover, the map $(ad\, c_\kappa)^{-1} c_\kappa$ is B_κ°-equivariant.

Proof. We put $c_1' = c_1 \circ \exp\left(\frac{\pi i}{4} h\right)$ and $\tilde{c}_1 = c_1'^{-1} c_1 c_1'$. Then, in view of the diagram (4. 5), we have an (H_1)-homomorphism

$$\tilde{\kappa} = \kappa \circ ad\, c_1' : \mathfrak{su}(1, 1) \longrightarrow \mathfrak{g},$$

and the corresponding group homomorphism $SU(1, 1) \rightarrow G$, which we denote also by $\tilde{\kappa}$, maps $\tilde{c}_1$ to c_κ. One has

$$\begin{aligned} \tilde{c}_1 &= \exp\left(-\frac{\pi i}{4} h\right) c_1 \exp\left(\frac{\pi i}{4} h\right) \\ &= \frac{1}{\sqrt{2}} \begin{pmatrix} 1 & 1 \\ -1 & 1 \end{pmatrix} \doteq \begin{pmatrix} 1 & 1 \\ 0 & 1 \end{pmatrix} \begin{pmatrix} \sqrt{2} & 0 \\ 0 & \sqrt{2}^{-1} \end{pmatrix} \begin{pmatrix} 1 & 0 \\ -1 & 1 \end{pmatrix}. \end{aligned} \tag{7. 7}$$

Therefore the corresponding linear fractional transformation $\tilde{c}_1 : z \mapsto (z+1)(-z+1)^{-1}$ is surely defined on the unit disc, and one has $\tilde{c}_1(0) = 1 (= e_+)$.

Now, to prove the first assertion of the Theorem, we have to show that

$$c_\kappa G^\circ \subset P_+ K_C^\circ P_-. \tag{7. 8}$$

As B_κ is parabolic, one has $G^\circ = B_\kappa^\circ K^\circ$ by Iwasawa decomposition. Hence one has

$$c_\kappa G^\circ = c_\kappa B_\kappa^\circ K^\circ = (c_\kappa \tilde{V}_\kappa c_\kappa^{-1})(c_\kappa G_\kappa^{(2)\circ} c_\kappa^{-1})(c_\kappa G_\kappa^{(1)\circ})K^\circ.$$

Hence to prove (7. 8) it is enough to prove the following three relations

$$(7.9\text{ a})\qquad c_\kappa \tilde{V}_\kappa c_\kappa^{-1} \subset P_+ K_c^\circ,$$

$$(7.9\text{ b})\qquad c_\kappa G_\kappa^{(2)\circ} c_\kappa^{-1} \subset K_c^\circ,$$

$$(7.9\text{ c})\qquad c_\kappa G_\kappa^{(1)\circ} \subset P_+ K_c^\circ P_-.$$

(7. 9 a, b) follow from (3. 21) and (3. 25). Since the inclusion map $G_\kappa^{(1)} \to G$ satisfies (H_1), every element $g \in G_\kappa^{(1)\circ}$ can be written as $g = g_+ g_0 g_-$ with

$$g_0 \in K_{\kappa,c}^{(1)} = K_c^\circ \cap G_{\kappa,c}^{(1)}, \qquad g_\pm \in P_{\kappa,\pm} = P_\pm \cap G_{\kappa,c}^{(1)}.$$

On the other hand, (7. 7) gives $\tilde{c}_1 \in P_+^1 K_c^1 P_-^1$ and hence $c_\kappa = \tilde{\kappa}(\tilde{c}_1) \in P_+ K_c^\circ P_-$. Therefore

$$c_\kappa g = g_+ c_\kappa (g_0 g_-) \in P_+ K_c^\circ P_-,$$

which proves (7. 9 c). Thus we can conclude that c_κ is defined on $\mathcal{D}$ and maps $\mathcal{D}$ biholomorphically onto $c_\kappa(\mathcal{D})$. In particular, since the map $\tilde{\kappa} : \mathcal{D}^1 \to \mathcal{D}$ is equivariant, one has

$$c_\kappa(o) = \tilde{\kappa}(\tilde{c}_1(0)) = \tilde{\kappa}(e_+) = o_\kappa.$$

In view of the diagram (4. 5), this implies (7. 6).

We know that the group B_κ° is transitive on both $\mathcal{D}$ and $\mathcal{S}_\kappa$, and one has $o \in \mathcal{D}$ and $ie_\kappa \in \mathcal{S}_\kappa$. Hence, to complete the proof, it suffices to show that the map $(ad\, c_\kappa)^{-1} c_\kappa$ is B_κ°-equivariant. Let $z \in \mathcal{D}$, $(ad\, c_\kappa)^{-1} c_\kappa(z) = (u, w, t) \in \mathcal{S}_\kappa$, and put

$$z' = g(z), \qquad (u', w', t') = g(u, w, t)$$

for $g \in B_\kappa^\circ$. We have to show that $(ad\, c_\kappa)^{-1} c_\kappa(z') = (u', w', t')$. According to (5. 33), the equation $(u', w', t') = g(u, w, t)$ means

$$(7.10)\qquad g \cdot \exp(u+w+t) \in \exp(u'+w'+t') C(X_\kappa, H_0^{(1)})_c \tilde{P}_{\kappa,-},$$

where $\tilde{P}_{\kappa,-} = \exp(V_{\kappa,-} + \mathfrak{p}_{\kappa,-})$. By (1. 10), (1. 8) and Lemma II. 4. 1, one has

$$(7.11\text{ a})\qquad c_\kappa C_{G_c}(X_\kappa, H_0^{(1)}) c_\kappa^{-1} = C_{G_c}(H_\kappa, H_0^{(1)}) \subset C_{G_c}(H_0) = K_c^\circ.$$

On the other hand, by Proposition 3. 5, one has

$$(7.11\text{ b})\qquad c_\kappa \tilde{P}_{\kappa,-} c_\kappa^{-1} \subset K_c^\circ P_-.$$

Hence (7. 10) implies

$$(7.10')\qquad g \cdot \exp(u+w+t) \in \exp(u'+w'+t') c_\kappa^{-1} K_c^\circ P_- c_\kappa,$$

or

$$c_\kappa g c_\kappa^{-1} \exp(c_\kappa(z)) \in \exp((ad\, c_\kappa)(u'+w'+t')) K_c^\circ P_-.$$

But, as $z' = g(z)$, one has

$$c_\kappa g c_\kappa^{-1} \exp(c_\kappa(z)) \in \exp(c_\kappa g(z)) K_c^\circ P_- = \exp(c_\kappa(z')) K_c^\circ P_-.$$

From these two relations, one obtains

$$c_\kappa(z') = (ad\, c_\kappa)(u'+w'+t'), \qquad \text{q. e. d.}$$

From the above proof, we see that a natural G°-action on $\mathscr{S}_\kappa$ is defined by (7. 10′), or equivalently by

$$(u, w, t) \longmapsto (ad\, c_\kappa)^{-1}(c_\kappa g c_\kappa^{-1}((ad\, c_\kappa)(u+w+t))), \tag{7.12}$$

which makes the correspondence $z \mapsto (ad\, c_\kappa)^{-1} c_\kappa(z)$ a G°-equivariant isomorphism of $\mathscr{D}$ onto $\mathscr{S}_\kappa$.

We note that, if $\mathscr{D}_\kappa$ reduces to a point, i. e., if $\mathfrak{g}_\kappa^{(1)}$ is compact, then $\mathscr{S}_\kappa$ becomes a "Siegel domain" (of the second kind). When $\mathfrak{g}$ is simple, this is the case if and only if κ is the *last* homomorphism (i. e., $m(\kappa)=r$). Moreover, as we remarked in § 1, this $\mathscr{S}_\kappa$ is a tube domain if and only if κ satisfies the condition (H_2).

Exercises

1. Prove the relation (7. 4).

2. Let $\mathscr{S}_\kappa$ be as defined in the text and let π_κ denote the canonical projection $\mathscr{S}_\kappa \to \mathscr{D}_\kappa$. The fiber

$$\pi_\kappa^{-1}(o) = \{(u, w, o) \mid \mathrm{Im}\, u - 2iA(\bar{w}, w) \in \Omega\}$$

is then a Siegel domain (of the second kind).

2. 1) Show that the group $G_\kappa^{(2)\circ}\tilde{V}_\kappa$ acts transitively on $\pi_\kappa^{-1}(o)$.

2. 2) Let $z_\xi = (\exp \xi X_\kappa)o$ $(\xi \in \boldsymbol{R})$. Show that

$$(ad\, c_\kappa)^{-1} c_\kappa(z_\xi) = (ie^{2\xi} e_\kappa, 0, o).$$

2. 3) Show that any half-line of the form

$$(u_0 + i\lambda u_1, w_0, o) \qquad (\lambda > 0),$$

where $u_0 \in U_C$, $u_1 \in \Omega$, $w_0 \in V_+$, $\mathrm{Im}\, u_0 = 2iA(\bar{w}_0, w_0)$, is a geodesic in $\pi_\kappa^{-1}(o)$ with respect to the Bergman metric.

Remark. 2. 3) shows that $\pi_\kappa^{-1}(o)$ is a union of geodesics in $\mathscr{S}_\kappa$. However, in general, $\pi_\kappa^{-1}(o)$ itself is not totally geodesic. In fact, we shall see in V, § 4 that each fiber of $\mathscr{S}_\kappa \to \mathscr{D}_\kappa$ is "quasi-symmetric" but not necessarily symmetric.

3. 3. 1) Show that the action of c_κ on $\mathscr{S}_\kappa$ is given by

$$\begin{aligned} z = (u, w, t) &\longmapsto ie_\kappa + ad(\exp(\sqrt{2}\, X_\kappa))(1 - z \,\Box\, ie_\kappa)^{-1} z \\ &= (i(e_\kappa - iu)(e_\kappa + iu)^{-1},\ 2\sqrt{2}\, R_{(e_\kappa+iu)^{-1}} w,\ t - 2i\{w, e_\kappa, R_{(e_\kappa+iu)^{-1}} w\}). \end{aligned} \tag{7.13}$$

Hint. Observe first that

$$c_\kappa = \tilde{\kappa}(\tilde{c}_1) = \exp(o_\kappa) \exp\left(-\frac{i}{2}(\log 2) H_\kappa\right) \exp(-\bar{o}_\kappa)$$

and then consider the holomorphic actions of each factor on $\mathscr{D}$ and $\mathscr{S}_\kappa$. Use (II. 5. 23) and Prop. 3. 4 (cf. Loos [9], § 10).

3. 2) Show that the action of the one-parameter group $\{\exp(\lambda H_0)\}$ on $\mathscr{S}_\kappa$ is given by

$$z = (u, w, t) \longmapsto (u', w', t'),$$

$$\begin{cases} u' = \left(\left(\cos\dfrac{\lambda}{2}\right)u + \left(\sin\dfrac{\lambda}{2}\right)e_\kappa\right)\left(-\left(\sin\dfrac{\lambda}{2}\right)u + \left(\cos\dfrac{\lambda}{2}\right)e_\kappa\right)^{-1}, \\ w' = 2e^{\frac{i\lambda}{2}} R\left(\left(-\left(\sin\dfrac{\lambda}{2}\right)u + \left(\cos\dfrac{\lambda}{2}\right)e_\kappa\right)^{-1}\right)w, \\ t' = e^{i\lambda}\left(t + 2\sin\dfrac{\lambda}{2}\left\{w, e_\kappa, R\left(\left(-\left(\sin\dfrac{\lambda}{2}\right)u + \left(\cos\dfrac{\lambda}{2}\right)e_\kappa\right)^{-1}\right)w\right\}\right). \end{cases} \tag{7.14}$$

In particular, the symmetry of $\mathscr{S}_\kappa$ at ie_κ is given by

$$(u, w, t) \longmapsto (-u^{-1},\ -2iR_{u^{-1}} w,\ -t + 2\{w, e_\kappa, R_{u^{-1}} w\}). \tag{7.15}$$

§ 8. Boundary components of a symmetric domain.

In this section, we study the structure of the boundary of a symmetric domain $\mathcal{D}$ in its bounded domain realization in $\mathfrak{p}_+$. We denote by $\bar{\mathcal{D}}$ the closure of $\mathcal{D}$ in $\mathfrak{p}_+$ and put $\partial\mathcal{D}=\bar{\mathcal{D}}-\mathcal{D}$. We retain the notation in the preceding sections. G_κ will denote the (Zariski connected) $\boldsymbol{R}$-subgroup of G corresponding to $\mathfrak{g}_\kappa$, i. e., the non-compact semi-simple part of $G_\kappa^{(1)}$, and $\mathcal{D}_\kappa=G_\kappa^\circ\cdot o\subset\mathfrak{p}_{\kappa,+}$ is the associated symmetric domain. Since the inclusion map $\mathfrak{g}_\kappa\to\mathfrak{g}$ satisfies (H_1), $\mathcal{D}_\kappa$ is a totally geodesic complex submanifold of $\mathcal{D}$ (with respect to the Bergman metric). It is clear that the holomorphic action of G° is defined on $\bar{\mathcal{D}}$ and one has $G^\circ(\bar{\mathcal{D}})=\bar{\mathcal{D}}$.

Lemma 8. 1. *On the subspace $\mathfrak{p}_{\kappa,+}$ the "Cayley transformation" (the holomorphic action of c_κ) is given by a parallel translation:*

(8. 1) $$c_\kappa(t) = o_\kappa+t \qquad (t\in\mathfrak{p}_{\kappa,+})$$

and one has

(8. 2) $$c_\kappa(\mathcal{D}_\kappa) = o_\kappa+\mathcal{D}_\kappa\subset\partial\mathcal{D}.$$

Proof. We know that $c_\kappa(o)=o_\kappa$, and the Cayley element c_κ commutes element-wise with G_κ. Hence for $t\in\mathfrak{p}_{\kappa,+}$ one has

$$c_\kappa(\exp t) = (\exp t)c_\kappa \in P_+K_C^\circ P_-,$$

i. e., $c_\kappa(t)$ is defined, and

$$\begin{aligned}\exp(c_\kappa(t)) &= (c_\kappa\cdot\exp t)_+ = \exp o_\kappa\cdot\exp t\\ &= \exp(o_\kappa+t),\end{aligned}$$

which proves (8. 1). It follows that

$$\begin{aligned}G_\kappa^\circ\cdot o_\kappa &= G_\kappa^\circ(c_\kappa(o)) = c_\kappa(G_\kappa^\circ\cdot o) = c_\kappa(\mathcal{D}_\kappa)\\ &= o_\kappa+\mathcal{D}_\kappa.\end{aligned}$$

Therefore, to complete the proof of (8. 2), it is enough to show that $o_\kappa\in\partial\mathcal{D}$, because the boundary $\partial\mathcal{D}$ is G°-stable. In the case of the unit disc $\mathcal{D}^1=\{\zeta\mid|\zeta|<1\}$, one has

$$\lim_{\lambda\to\infty}(\exp\lambda h')(0) = \lim_{\lambda\to\infty}\operatorname{th}\lambda = 1\in\partial\mathcal{D}^1.$$

Applying the equivariant map $\tilde{\kappa}$: $(SU(1,1),\mathcal{D}^1)\to(G,\mathcal{D})$ to this equation and noting that $\tilde{\kappa}(0)=o$, $\tilde{\kappa}(1)=o_\kappa$, $\tilde{\kappa}(h')=X_\kappa$, one has

$$\lim_{\lambda\to\infty}(\exp\lambda X_\kappa)(o) = o_\kappa,$$

which shows that o_κ is a limit point of a geodesic in $\mathcal{D}$. Hence one has $o_\kappa\in\partial\mathcal{D}$, q. e. d.

We put

(8. 3) $$\mathcal{F}_\kappa = c_\kappa(\mathcal{D}_\kappa) = G_\kappa^\circ\cdot o_\kappa$$

and call it a (proper) *boundary component* of $\mathcal{D}$ corresponding to the homomorphism

κ. ($\mathcal{D}$ itself may be counted as an improper boundary component corresponding to the trivial homomorphism.) We will show that the boundary $\partial\mathcal{D}$ is a disjoint union of all (proper) boundary components $\mathcal{F}_\kappa$ corresponding to all possible (non-trivial) (H_1)-homomorphisms $\kappa : \mathfrak{g}^1 \to (\mathfrak{g}, H_0)$.

For that purpose, we first observe that, from $G^\circ = K^\circ(\exp \mathfrak{a}^+)K^\circ$ ((I. 5. 8)), one has $\mathcal{D} = K^\circ(\exp \mathfrak{a}^+)o$ and so

$$\bar{\mathcal{D}} = K^\circ\overline{(\exp \mathfrak{a}^+)o}. \tag{8. 4}$$

Clearly the study of the boundary $\partial\mathcal{D}$ can be reduced to the case where $\mathcal{D}$ is irreducible. Hence, for simplicity, we assume for a while that $\mathfrak{g}$ is simple and introduce the following notation. Let $(\kappa_1, \cdots, \kappa_r)$ be a maximal system of mutually commutative (H_1)-homomorphisms $\mathfrak{g}^1 \to (\mathfrak{g}, H_0)$ and set

$$X_i = X_{\kappa_i} \ (\in \mathfrak{p}), \qquad o_i = o_{\kappa_i} \ (\in \mathfrak{p}_+)\,; \tag{8. 5}$$

then $X_i = o_i + \bar{o}_i = 2\,\mathrm{Re}\, o_i$. $\mathfrak{a} = \{X_1, \cdots, X_r\}_{\boldsymbol{R}}$ is a maximal abelian subalgebra of $\mathfrak{g}$ contained in $\mathfrak{p}$, and

$$\mathfrak{a}^+ = \left\{\sum_{i=1}^{r} \xi_i X_i \,\middle|\, \xi_1 \geq \cdots \geq \xi_r \geq 0\right\}$$

is a Weyl chamber in $\mathfrak{a}$ (§ 4). We denote the b-th canonical homomorphism belonging to $\mathfrak{a}^+$ by $\kappa^{(b)}$ and set $X^{(b)} = X_{\kappa^{(b)}}$, $o^{(b)} = o_{\kappa^{(b)}}$, i. e.,

$$X^{(b)} = \sum_{i=1}^{b} X_i, \qquad o^{(b)} = \sum_{i=1}^{b} o_i. \tag{8. 6}$$

(For $b=0$, one has $\kappa^{(0)}=0$, and $X^{(0)}=0$, $o^{(0)}=o$.) We also write $\mathfrak{g}_b$, G_b, etc. for $\mathfrak{g}_{\kappa^{(b)}}$, $G_{\kappa^{(b)}}$, etc. Then $(\kappa_{b+1}, \cdots, \kappa_r)$ is a maximal system of mutually commutative (H_1)-homomorphisms for $\mathfrak{g}_b$ and one has $X_i \in \mathfrak{p}_b$, $o_i \in \mathfrak{p}_{b,+}$ for $b+1 \leq i \leq r$. We set

$$\mathfrak{a}_b = \{X_{b+1}, \cdots, X_r\}_{\boldsymbol{R}}, \qquad \mathfrak{a}_b^+ = \mathfrak{a}_b \cap \mathfrak{a}^+. \tag{8. 7}$$

Now, since

$$\exp(\textstyle\sum \lambda_i X_i)o = \sum (\mathrm{th}\,\lambda_i) o_i \qquad (\lambda_i \in \boldsymbol{R}),$$

one has

$$\overline{(\exp \mathfrak{a}^+)o} = \left\{\sum_{i=1}^{r} \xi_i o_i \,\middle|\, 1 \geq \xi_1 \geq \cdots \geq \xi_r \geq 0\right\}. \tag{8. 8}$$

On the other hand, by Lemma 8. 1 applied to $\kappa = \kappa^{(b)}$, one has

$$\begin{aligned}(\exp \mathfrak{a}_b^+)o^{(b)} &= c_{\kappa^{(b)}}((\exp \mathfrak{a}_b^+)o) = o^{(b)} + (\exp \mathfrak{a}_b^+)o \\ &= \left\{\sum_{i=1}^{r} \xi_i o_i \,\middle|\, 1 = \xi_1 = \cdots = \xi_b > \xi_{b+1} \geq \cdots \geq \xi_r \geq 0\right\}.\end{aligned} \tag{8. 9}$$

Hence we obtain

$$\overline{(\exp \mathfrak{a}^+)o} = \coprod_{b=0}^{r} (\exp \mathfrak{a}_b^+)o^{(b)} \quad \text{(disjoint union)}. \tag{8. 10}$$

[We note that the limit point of the geodesic $\exp(\lambda X)o$ $(\lambda \to \infty)$ for $X = \sum \xi_i X_i$, $\xi_1 \geq \cdots \geq \xi_b > 0$, $\xi_{b+1} = \cdots = \xi_r = 0$, is equal to $o^{(b)}$. Thus the set of all limit points of geodesics passing through o coincides with the set of all idempotents $\{o_\kappa\}$ in the

JTS $\mathfrak{p}_+$.]

Lemma 8. 2. *When $\mathfrak{g}$ is simple, every element z in $\mathfrak{p}_+$ can be expressed in the form*

$$z = (ad\, k)\left(\sum_{i=1}^{r} \xi_i o_i\right) \tag{8. 11}$$

with $k \in K^\circ$, $\xi_i \in \boldsymbol{R}$, $\xi_1 \geq \cdots \geq \xi_r \geq 0$. The real numbers $\xi_1, \cdots, \xi_r$ are uniquely determined by z.

Proof (cf. Sophus Lie [1], Exp. 23). We use the results in I, § 5. Let $z \in \mathfrak{p}_+$. From $\mathfrak{p} = (ad\, K^\circ)\mathfrak{a}^+$, one has Re $z = (ad\, k)X$ with $k \in K^\circ$, $X \in \mathfrak{a}^+$. Writing $X = \frac{1}{2}\sum \xi_i X_i$ and taking the $\mathfrak{p}_+$-part, one obtains (8. 11). To prove the uniqueness, suppose one has another expression Re $z = (ad\, k')X'$ with $k' \in K^\circ$, $X' \in \mathfrak{a}^+$. It suffices to show $X = X'$. Put $k_1 = k^{-1}k'$. Then $\mathfrak{a}$ and $(ad\, k_1)\mathfrak{a}$ are both maximal abelian subalgebras of $\mathfrak{c}(X)$, contained in its $\mathfrak{p}$-part. Therefore, by the conjugacy of such subalgebras, there exists $k_2 \in C_G(X) \cap K^\circ$ such that $(ad\, k_1)\mathfrak{a} = (ad\, k_2)\mathfrak{a}$, i. e., $k_1^{-1}k_2 \in N_G(\mathfrak{a})$. It follows that $s = ad(k_1^{-1}k_2)|\mathfrak{a} \in W$ and $sX = (ad\, k_1)^{-1}X = X'$. Since X and X' are both in the Weyl chamber $\mathfrak{a}^+$, one has $X = X'$, q. e. d.

Lemma 8. 3. *Let $k \in C(X_\kappa) \cap K$. Then, for some $\varepsilon = \pm 1$, one has*

$$(ad\, k)H_0 = \varepsilon H_0, \qquad (ad\, k)H_0^{(1)} = \varepsilon H_0^{(1)}, \qquad (ad\, k)e_\kappa = \varepsilon e_\kappa.$$

Proof. Since $\mathfrak{g}$ is simple, one has $(ad\, k)H_0 = \varepsilon H_0$ for some $\varepsilon = \pm 1$. Then, from $[H_0, X_\kappa] = \frac{1}{2}[H_\kappa, X_\kappa] = -Y_\kappa$ and $(ad\, k)X_\kappa = X_\kappa$, it follows that $(ad\, k)Y_\kappa = \varepsilon Y_\kappa$ and so $(ad\, k)H_\kappa = \varepsilon H_\kappa$. Since $H_0^{(1)} = H_0 - \frac{1}{2}H_\kappa$ and $e_\kappa = \frac{1}{2}(Y_\kappa + H_\kappa)$, one has $(ad\, k)H_0^{(1)} = \varepsilon H_0^{(1)}$ and $(ad\, k)e_\kappa = \varepsilon e_\kappa$, q. e. d.

Lemma 8. 4. *When $H_0^{(1)} \neq 0$ (i. e., when κ does not satisfy* (H$_2$)*), one has*

$$C(X_\kappa) \cap K^\circ = C(X_\kappa, H_0^{(1)}) \cap K, \tag{8. 12 a}$$

$$C(X_\kappa) \cap G^\circ = G_\kappa^{(1)\circ} \cdot C(X_\kappa, H_0^{(1)}). \tag{8. 12 b}$$

Proof. (8. 12 a) follows from Lemma 8. 3 and Lemma II. 4. 1. To prove (8. 12 b), let $g \in C(X_\kappa) \cap G^\circ$. Then, since $C(X_\kappa)$ is stable under θ, there exist $g_1 \in G_\kappa^{(1)\circ}$, $g_2 \in G_\kappa^{(2)\circ} \subset C(X_\kappa, H_0^{(1)})$ such that $(g_1 g_2)^{-1} g \in K^\circ$. By (8. 12 a), this implies $(g_1 g_2)^{-1} g \in C(X_\kappa, H_0^{(1)})$, which shows that g belongs to the right-hand side of (8. 12 b). To prove the converse, it suffices to show that $C(X_\kappa, H_0^{(1)}) \subset G^\circ$. Let $g \in C(X_\kappa, H_0^{(1)})$. Then, since $\mathfrak{c}(X_\kappa, H_0^{(1)}) = \mathfrak{k}_\kappa^{(1)} \oplus \mathfrak{g}_\kappa^{(2)}$, there exists $g_2 \in G_\kappa^{(2)\circ}$ such that $g_2^{-1} g \in C(X_\kappa, H_0^{(1)}) \cap K$. By (8. 12 a) it follows that $g_2^{-1} g \in K^\circ$ and so $g \in G^\circ$, q. e. d.

Proposition 8. 5. *The stabilizer $(G^\circ)_{o_\kappa}$ of o_κ in G° is given by*

$$(G^\circ)_{o_\kappa} = \begin{cases} C(X_\kappa, H_0^{(1)}) \cdot \tilde{V}_\kappa & \text{if } H_0^{(1)} \neq 0, \\ (C(X_\kappa) \cap G^\circ) \cdot \tilde{V}_\kappa & \text{if } H_0^{(1)} = 0. \end{cases} \tag{8. 13}$$

Proof. Since $o_\kappa = c_\kappa(o)$, one has $(G^\circ)_{o_\kappa} = G^\circ \cap c_\kappa(K_C^\circ P_-)c_\kappa^{-1}$. One has

$$U_\kappa \subset (ad\, c_\kappa)\mathfrak{p}_-^{[2]} \qquad \text{(by (2. 7), (1. 10))},$$
$$V_\kappa \subset (ad\, c_\kappa)(\mathfrak{k}_C^{[1]} + \mathfrak{p}_-^{[1]}) \qquad \text{(by (3. 7), (3. 1))},$$

so that $\tilde{V}_\kappa \subset c_\kappa(K_C^\circ P_-)c_\kappa^{-1}$. On the other hand, analogously to (7. 11 a), one obtains $C(X_\kappa, H_0^{(1)})_C \subset c_\kappa K_C^\circ c_\kappa^{-1}$. Hence it follows that

$$C(X_\kappa, H_0^{(1)}) \cdot \tilde{V}_\kappa \subset c_\kappa(K_C^\circ P_-)c_\kappa^{-1}. \tag{8. 14}$$

Since by (8. 12 b) one has $C(X_\kappa, H_0^{(1)}) \subset G^\circ$ when $H_0^{(1)} \neq 0$, this proves that the right-hand side of (8. 13) is contained in the stabilizer $(G^\circ)_{o_\kappa}$. Conversely, let $g \in (G^\circ)_{o_\kappa}$ and write $g = kg_1$ with $k \in K^\circ$, $g_1 \in B_\kappa^\circ = G_\kappa^{(1)\circ} G_\kappa^{(2)\circ} \tilde{V}_\kappa$. We want to show that $g \in C(X_\kappa, H_0^{(1)}) \cdot \tilde{V}_\kappa$. From what we have proved above, we may assume that $g_1 \in G_\kappa^{(1)\circ}$. We may further assume that $\kappa = \kappa^{(b)}$. Under these assumptions, one can write $g_1 = k_1 a_1 k_2$ with $k_1, k_2 \in K_\kappa^{(1)\circ}$, $a_1 \in \exp \mathfrak{a}_b^+$. Then from $kg_1 o^{(b)} = o^{(b)}$ one has

$$kk_1a_1k_2o^{(b)} = kk_1a_1o^{(b)} = o^{(b)}.$$

If $a_1 = \exp\left(\sum_{i=b+1}^{r} \lambda_i X_i\right)$, then

$$a_1 o^{(b)} = o^{(b)} + \sum_{i=b+1}^{r} (\text{th}\, \lambda_i) o_i.$$

Hence, by the uniqueness in Lemma 8. 2, one has $\lambda_{b+1} = \cdots = \lambda_r = 0$, i. e., $a_1 = 1$, and hence $g \in K^\circ$. But then $g(o^{(b)}) = (ad\, g)o^{(b)} = o^{(b)}$ implies $(ad\, g)X^{(b)} = X^{(b)}$, i. e., $g \in C(X^{(b)})$. When $H_0^{(1)} \neq 0$, one has by (8. 12 a) $g \in C(X^{(b)}) \cap K^\circ \subset C(X^{(b)}, H_0^{(1)})$, as was to be shown, q. e. d.

Corollary 8. 6. *For $g \in G^\circ$, one has*

$$g\mathcal{F}_\kappa = \mathcal{F}_\kappa \Longleftrightarrow g \in B_\kappa = C(X_\kappa) \cdot \tilde{V}_\kappa, \tag{8. 15 a}$$

$$g(z) = z \quad \textit{for all } z \in \mathcal{F}_\kappa \Longleftrightarrow g \in C(X_\kappa, \mathfrak{g}_\kappa) \cdot \tilde{V}_\kappa. \tag{8. 15 b}$$

Proof. When $\mathfrak{g}_\kappa = \{0\}$, one has $\mathcal{F}_\kappa = \{o_\kappa\}$ and, if moreover, $H_0^{(1)} \neq 0$, one has $C(X_\kappa, H_0^{(1)}) = C(X_\kappa) \cap G^\circ$ by (8. 12 b). Hence (8. 15 a, b) are obvious by (8. 13). Therefore we assume $\mathfrak{g}_\kappa \neq \{0\}$ and so $H_0^{(1)} \neq 0$. Then, since $\mathcal{F}_\kappa = G_\kappa^{(1)\circ} \cdot o_\kappa$, (8. 15 a) follows immediately from Proposition 8. 5 and (8. 12 b). The $\Leftarrow$ part of (8. 15 b) is also obvious in view of the relation $C(X_\kappa, \mathfrak{g}_\kappa) \subset C(X_\kappa, H_0^{(1)})$. [In fact, when $\mathfrak{g}_\kappa \neq \{0\}$, one has $C(X_\kappa, \mathfrak{g}_\kappa) \cap K \subset C(X_\kappa, H_0^{(1)})$ by Lemma 8. 3 and hence $C(X_\kappa, \mathfrak{g}_\kappa) = G_\kappa^{(2)\circ} \cdot (C(X_\kappa, \mathfrak{g}_\kappa) \cap K) \subset C(X_\kappa, H_0^{(1)})$.] To prove the converse, suppose that $gz = z$ for all $z \in \mathcal{F}_\kappa$. Then, again from $\mathcal{F}_\kappa = G_\kappa^{(1)\circ} \cdot o_\kappa$, one has $g_1^{-1} g g_1 \in (G^\circ)_{o_\kappa}$ for all $g_1 \in G_\kappa^{(1)\circ}$; in particular, $g \in (G^\circ)_{o_\kappa}$. According to (8. 13), write $g = h\tilde{v}$ with $h \in C(X_\kappa, H_0^{(1)})$, $\tilde{v} \in \tilde{V}_\kappa$. Then, for all $g_1 \in G_\kappa^{(1)\circ}$, one has $g_1^{-1} h g_1 \in C(X_\kappa, H_0^{(1)})$ and so $h \in C(X_\kappa, (ad\, g_1) H_0^{(1)})$. It follows by Corollary II. 2. 7 that $(ad\, h)|\mathfrak{g}_\kappa = id$, i. e., $h \in C(X_\kappa, \mathfrak{g}_\kappa)$, q. e. d.

Theorem 8. 7. *Let $\mathfrak{g}$ be a simple Lie algebra of hermitian type of $\boldsymbol{R}$-rank r and $\mathcal{D}$ the associated irreducible symmetric domain in $\mathfrak{p}_+$. Then $\bar{\mathcal{D}}$ decomposes into $r+1$ G°-orbits:*

(8. 16) $$\bar{\mathcal{D}} = \coprod_{b=0}^{r} G^\circ o^{(b)}.$$

Each orbit $G^\circ o^{(b)}$ is a disjoint union of boundary components $\mathcal{F}_\kappa$ corresponding to the (H$_1$)-*homomorphisms $\kappa: \mathfrak{g}^1 \to (\mathfrak{g}, H_0)$ with $m(\kappa)=b$:*

(8. 17) $$G^\circ o^{(b)} = \coprod_{m(\kappa)=b} \mathcal{F}_\kappa.$$

Proof. From (8. 4) and (8. 10) we obtain

$$\bar{\mathcal{D}} = \bigcup_{b=0}^{r} G^\circ o^{(b)}.$$

Hence, to establish (8. 16), we have only to show that these orbits are mutually disjoint. Suppose there were two indices $b \neq b'$ and $g \in G^\circ$ such that $go^{(b)} = o^{(b')}$. We write $g = kg_1$ with $k \in K^\circ$, $g_1 \in B_b^\circ$; as in the proof of Proposition 8. 5, we may assume that $g_1 \in G_b^{(1)\circ}$, [B_b and $G_b^{(1)}$ standing for B_κ and $G_\kappa^{(1)}$ ($\kappa = \kappa^{(b)}$)]. Then, writing $g_1 = k_1 a_1 k_2$ as there, we obtain the relation

$$kk_1 a_1 o^{(b)} = o^{(b')},$$

which contradicts the uniqueness of the ξ_i's in Lemma 8. 2. This proves (8. 16). Next, by Proposition 8. 5 and Corollary 8. 6, one has $\mathcal{F}_b \approx (B_b \cap G^\circ)/(G^\circ)_{o^{(b)}}$ and so

$$G^\circ o^{(b)} \approx G^\circ/(G^\circ)_{o^{(b)}} \approx \coprod_{g \in G^\circ/(B_b \cap G^\circ)} g\mathcal{F}_b,$$

where g ranges over a full set of representatives for $G^\circ/(B_b \cap G^\circ)$. Since $G^\circ = K^\circ B_b^\circ$, the representatives g can be chosen from K°. For $k \in K^\circ$, the homomorphism $\kappa = (ad\, k)\kappa^{(b)}$ satisfies (H$_1$) with respect to the fixed H-elements H_0^1 and H_0, and one has $\mathcal{F}_\kappa = k\mathcal{F}_b$. [More generally, for any $g \in G^\circ$, $\mathcal{F}_\kappa$ with $\kappa = (ad\, g)\kappa_b$ can be identified with $g(\mathcal{F}_b)$, cf. Exerc. 2.] Conversely, by Proposition 4. 3, every (H$_1$)-homomorphism $\kappa: \mathfrak{g}^1 \to (\mathfrak{g}, H_0)$ with $m(\kappa)=b$ can be written in the form $\kappa = (ad\, k)\kappa^{(b)}$ with $k \in K^\circ$. Moreover, if $k\mathcal{F}_b = k'\mathcal{F}_b$ for $k, k' \in K^\circ$, then one has by (8. 15 a) $k^{-1}k' \in B_b \cap K^\circ = C(X^{(b)}, H_0) \subset C(\kappa^{(b)}(\mathfrak{g}^1))$ (see the proof of Lemma 8. 3) and hence $(ad\, k)\kappa^{(b)} = (ad\, k')\kappa^{(b)}$. Thus, when the H-element H_0 is fixed, there exists a bijective correspondence between the (H$_1$)-homomorphisms $\kappa: \mathfrak{g}^1 \to (\mathfrak{g}, H_0)$ with $m(\kappa)=b$ and the boundary components $\mathcal{F}_\kappa$ contained in $G^\circ o^{(b)}$. This proves (8. 17), q. e. d.

As shown above, the orbit $G^\circ o^{(b)}$ has the following fiber structure:

(8. 18) $$\begin{array}{ccc} G^\circ o^{(b)} & \approx & G^\circ/(G^\circ)_{o^{(b)}} \\ \Big\downarrow & \text{(fiber } \mathcal{F}_b \approx G_b^\circ/K_b^\circ) & \\ K^\circ o^{(b)} \approx K^\circ/(C(X_\kappa) \cap K^\circ) & \approx & G^\circ/(B_b \cap G^\circ) \end{array}$$

where, $\mathcal{F}_b$, G_b, etc. stand for $\mathcal{F}_\kappa$, G_κ, etc. with $\kappa = \kappa^{(b)}$.

$\mathcal{F}_b$ is called the b-th boundary component belonging to $(\mathfrak{a}, \Delta)$. In general, $\mathcal{F}_\kappa$ with $m(\kappa)=b$ is called *a b-th boundary component.* ($\mathcal{D}$ itself may be counted as the unique 0-th boundary component.) An r-th (i. e., last) boundary component, which consists of a single point o_κ, is also called a "point boundary component".

From (8. 18) one has

$$\text{(8.19)}\qquad \begin{aligned}\dim G^\circ o^{(b)} &= (\boldsymbol{R}\text{-})\dim \mathcal{D}_b+\dim \tilde{V}_b,\\ \dim K^\circ o^{(b)} &= \dim \tilde{V}_b,\end{aligned}$$

so that the (real) codimension of the orbit $G^\circ o^{(b)}$ in $\mathfrak{p}_+$ coincides with $\dim U_b$. In particular, $G^\circ o^{(1)}$ is a smooth *hypersurface* in $\mathfrak{p}_+$, since $\dim U_1=1$ (cf. § 4).

Proposition 8. 8. *For two* (H_1)*-homomorphisms* κ *and* $\kappa' : \mathfrak{sl}_2(\boldsymbol{R})\to(\mathfrak{g}, H_0)$, *where* $\mathfrak{g}$ *is simple, the following three conditions are equivalent.*

(a) *There exists a third* (H_1)*-homomorphism* $\kappa'' : \mathfrak{sl}_2(\boldsymbol{R})\to(\mathfrak{g}, H_0)$ *such that* κ *is the commutative sum of* κ' *and* κ'' : $\kappa=\kappa'+\kappa''$.

(b) *There exists* $k\in K^\circ$ *and integers* $b\geq b'$ *such that*

$$\kappa = (ad\, k)\cdot\kappa^{(b)},\qquad \kappa' = (ad\, k)\cdot\kappa^{(b')}.$$

(c) *One has* $\mathcal{F}_\kappa\subset\overline{\mathcal{F}}_{\kappa'}$.

Proof. (a)⇔(b) If the condition (a) is satisfied and $m(\kappa')=b'$, $m(\kappa'')=b''$, then by the proof of Proposition 4. 3 there exists $k\in K^\circ$ such that

$$\kappa' = (ad\, k)\sum_{i=1}^{b'}\kappa_i,\qquad \kappa'' = (ad\, k)\sum_{i=b'+1}^{b'+b''}\kappa_i,$$

whence follows (b). The converse is obvious.

(a)⇔(c) If the condition (a) is satisfied, $\mathfrak{g}_\kappa$ and $\kappa''(\mathfrak{g}^1)$ are contained in $\mathfrak{g}_{\kappa'}$, and $o_{\kappa''}+\mathcal{D}_\kappa$ is a boundary component of $\mathcal{D}_{\kappa'}$ corresponding to the (H_1)-homomorphism $\kappa'' : \mathfrak{g}^1\to(\mathfrak{g}_{\kappa'}, H_0-\frac{1}{2}H_{\kappa'})$. Thus

$$\text{(8.20)}\qquad o_{\kappa''}+\mathcal{D}_\kappa\subset\overline{\mathcal{D}}_{\kappa'}.$$

Applying $c_{\kappa'}$, which is a translation $z\mapsto o_{\kappa'}+z$ on $\mathfrak{p}_{\kappa',+}$, and noting that $o_\kappa=o_{\kappa'}+o_{\kappa''}$, one has

$$o_\kappa+\mathcal{D}_\kappa\subset o_{\kappa'}+\overline{\mathcal{D}}_{\kappa'},$$

i. e., $\mathcal{F}_\kappa\subset\overline{\mathcal{F}}_{\kappa'}$. Conversely, any (H_1)-homomorphism $\kappa'' : \mathfrak{g}^1\to(\mathfrak{g}_{\kappa'}, H_0-\frac{1}{2}H_{\kappa'})$ may be viewed as an (H_1)-homomorphism $\mathfrak{g}^1\to(\mathfrak{g}, H_0)$ and is commutative with κ'. Since $\partial\mathcal{D}_{\kappa'}$ is the union of the boundary components (of $\mathcal{D}_{\kappa'}$) corresponding to all such homomorphisms κ'', any boundary component (of $\mathcal{D}$) contained in $\partial\mathcal{F}_{\kappa'}$ must be of the form $\mathcal{F}_\kappa$ with $\kappa=\kappa'+\kappa''$. Hence (c) implies (a), q. e. d.

It follows from the above proof that $\overline{\mathcal{F}}_{\kappa'}$ is the union of all boundary components of $\mathcal{D}$ contained in $\overline{\mathcal{F}}_{\kappa'}$.

Corollary 8. 9. *One has*

$$\text{(8.21)}\qquad \overline{G^\circ o^{(b)}} = \coprod_{b'\geq b} G^\circ o^{(b')}.$$

In particular, $G^\circ o^{(r)}$ *is the unique closed orbit contained in* $\overline{\mathcal{D}}$, *whereas* $G^\circ o^{(1)}$ *is relatively open and everywhere dense in* $\partial\mathcal{D}$.

This follows immediately from Theorem 8. 7, (8. 18) and Proposition 8. 8.

Next we want to obtain an analytic characterization of boundary components. We start with

Lemma 8. 10. *One has*

(8. 22)
$$\bar{\mathcal{D}} \cap (o_\kappa + \mathfrak{p}_{\kappa,+}) = \mathcal{F}_\kappa.$$

Proof. Suppose $z \in \bar{\mathcal{D}}$, $z = o_\kappa + z'$, $z' \in \mathfrak{p}_{\kappa,+}$. We may (hence shall) assume $\kappa = \kappa^{(b)}$. Then one can write

$$z = (ad\, k)\Big(\sum_{i=1}^{r} \xi_i o_i\Big),$$
$$z' = (ad\, k')\Big(\sum_{i=b+1}^{r} \xi'_i o_i\Big)$$

with $k \in K^\circ$, $1 \geq \xi_1 \geq \cdots \geq \xi_r \geq 0$, $k' \in K_b^\circ$, $\xi'_{b+1} \geq \cdots \geq \xi'_r \geq 0$. Hence one has

$$z = (ad\, k')\Big(\sum_{i=1}^{b} o_i + \sum_{i=b+1}^{r} \xi'_i o_i\Big)$$

and by the uniqueness of the expression (Lem. 8. 2) the r-tuple $(\xi_1, \cdots, \xi_r)$ must coincide with $(1, \cdots, 1, \xi'_{b+1}, \cdots, \xi'_r)$ up to the order. Therefore $\xi'_{b+1} \leq 1$, which implies $z' \in \bar{\mathcal{D}}_b$, and hence $z \in \mathcal{F}_b$, q. e. d.

We note that the (affine) subspace $o_\kappa + \mathfrak{p}_{\kappa,+}$ is perpendicular to the line spanned by the segment oo_κ. Hence the point o_κ is characterized as the "nearest" point on $\mathcal{F}_\kappa$ from the origin, i. e., $\{\|z\| \,|\, z \in \mathcal{F}_\kappa\}$ attains the minimum at $z = o_\kappa$ (see (II. 4. 20)). We put

$$\|o_1\| = \cdots = \|o_r\| = \delta > 0.$$

Then, if $m(\kappa) = b$, one has

(8. 23)
$$\|o_\kappa\| = \sqrt{b}\,\delta.$$

(Note that o_κ can also be characterized algebraically as the unique idempotent of the JTS $\mathfrak{p}_+$ contained in $\mathcal{F}_\kappa$.)

Let L be a (real) hyperplane in $\mathfrak{p}_+$ not passing through the origin. Among the two (open) half-spaces bounded by L, we denote the one containing the origin by L^- and the other one by L^+. A hyperplane L is called a *supporting hyperplane* of $\mathcal{D}$ if $\bar{\mathcal{D}} \cap L \neq \phi$ and $\mathcal{D} \subset L^-$. Since $\mathcal{D}$ is convex (Prop. II. 4. 6), the second condition is equivalent to saying that $\bar{\mathcal{D}} \cap L \subset \partial\mathcal{D}$.

Lemma 8. 11. *If L is a supporting hyperplane of $\mathcal{D}$, then there exists a uniquely determined boundary component $\mathcal{F}_\kappa$ such that*

$$\bar{\mathcal{D}} \cap L = \mathcal{F}_\kappa.$$

Proof. First we note that, for any supporting hyperplane L and any boundary component $\mathcal{F}_\kappa$, if $\mathcal{F}_\kappa \cap L \neq \phi$ then $o_\kappa + \mathfrak{p}_{\kappa,+} \subset L$. In fact, if $z_0 \in \mathcal{F}_\kappa \cap L$ and $o_\kappa + \mathfrak{p}_{\kappa,+} \not\subset L$, then any neighbourhood of z_0 in $o_\kappa + \mathfrak{p}_{\kappa,+}$ contained in $\mathcal{F}_\kappa$ would have points in common with both half-spaces $L^\pm$, which is absurd. It follows that $\bar{\mathcal{D}} \cap L$ is a union of boundary components. Let $\mathcal{F}_\kappa$ be a boundary component contained in $\bar{\mathcal{D}} \cap L$ for which $m(\kappa) = b$ is the smallest. We claim that $\mathcal{F}_\kappa$ is uniquely determined by this property. In fact, by what we remarked above, one has $\|z\| \geq \sqrt{b}\,\delta$ for all $z \in \bar{\mathcal{D}} \cap L$ and $\|o_\kappa\| = \sqrt{b}\,\delta$. If there were another $\kappa' \neq \kappa$ such that $\mathcal{F}_{\kappa'} \subset L$ and $m(\kappa') = b$, then $\|o_{\kappa'}\| = \sqrt{b}\,\delta$. But then, since $\bar{\mathcal{D}} \cap L$ is convex, the line segment $\{z(\lambda) = \lambda o_\kappa + (1-\lambda) o_{\kappa'}\ (0 \leq \lambda \leq 1)\}$ is contained in $\bar{\mathcal{D}} \cap L$ and one would have $\|z(\lambda)\| < \sqrt{b}\,\delta$ for $0 < \lambda < 1$, which is absurd. It follows that all boundary components in $\bar{\mathcal{D}} \cap L$ other than $\mathcal{F}_\kappa$ are contained in

$$\{z \in \mathfrak{p}_+ \mid \|z\| \geq \sqrt{b+1}\,\delta\}.$$

Therefore, again by the convexity of $\bar{\mathcal{D}} \cap L$, we can conclude that $\bar{\mathcal{D}} \cap L$ is contained in the subspace $o_\kappa + \mathfrak{p}_{\kappa,+}$. Hence by Lemma 8. 10 one has

$$\bar{\mathcal{D}} \cap L = \bar{\mathcal{D}} \cap (o_\kappa + \mathfrak{p}_{\kappa,+}) = \bar{\mathcal{F}}_\kappa, \qquad \text{q. e. d.}$$

Proposition 8. 12. *For a non-trivial* (H_1)*-homomorphism* $\kappa : \mathfrak{g}^1 \to (\mathfrak{g}, H_0)$, *let* L_κ *be the hyperplane in* $\mathfrak{p}_+$ *passing through* o_κ *and perpendicular to the line* oo_κ. *Then* L_κ *is a supporting hyperplane of* $\mathcal{D}$ *and one has*

(8. 24) $$\bar{\mathcal{D}} \cap L_\kappa = \bar{\mathcal{F}}_\kappa.$$

Proof. First we treat the case $m(\kappa) = 1$; clearly, we may assume that $\kappa = \kappa^{(1)}$. Since $G^\circ o^{(1)}$ is a smooth hypersurface in $\mathfrak{p}_+$, $o^{(1)}$ is a regular point on $\partial\mathcal{D}$. We know that the tangent space to $\mathcal{F}_1$ at $o^{(1)}$, i. e., $o^{(1)} + \mathfrak{p}_{1,+}$, is contained in L_κ. On the other hand, since $K^\circ o^{(1)}$ is on the sphere $\{\|z\| = \delta\}$, the tangent space to $K^\circ o^{(1)}$ at $o^{(1)}$ is also contained in L_κ. Therefore L_κ coincides with the tangent hyperplane to $\partial\mathcal{D}$ at $o^{(1)}$. Since $\mathcal{D}$ is convex, this implies that L_κ is a supporting hyperplane. By Lemma 8. 11, it is then clear that $\bar{\mathcal{D}} \cap L_\kappa = \bar{\mathcal{F}}_\kappa$. In the general case, we proceed by induction on $r = \operatorname{rank} \mathfrak{g}$. When $r = 1$, there is nothing more to prove. Let $r > 1$ and assume that the assertion of our Proposition is true for symmetric domains of rank $r-1$. Suppose that L_κ with $m(\kappa) > 1$ is not a supporting hyperplane; then $\mathcal{D} \cap L_\kappa^+ \neq \phi$. Since $\mathcal{D}$ is convex, this implies that for any neighbourhood N of o_κ one has $\partial\mathcal{D} \cap N \cap L_\kappa^+ \neq \phi$. As $G^\circ o^{(1)}$ is everywhere dense in $\partial\mathcal{D}$, there exists κ' with $m(\kappa') = 1$ such that $\bar{\mathcal{F}}_{\kappa'} \supset \mathcal{F}_\kappa$ and $\mathcal{F}_{\kappa'} \cap N \cap L_\kappa^+ \neq \phi$. But then $L'_\kappa = (o_{\kappa'} + \mathfrak{p}_{\kappa',+}) \cap L_\kappa$ is a hyperplane in $o_{\kappa'} + \mathfrak{p}_{\kappa',+}$, perpendicular to the line $o_{\kappa'} o_\kappa$. Hence, by induction assumption applied to the symmetric domain $\mathcal{F}_{\kappa'}$, L'_κ is a supporting hyperplane of $\mathcal{F}_{\kappa'}$. Thus, denoting by L'^-_κ the half-space (in $o_{\kappa'} + \mathfrak{p}_{\kappa',+}$) bounded by L'_κ and containing $o_{\kappa'}$, one has $\mathcal{F}_{\kappa'} \subset L'^-_\kappa$. Since $\|o_{\kappa'}\| \leq \|o_\kappa\|$, one has $L'^-_\kappa \subset L_\kappa^-$, and hence $\mathcal{F}_{\kappa'} \subset L_\kappa^-$, which is absurd. Therefore L_κ is a supporting hyperplane. Since o_κ is the nearest point on L_κ from the origin, (8. 24) follows immediately from Proposition 8. 8 and Lemma 8. 11, q. e. d.

Now we define the notion of holomorphic arc components. In general, let M be a complex manifold and S a subset of M. We say that two points x, y in S can be connected by a holomorphic arc in S, if there exists a finite chain $f_1, \cdots, f_s$ of holomorphic maps $f_i : \mathscr{D}^1 = \{|\zeta|<1\} \to M$ $(1 \leq i \leq s)$ such that

$$f_i(\mathscr{D}^1) \subset S, \quad f_i(\mathscr{D}^1) \cap f_{i+1}(\mathscr{D}^1) \neq \phi \quad (1 \leq i \leq s-1),$$
$$x \in f_1(\mathscr{D}^1), \quad y \in f_s(\mathscr{D}^1).$$

Clearly this is an equivalence relation in S. A *holomorphic arc component* of S is an equivalence class for this equivalence relation.

Theorem 8. 13. *A boundary component $\mathscr{F}_\kappa$ is a holomorphic arc component $\bar{\mathscr{D}}$ (in $\mathfrak{p}_+$).*

Proof. Since $\mathscr{F}_\kappa$ is holomorphically equivalent to a symmetric domain $\mathscr{D}_\kappa$, it is clear that any two points x, y in $\mathscr{F}_\kappa$ can be connected by a holomorphic arc. [Actually, there exists a single holomorphic map $f : \mathscr{D}^1 \to \mathscr{F}_\kappa$ such that $x, y \in f(\mathscr{D}^1)$.] Hence it is enough to show that, if $f : \mathscr{D}^1 \to \mathfrak{p}_+$ is a holomorphic map such that $f(\mathscr{D}^1) \subset \bar{\mathscr{D}}$, then $f(\mathscr{D}^1)$ is contained in one of the boundary components. Let $\mathscr{F}_\kappa$ be a boundary component such that $f(\mathscr{D}^1) \cap \mathscr{F}_\kappa \neq \phi$, for which $m(\kappa)$ is the largest. If $m(\kappa)=0$, then $f(\mathscr{D}^1) \subset \mathscr{D}$. If $m(\kappa) \geq 1$, then by Proposition 8. 12 one has $\bar{\mathscr{D}} \cap L_\kappa = \mathscr{F}_\kappa$. Let f_0 be a $\boldsymbol{C}$-linear function on $\mathfrak{p}_+$ such that $\operatorname{Re} f_0 = 1$ on L_κ and $f_0(o)=0$. Then one has

$$\operatorname{Re} f_0(z) \leq 1 \qquad \text{for all} \quad z \in \bar{\mathscr{D}}.$$

Hence $f_0 \circ f$ is a holomorphic function on $\mathscr{D}^1$ for which $\operatorname{Re}(f_0 \circ f)$ attains a maximum. Therefore, by the maximum modulus principle, $f_0 \circ f$ is constant. It follows that $f(\mathscr{D}^1) \subset L_\kappa$ and hence by (8. 24)

$$f(\mathscr{D}^1) \subset \bar{\mathscr{D}} \cap L_\kappa = \mathscr{F}_\kappa.$$

From our choice of κ, this implies that $f(\mathscr{D}^1) \subset \mathscr{F}_\kappa$, q. e. d.

As another application of the maximum modulus principle, we add

Theorem 8. 14. *The Bergman-Šilov boundary of $\mathscr{D}$ coincides with $G^\circ o^{(r)}$. It is the unique closed orbit of G° in $\bar{\mathscr{D}}$ and is the union of all point boundary components of $\mathscr{D}$.*

We have only to prove the first assertion. By definition, the "Bergman-Šilov boundary" $B(\mathscr{D})$ of $\mathscr{D}$ is the *smallest* closed subset S of $\partial\mathscr{D}$ having the property that for any ($\boldsymbol{C}$-valued) continuous function f on $\bar{\mathscr{D}}$, holomorphic in $\mathscr{D}$, there exists $z_0 \in S$ such that f attains the maximum at z_0, i. e., $|f(z_0)| = \operatorname{Sup}\{|f(z)| \ (z \in \bar{\mathscr{D}})\}$. It is clear that $B(\mathscr{D})$ is G°-stable. Hence it suffices to show that for any continuous function f on $\bar{\mathscr{D}}$, holomorphic in $\mathscr{D}$, there exists a point in $G^\circ o^{(r)}$ where f attains the maximum. Suppose f attains the maximum at $z_0 \in \bar{\mathscr{D}}$; we may assume that z_0 is in $(\exp \mathfrak{a}^+)o$. Let $\varphi : \boldsymbol{C}^r \to \mathfrak{p}_+$ be a $\boldsymbol{C}$-linear map defined by

$$\varphi(\zeta_1, \cdots, \zeta_r) = \sum_{i=1}^{r} \zeta_i o_i.$$

Then φ induces an equivariant holomorphic map $(\mathscr{D}^1)^r \to \mathscr{D}$ and

$$\overline{(\exp \mathfrak{a}^+)o} \subset \varphi((\overline{\mathscr{D}^1})^r) \subset \overline{\mathscr{D}}.$$

Hence $f \circ \varphi$ is continuous on $(\overline{\mathscr{D}^1})^r$, holomorphic in $(\mathscr{D}^1)^r$, and attains the maximum at $\varphi^{-1}(z_0)$. It is well-known (and easy to see) that $B((\mathscr{D}^1)^r) = (\partial\mathscr{D}^1)^r$. Hence there exists $z_0' \in \varphi((\partial\mathscr{D}^1)^r)$ such that $|f(z_0')| = |f(z_0)|$. This proves our assertion, because

$$\begin{aligned} \varphi((\partial\mathscr{D}^1)^r) &= \varphi((G^1)^r(1, \cdots, 1)) \\ &\subset G^\circ \varphi(1, \cdots, 1) = G^\circ o^{(r)}. \end{aligned}$$

So far we have assumed that $\mathfrak{g}$ is simple (and non-compact). In the general case, let

(8. 25) $$\mathfrak{g} = \mathfrak{g}_0 \oplus \left(\bigoplus_{i=1}^{s} \mathfrak{g}_i\right), \qquad \boldsymbol{R}\text{-rank}\, \mathfrak{g}_i = r_i,$$

where $\mathfrak{g}_0$ is compact and $\mathfrak{g}_i\,(1 \le i \le s)$ is simple, non-compact. Then one has the corresponding direct product decomposition

(8. 26) $$\mathscr{D} = \mathscr{D}_1 \times \cdots \times \mathscr{D}_s,$$

where $\mathscr{D}_i$ is an irreducible symmetric domain associated with $\mathfrak{g}_i$. Let π_i denote the projection map $\mathscr{D} \to \mathscr{D}_i$. For any (H_1)-homomorphism $\kappa : \mathfrak{g}^1 \to (\mathfrak{g}, H_0)$, the boundary component $\mathscr{F}_\kappa$ decomposes as

(8. 27) $$\mathscr{F}_\kappa = \mathscr{F}_{\kappa,1} \times \cdots \times \mathscr{F}_{\kappa,s},$$

where $\mathscr{F}_{\kappa,i}$ is a boundary component of $\mathscr{D}_i$ corresponding to the (H_1)-homomorphism $\pi_i \circ \kappa$. We define an s-tuple $\boldsymbol{m}(\kappa)$ by

(8. 28) $$\boldsymbol{m}(\kappa) = (b_1, \cdots, b_s), \qquad b_i = m(\pi_i \circ \kappa).$$

For a fixed $(\mathfrak{a}, \Delta)$, we define the canonical homomorphism $\kappa^{(\boldsymbol{b})}$ with $\boldsymbol{b} = (b_1, \cdots, b_s)$ $(0 \le b_i \le r_i)$ to be the unique (H_1)-homomorphism $\mathfrak{g}^1 \to (\mathfrak{g}, H_0)$ such that $\pi_i \circ \kappa^{(\boldsymbol{b})}$ is the b_i-th canonical (H_1)-homomorphism $\mathfrak{g}^1 \to (\mathfrak{g}_i, \pi_i(H_0))$. The corresponding boundary component $\mathscr{F}_{\boldsymbol{b}}$ is called a *standard boundary component of type* $\boldsymbol{b}$. We also write $o^{(\boldsymbol{b})}$ for o_κ when $\kappa = \kappa^{(\boldsymbol{b})}$. Then, from Theorem 8. 7, it is clear that $\overline{\mathscr{D}}$ decomposes into $\prod_{i=1}^{s} (r_i + 1)$ G°-orbits $G^\circ o^{(\boldsymbol{b})}$, and each orbit $G^\circ o^{(\boldsymbol{b})}$ is a disjoint union of the boundary components $\mathscr{F}_\kappa$ with $\boldsymbol{m}(\kappa) = \boldsymbol{b}$; in general, such a boundary component is called *of type* $\boldsymbol{b}$. The expressions for the stabilizers of $\mathscr{F}_\kappa$ and o_κ given in Proposition 8. 5, Corollary 8. 6 hold without change. Proposition 8. 8 remains valid, if we replace (b) by

(b′) There exist $k \in K^\circ$ and $\boldsymbol{b} = (b_i)$, $\boldsymbol{b}' = (b_i') \in \boldsymbol{Z}^s$ such that $0 \le b_i' \le b_i \le r_i\,(1 \le i \le s)$ and $\kappa = (ad\, k) \cdot \kappa^{(\boldsymbol{b})}$, $\kappa' = (ad\, k) \cdot \kappa^{(\boldsymbol{b}')}$.

We write $\boldsymbol{b}' \le \boldsymbol{b}$ if $b_i' \le b_i\,(1 \le i \le s)$, and put $\boldsymbol{0} = (0, \cdots, 0)$, $\boldsymbol{r} = (r_1, \cdots, r_s)$. Then, in place of (8. 16), (8. 17), (8. 21), we have

(8. 16′) $$\overline{\mathscr{D}} = \coprod_{\boldsymbol{0} \le \boldsymbol{b} \le \boldsymbol{r}} G^\circ o^{(\boldsymbol{b})},$$

(8. 17′) $$G^\circ o^{(b)} = \coprod_{m(\kappa)=b} \mathcal{F}_\kappa,$$

(8. 21′) $$\overline{G^\circ o^{(b)}} = \coprod_{b \le b' \le r} G^\circ o^{(b')}.$$

In particular, $G^\circ o^{(r)}$ is the unique closed G°-orbit in $\bar{\mathcal{D}}$, whereas the union $\bigcup G^\circ o^{(b)}$ for $\boldsymbol{b}=(1,0,\cdots,0), (0,1,0,\cdots,0), \cdots, (0,\cdots,0,1)$ is relatively open and everywhere dense in $\partial\mathcal{D}$. Corollary 8. 9∼Theorem 8. 13 remain valid without any change. Note that a holomorphic arc component of the direct product $\bar{\mathcal{D}}=\prod \bar{\mathcal{D}}_i$ is given by the direct product of the holomorphic arc components of each $\bar{\mathcal{D}}_i$. Similarly, for the Bergman-Šilov boundary, one has

$$B(\mathcal{D}) = \prod_{i=1}^{s} B(\mathcal{D}_i).$$

Thus $B(\mathcal{D})$ coincides with the unique compact orbit $G^\circ o^{(r)}$ and is the union of all point boundary components of $\mathcal{D}$.

Remark 1. In the notation of Corollary 1. 6, one has

$$\mathfrak{g}^{[1]} = \{0\} \Longleftrightarrow V_\kappa = \{0\} \Longleftrightarrow \mathcal{S}_\kappa = U_\kappa + i\Omega_\kappa \text{ (tube domain)},$$
$$\mathfrak{g}_\kappa = \{0\} \Longleftrightarrow \mathcal{D}_\kappa = \{o\} \Longleftrightarrow \mathcal{F}_\kappa = \{o_\kappa\} \text{ (point boundary component)}.$$

Thus the conditions (ii) and (iii) in Corollary 1. 6 are equivalent, in virtue of the uniqueness of the Siegel domain realization (V, § 3, Rem. 2; see also V, § 5).

Remark 2. We have seen that, when the origin o of the symmetric domain $\mathcal{D}$ is fixed, there are one-to-one correspondences between the following objects:

(a) (non-trivial) (H_1)-homomorphisms $\kappa: \mathfrak{g}^1 \to (\mathfrak{g}, H_0)$,

(b) (proper) boundary components $\mathcal{F}_\kappa \subset \partial\mathcal{D}$,

(c) (non-zero) idempotents o_κ in the JTS $\mathfrak{p}_+$.

In fact, the correspondence $\kappa \leftrightarrow o_\kappa$ was given in Proposition 4. 1. By Theorems 8. 7 and 8. 13, $\mathcal{F}_\kappa$ is the unique holomorphic arc component containing o_κ, while o_κ is determined as the nearest point in $\mathcal{F}_\kappa$ from the origin o; o_κ is also characterized as the unique limit point of a geodesic passing through o and converging to a point in $\mathcal{F}_\kappa$ (though the geodesic itself is not unique in general). Note also that o_κ and X_κ determine each other uniquely by the relations $X_\kappa = 2\,\mathrm{Re}\, o_\kappa$ and $o_\kappa =$ (the $\mathfrak{p}_+$-part of X_κ). Moreover, by Corollary 8. 6, $\mathfrak{b}_\kappa$ is the Lie algebra of the stabilizer of $\mathcal{F}_\kappa$, and X_κ is the unique element in $\mathfrak{p}$ such that $\mathfrak{b}_\kappa = \mathfrak{g}(ad\, X_\kappa; 0, 1, 2)$. The correspondences $\kappa \to X_\kappa \to \mathfrak{b}_\kappa \leftrightarrow \mathcal{F}_\kappa$ are given independently of the choice of the origin o, but the other correspondences are not. In Exercises 1∼4, we will examine what happens to these correspondences if we change the origin o.

Remark 3. When the $\boldsymbol{R}$-group G is defined over a field $F(\subset \boldsymbol{R})$, the following implications are obvious:

$$\kappa: \mathfrak{g}^1 \to \mathfrak{g} \text{ is defined over } F \Longrightarrow X_\kappa \text{ is } F\text{-rational (i. e., } X_\kappa \in \mathfrak{g}_F)$$
$$\Longrightarrow \mathfrak{b}_\kappa \text{ is defined over } F.$$

Conversely, it can be shown that, when $\mathfrak{b}_\kappa$ is defined over F, there exists a homomorphism $\kappa': \mathfrak{g}^1 \to \mathfrak{g}$ satisfying (H_1) (possibly with respect to a different H-element) and defined over F such that $\mathfrak{b}_\kappa = \mathfrak{b}_{\kappa'}$ (Exerc. 3, 4). In other words, for a given boundary component, the above three conditions become equivalent for a "flexible" origin o. The boundary component $\mathcal{F}_\kappa$ is called *F-rational* if $\mathfrak{b}_\kappa$ is defined over F.

Exercises

1. We say that a homomorphism $\kappa: \mathfrak{g}^1 \to \mathfrak{g}$ "gets along" with the origin o if κ satisfies the condition (H_1) with respect to H_0^1 and the H-element H_0 corresponding to o. Fix such a pair (κ, o). Show that κ gets along with $o'(\in \mathcal{D})$ if and only if $o' \in \mathcal{D}_\kappa = G_\kappa^\circ o$. (Use II, § 8, Exerc. 1.)

2. Let $o' = g(o)$ with $g \in G^\circ$ and let ψ and ψ' denote the Harish-Chandra embeddings of $\mathcal{D}$ (into $\mathfrak{p}_+$ and $\mathfrak{p}'_+ = (ad\, g)\mathfrak{p}_+$) at o and o'. Prove the following.

2. 1) For $z \in \mathcal{D}$, one has

$$\psi'(z) = (ad\, g)\psi(g^{-1}(z)).$$

2. 2) The map $\psi'\psi^{-1}$ can be extended to a homeomorphism φ of $\overline{\psi(\mathcal{D})}$ onto $\overline{\psi'(\mathcal{D})}$. When κ gets along with o, the homomorphism $\kappa' = (ad\, g)\kappa$ gets along with o', and φ maps $g(o_\kappa)$ and $g(\mathcal{F}_\kappa)$ (in $\mathfrak{p}_+$) to $o_{\kappa'}$ and $\mathcal{F}_{\kappa'}$ (in $\mathfrak{p}'_+$), respectively. [This legitimates the identifications $g(o_\kappa) = o_{\kappa'}, g(\mathcal{F}_\kappa) = \mathcal{F}_{\kappa'}$.]

3. Prove the following.

3. 1) Let X be a semi-simple element in $\mathfrak{g}$ such that the set of eigenvalues of $ad\, X$ in each simple factor $\mathfrak{g}_i$ of $\mathfrak{g}$ is either $\{0, \pm 1, \pm 2\}$ or $\{0, \pm 2\}$ or $\{0\}$. Then there exists an (H_1)-homomorphism $\kappa: \mathfrak{g}^1 \to \mathfrak{g}$ such that $X = X_\kappa$.

3. 2) Let κ be a non-trivial (H_1)-homomorphism $\mathfrak{g}^1 \to \mathfrak{g}$. Then one has

$$C(X_\kappa) \cap K^\circ = C(\kappa(\mathfrak{g}^1)) \cap K, \tag{8. 12$'$ a}$$

$$C(X_\kappa) \cap G^\circ = C(\kappa(\mathfrak{g}^1)) \cdot G_\kappa^{(2)\circ}. \tag{8. 12$'$ b}$$

(The proof is quite analogous to that of Lemma 8. 4.)

3. 3) Let $\kappa': \mathfrak{g}^1 \to \mathfrak{g}$ be another (H_1)-homomorphism such that $X_\kappa = X_{\kappa'}$. Then there exists $g \in G_\kappa^{(2)\circ}$ such that $\kappa' = (ad\, g)\kappa$. (Use Prop. 4. 3 and (8. 12$'$ b).) Then one has $e_{\kappa'} = (ad\, g)e_\kappa \in \Omega_\kappa$, and the map $\kappa' \mapsto e_{\kappa'}$ gives a bijective correspondence between the set of (H_1)-homomorphisms κ' with $X_\kappa = X_{\kappa'}$ and the cone Ω_κ. (Note that $\mathfrak{b}_\kappa$, $\mathfrak{c}(X_\kappa)$, U_κ, V_κ are all defined by X_κ alone. The above result shows that the cone Ω_κ, too, is determined only by X_κ.)

3. 4) When X_κ is F-rational, the (H_1)-homomorphism κ' in the above correspondence is defined over F if and only if $e_{\kappa'}$ is F-rational. (Since U_κ is defined over F, an F-rational element $e_{\kappa'}$ in Ω_κ surely exists.)

4. Prove the following.

4. 1) Let $\mathfrak{b}$ be a parabolic subalgebra of $\mathfrak{g}$ such that, for each i, $\mathfrak{b} \cap \mathfrak{g}_i$ is either maximal in $\mathfrak{g}_i$ or $= \mathfrak{g}_i$. Then there exists an (H_1)-homomorphism $\kappa: \mathfrak{g}^1 \to \mathfrak{g}$ such that $\mathfrak{b} = \mathfrak{b}_\kappa (= \mathfrak{g}(ad\, X_\kappa; 0, 1, 2))$.

4. 2) Let κ' be another (H_1)-homomorphism $\mathfrak{g}^1 \to \mathfrak{g}$ such that $\mathfrak{b} = \mathfrak{b}_{\kappa'}$. Then there exists $g \in B_\kappa \cap G^\circ$ such that $\kappa' = (ad\, g)\kappa$ (again by Prop. 4. 3 and by the property of parabolic subalgebras (I, § 5)). It follows that $\mathcal{F}_\kappa = \mathcal{F}_{\kappa'}$ (in the sense of Exerc. 2). Thus the boundary component $\mathcal{F}_\kappa$ is uniquely determined only by $\mathfrak{b}_\kappa$. (Conversely, $\mathfrak{b}_\kappa$ is determined as the Lie algebra of the stabilizer of $\mathcal{F}_\kappa$.) On the other hand, writing $g = \tilde{v} g_1$ with $\tilde{v} \in \tilde{V}_\kappa$ and $g_1 \in C(X_\kappa) \cap G^\circ$, one has $X_{\kappa'} = (ad\, \tilde{v})X_\kappa$, and so $\mathfrak{c}(X_{\kappa'}) = (ad\, \tilde{v})\mathfrak{c}(X_\kappa)$. The map $X_{\kappa'} \mapsto \mathfrak{c}(X_{\kappa'})$ gives a bijective correspondence between the set of $X_{\kappa'}$ such that $\mathfrak{b}_\kappa = \mathfrak{b}_{\kappa'}$ onto the set of the reductive parts of $\mathfrak{b}_\kappa$.

4. 3) When $\mathfrak{b}_\kappa$ is defined over F, the element $X_{\kappa'}$ in the above correspondence is F-rational if and only if $\mathfrak{c}(X_{\kappa'})$ is defined over F. [The existence of an F-rational element $X_{\kappa'}$ follows easily from the triviality of the Galois cohomology $H^1(\mathrm{Gal}(\bar{F}/F), (\tilde{V}_\kappa)_{\bar{F}})$. For another proof, see Satake [7], p. 240.]

5. Let (V, A) be a symplectic space of dimension $2n$ and let V_1 be a (real) totally isotropic subspace of V of dimension b. Let $\mathcal{F}(V_1)$ denote the set of all maximal totally isotropic complex subspaces W of V_C satisfying the following two conditions:

(8.29) $h_A|W$ is negative semi-definite.

(8.30) $W \cap \overline{W} = V_{1C}$.

Prove the following.

5.1) Let $V_1^\perp$ be the annihilator of V_1 in V with respect to A and let A_1 be the alternating bilinear form on the quotient space $V_1^\perp/V_1$ induced by A. Then A_1 is non-degenerate and the correspondence $W \mapsto W/V_{1C}$ gives an (analytic) isomorphism of $\mathcal{F}(V_1)$ (viewed as a submanifold of the Grassmannian $\mathcal{G}_n(V_C)$) onto the Siegel space $\mathfrak{S}(V_1^\perp/V_1, A_1)$.

5.2) Take a (real) symplectic basis $(e_1, \cdots, e_{2n})$ of V such that $V_1 = \{e_{n-b+1}, \cdots, e_n\}_{\boldsymbol{R}}$ and identify $\mathfrak{sp}(V, A)$ with $\mathfrak{sp}(2n, \boldsymbol{R})$. Then the stabilizer of V_1 in $\mathfrak{sp}(V, A)$ and $\mathcal{F}(V_1)$ can be identified, respectively, with the parabolic subalgebra $\mathfrak{b}_\kappa$ given in § 3, Exerc. 2 and the corresponding boundary component $\mathcal{F}_\kappa$. By the map $V_1 \mapsto \mathcal{F}(V_1)$, the set of all totally isotropic subspaces V_1 of dimension b in V is in a one-to-one correspondence with the set of all b-th boundary components of the Siegel space $\mathfrak{S}(V, A)$. Give a similar geometric interpretation of boundary components for other types of symmetric domains. (Cf. Satake [9] for the domain of type (IV).)

6. Show that for a symmetric bounded domain $\mathcal{D}$ the real dimension of the Bergman-Šilov boundary $B(\mathcal{D})$ is not greater than the complex dimension of $\mathcal{D}$ and the equality holds if and only if $\mathcal{D}$ is of tube type.

§ 9. The relations between two Siegel domain realizations.

We begin with a preliminary observation on the representations of $\mathfrak{g}^1 = \mathfrak{sl}_2(\boldsymbol{R})$.

Suppose there are given two mutually commutative representations ρ and ρ' of $\mathfrak{g}^1$ on a (finite-dimensional) real vector space V, and let

$$V = \bigoplus_\nu V_\rho^{[\nu]} = \bigoplus_\nu V_{\rho'}^{[\nu]}$$

be the corresponding primary decompositions of V, where as in § 1 $V_\rho^{[\nu]}$ (resp. $V_{\rho'}^{[\nu]}$) denotes the primary component of (V, ρ) (resp. (V, ρ')) belonging to the irreducible representation ρ_ν of $\mathfrak{g}^1$. Put $V^{[\lambda,\mu]} = V_\rho^{[\lambda]} \cap V_{\rho'}^{[\mu]}$. Then $V^{[\lambda,\mu]}$, being stable under both ρ and ρ', is also stable under $\rho + \rho'$.

Lemma 9.1. *Suppose $V^{[\lambda,\mu]} \neq \{0\}$. Then the irreducible representation ρ_ν is contained in $(V^{[\lambda,\mu]}, \rho+\rho')$ if and only if ρ_ν is contained in $\rho_\lambda \otimes \rho_\mu$ (as group representation).*

Proof. The restriction of ρ on $V^{[\lambda,\mu]}$ is a primary representation of $\mathfrak{g}^1$ belonging to ρ_λ. Hence one may write

(*) $$V^{[\lambda,\mu]} = V_\lambda \otimes W, \qquad \rho | V^{[\lambda,\mu]} = \rho_\lambda \otimes 1_W,$$

where $V_\lambda = \boldsymbol{R}^{\lambda+1}$ is the representation-space for ρ_λ. Then, since ρ and ρ' are commutative, there exists a representation ρ'_W of $\mathfrak{g}^1$ on W such that

$$\rho' | V^{[\lambda,\mu]} = 1_{V_\lambda} \otimes \rho'_W.$$

Since $\rho' | V^{[\lambda,\mu]}$ is a primary representation (belonging to ρ_μ), so is ρ'_W. Hence, by the same reason as above, one has a tensor product decomposition

(**) $$W = V_\mu \otimes W', \qquad \rho'_W = \rho_\mu \otimes 1_{W'}.$$

By (*), (**), one has

$$V^{[\lambda,\mu]} = (V_\lambda \otimes V_\mu) \otimes W',$$
$$(\rho+\rho')|V^{[\lambda,\mu]} = (\rho_\lambda \otimes 1_{V_\mu} + 1_{V_\lambda} \otimes \rho_\mu) \otimes 1_{W'},$$

where $\rho_\lambda \otimes 1_{V_\mu} + 1_{V_\lambda} \otimes \rho_\mu$ is the Lie algebra representation of $\mathfrak{sl}_2(\boldsymbol{R})$ corresponding to the group representation $\rho_\lambda \otimes \rho_\mu$ of $SL_2(\boldsymbol{R})$. The assertion of the Lemma is now obvious, q. e. d.

One has, for instance,

$$\rho_1 \otimes \rho_1 = \rho_0 \oplus \rho_2. \tag{9. 1}$$

In general, it is well-known (and easy to see) that $\rho_{\lambda+\mu}$ is contained in $\rho_\lambda \otimes \rho_\mu$. Hence, by Lemma 9. 1, if $V^{[\lambda,\mu]} \neq \{0\}$, then one has $V^{[\lambda,\mu]} \cap V^{[\lambda+\mu]}_{\rho+\rho'} \neq \{0\}$.

Now, let $(\mathfrak{g}, H_0)$ be a semi-simple Lie algebra of hermitian type, and let κ', κ'' be two mutually commutative (H_1)-homomorphisms $\mathfrak{g}^1 \to (\mathfrak{g}, H_0)$; then $\kappa = \kappa' + \kappa''$ is also an (H_1)-homomorphism (with respect to the same H-elements). For simplicity, we write $\mathfrak{g}^{[\nu]}_\kappa$, etc. for $\mathfrak{g}^{[\nu]}_{ad\,\mathfrak{g}\circ\kappa}$, etc. and set $\mathfrak{g}^{[\lambda,\mu]} = \mathfrak{g}^{[\lambda]}_{\kappa'} \cap \mathfrak{g}^{[\mu]}_{\kappa''}$. Then from Lemma 9. 1, Lemma 1. 2, and (9. 1), one has

$$\mathfrak{g}^{[\lambda,\mu]} = \{0\} \qquad \text{if} \quad \lambda+\mu > 2, \tag{9. 2 a}$$
$$\mathfrak{g}^{[\nu,0]} + \mathfrak{g}^{[0,\nu]} \subset \mathfrak{g}^{[\nu]}_\kappa \qquad (\nu=0, 1, 2), \tag{9. 2 b}$$
$$\mathfrak{g}^{[1,1]} \subset \mathfrak{g}^{[0]}_\kappa + \mathfrak{g}^{[2]}_\kappa. \tag{9. 2 c}$$

We set $\mathfrak{g}^{[1,1]}_\nu = \mathfrak{g}^{[1,1]} \cap \mathfrak{g}^{[\nu]}_\kappa$ ($\nu=0, 2$). Then we obtain

$$\begin{cases} \mathfrak{g}^{[0]}_\kappa = \mathfrak{g}^{[0,0]} + \mathfrak{g}^{[1,1]}_0, \\ \mathfrak{g}^{[1]}_\kappa = \mathfrak{g}^{[0,1]} + \mathfrak{g}^{[1,0]}, \\ \mathfrak{g}^{[2]}_\kappa = \mathfrak{g}^{[0,2]} + \mathfrak{g}^{[1,1]}_2 + \mathfrak{g}^{[2,0]}. \end{cases} \tag{9. 3}$$

On the other hand, it is clear from the definitions that

$$\begin{cases} \mathfrak{g}^{[0]}_{\kappa'} = \mathfrak{g}^{[0,0]} + \mathfrak{g}^{[0,1]} + \mathfrak{g}^{[0,2]}, \\ \mathfrak{g}^{[1]}_{\kappa'} = \mathfrak{g}^{[1,1]}_0 + \mathfrak{g}^{[1,0]} + \mathfrak{g}^{[1,1]}_2, \\ \mathfrak{g}^{[2]}_{\kappa'} = \mathfrak{g}^{[2,0]}. \end{cases} \tag{9. 4}$$

Now, by the assumption, one has

$$X_\kappa = X_{\kappa'} + X_{\kappa''}, \qquad [X_{\kappa'}, X_{\kappa''}] = 0, \tag{9. 5}$$
$$c_\kappa = c_{\kappa'} c_{\kappa''} = c_{\kappa''} c_{\kappa'}.$$

Also, from $[\kappa'(\mathfrak{g}^1), \kappa''(\mathfrak{g}^1)] = 0$, one obtains $\kappa''(\mathfrak{g}^1) \subset \mathfrak{g}^{[0]}_{\kappa'}$. More precisely, since $\kappa''(\mathfrak{g}^1)$ is semi-simple, non-compact, one has $\kappa''(\mathfrak{g}^1) \subset \mathfrak{g}_{\kappa'}$. Hence $(ad\, c_{\kappa''})|\mathfrak{g}_{\kappa',c}$ may be viewed as the Cayley element in Ad $\mathfrak{g}_{\kappa',c}$ corresponding to the (H_1)-homomorphism $\kappa'': \mathfrak{g}^1 \to \mathfrak{g}_{\kappa'}$. By Proposition 8. 8, for the corresponding boundary components, one has $\mathscr{F}_\kappa \subset \partial \mathscr{F}_{\kappa'}$. In particular, $\mathfrak{p}_{\kappa,+} \subset \mathfrak{p}_{\kappa',+}$ and hence $\mathfrak{p}_\kappa \subset \mathfrak{p}_{\kappa'}$, where $\mathfrak{p}_\kappa = \mathfrak{g}^{[0]}_\kappa \cap \mathfrak{p}$, etc. In view of (9. 3), (9. 4), this implies that $\mathfrak{g}^{[1,1]}_0 \cap \mathfrak{p} = \{0\}$, i. e., $\mathfrak{g}^{[1,1]}_0$ is a compact ideal of $\mathfrak{g}^{[0]}_\kappa$. Since $\mathfrak{g}_\kappa$ and $\mathfrak{g}_{\kappa'}$ are subalgebras of $\mathfrak{g}$ generated by $\mathfrak{p}_\kappa$ and $\mathfrak{p}_{\kappa'}$, respectively, we have

(9. 6) $$\mathfrak{g}_\kappa \subset \mathfrak{g}_{\kappa'}.$$

(The inclusion $\mathfrak{g}_\kappa^{[1]} \subset \mathfrak{g}_{\kappa'}^{[1]}$ is also true, as is seen from the list in § 4.)

We set

$$\mathfrak{p}^{[\lambda,\mu]} = \mathfrak{g}^{[\lambda,\mu]} \cap \mathfrak{p}, \qquad \mathfrak{p}_\nu^{[1,1]} = \mathfrak{g}_\nu^{[1,1]} \cap \mathfrak{p} \quad (\nu=0, 2).$$

Then, since $\mathfrak{g}_\kappa^*$, $\mathfrak{g}_{\kappa'}^*$, $\mathfrak{g}_{\kappa''}^*$ are subalgebras of $\mathfrak{g}$ generated by $\mathfrak{p}^{[0,2]}+\mathfrak{p}_2^{[1,1]}+\mathfrak{p}^{[2,0]}$, $\mathfrak{p}^{[2,0]}$, $\mathfrak{p}^{[0,2]}$, respectively, one has

(9. 7) $$\mathfrak{g}_{\kappa'}^* \oplus \mathfrak{g}_{\kappa''}^* \subset \mathfrak{g}_\kappa^*$$

and so

(9. 8) $$\mathfrak{k}_{\kappa'}^* \oplus \mathfrak{k}_{\kappa''}^* \subset \mathfrak{k}_\kappa^*.$$

As $\gamma_{\kappa'}(=c_{\kappa'}^2)$ and $\gamma_{\kappa''}(=c_{\kappa''}^2)$ are commutative and the restrictions $\gamma_{\kappa'}|\mathfrak{k}_{\kappa''}^*$ and $\gamma_{\kappa''}|\mathfrak{k}_{\kappa'}^*$ are trivial, $\gamma_\kappa|\mathfrak{k}_\kappa^*$ induces $\gamma_{\kappa'}|\mathfrak{k}_{\kappa'}^*$ and $\gamma_{\kappa''}|\mathfrak{k}_{\kappa''}^*$, and these three are Cartan involutions of $\mathfrak{k}_\kappa^*$, $\mathfrak{k}_{\kappa'}^*$, and $\mathfrak{k}_{\kappa''}^*$, respectively. By definition, the corresponding non-compact duals are $\mathfrak{g}_\kappa^{(2)}$, $\mathfrak{g}_{\kappa'}^{(2)}$, and $\mathfrak{g}_{\kappa''}^{(2)}$. Hence one has by (9. 8)

(9. 9) $$\mathfrak{g}_{\kappa'}^{(2)} \oplus \mathfrak{g}_{\kappa''}^{(2)} \subset \mathfrak{g}_\kappa^{(2)}.$$

By (9. 4) we have $\mathfrak{p}^{[0,2]} \subset \mathfrak{p}_{\kappa'}$, and hence $\mathfrak{g}_{\kappa''}^* \subset \mathfrak{g}_{\kappa'}$. In sum, we obtain the following inclusion relations between semi-simple subalgebras of hermitian type in $\mathfrak{g}$:

(9. 10)
$$\begin{array}{ccccc} \mathfrak{g} & \supset & \mathfrak{g}_\kappa^* & \supset & \mathfrak{g}_{\kappa'}^* \\ \cup & & \cup & & \\ \mathfrak{g}_{\kappa'} & \supset & \mathfrak{g}_{\kappa''}^* & & \\ \cup & & & & \\ \mathfrak{g}_\kappa & & & & \end{array}$$

Next we consider the parabolic subalgebras corresponding to κ and κ'. Put

$$\mathfrak{g}^{(\lambda,\mu)} = \mathfrak{g}(ad\, X_{\kappa'}\,;\lambda) \cap \mathfrak{g}(ad\, X_{\kappa''}\,;\mu) \qquad (\lambda, \mu=0, \pm 1, \pm 2).$$

Then by (9. 2 a) and (9. 5)

(9. 11)
$$\begin{cases} \mathfrak{c}(X_\kappa) = \mathfrak{g}(ad\, X_\kappa\,; 0) = \mathfrak{g}^{(0,0)}+\mathfrak{g}^{(1,-1)}+\mathfrak{g}^{(-1,1)}, \\ V_\kappa = \mathfrak{g}(ad\, X_\kappa\,; 1) = \mathfrak{g}^{(0,1)}+\mathfrak{g}^{(1,0)}, \\ U_\kappa = \mathfrak{g}(ad\, X_\kappa\,; 2) = \mathfrak{g}^{(0,2)}+\mathfrak{g}^{(1,1)}+\mathfrak{g}^{(2,0)} \end{cases}$$

and

(9. 12)
$$\begin{cases} \mathfrak{c}(X_{\kappa'}) = \mathfrak{g}^{(0,0)}+\mathfrak{g}^{(0,\pm 1)}+\mathfrak{g}^{(0,\pm 2)}, \\ V_{\kappa'} = \mathfrak{g}^{(1,-1)}+\mathfrak{g}^{(1,0)}+\mathfrak{g}^{(1,1)}, \\ U_{\kappa'} = \mathfrak{g}^{(2,0)}. \end{cases}$$

We introduce the following notations:

(9. 13) $$V_{\kappa',\kappa''} = \mathfrak{g}^{(0,1)}, \qquad U_{\kappa',\kappa''} = \mathfrak{g}^{(1,1)}.$$

Then the second and third equations in (9. 11) can be rewritten as

(9. 14)
$$\begin{cases} V_\kappa = V_{\kappa',\kappa''} + V_{\kappa'',\kappa'}, \\ U_\kappa = U_{\kappa''} + U_{\kappa',\kappa''} + U_{\kappa'}. \end{cases}$$

In terms of these notations, we shall now compare the two Siegel domain realizations

$$(ad\, c_\kappa)^{-1} c_\kappa : \mathscr{D} \longrightarrow \mathscr{S}_\kappa \quad \text{and}$$
$$(ad\, c_{\kappa'})^{-1} c_{\kappa'} : \mathscr{D} \longrightarrow \mathscr{S}_{\kappa'}.$$

From (9. 3) and (9. 4), noting that $\mathfrak{p}_0^{[1,1]} = \{0\}$, we have

$$\begin{cases} \mathfrak{p}_{\kappa,+}^{[0]} = \mathfrak{p}_+^{[0,0]} \\ \qquad\qquad + \\ \mathfrak{p}_{\kappa,+}^{[1]} = \mathfrak{p}_+^{[0,1]} + \mathfrak{p}_+^{[1,0]} \\ \qquad\qquad + \qquad + \\ \mathfrak{p}_{\kappa,+}^{[2]} = \mathfrak{p}_+^{[0,2]} + \mathfrak{p}_+^{[1,1]} + \mathfrak{p}_+^{[2,0]} \end{cases}$$
$$\qquad\qquad \| \qquad\quad \| \qquad\quad \|$$
$$\qquad\qquad \mathfrak{p}_{\kappa',+}^{[0]} \quad \mathfrak{p}_{\kappa',+}^{[1]} \quad \mathfrak{p}_{\kappa',+}^{[2]}$$

First we apply $(ad\, c_{\kappa'})^{-1}$ to these equations. Then the first vertical equation remains unchanged, while the second and third ones become

(9. 15)
$$\begin{aligned} (ad\, c_{\kappa'})^{-1}\mathfrak{p}_+^{[1,0]} + (ad\, c_{\kappa'})^{-1}\mathfrak{p}_+^{[1,1]} &= (ad\, c_{\kappa'})^{-1}\mathfrak{p}_{\kappa',+}^{[1]} = V_{\kappa',+}, \\ (ad\, c_{\kappa'})^{-1}\mathfrak{p}_+^{[2,0]} &= (ad\, c_{\kappa'})^{-1}\mathfrak{p}_{\kappa',+}^{[2]} = U_{\kappa',C}. \end{aligned}$$

Applying further $(ad\, c_{\kappa''})^{-1}$ and using (9. 5), one obtains

(9. 16)
$$\begin{cases} \mathfrak{p}_{\kappa,+} = \quad \mathfrak{p}_+^{[0,0]} \\ \qquad\qquad + \\ V_{\kappa,+} = (ad\, c_{\kappa''})^{-1}\mathfrak{p}_+^{[0,1]} + (ad\, c_{\kappa'})^{-1}\mathfrak{p}_+^{[1,0]} \\ \qquad\qquad + \qquad\qquad\qquad + \\ U_{\kappa,C} = \quad U_{\kappa'',C} \quad + (ad\, c_\kappa)^{-1}\mathfrak{p}_+^{[1,1]} + U_{\kappa',C} \end{cases}$$
$$\qquad\qquad \| \qquad\qquad\qquad \|$$
$$(ad\, c_{\kappa''})^{-1}\mathfrak{p}_{\kappa',+} \quad (ad\, c_{\kappa''})^{-1} V_{\kappa',+}$$

Comparing this with (9. 14), one obtains the relations

(9. 17)
$$\begin{cases} (V_{\kappa',\kappa''})_+ = (ad\, c_{\kappa''})^{-1}\mathfrak{p}_+^{[0,1]}, \\ (U_{\kappa',\kappa''})_C = (ad\, c_\kappa)^{-1}\mathfrak{p}_+^{[1,1]}, \end{cases}$$

where $(V_{\kappa',\kappa''})_+ = (V_{\kappa',\kappa''})_C(I_0; i)$ (Exerc. 1). Thus one obtains

(9. 18)
$$\begin{cases} (ad\, c_{\kappa''})^{-1}\mathfrak{p}_{\kappa',+} = U_{\kappa'',C} + (V_{\kappa',\kappa''})_+ + \mathfrak{p}_{\kappa,+}, \\ (ad\, c_{\kappa''})^{-1} V_{\kappa',+} = (U_{\kappa',\kappa''})_C + (V_{\kappa'',\kappa'})_+. \end{cases}$$

Proposition 9. 2. *Let κ', κ'' be two mutually commutative* (H_1)*-homomorphisms $\mathfrak{sl}_2(\boldsymbol{R}) \to (\mathfrak{g}, H_0)$ and $\kappa = \kappa' + \kappa''$. Then the Cayley transformation $c_{\kappa''}$ is "defined" on $\mathscr{S}_{\kappa'} = (ad\, c_{\kappa'})^{-1} c_{\kappa'}(\mathscr{D})$ and one has*

(9. 19)
$$(ad\, c_\kappa)^{-1} c_\kappa = (ad\, c_{\kappa''})^{-1} c_{\kappa''} \circ (ad\, c_{\kappa'})^{-1} c_{\kappa'} \qquad \text{on} \quad \mathscr{D}.$$

Proof. The proof is similar to that of Theorem 7. 1. Let $z \in \mathcal{D}$ and

$$(ad\, c_\kappa)^{-1} c_\kappa(z) = (u, w, t) \in \mathcal{S}_\kappa,$$
$$(ad\, c_{\kappa'})^{-1} c_{\kappa'}(z) = (u', w', t') \in \mathcal{S}_{\kappa'}.$$

By the definition, this means

$$c_\kappa(\exp z) \in \exp((ad\, c_\kappa)(u+w+t))K_C^\circ P_-,$$

or

$$\exp z \in \exp(u+w+t)c_\kappa^{-1}K_C^\circ P_- \tag{9. 20}$$

and similarly

$$\exp z \in \exp(u'+w'+t')c_{\kappa'}^{-1}K_C^\circ P_-. \tag{9. 20'}$$

Since $c_{\kappa''}$ is a Cayley element in $(G_{\kappa'})_C$, one has by (7. 8)

$$c_{\kappa''} G_{\kappa'}^\circ \subset P_{\kappa',+} K_{\kappa',C}^\circ P_{\kappa',-}$$

and hence

$$c_{\kappa''} B_{\kappa'}^\circ \subset \tilde{P}_{\kappa',+} \tilde{K}_{\kappa',C}^\circ \tilde{P}_{\kappa',-},$$

where $\tilde{P}_{\kappa',\pm} = P_{\kappa',\pm}(\exp V_{\kappa',\pm})$, $\tilde{K}_{\kappa',C}^\circ = (K_{\kappa'}^{(1)\circ})_C (G_{\kappa'}^{(2)\circ})_C (\exp U_{\kappa',C})$. This shows that $c_{\kappa''}(u', w', t')$ is defined. If we set

$$(ad\, c_{\kappa''})^{-1} c_{\kappa''}(u', w', t') = (u'', w'', t'') \in U_{\kappa',C} \times (ad\, c_{\kappa''})^{-1} V_{\kappa',+} \times (ad\, c_{\kappa''})^{-1} \mathfrak{p}_{\kappa',+},$$

i. e.,

$$\exp(u'+w'+t') \in \exp(u''+w''+t'') c_{\kappa''}^{-1} C(X_{\kappa'}, H_{0,\kappa'}^{(1)})_C \tilde{P}_{\kappa',-}, \tag{9. 20''}$$

where $H_{0,\kappa'}^{(1)} = H_0 - \frac{1}{2} H_{\kappa'}$, then we have by (7. 11 a, b) and (9. 5)

$$\begin{aligned} &\exp(u'+w'+t') c_{\kappa'}^{-1} K_C^\circ P_- \\ &\subset \exp(u''+w''+t'') c_{\kappa''}^{-1} C(X_{\kappa'}, H_{0,\kappa'}^{(1)})_C \tilde{P}_{\kappa',-} \cdot c_{\kappa'}^{-1} K_C^\circ P_- \\ &= \exp(u''+w''+t'') c_\kappa^{-1} K_C^\circ P_-. \end{aligned}$$

Comparing this with (9. 20), (9. 20'), we obtain

$$u''+w''+t'' = u+w+t \ (\in \mathcal{S}_\kappa), \qquad \text{q. e. d.}$$

The relation between (u, w, t) and (u', w', t') is obtained as follows. First, by (5. 34) one has

$$\begin{cases} u'' = u' - 2A_{\kappa'}(w', c_{\kappa''}^{-1} J'' w'), \\ w'' = (ad\, c_{\kappa''})^{-1} J'' w', \\ t'' = (ad\, c_{\kappa''})^{-1} c_{\kappa''}(t'), \end{cases} \tag{9. 21}$$

where $J'' = J(c_{\kappa''}, t')$. In particular, t'' belongs to the Siegel domain

$$\mathcal{S}_{\kappa',\kappa''} = (ad\, c_{\kappa''})^{-1} c_{\kappa''}(\mathcal{D}_{\kappa'}) \ (\subset U_{\kappa'',C} \oplus (V_{\kappa',\kappa''})_+ \oplus \mathfrak{p}_{\kappa,+}),$$

which is the Siegel domain realization of $\mathcal{D}_{\kappa'}$ corresponding to $\kappa'' : \mathfrak{g}^1 \to \mathfrak{g}_{\kappa'}$. Next, according to (9. 16) and (9. 18), one has

(9.22)
$$\left\{\begin{array}{l} t = \; t_2 \\ \qquad + \\ w = w_2 + w_1 \\ \qquad + \quad + \\ u = u_2 + u_{12} + u_1 \end{array}\right.$$
$$\begin{array}{ccc} \| & \| & \| \\ t'' & w'' & u'' \end{array}$$

where

$$t_2 \in \mathcal{D}_\kappa, \qquad w_2 \in (V_{\kappa',\kappa''})_+, \qquad w_1 \in (V_{\kappa'',\kappa'})_+,$$
$$u_2 \in U_{\kappa'',+}, \qquad u_{12} \in (U_{\kappa',\kappa''})_C, \qquad u_1 \in U_{\kappa',C}.$$

Combining (9.21) and (9.22), we obtain the relation between (u, w, t) and (u', w', t').

Let us consider the special case where $\mathcal{S}_\kappa$ is a tube domain. In this case, one has $\mathfrak{g}^{[0,1]}=\mathfrak{g}^{[1,0]}=\{0\}$ (hence by (9.17) $V_{\kappa',\kappa''}=V_{\kappa'',\kappa'}=\{0\}$) and $\mathfrak{g}^{[0,0]}$ is compact. Hence one has

$$\begin{array}{llll} \mathfrak{g}_{\kappa''}^{[2]} = \mathfrak{g}^{[0,2]} & & & \\ \mathfrak{g}_{\kappa''}^{[1]} = & & \mathfrak{g}^{[1,1]} & \\ & + & & \\ \mathfrak{g}_{\kappa''}^{[0]} = & \mathfrak{g}^{[0,0]} & & + \; \mathfrak{g}^{[2,0]} \\ & \| & \| & \quad \| \\ & \mathfrak{g}_{\kappa'}^{[0]} & \mathfrak{g}_{\kappa'}^{[1]} & \quad \mathfrak{g}_{\kappa'}^{[2]} \end{array}$$

and

(9.23)
$$\mathfrak{g}_\kappa^* = \mathfrak{g}, \qquad \mathfrak{g}_{\kappa'}^* = \mathfrak{g}_{\kappa''}, \qquad \mathfrak{g}_{\kappa''}^* = \mathfrak{g}_{\kappa'}.$$

The Siegel domain

$$\mathcal{S}_{\kappa'} = \left\{(u', w', t') \in U_{\kappa',C} \times V_{\kappa',+} \times \mathcal{D}_{\kappa'} \,\middle|\, \operatorname{Im} u' - \frac{1}{2i}\mathcal{K}_{U_{\kappa'}}[(w', t')] \in \Omega_{\kappa'}\right\}$$

is mapped to the tube domain $\mathcal{S}_\kappa = U_\kappa + i\Omega_\kappa$ by the Cayley transformation $(ad\, c_{\kappa''})^{-1} c_{\kappa''}$. Let $(u', w', t') \in \mathcal{S}_{\kappa'}$ and let $u = (ad\, c_{\kappa''})^{-1} c_{\kappa''}(u', w', t')$. Then, in the general notation, one has

$$u = u_1 + u_{12} + u_2 \qquad (t_1 = w_1 = w_2 = 0),$$

where

(9.24)
$$\left\{\begin{array}{l} u_1 = u'' = u' - 2A_{\kappa'}(w', c_{\kappa''}^{-1} J'' w') \in U_{\kappa',C}, \\ u_{12} = w'' = (ad\, c_{\kappa''})^{-1} J'' w' \in (U_{\kappa',\kappa''})_C, \\ u_2 = t'' = (ad\, c_{\kappa''})^{-1} c_{\kappa''}(t') \in \mathcal{S}_{\kappa',\kappa''} \subset U_{\kappa'',C}. \end{array}\right.$$

In this case, $\mathcal{S}_{\kappa',\kappa''}$ is also a tube domain, i.e., $U_{\kappa''} + i\Omega_{\kappa''}$. Translating the condition for $(u', w', t') \in \mathcal{S}_{\kappa'}$, one obtains

(9.25)
$$\operatorname{Im} u \in \Omega_\kappa \Longleftrightarrow \left\{\begin{array}{l} \operatorname{Im} u_1 - 2A_{\kappa'}((\operatorname{Im} u_2)^{-1} \operatorname{Im} u_{12}, \operatorname{Im} u_{12}) \in \Omega_{\kappa'}, \\ \operatorname{Im} u_2 \in \Omega_{\kappa''}, \end{array}\right.$$

where $\operatorname{Im} u_2 \in \mathfrak{g}^{[0,2]}$ is viewed as a linear map $\mathfrak{g}^{[1,-1]} \to \mathfrak{g}^{[1,1]}$ (Exerc. 2, cf. also § 6, Exerc.

2).

In the general case, the situation is as shown in the following diagram where (unmarked) arrows indicate fiber structures:

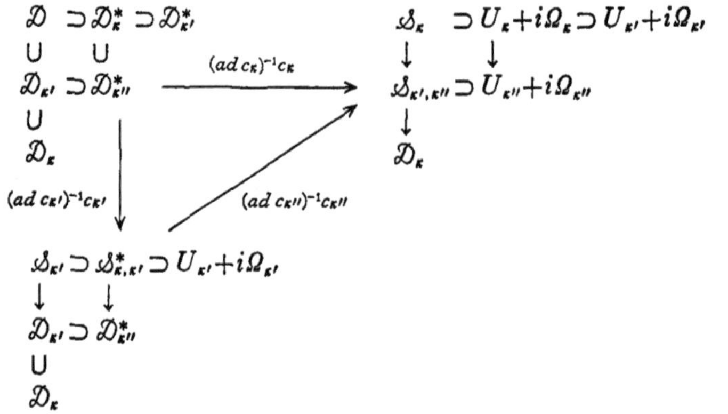

$\mathscr{D}^*_\kappa$, etc. are the symmetric domains associated with $\mathfrak{g}^*_\kappa$, etc. and $\mathscr{S}^*_{\kappa,\kappa'}$ is the Siegel domain realization of $\mathscr{D}^*_\kappa$ in $U_{\kappa',C} \oplus (ad\, c_{\kappa''})(U_{\kappa',\kappa''})_C \oplus \mathfrak{p}^{[2]}_{\kappa'',+}$ obtained by $c_{\kappa'}$. In the above diagram, the part involving $\mathscr{D}^*_\kappa$, $\mathscr{D}^*_{\kappa'}$, $\mathscr{D}^*_{\kappa''}$ reduces to the special case considered above, for $\mathscr{D}^*_\kappa$ is always of tube type and the restriction $c_\kappa|\mathfrak{g}^*_\kappa$ is a "full" Cayley transformation.

Exercises

1. Prove the relation (9. 17).
2. Prove the equivalence (9. 25).

§ 10. The correspondence of boundary components under equivariant holomorphic maps.

Let $(\mathfrak{g}, H_0)$ and $(\mathfrak{g}', H'_0)$ be two semi-simple Lie algebras of hermitian type and $\mathscr{D}$ and $\mathscr{D}'$ the associated symmetric bounded domains. Suppose there are given (H_1)-homomorphisms $\rho : (\mathfrak{g}, H_0) \to (\mathfrak{g}', H'_0)$ and $\kappa : \mathfrak{g}^1 \to (\mathfrak{g}, H_0)$. We fix $\boldsymbol{R}$-groups G and G' with Lie algebras $\mathfrak{g}$ and $\mathfrak{g}'$, respectively, and assume that ρ can be lifted to an $\boldsymbol{R}$-homomorphism $\rho_G : G \to G'$. We denote by ρ_+ and $\rho_{\mathscr{D}}$ the restrictions of ρ_C on $\mathfrak{p}_+$ and $\mathscr{D}$; then $\rho_{\mathscr{D}}$ is a (strongly) equivariant holomorphic map of $\mathscr{D}$ into $\mathscr{D}'$ (II, § 8). It is clear that $\kappa' = \rho \circ \kappa : \mathfrak{g}^1 \to (\mathfrak{g}', H'_0)$ is also an (H_1)-homomorphism. The objects relative to $\mathfrak{g}'$, H'_0, κ' will be denoted by the corresponding symbols with a prime. For instance, $X'_{\kappa'} = \kappa'(h)$ and

$$e'_{\kappa'} = \frac{1}{2}(H'_{\kappa'} + Y'_{\kappa'}), \qquad o'_{\kappa'} = \frac{1}{2}(X'_{\kappa'} + iY'_{\kappa'}), \qquad \text{etc.};$$

c_κ and $c'_{\kappa'}$ are the "Cayley elements" in G_C and G'_C, respectively.

From the definitions, it is clear that

(10. 1) $$\rho(X_\kappa)=X'_{\kappa'},\quad \rho(Y_\kappa)=Y'_{\kappa'},\quad \rho(H_\kappa)=H'_{\kappa'}$$

and hence

(10. 2) $$\rho(e_\kappa)=e'_{\kappa'},\quad \rho_+(o_\kappa)=o'_{\kappa'},\quad \rho_{G_C}(c_\kappa)=c'_{\kappa'}.$$

It follows also from the definitions that $\rho(\mathfrak{g}^{[\nu]})\subset\mathfrak{g}'^{[\nu]}$ $(\nu=0, 1, 2)$ and

(10. 3) $$\begin{cases}\rho(\mathfrak{c}_{\mathfrak{g}}(X_\kappa))\subset\mathfrak{c}_{\mathfrak{g}'}(X'_{\kappa'}),\\ \rho(U_\kappa)\subset U'_{\kappa'},\\ \rho(V_\kappa)\subset V'_{\kappa'}.\end{cases}$$

Thus one has $\rho(\mathfrak{b}_\kappa)\subset\mathfrak{b}'_{\kappa'}$. Since $\mathfrak{g}_\kappa$ (resp. $\mathfrak{g}'_{\kappa'}$) is the non-compact part of $\mathfrak{g}^{[0]}$ (resp. $\mathfrak{g}'^{[0]}$), one has

(10. 4) $$\rho(\mathfrak{g}_\kappa)\subset\mathfrak{g}'_{\kappa'}.$$

(But, in general, $\rho(\mathfrak{l}_2)\subset\mathfrak{l}'_2$ is not true.) Let G_κ, $G'_{\kappa'}$ denote the $\boldsymbol{R}$-subgroups of G, G' corresponding to $\mathfrak{g}_\kappa$, $\mathfrak{g}'_{\kappa'}$, respectively. Then, since $\mathcal{D}_\kappa$, $\mathcal{F}_\kappa$ (resp. $\mathcal{D}'_{\kappa'}$, $\mathcal{F}'_{\kappa'}$) are G°_κ-orbits (resp. $G'^\circ_{\kappa'}$-orbits) of o, o_κ (resp. o', $o'_{\kappa'}$), one has

(10. 5) $$\rho_{\mathcal{D}}(\mathcal{D}_\kappa)\subset\mathcal{D}'_{\kappa'},\quad \rho_+(\mathcal{F}_\kappa)\subset\mathcal{F}'_{\kappa'}.$$

Clearly, the pair $(\rho|\mathfrak{g}_\kappa, \rho_+|\mathcal{D}_\kappa)$ is (strongly) equivariant.

Lemma 10. 1. *Let* $\rho:(\mathfrak{g}, H_0)\to(\mathfrak{g}', H'_0)$, $\kappa:\mathfrak{g}^1\to(\mathfrak{g}, H_0)$ *be* (H_1)*-homomorphisms and let* $\kappa'=\rho\circ\kappa$. *Then one has*

(10. 6) $$\rho([H_0^{(1)}, X])=[H_0'^{(1)}, \rho(X)]$$

for all $X\in\mathfrak{g}$, *where* $H_0^{(1)}=H_0-\frac{1}{2}H_\kappa$, $H_0'^{(1)}=H'_0-\frac{1}{2}H'_{\kappa'}$.

Proof. By the assumptions, one has

$$\rho(H_0^{(1)})-H_0'^{(1)}=\rho(H_0)-H'_0\in\mathfrak{c}_{\mathfrak{g}'}(\rho(\mathfrak{g})),$$

which implies (10. 6), q. e. d.

Corollary 10. 2. ρ *induces an* (H_1)*-homomorphism* $(\mathfrak{g}^{[0]}, H_0^{(1)})\to(\mathfrak{g}'^{[0]}, H_0'^{(1)})$.

Corollary 10. 3. $\rho|V_\kappa$ *is* $\boldsymbol{C}$*-linear.*

These follow from Proposition 1. 3, Lemma 3. 1, (10. 3) and Lemma 10. 1. Corollary 10. 3 implies that $\rho_C(V_{\kappa,+})\subset V'_{\kappa',+}$.

Corollary 10. 4. *One has* $\rho(\mathfrak{c}^*(\kappa))\subset\mathfrak{c}^*(\kappa')$ (*cf.* § 1, *Exerc.* 1).

By the definition, one has $\mathfrak{c}^*(\kappa)=\mathfrak{c}_{\mathfrak{g}}(H_0^{(1)})$, $\mathfrak{c}^*(\kappa')=\mathfrak{c}_{\mathfrak{g}'}(H_0'^{(1)})$. Hence Corollary 10. 4 follows from Lemma 10. 1 immediately.

Since $\mathfrak{g}^*_\kappa$ (resp. $\mathfrak{g}'^*_{\kappa'}$) is the non-compact part of $\mathfrak{c}^*(\kappa)$ (resp. $\mathfrak{c}^*(\kappa')$), one has $\rho(\mathfrak{g}^*_\kappa)\subset\mathfrak{g}'^*_{\kappa'}$. Since, by the definition, $\mathfrak{g}^{(2)}_\kappa$ (resp. $\mathfrak{g}'^{(2)}_{\kappa'}$) is the real part of $(ad\, c_\kappa)^{-1}(\mathfrak{k}^*_\kappa)_C$ (resp.

$(ad\, c'_{\kappa'})^{-1}(\mathfrak{k}'^{*}_{\kappa'})_C)$, one has

$$\rho(\mathfrak{g}^{(2)}_{\kappa}) \subset \mathfrak{g}'^{(2)}_{\kappa'}. \tag{10.7}$$

We denote by $G^{(2)}_{\kappa}$, $G'^{(2)}_{\kappa'}$ the $\boldsymbol{R}$-subgroups of G, G' corresponding to $\mathfrak{g}^{(2)}_{\kappa}$, $\mathfrak{g}'^{(2)}_{\kappa'}$. Then, since Ω_{κ} (resp. $\Omega'_{\kappa'}$) is a $G^{(2)\circ}_{\kappa}$-orbit (resp. $G'^{(2)\circ}_{\kappa'}$-orbit) of e_{κ} (resp. $e'_{\kappa'}$), one has

$$\rho(\Omega_{\kappa}) \subset \Omega'_{\kappa'}. \tag{10.8}$$

Clearly $(\rho|\mathfrak{g}^{(2)}_{\kappa}, \rho|\Omega_{\kappa})$ is an equivariant pair.

Now the Siegel domain expressions of $\mathcal{D}$ and $\mathcal{D}'$ corresponding to κ and κ' are given, respectively, by

$$\mathcal{S}_{\kappa} = (ad\, c_{\kappa})^{-1}c_{\kappa}(\mathcal{D}) \subset U_{\kappa,C} \times V_{\kappa,+} \times \mathfrak{p}_{\kappa,+}, \quad \text{and}$$
$$\mathcal{S}'_{\kappa'} = (ad\, c'_{\kappa'})^{-1}c'_{\kappa'}(\mathcal{D}') \subset U'_{\kappa',C} \times V'_{\kappa',+} \times \mathfrak{p}'_{\kappa',+}.$$

From what we mentioned above, we see that ρ_C maps $(ad\, c_{\kappa})^{-1}\mathfrak{p}_+$ into $(ad\, c'_{\kappa'})^{-1}\mathfrak{p}'_+$, preserving the above decompositions. Moreover, one has

$$\rho_C((ad\, c_{\kappa})^{-1}c_{\kappa}(z)) = (ad\, c'_{\kappa'})^{-1}c'_{\kappa'}(\rho_{\mathcal{D}}(z))$$

for $z \in \mathcal{D}$. Hence, in view of the result of § 7, if we put $\rho_{\mathcal{S}} = \rho_C|\mathcal{S}_{\kappa}$, $\rho_{\mathcal{S}}$ is a (strongly) equivariant holomorphic map of $\mathcal{S}_{\kappa}$ into $\mathcal{S}'_{\kappa'}$. Thus we obtain

Theorem 10. 5. *Let* $\rho : (\mathfrak{g}, H_0) \to (\mathfrak{g}', H'_0)$, $\kappa : \mathfrak{sl}_2(\boldsymbol{R}) \to (\mathfrak{g}, H_0)$ *be* (H_1)*-homomorphisms, and let* $\kappa' = \rho\circ\kappa$. *Then* ρ_C *maps the subspaces* $U_{\kappa,C}$, $V_{\kappa,+}$, $\mathfrak{p}_{\kappa,+}$ *into* $U'_{\kappa',C}$, $V'_{\kappa',+}$, $\mathfrak{p}'_{\kappa',+}$, *respectively, inducing (strongly) equivariant maps* $(\Omega_{\kappa}, e_{\kappa}) \to (\Omega'_{\kappa'}, e'_{\kappa'})$, $\mathcal{D}_{\kappa} \to \mathcal{D}'_{\kappa'}$, *and* $(\mathcal{S}_{\kappa}, ie_{\kappa}) \to (\mathcal{S}'_{\kappa'}, ie'_{\kappa'})$. *Moreover, one has*

$$\rho_C(\mathcal{L}_{\varphi(t)}(w, w)) = \mathcal{L}_{\varphi'(\rho_C(t))}(\rho_C(w), \rho_C(w)) \tag{10.9}$$

for all $t \in \mathcal{D}_{\kappa}$, $w \in V_{\kappa,+}$.

The last assertion follows immediately from the definition in § 6. (We leave the detail for an exercise of the reader.) The conclusion in Theorem 10. 5 can be expressed by saying that ρ_C induces a (strongly) equivariant holomorphic map $\rho_{\mathcal{S}}$: $\mathcal{S}_{\kappa} \to \mathcal{S}'_{\kappa'}$ preserving the structure of Siegel domain of the third kind. In particular, when ρ is an (H_2)-isomorphism, the $\boldsymbol{C}$-linear map $\rho_C|(ad\, c_{\kappa})^{-1}\mathfrak{p}_+$ and all the induced maps mentioned in Theorem 10. 5 become isomorphisms, so that $\mathcal{S}_{\kappa}$ and $\mathcal{S}'_{\kappa'}$ are "linearly equivalent".

Corollary 10. 6. *Let* κ *and* κ' *be two* (H_1)*-homomorphisms* $\mathfrak{sl}_2(\boldsymbol{R}) \to (\mathfrak{g}, H_0)$ *with* $\boldsymbol{m}(\kappa) = \boldsymbol{m}(\kappa')$ *(cf. § 8). Then the corresponding Siegel domains of the third kind* $\mathcal{S}_{\kappa}$ *and* $\mathcal{S}_{\kappa'}$ *are linearly equivalent.*

Proof. If $\boldsymbol{m}(\kappa) = \boldsymbol{m}(\kappa')$, then by Proposition 4. 3 (applied to each simple component of $\mathfrak{g}$) one has $\kappa' = \rho\circ\kappa$ with $\rho \in \mathrm{Inn}(\mathfrak{g})$. If κ satisfies (H_1) with respect to H_0, then so does κ' with respect to $H'_0 = \rho(H_0)$. Therefore by Theorem 10. 5 $\mathcal{S}_{\kappa}$ and $\mathcal{S}_{\kappa'}$ are linearly equivalent, q. e. d.

Now let us assume for a moment that $\mathfrak{g}$ is simple (and non-compact). Extending the definition given in § 4, we define the "multiplicity" $m(\rho)$ by setting

$$m(\rho) = m(\rho\circ\kappa_1),$$

where κ_1 is any (H_1)-homomorphism $\mathfrak{g}^1\to\mathfrak{g}$ with $m(\kappa_1)=1$. This definition is independent of the choice of κ_1. In fact, if κ_2 is another (H_1)-homomorphism $\mathfrak{g}^1\to\mathfrak{g}$ with $m(\kappa_2)=1$, then by Proposition 4. 3 there exists $\varphi\in\mathrm{Inn}(\mathfrak{g})$ such that $\kappa_2=\varphi\circ\kappa_1$. Let $\varphi=ad\,g$ with $g\in G^\circ$. Then one has

$$\rho\circ\kappa_2 = \rho\circ(ad\,g)\circ\kappa_1 = (ad\,\rho_G(g))\circ\rho\circ\kappa_1,$$

which, by the same proposition, implies $m(\rho\circ\kappa_2)=m(\rho\circ\kappa_1)$.

Lemma 10. 7. 1) *Assuming* $\mathfrak{g}$ *to be simple, suppose that the* (H_1)*-homomorphism* ρ: $(\mathfrak{g}, H_0)\to(\mathfrak{g}', H_0')$ *is a commutative sum of two* (H_1)*-homomorphisms* ρ_1 *and* ρ_2 *(with respect to the same H-elements). Then one has*

$$m(\rho) = m(\rho_1)+m(\rho_2). \tag{10. 10}$$

2) *Assume that both* $\mathfrak{g}$ *and* $\mathfrak{g}'$ *are simple and let* ρ' *be an* (H_1)*-homomorphism* $(\mathfrak{g}', H_0')\to(\mathfrak{g}'', H_0'')$*, where* $\mathfrak{g}''$ *is a third semi-simple Lie algebra of hermitian type. Then one has*

$$m(\rho'\circ\rho) = m(\rho')m(\rho). \tag{10. 11}$$

Proof. 1) Let κ_1 be an (H_1)-homomorphism $\mathfrak{g}^1\to(\mathfrak{g}, H_0)$ with $m(\kappa_1)=1$. Then $\rho\circ\kappa_1$ is a commutative sum of two (H_1)-homomorphisms $\rho_1\circ\kappa_1$ and $\rho_2\circ\kappa_1:\mathfrak{g}^1\to\mathfrak{g}'$. Hence by the definition one has

$$\begin{aligned} m(\rho) &= m(\rho\circ\kappa_1) = m(\rho_1\circ\kappa_1)+m(\rho_2\circ\kappa_1) \\ &= m(\rho_1)+m(\rho_2). \end{aligned}$$

2) Let κ_1 be as above, and let $m(\rho)=m$. Then $\rho\circ\kappa_1$ can be expressed as a commutative sum of m (H_1)-homomorphisms $\kappa_i':\mathfrak{g}^1\to(\mathfrak{g}', H_0')$ $(1\leq i\leq m)$ with $m(\kappa_i')=1$. Hence by the definition and (10. 10) one has

$$\begin{aligned} m(\rho'\circ\rho) &= m(\rho'\circ\rho\circ\kappa_1) = \sum_{i=1}^{m} m(\rho'\circ\kappa_i') \\ &= m\cdot m(\rho'), \end{aligned}$$

q. e. d.

In the general case, we decompose $\mathfrak{g}$ and $\mathfrak{g}'$ into the direct sum of ideals:

$$\mathfrak{g} = \bigoplus_{i=0}^{s}\mathfrak{g}_i, \quad \mathfrak{g}' = \bigoplus_{j=0}^{s'}\mathfrak{g}_j', \tag{10. 12}$$

where $\mathfrak{g}_0$, $\mathfrak{g}_0'$ are compact and $\mathfrak{g}_i$, $\mathfrak{g}_j'$ $(1\leq i\leq s,\ 1\leq j\leq s')$ are simple, non-compact. Let ι_i and π_i $(1\leq i\leq s)$ denote the canonical injections $\mathfrak{g}_i\to\mathfrak{g}$ and the canonical projections $\mathfrak{g}\to\mathfrak{g}_i$, respectively, and define ι_j' and π_j' $(1\leq j\leq s')$ similarly. Put $m_{ji}=m(\pi_j'\circ\rho\circ\iota_i)$ $(1\leq i\leq s,\ 1\leq j\leq s')$ and denote by $M(\rho)$ the $s'\times s$ integral matrix whose (j, i) component is given by m_{ji}, i. e.,

$$M(\rho) = (m(\pi_j'\circ\rho\circ\iota_i))_{ji}. \tag{10. 13}$$

Then we have

Proposition 10. 8. *Let* $(\mathfrak{g}, H_0)$ *and* $(\mathfrak{g}', H_0')$ *be two semi-simple Lie algebras of hermitian type and* $\rho : (\mathfrak{g}, H_0) \to (\mathfrak{g}', H_0')$ *an* (H_1)*-homomorphism.*

1) *If* ρ *is a commutative sum of two* (H_1)*-homomorphisms* ρ_1, ρ_2 *(with respect to the same H-elements), then one has*

$$M(\rho) = M(\rho_1) + M(\rho_2). \tag{10. 14}$$

2) *Let* $(\mathfrak{g}'', H_0'')$ *be a third semi-simple Lie algebra of hermitian type and* $\rho' : (\mathfrak{g}', H_0') \to (\mathfrak{g}'', H_0'')$ *be an* (H_1)*-homomorphism. Then one has*

$$M(\rho' \circ \rho) = M(\rho') M(\rho). \tag{10. 15}$$

Proof. 1) For each $1 \leq i \leq s$, $1 \leq j \leq s'$, the homomorphism $\pi'_j \circ \rho \circ \iota_i$ is a commutative sum of $\pi'_j \circ \rho_1 \circ \iota_i$ and $\pi'_j \circ \rho_2 \circ \iota_i$. Hence by Lemma 10. 7, 1) one has

$$m(\pi'_j \circ \rho \circ \iota_i) = m(\pi'_j \circ \rho_1 \circ \iota_i) + m(\pi'_j \circ \rho_2 \circ \iota_i),$$

which gives (10. 14).

2) Let $\mathfrak{g}'' = \bigoplus_{k=0}^{s''} \mathfrak{g}''_k$ be the decomposition similar to (10. 12) and π''_k the canonical projection $\mathfrak{g}'' \to \mathfrak{g}''_k$. Then each homomorphism $\pi''_k \circ \rho' \circ \rho \circ \iota_i$ $(1 \leq i \leq s,\ 1 \leq k \leq s'')$ is a commutative sum of the form

$$\pi''_k \circ \rho' \circ \rho \circ \iota_i = \sum_{j=1}^{s'} (\pi''_k \circ \rho' \circ \iota'_j) \circ (\pi'_j \circ \rho \circ \iota_i).$$

Hence by Lemma 10. 7, 1), 2) one has

$$m(\pi''_k \circ \rho' \circ \rho \circ \iota_i) = \sum_{j=1}^{s'} m(\pi''_k \circ \rho' \circ \iota'_j) m(\pi'_j \circ \rho \circ \iota_i),$$

which gives (10. 15), q. e. d.

Corollary 10. 9. *Let* $\rho : (\mathfrak{g}, H_0) \to (\mathfrak{g}', H_0')$ *be an* (H_1)*-homomorphism and suppose that* $\rho_+(\mathcal{F}) \subset \mathcal{F}'$*, where* $\mathcal{F}$ *and* $\mathcal{F}'$ *are boundary components of* $\mathcal{D}$ *and* $\mathcal{D}'$ *of type* $\boldsymbol{b}$ *and* $\boldsymbol{b}'$*, respectively. Then one has*

$$\boldsymbol{b}' = M(\rho) \boldsymbol{b}. \tag{10. 16}$$

Proof. Let κ be an (H_1)-homomorphism $\mathfrak{g}^1 \to (\mathfrak{g}, H_0)$ such that $\mathcal{F} = \mathcal{F}_\kappa$, and let $\kappa' = \rho \circ \kappa$. Then, since $\rho_+(\mathcal{F})$ is contained in both $\mathcal{F}'$ and $\mathcal{F}'_{\kappa'}$, one has $\mathcal{F}' = \mathcal{F}'_{\kappa'}$. Therefore, by the definition, one has $\boldsymbol{b} = M(\kappa)$, $\boldsymbol{b}' = M(\kappa')$, both sides being viewed as column vectors of dimension s and s' (cf. § 8, where $M(\kappa)$ is denoted by $\boldsymbol{m}(\kappa)$). Hence our result follows from (10. 15), q. e. d.

Remark. In the situation of Corollary 10. 9, suppose further that $\mathfrak{g}$ and $\mathfrak{g}'$ are given F-structures, $\mathcal{F}$ is F-rational, and ρ is defined over F, where F is a subfield of $\boldsymbol{R}$. Then, by Remark 3 at the end of § 8, we may assume that $\mathcal{F} = \mathcal{F}_\kappa$ with κ defined over F (for a suitably chosen origin o). Then $\kappa' = \rho \circ \kappa$ is defined over F, and therefore $\mathcal{F}' = \mathcal{F}'_{\kappa'}$ is also F-rational.

We say that a homomorphism $\rho : \mathfrak{g} \to \mathfrak{g}'$ *satisfies the condition* (H_2') (with respect to H_0, κ_0 and H_0', κ_0') if it satisfies (H_1) (with respect to H_0 and H_0') and the $\boldsymbol{C}$-linear map ρ_+ maps a point boundary component $\mathcal{F}_{\kappa_0}$ of $\mathcal{D}$ into a point boundary component $\mathcal{F}'_{\kappa_0'}$ of $\mathcal{D}'$. (This condition implies that all point boundary components of $\mathcal{D}$ are mapped to point boundary components of $\mathcal{D}'$.) Recall that a boundary component $\mathcal{F}$ of type $\boldsymbol{b}=(b_i)$ is a point boundary component if and only if one has $b_i=r_i=\text{rank}\,\mathfrak{g}_i$ for all $1\leq i\leq s$.

Corollary 10. 10. *Let* $\rho:(\mathfrak{g},H_0)\to(\mathfrak{g}',H_0')$ *be an* (H_1)*-homomorphism, and* $M(\rho)=(m_{ji})$. *Then one has*

$$\sum_{i=1}^{s} m_{ji}r_i \leq r'_j \qquad (1\leq j\leq s'), \tag{10. 17}$$

where $r_i=\text{rank}\,\mathfrak{g}_i$ *and* $r'_j=\text{rank}\,\mathfrak{g}'_j$. *The equality sign holds if and only if* ρ *satisfies* (H_2').

Proof. Using the same notation as in the proof of Corollary 10. 9, assume that $\mathcal{F}$ is a point boundary component. Then, since $b_i=r_i\,(1\leq i\leq s)$, one has by (10. 16)

$$\sum m_{ji}r_i = b'_j \leq r'_j,$$

i. e., (10. 17). Here the second equality sign holds for all $1\leq j\leq s'$, if and only if $\mathcal{F}'$ is a point boundary component, which, by the definition, amounts to saying that ρ satisfies (H_2'), q. e. d.

Lemma 10. 11. *Let* $(\mathfrak{g},H_0)$, $(\mathfrak{g}',H_0')$, $(\mathfrak{g}'',H_0'')$ *be semi-simple Lie algebras of hermitian type and let* $\rho:\mathfrak{g}\to\mathfrak{g}'$, $\rho':\mathfrak{g}'\to\mathfrak{g}''$ *be Lie algebra homomorphisms. If* ρ *and* ρ' *satisfy the condition* (H_2) *(with respect to the given H-elements), then so does* $\rho'\circ\rho$. *Conversely, if* $\rho'\circ\rho$ *satisfies* (H_2) *and* ρ' *satisfies* (H_1), *then* ρ' *satisfies also* (H_2); *if moreover* ρ' *is injective, then* ρ *satisfies* (H_2).

Proof. The first assertion is obvious. To prove the second one, suppose that $\rho'\circ\rho$ satisfies (H_2), i. e., $\rho'\rho(H_0)=H_0''$, and ρ' satisfies (H_1). Then one has

$$\rho'(H_0'-\rho(H_0)) = \rho'(H_0')-H_0'' \in \mathfrak{c}_{\mathfrak{g}''}(\rho'(\mathfrak{g}')).$$

This implies that $\rho'(H_0'-\rho(H_0))$ is in the center of $\rho'(\mathfrak{g}')$, which reduces to zero, because $\mathfrak{g}'$ is semi-simple. Hence one has $\rho'(H_0')=H_0''$, i. e., ρ' satisfies (H_2). If ρ' is injective, it also follows that $H_0'=\rho(H_0)$, i. e., ρ satisfies (H_2), q. e. d.

Proposition 10. 12. *Let* $\mathfrak{g}$ *and* $\mathfrak{g}'$ *be semi-simple Lie algebras of hermitian type,* $\mathcal{D}$ *and* $\mathcal{D}'$ *the associated symmetric bounded domains, and* $\rho:\mathfrak{g}\to\mathfrak{g}'$ *an* (H_1)*-homomorphism. If* $\mathcal{D}$ *is of tube type and* ρ *satisfies* (H_2), *then* $\mathcal{D}'$ *is also of tube type and* ρ *satisfies* (H_2'). *Conversely, if* $\mathcal{D}'$ *is of tube type and* ρ *satisfies* (H_2'), *then* ρ *satisfies* (H_2); *if moreover* ρ *is injective, then* $\mathcal{D}$ *is of tube type.*

Proof. Let $\kappa_0:\mathfrak{g}^1\to\mathfrak{g}$ be an (H_1)-homomorphism such that $\mathcal{F}_{\kappa_0}$ is a point boundary component of $\mathcal{D}$. We know that $\mathcal{D}$ is of tube type if and only if κ_0 satisfies (H_2) (Cor. 1. 6). Hence, if $\mathcal{D}$ is of tube type and ρ satisfies (H_2), then by Lemma 10. 11

$\kappa_0' = \rho \circ \kappa_0$ satisfies (H_2), so that the corresponding boundary component $\mathcal{F}'_{\kappa_0'}$ reduces to a point and ρ satisfies (H_2'). Conversely, if $\mathcal{D}'$ is of tube type and ρ satisfies (H_2'), then $\mathcal{F}'_{\kappa_0'}$ is a point boundary component and hence (again by Cor. 1. 6) $\kappa_0' = \rho \circ \kappa_0$ satisfies (H_2). Hence by Lemma 10. 11 ρ also satisfies (H_2). If moreover ρ is injective, then κ_0 satisfies (H_2) and hence $\mathcal{D}$ is of tube type, q. e. d.

We note that, when ρ satisfies (H_2') with respect to κ_0 and κ_0', ρ_c induces a strongly equivariant holomorphic map $\mathcal{S}_{\kappa_0} \rightarrow \mathcal{S}'_{\kappa_0'}$ preserving the structure of Siegel domain (of the second kind) by Theorem 10. 5.

For more on the matrix $M(\rho)$ and applications, the reader is referred to Satake [7], [8].

Chapter IV

Equivariant Holomorphic Maps of a Symmetric Domain into a Siegel Space

§ 1. Fully reducible representations.

Let G be a group. A (linear) representation of G over F is a pair (V, ρ) formed of a finite-dimensional vector space V over F and a group homomorphism $\rho: G \to GL(V)$. Sometimes ρ alone is called a representation and V is referred to as a representation-space. When we are dealing with an F-group G, it will tacitly be assumed that ρ is an F-homomorphism of F-group (called sometimes an "F-representation").

Irreducible representations, direct sums, etc. are defined in the usual manner. A representation (V, ρ) of G is called *fully reducible* if it is a direct sum of irreducible representations, or equivalently, if for any G-stable subspace W of V there exists a G-stable subspace W' of V such that $V = W \oplus W'$. When an irreducible representation (V_1, ρ_1) is equivalent to a subrepresentation of (V, ρ), we will say that (V_1, ρ_1) is contained in (V, ρ). A representation is called *primary* if it is a direct sum of mutually equivalent irreducible representations, and, if (V_1, ρ_1) is an irreducible representation contained in a primary representation (V, ρ), we say that (V, ρ) is *belonging to* (V_1, ρ_1).

Let (V, ρ) be a fully reducible representation and (V_1, ρ_1) an irreducible representation of G over F. We set

$$(1.1) \qquad \begin{aligned} D_1 &= \mathrm{End}_G(V_1), \qquad F_1 = \mathrm{Cent}\, D_1, \\ U_1 &= \mathrm{Hom}_G(V_1, V), \end{aligned}$$

where $\mathrm{End}_G(V_1)$ and $\mathrm{Hom}_G(V_1, V)$ denote the algebra over F of all G-endomorphisms of V_1 and the space of all G-homomorphisms of V_1 into V, respectively. Then, by Schur's lemma, D_1 is a division algebra over F and so F_1 is a finite extension field of F. V_1 (resp. U_1) has a natural structure of a left (resp. right) D_1-module, the right action of D_1 on U_1 being defined by composition. Denoting by $\bar{D}_1$ a division algebra anti-isomorphic to D_1 and fixing an anti-isomorphism $\alpha \mapsto \bar{\alpha}$ of D_1 onto $\bar{D}_1$, we can convert the right D_1-module U_1 into a left $\bar{D}_1$-module by setting

$$\bar{\alpha} u = u \circ \alpha \qquad \text{for} \quad \alpha \in D_1, \ u \in U_1.$$

We denote by $U_1 \otimes_{D_1} V_1$ the "tensor product" of U_1 and V_1 over D_1, i. e., the factor space of the usual tensor product $U_1 \otimes_F V_1$ modulo the subspace spanned by

$$\bar{\alpha}u \otimes v - u \otimes \alpha v \qquad (\alpha \in D_1,\ u \in U_1,\ v \in V_1),$$

where $\otimes$ stands for $\otimes_F$. If we put

$$(1.2) \qquad \begin{array}{ll} [D_1 : F_1] = r_1^2, & [F_1 : F] = d_1, \\ \dim_{D_1} U_1 = m_1, & \dim_{D_1} V_1 = n_1, \end{array}$$

then it is clear that

$$(1.3) \qquad \dim_F (U_1 \otimes_{D_1} V_1) = d_1 r_1^2 m_1 n_1.$$

Lemma 1. 1. (i) *Let $V^{[1]}$ denote the sum of all irreducible G-stable subspaces of V, which are G-isomorphic to V_1. Then one has a natural isomorphism*

$$(1.4) \qquad V^{[1]} \cong U_1 \otimes_{D_1} V_1.$$

(ii) *There exists a natural isomorphism:*

$$(1.5) \qquad \mathrm{End}_{F_1}(V^{[1]}) \cong \mathrm{End}_{\bar{D}_1}(U_1) \otimes_{F_1} \mathrm{End}_{D_1}(V_1).$$

[*Note that one has* $\mathrm{End}_{\bar{D}_1}(U_1) \cong \mathcal{M}_{m_1}(D_1)$, $\mathrm{End}_{D_1}(V_1) \cong \mathcal{M}_{n_1}(\bar{D}_1)$, *and* (1. 3), (1. 4) *give* $\mathrm{End}_{F_1}(V^{[1]}) \cong \mathcal{M}_{r_1^2 m_1 n_1}(F_1)$.]

Proof. Consider an F-bilinear map

$$U_1 \times V_1 \ni (u, v) \longmapsto u(v) \in V.$$

Then, since $u(\alpha v) = (u \circ \alpha)(v) = (\bar{\alpha}u)(v)$ for $\alpha \in D_1$, this gives rise to an F-linear map $\Phi_1 : U_1 \otimes_{D_1} V_1 \to V$, and the image of Φ_1 clearly coincides with $V^{[1]}$. Hence it suffices to prove that Φ_1 is injective. Since $V^{[1]}$ is fully reducible, we can find non-zero elements $\varphi_1, \cdots, \varphi_{m_1}$ of U_1 such that $V = \bigoplus_{i=1}^{m_1} \varphi_i(V_1)$. Then every $v \in V^{[1]}$ can be written uniquely in the form

$$v = \sum_{i=1}^{m_1} \varphi_i(v_i)$$

with $v_i \in V_1$. Call ψ_i the map $v \mapsto v_i$. Then it is clear that $\psi_i \in \mathrm{Hom}_G(V, V_1)$ and one has the following relations

$$(1.6) \qquad \sum_{i=1}^{m_1} \varphi_i \circ \psi_i = 1_{V^{[1]}}, \qquad \psi_i \circ \varphi_j = \delta_{ij} 1_{V_1}.$$

It follows that $(\varphi_1, \cdots, \varphi_{m_1})$ is a $\bar{D}_1$-basis of U_1 [and $(\psi_1, \cdots, \psi_{m_1})$ is its dual basis of $\mathrm{Hom}_G(V, V_1)$ over D_1]. Indeed, from (1. 6) one has for $u \in U_1$ and $\alpha_i \in D_1$

$$u = \sum \varphi_i \circ \alpha_i \Longleftrightarrow \alpha_i = \psi_i \circ u,$$

which proves our assertion. It follows that every element in $U_1 \otimes_{D_1} V_1$ can be written uniquely in the form $\sum \varphi_i \otimes_{D_1} v_i$ with $v_i \in V_1$, and one has $\Phi_1(\sum \varphi_i \otimes_{D_1} v_i) = \sum \varphi_i(v_i)$. Hence Φ_1 is injective.

(ii) For simplicity, we identify $V^{[1]}$ with $U_1 \otimes_{D_1} V_1$ through the isomorphism (1. 4) and consider $U_1, V_1, V^{[1]}$ as vector spaces over F_1. Then one has

$$V^{[1]} = U_1 \otimes_{D_1} V_1 = (U_1 \otimes_{F_1} V_1)/N_1,$$

where N_1 is the subspace of $U_1 \otimes_{F_1} V_1$ generated by

$$\bar{\alpha}u\otimes_{F_1}v-u\otimes_{F_1}\alpha v \qquad (u\in U_1,\ v\in V_1,\ \alpha\in D_1).$$

As is well-known, $\mathrm{End}_{F_1}(U_1\otimes_{F_1}V_1)$ can naturally be identified with $\mathrm{End}_{F_1}(U_1)\otimes_{F_1}\mathrm{End}_{F_1}(V_1)$. For $\varphi\in\mathrm{End}_{\bar{D}_1}(U_1)$ and $\psi\in\mathrm{End}_{D_1}(V_1)$, the endomorphism $\varphi\otimes_{F_1}\psi$, leaving N_1 invariant, induces an F_1-linear endomorphism of $V^{[1]}$. Thus one has a natural map

$$\mathrm{End}_{\bar{D}_1}(U_1)\otimes_{F_1}\mathrm{End}_{D_1}(V_1)\longrightarrow\mathrm{End}_{F_1}(V^{[1]}).$$

We will show that this is actually an isomorphism. Since the dimensions of both spaces are equal, it suffices to show that this map is surjective. Let $(\varphi_1,\cdots,\varphi_{m_1})$ and $(e_1,\cdots,e_{n_1})$ be bases of U_1 and V_1 over $\bar{D}_1$ and D_1, respectively. Then every element in $V^{[1]}$ can be written uniquely in the form

$$\sum_{i,j}\varphi_i\otimes_{D_1}\xi_{ij}e_j \qquad (\xi_{ij}\in D_1).$$

For $\alpha\in D_1$ and $1\leq i,j\leq m$, we denote by $\varphi_{ij}^{\bar{\alpha}}$ the unique element in $\mathrm{End}_{\bar{D}_1}(U_1)$ such that

$$\varphi_{ij}^{\bar{\alpha}}(\varphi_k)=\delta_{jk}\bar{\alpha}\varphi_i.$$

Similarly, we define $\psi_{ij}^{\alpha}\in\mathrm{End}_{D_1}(V_1)$ by the relation

$$\psi_{ij}^{\alpha}(\psi_k)=\delta_{jk}\alpha\psi_i.$$

Then one has

$$(\varphi_{ij}^{\bar{\alpha}}\otimes_{F_1}\psi_{kl}^{\beta})(\varphi_p\otimes_{D_1}\xi e_q)=\delta_{jp}\delta_{lq}\varphi_i\otimes_{D_1}(\alpha\xi\beta)e_k.$$

Since any linear transformation of D_1 over F_1 is a linear combination of the transformations of the form $\xi\mapsto\alpha\xi\beta$, our assertion follows, q. e. d.

The subspace $V^{[1]}$ is called the *primary component* of V belonging to the irreducible representation (V_1,ρ_1). By definition, it is the largest primary G-stable subspace of V containing (V_1,ρ_1). In the following, we consider the natural isomorphisms (1. 4), (1. 5) as identifications. Let $\rho^{[1]}$ denote the restriction of ρ on $V^{[1]}$. Then one has

$$\rho^{[1]}(g)=1_{U_1}\otimes_{F_1}\rho_1(g) \qquad (g\in G). \tag{1.7}$$

On the other hand, one has

$$\mathrm{End}_G(V^{[1]})=\mathrm{End}_{\bar{D}_1}(U_1)\otimes_{F_1}1_{V_1}. \tag{1.8}$$

Thus $\mathrm{End}_{\bar{D}_1}(U_1)$ is essentially the centralizer ("commutor algebra") of $\rho^{[1]}(G)$ in $\mathrm{End}(V^{[1]})$. In the following, when there is no fear of confusion, $\mathrm{End}_{\bar{D}_1}(U_1)$ and $\mathrm{End}_{D_1}(V_1)$ will also be identified with the corresponding subalgebras of $\mathrm{End}_{F_1}(V^{[1]})$.

Let $\{(V_i,\rho_i)\}$ be a complete set of representatives of equivalence classes of irreducible representations of G over F. For each i, let $V^{[i]}$ be the primary component of V belonging to (V_i,ρ_i) ; one has $V^{[i]}=\{0\}$ for almost all i for which (V_i,ρ_i) is not contained in V. One then obtains the direct sum decomposition

$$V=\bigoplus_i V^{[i]}, \tag{1.9}$$

called the *primary decomposition* of V.

Example 1. For $F=\boldsymbol{R}$, there are three types of primary representations:

(1.10 a) $$V^{[1]}=U_1\otimes_{\boldsymbol{R}}V_1 \qquad (D_1=F_1=\boldsymbol{R}),$$

(1.10 b) $$V^{[1]}=U_1\otimes_{\boldsymbol{H}}V_1 \qquad (D_1=\boldsymbol{H},\ F_1=\boldsymbol{R}),$$

(1.10 c) $$V^{[1]}=U_1\otimes_{\boldsymbol{C}}V_1 \qquad (D_1=F_1=\boldsymbol{C}).$$

According to the cases, the irreducible representation ρ_1 or the corresponding primary representation $\rho^{[1]}$ will be called of ***R**-type, **H**-type*, or ***C**-type*.

Example 2. Suppose $\rho_1(G)$ is abelian. Then one has $\rho_1(G)\subset D_1$ and hence $\rho_1(G)\subset F_1$. Therefore, $\dim_{F_1}V_1=1$ and $D_1=F_1$.

For a later use, let us consider a primary representation of the direct product $G=G_1\times G_2$. Generalizing the situation slightly, we assume that there are two (normal) subgroups G_1, G_2 of G such that one has $G=G_1G_2$ and $g_1g_2=g_2g_1$ for all $g_1\in G_1$, $g_2\in G_2$. Let (V,ρ) be a primary representation of G over F and assume that $\rho|G_i\ (i=1,2)$ are both fully reducible. Then $(V,\rho|G_i)\ (i=1,2)$ are also primary, for otherwise two distinct components of them, being G-stable, would contain inequivalent irreducible representations of G. Let (V_{1i},ρ_{1i}) be irreducible representations of G_i contained in $(V,\rho|G_i)$ and put

(1.11) $$D_{1i}=\mathrm{End}_{G_i}(V_{1i}),\qquad F_{1i}=\mathrm{Cent}\,D_{1i}\qquad (i=1,2),$$ $$U_{11}=\mathrm{Hom}_{G_1}(V_{11},V).$$

Then, since $\rho(G_2)\subset\mathrm{End}_{G_1}(V)\cong\mathrm{End}_{\bar{D}_{11}}(U_{11})$, U_{11} can be viewed as a representation-space of G_2, which is also primary and belonging to ρ_{12}. Hence, if we set

(1.12) $$W=\mathrm{Hom}_{G_2}(V_{12},U_{11}),$$

then one has

(1.13) $$V=U_{11}\otimes_{D_{11}}V_{11}=(W\otimes_{D_{12}}V_{12})\otimes_{D_{11}}V_{11},$$

and the representation ρ on V is given by

(1.14) $$\rho(g_1g_2)=(1_W\otimes_{F_{12}}\rho_{12}(g_2))\otimes_{F_{11}}\rho_{11}(g_1)\qquad (g_1\in G_1,\ g_2\in G_2).$$

Since U_{11} has a structure of $\bar{D}_{11}$-G_2-module, W has a structure of $\bar{D}_{11}$-$\bar{D}_{12}$-module (where the actions of $\bar{D}_{11}$ and $\bar{D}_{12}$ commute), or what amounts to the same thing, that of a $\bar{D}_{11}\otimes\bar{D}_{12}$-module. The action of $\bar{D}_{11}\otimes\bar{D}_{12}$ on W is given by

$$(\bar{\alpha}\otimes\bar{\beta})_W:\ w\longmapsto(\bar{\alpha})_{U_{11}}\circ w\circ\beta,$$

$(\bar{\alpha})_{U_{11}}$ denoting the action of $\bar{\alpha}\in\bar{D}_{11}$ on U_{11}. Clearly, if $g\in G_1\cap G_2$, then $\rho_{1i}(g)\in F_{1i}$ $(i=1,2)$ and one has

$$(\overline{\rho_{11}(g)})_{U_{11}}\circ w=w\circ\rho_{12}(g),$$

or

(1.15) $$(\overline{\rho_{11}(g)}\otimes\overline{\rho_{12}(g)^{-1}})_W=1_W\qquad (g\in G_1\cap G_2).$$

Conversely, given any $\bar{D}_{11}\otimes\bar{D}_{12}$-module W satisfying the condition (1.15), the space V given by the tensor product (1.13) can be made into a (primary) representation-space of G by (1. 14). It follows that G-stable subspaces of V are in one-to-one correspondence with $\bar{D}_{11}\otimes\bar{D}_{12}$-submodules of W. In particular, (V, ρ) is irreducible if and only if W is a simple $\bar{D}_{11}\otimes\bar{D}_{12}$-module and, if $\rho\sim\rho_1$, one has

$$(1.16)\qquad D_1 = \mathrm{End}_G(V_1) \cong \mathrm{End}_{\bar{D}_{11}\otimes\bar{D}_{12}}(W).$$

As is well-known, W is then isomorphic (as a $\bar{D}_{11}\otimes\bar{D}_{12}$-module) to a simple left ideal of a simple component of the semi-simple algebra $\bar{D}_{11}\otimes\bar{D}_{12}$, simple component which is equivalent to $\bar{D}_1$ in the sense of Brauer.

The space W can be given a more convenient interpretation as follows. Put

$$(1.17)\qquad \tilde{W} = \mathrm{Hom}_{G_1\times G_2}(V_{11}\otimes V_{12}, V).$$

Then one has a natural isomorphism $W\cong\tilde{W}$ given by the correspondence $w\leftrightarrow\tilde{w}$ defined by the relation

$$(1.18)\qquad \tilde{w}(v_1\otimes v_2) = (w(v_2))(v_1) = (w\otimes_{D_{12}}v_2)\otimes_{D_{11}}v_1 \qquad (v_1\in V_1,\ v_2\in V_2).$$

This can be proved by a standard argument, which is left to the reader. In the following, we identify W with $\tilde{W}$ by this correspondence. The action of $\bar{D}_{11}\otimes\bar{D}_{12}$ on W is then given by

$$(1.19)\qquad (\bar{\alpha}\otimes\bar{\beta})_W(w) = w\circ(\alpha\otimes\beta) \qquad (\alpha\in D_{11},\ \beta\in D_{12}).$$

The above results can also be applied (with an obvious modification) to the case where G is an almost direct product of two reductive F-groups G_1 and G_2. For simplicity, we will refer to these cases as the "direct product case".

Example 3. For $F=\boldsymbol{R}$, $D_{11}\otimes D_{12}$ is simple except for the case $D_{11}\cong D_{12}\cong\boldsymbol{C}$, in which case one has $D_{11}\otimes D_{12}\cong\boldsymbol{C}\oplus\boldsymbol{C}$. When $D_{11}\cong\boldsymbol{H}$, a simple $\bar{D}_{11}\otimes\bar{D}_{12}$-module W, viewed as a $\bar{D}_1$-module, is $\cong\boldsymbol{H}, \boldsymbol{C}^2, \boldsymbol{R}^4$ according as $D_{12}\cong\boldsymbol{R}, \boldsymbol{C}, \boldsymbol{H}$.

Finally, we consider the representations by dual spaces. Retaining the notation U_1, V_1, D_1, etc. for a given primary representation (V, ρ), let V^*, U_1^*, V_1^* denote the dual spaces of V, U_1, V_1 over F and consider V^* and V_1^* as representation-spaces for the contragradient representations ${}^t\rho^{-1}$, ${}^t\rho_1^{-1}$ of ρ, ρ_1, respectively.

Lemma 1. 2. (i) *One has a natural identification:*

$$(1.20)\qquad U_1^* = \mathrm{Hom}_G(V_1^*, V^*).$$

(ii) *In the "direct product" case* ($G=G_1G_2$, $V=(W\otimes_{D_{12}}V_{12})\otimes_{D_{11}}V_{11}$), *one has a natural identification:*

$$(1.21)\qquad W^* = \mathrm{Hom}_{G_1\times G_2}(V_{11}^*\otimes V_{12}^*, V^*).$$

Proof. (i) $\mathrm{Hom}(V_1, V)$ and $\mathrm{Hom}(V_1^*, V^*)$ are mutually dual with respect to the inner product defined as follows:

$$\langle\varphi, \psi\rangle = \mathrm{tr}_V(\varphi\circ{}^t\psi) = \mathrm{tr}_{V_1}({}^t\psi\circ\varphi)$$
$$(\varphi\in\mathrm{Hom}(V_1, V),\quad \psi\in\mathrm{Hom}(V_1^*, V^*)),$$

where tr_V and tr_{V_1} denote the traces of linear transformations on V and V_1, respectively. Hence it is enough to show that the restriction of this inner product on $\mathrm{Hom}_G(V_1, V)\times\mathrm{Hom}_G(V_1^*, V^*)$ is non-degenerate. Let $\psi\in\mathrm{Hom}_G(V_1^*, V^*)$ and suppose that $\langle u, \psi\rangle=0$ for all $u\in U_1=\mathrm{Hom}_G(V_1, V)$. Then one has

$$\langle\bar{\alpha}u, \psi\rangle=\mathrm{tr}_{V_1}({}^t\psi\circ u\circ\alpha)=0 \qquad \text{for all } \alpha\in D_1.$$

Since D_1 is self-dual with respect to the inner product defined by tr_{V_1}, this implies that ${}^t\psi\circ u=0$ for all $u\in U_1$, whence follows that ${}^t\psi=0$ and so $\psi=0$, which proves our assertion.

(ii) By (i) U_{11}^* can be identified with $\mathrm{Hom}_{G_1}(V_{11}^*, V^*)$ and W^* with $\mathrm{Hom}_{G_2}(V_{12}^*, U_{11}^*)$, and the duality between W and W^* is given by the inner product

$$\langle w, w^*\rangle=\mathrm{tr}_{V_{12}}({}^tw^*\circ w) \qquad (w\in W,\ w^*\in W^*).$$

A straightforward computation shows that this coincides with

$$\langle\tilde{w}, \tilde{w}^*\rangle=\mathrm{tr}_{V_{11}\otimes V_{12}}({}^t\tilde{w}^*\circ\tilde{w}),$$

where $\tilde{w}$ and $\tilde{w}^*$ are the elements of $\mathrm{Hom}_{G_1\times G_2}(V_{11}\otimes V_{12}, V)$ and $\mathrm{Hom}_{G_1\times G_2}(V_{11}^*\otimes V_{12}^*, V^*)$ corresponding to w and w^*, respectively. This proves our assertion, q. e. d.

We note that $\mathrm{End}_G(V_1^*)={}^tD_1$ is a division algebra anti-isomorphic to D_1, so that we may identify it with $\bar{D}_1$ by the correspondence ${}^t\alpha\leftrightarrow\bar{\alpha}$ (which depends on the realization of D_1 as $\mathrm{End}_G(V_1)$).

Exercises

1. For a given irreducible representation (V_1, ρ_1), show that the category of finite-dimensional $\bar{D}_1$-modules U and that of finite-dimensional primary representation-spaces of G over F belonging to (V_1, ρ_1) are equivalent, the equivalence being given by the functor $U\mapsto U\otimes_{D_1}V_1$.

2. The notation being as in the text, let $\boldsymbol{F}$ be an algebraically closed extension of F and fix an injection of F_1 into $\boldsymbol{F}$. Then one has an isomorphism $\bar{D}_1\otimes_{F_1}\boldsymbol{F}\cong\mathcal{M}_{r_1}(\boldsymbol{F})$, which gives rise to a representation of G in $\boldsymbol{F}$ by the following scheme:

$$G\xrightarrow{\rho_1}\mathrm{End}_{D_1}(V_1)\cong\mathcal{M}_{n_1}(\bar{D}_1)\longrightarrow\mathcal{M}_{r_1n_1}(\boldsymbol{F}).$$

Show that this is an absolutely irreducible representation of G and that, over $\boldsymbol{F}$, the representation ρ_1 decomposes into a direct sum of d_1 distinct (but mutually conjugate) absolutely irreducible representations of dimension r_1n_1, each repeated r_1 times, corresponding to the d_1 distinct injections of F_1 into $\boldsymbol{F}$ (cf. § 3).

§ 2. Invariant alternating bilinear forms.

In this section, we explain a general procedure of constructing an invariant alternating bilinear form for a self-dual primary representation. Let (V, ρ) be a primary representation of a group G belonging to an irreducible representation (V_1, ρ_1) over F, and we retain the notation introduced in § 1. We first recall basic facts on the correspondence between involutions and hermitian fcrms. Suppose there is given an "involution" ι_0 of the division algebra D_1 (i. e., an anti-automorphism of D_1 of order ≤ 2) and let $\varepsilon=\pm1$. By a D_1-valued ε-hermitian form h' on

$U_1 \times U_1$ (with respect to ι_0), or a (D_1, ε)-*hermitian form* h' on $U_1 \times U_1$ for short, we mean an F-bilinear map $h' : U_1 \times U_1 \to D_1$ satisfying the following conditions:

$$(2.1)\qquad \begin{aligned} h'(x, y\alpha) &= h'(x, y)\alpha, \\ h'(y, x) &= \varepsilon h'(x, y)^{\iota_0} \qquad (x, y \in U_1,\ \alpha \in D_1). \end{aligned}$$

h' is called *non-degenerate* if $\mathrm{tr}_{D_1/F}(h'(x, y))$ is non-degenerate (as an F-bilinear form), where $\mathrm{tr}_{D_1/F}$ denotes the reduced trace. Let F_{10} be the fixed field of ι_0 in F_1. The involution ι_0 is called *of the first kind*, if $\iota_0|F_1$ is the identity, i. e., if $F_{10}=F_1$; and *of the second kind* otherwise, i. e., if $[F_1 : F_{10}]=2$.

Lemma 2. 1. *Suppose there is given an involution ι_0 of D_1. Then, to an involution ι of* $\mathrm{End}_{D_1}(U_1)$ *such that $\iota|F_1=\iota_0|F_1$, there corresponds a (D_1, ε)-hermitian form h' on $U_1 \times U_1$ such that*

$$(2.2)\qquad h'(x, \varphi y) = h'(\varphi^{\iota} x, y)$$

for all $x, y \in U_1$, $\varphi \in \mathrm{End}_{D_1}(U_1)$. The sign ε is uniquely determined by ι if ι_0 is of the first kind, but is arbitrary if ι_0 is of the second kind. A hermitian form h' having the above properties is uniquely determined by ι up to a scalar multiplication by a non-zero element in F_{10}. Conversely, any non-degenerate (D_1, ε)-hermitian form h' on $U_1 \times U_1$ determines an involution ι by the relation (2. 2).

A sketch of proof. Fixing a $\bar{D}_1$-basis $(e_1, \cdots, e_{m_1})$ of U_1, we have an algebra isomorphism $M : \mathrm{End}_{D_1}(U_1) \xrightarrow{\cong} \mathcal{M}_{m_1}(D_1)$. For $a=(\alpha_{ij}) \in \mathcal{M}_{m_1}(D_1)$, we put ${}^t a^{\iota_0}=(\alpha_{ji}^{\iota_0})$. Then by a theorem of Skolem-Noether, there exists $L \in GL_{m_1}(D_1)$, uniquely determined modulo $F_1^{\times}$, such that

$$(2.3)\qquad M(\varphi^{\iota}) = L^{-1}\,{}^t M(\varphi)^{\iota_0} L \qquad \text{for all } \varphi \in \mathrm{End}_{D_1}(U_1).$$

Then L satisfies the condition

$$(2.4)\qquad {}^t L^{\iota_0} = \zeta L \qquad \text{with } \zeta \in F_1,\ \zeta^{\iota_0}\zeta = 1.$$

If ι_0 is of the first kind, one has $\zeta=\pm 1$ (which is uniquely determined by ι independently of the choice of L). If ι_0 is of the second kind, there exists, by Hilbert's lemma, an $\eta \in F_1^{\times}$ such that $\pm\zeta=\eta\eta^{-\iota_0}$ (for any choice of $\pm$). Replacing L by ηL, one may assume that $\zeta=\pm 1$. In either case, once $\zeta=\pm 1$ is fixed, L is uniquely determined modulo $F_{10}^{\times}$. We then define h' by

$$(2.5)\qquad h'(x, y) = {}^t(\xi_i^{\iota_0}) L(\eta_i),$$

where (ξ_i), $(\eta_i) \in \mathcal{M}_{m_1,1}(D_1)$ are the coordinate (column) vectors of x, y with respect to the fixed basis (e_i). It is easy to check that h' is a (D_1, ε)-hermitian form (with $\varepsilon=\zeta$) satisfying all the requirements and that all such forms h' are obtained in this manner. The uniqueness of h' follows from that of L. The last assertion of the Lemma is trivial, q. e. d.

Next we prove a lemma on an irreducible representation (V_1, ρ_1).

Lemma 2. 2. *Suppose ρ_1 is "self-dual", i. e., $\rho_1 \sim {}^t\rho_1^{-1}$. Then there exists a unique involution ι_1 of* $\mathrm{End}_{D_1}(V_1)$ *such that one has*

$$\rho_1(g)^{\iota_1} = \rho_1(g)^{-1}. \tag{2. 6}$$

This involution can be extended to an involution of $\mathrm{End}(V_1)$.

Proof. By the assumption there exists a G-isomorphism $\psi_1 : V_1 \to V_1^*$, where V_1^* is the dual space of V_1 viewed as a representation-space of ${}^t\rho_1^{-1}$. Then one has

$${}^t\rho_1(g)^{-1} = \psi_1 \circ \rho_1(g) \circ \psi_1^{-1} \qquad (g \in G). \tag{2. 7}$$

Identifying $\mathrm{End}_G(V_1^*) = {}^tD_1$ with $\bar{D}_1$ in a natural manner, we consider V_1^* as a $\bar{D}_1$-module. For $\varphi \in \mathrm{End}_{D_1}(V_1)$ we set

$$\varphi^{\iota_1} = \psi_1^{-1} \circ {}^t\varphi \circ \psi_1. \tag{2. 8}$$

Then, since ${}^t\varphi \in \mathrm{End}_{\bar{D}_1}(V_1^*)$ and $\psi_1 \alpha \psi_1^{-1} \in \bar{D}_1$ for any $\alpha \in D_1$, one has $\varphi^{\iota_1} \in \mathrm{End}_{D_1}(V_1)$. Moreover, since ${}^t\psi_1^{-1}\psi_1 \in D_1^\times$, one obtains

$$\varphi^{\iota_1\iota_1} = ({}^t\psi_1^{-1}\psi_1)^{-1}\varphi({}^t\psi_1^{-1}\psi_1) = \varphi.$$

Thus ι_1 is an involution of $\mathrm{End}_{D_1}(V_1)$ and by (2. 7) the condition (2. 6) is satisfied. To prove the uniqueness, let ι_1' be any involution of $\mathrm{End}_{D_1}(V_1)$ satisfying (2. 6). Then $\sigma = \iota_1'\iota_1$ is an (algebra) automorphism of $\mathrm{End}_{D_1}(V_1)$. Extending σ to an automorphism of $\mathrm{End}(V_1)$, one sees by Skolem-Noether's theorem that there exists $\alpha_1 \in GL(V_1)$ such that $\varphi^\sigma = \alpha_1^{-1}\varphi\alpha_1$ $(\varphi \in \mathrm{End}(V_1))$. But then by (2. 6) one has $\alpha_1 \in D_1$ and so $\sigma = id$ on $\mathrm{End}_{D_1}(V_1)$, i. e., $\iota_1' = \iota_1$.

Now by a theorem of Albert [5], there exists an involution ι_0 of D_1 such that $\iota_1|F_1 = \iota_0|F_1$. We denote by $\bar{\iota}_0$ the involution of $\bar{D}_1$ defined by

$$\bar{\alpha}^{\bar{\iota}_0} = \overline{\alpha^{\iota_0}} \qquad (\alpha \in D_1). \tag{2. 9}$$

Then by Lemma 2. 1, applied to $\mathrm{End}_{D_1}(V_1)$, ι_1, and $\bar{\iota}_0$, there exists a non-degenerate $(\bar{D}_1, \varepsilon)$-hermitian form h on $V_1 \times V_1$ with respect to $\bar{\iota}_0$ such that

$$h(v, \varphi v') = h(\varphi^{\iota_1} v, v') \qquad (v, v' \in V, \ \varphi \in \mathrm{End}_{D_1}(V_1)). \tag{2. 10}$$

This h is G-invariant by (2. 6) and is uniquely characterized by these properties modulo $F_{10}^\times$. If we define $\psi_1' \in \mathrm{Hom}(V_1, V_1^*)$ by

$$\mathrm{tr}_{\bar{D}_1/F}\, h(v, v') = \langle v, \psi_1'(v') \rangle, \tag{2. 11}$$

then ψ_1' satisfies the relation $\varphi^{\iota_1} = \psi_1'^{-1}\,{}^t\varphi\psi_1'$ for $\varphi \in \mathrm{End}_{D_1}(V_1)$; in particular, by (2. 6) it is a G-isomorphism. Hence we may assume $\psi_1 = \psi_1'$. Then this ψ_1 satisfies further relations

$$(\alpha^{\iota_0})_{V_1} = \psi_1^{-1}\,{}^t\alpha_{V_1}\psi_1 \qquad (\alpha \in D_1), \tag{2. 12}$$

$${}^t\psi_1 = \varepsilon\psi_1, \tag{2. 13}$$

where α_{V_1} denotes the action of α on V_1. Therefore, extending the definition of ι_1 by (2. 8) for an arbitrary $\varphi \in \mathrm{End}(V_1)$, we obtain an involution of $\mathrm{End}(V_1)$ (of the first kind) such that $\iota_1|D_1 = \iota_0$ (or more precisely, $(\alpha_{V_1})^{\iota_1} = (\alpha^{\iota_0})_{V_1}$ for $\alpha \in D_1$). This proves the last assertion of the Lemma, q. e. d.

From the above proof, it is clear that to give an involution ι_1 of $\mathrm{End}(V_1)$ satisfying (2. 6) is equivalent to giving $\iota_1|\mathrm{End}_{D_1}(V_1)$ together with an involution $\iota_0=\iota_1|D_1$.

Going back to the primary representation (V, ρ), suppose there is a G-invariant non-degenerate alternating bilinear form A on $V\times V$. For simplicity, we denote the corresponding G-isomorphism $V\to V^*$ by the same letter A, i. e., we put

$$(2.14)\qquad A(x,y)=\langle x, Ay\rangle \qquad (x,y\in V).$$

Then one has ${}^t\rho(g)^{-1}=A\rho(g)A^{-1}$. Thus ρ is self-dual, and so is ρ_1. Hence, by Lemma 2. 2, there exists an involution ι_1 of $\mathrm{End}(V_1)$ satisfying (2. 6) and a hermitian form h on $V_1\times V_1$ (with respect to $\iota_0=\iota_1|D_1$) satisfying (2. 10). On the other hand, putting

$$(2.15)\qquad \varphi^\iota=A^{-1}\,{}^t\varphi A \qquad \text{for}\quad \varphi\in\mathrm{End}(V),$$

we obtain an involution of $\mathrm{End}(V)$ satisfying the condition $\rho(g)^\iota=\rho(g)^{-1}$ $(g\in G)$. It follows that ι leaves $\mathrm{End}_G(V)=\mathrm{End}_{\bar D_1}(U_1)$ stable. Hence the center F_1, $\mathrm{End}_{F_1}(V)$, and $\mathrm{End}_{D_1}(V_1)$ are also stable under ι. By the uniqueness of ι_1, the restriction of ι on $\mathrm{End}_{D_1}(V_1)$ must coincide with ι_1.

Now, applying Lemma 2. 1 to the restriction of ι on $\mathrm{End}_{\bar D_1}(U_1)$, we obtain a non-degenerate (D_1, ε')-hermitian form h' on $U_1\times U_1$ (with respect to $\bar\iota_0$) such that

$$h'(u,\varphi u')=h'(\varphi^\iota u, u') \qquad \text{for}\quad u,u'\in U_1,\ \varphi\in\mathrm{End}_{\bar D_1}(U_1).$$

Put

$$\tilde A(u,u',v,v')=\mathrm{tr}_{D_1/F}(h'(u,u')\overline{h(v,v')}) \qquad (u,u'\in U_1,\ v,v'\in V_1).$$

Then, since $\tilde A(u,u',v,v')$ depends only on $u\otimes_{D_1}v$, $u'\otimes_{D_1}v'$, there exists a unique bilinear form A' on $V\times V$ such that one has

$$\tilde A(u,u',v,v')=A'(u\otimes_{D_1}v,\ u'\otimes_{D_1}v').$$

It is clear that A' is $\varepsilon\varepsilon'$-symmetric and satisfies a relation similar to (2. 15). Hence by Skolem-Noether's theorem $A^{-1}A'$, viewed as an endomorphism of V, belongs to $F_1^\times$, i. e., one has

$$A'(x,y)=A(x,\alpha_1 y)$$

for some $\alpha_1\in F_1^\times$. Since A is alternating, one has $\alpha_1^\iota=-\varepsilon\varepsilon'\alpha_1$. Hence, replacing h' by $\alpha_1 h'$ if necessary, one may assume that $A=A'$ and $\varepsilon\varepsilon'=-1$.

Conversely, it is clear that for any non-degenerate $(D_1, -\varepsilon)$-hermitian form h' on $U_1\times U_1$ one obtains a non-degenerate G-invariant alternating bilinear form A on $V\times V$ by setting

$$(2.16)\qquad A(u\otimes_{D_1}v,\ u'\otimes_{D_1}v')=\mathrm{tr}_{D_1/F}(h'(u,u')\overline{h(v,v')})$$

for $u,u'\in U_1$, $v,v'\in V_1$. Note that for the existence of such an h' we have to assume that $\dim U_1$ is even if $D_1=F$ and $\varepsilon=1$. Summing up, we obtain the following theorem.

Theorem 2. 3. *Let (V, ρ) be a primary representation of a group G and let the notation V_1, U_1, D_1, etc. be as in § 1. Suppose that ρ_1 is self-dual, and let ι_1 be an involution of* $\mathrm{End}(V_1)$ *satisfying* (2. 6) *and h the corresponding non-degenerate G-invariant $(\bar{D}_1, \varepsilon)$-hermitian form on $V_1\times V_1$ (with respect to $\iota_0=\iota_1|D_1$) (cf. Lem. 2. 2). Then non-degenerate G-invariant alternating bilinear forms A on $V\times V$ are in one-to-one correspondence with non-degenerate $(D_1, -\varepsilon)$-hermitian forms h' on $U_1\times U_1$ (with respect to $\bar{\iota}_0$) by the relation* (2. 16).

For simplicity, when the relation (2. 16) holds, we write $A=\mathrm{tr}_{D_1/F}(h'\otimes h)$.

Next, let us consider the relation between invariant hermitian forms in the "direct product" case considered in § 1. Namely, let $G=G_1G_2$ (where G_1 and G_2 commute elementwise), and let (V_1, ρ_1) be an irreducible representation of G over F such that $\rho_1|G_i\,(i=1, 2)$ are both fully reducible. Then in the notation of § 1 one has

$$V_1=(W\otimes_{D_{12}}V_{12})\otimes_{D_{11}}V_{11}, \tag{2. 17}$$

where W is a simple $\bar{D}_{11}\otimes\bar{D}_{12}$-module. Suppose further that ρ_1 is self-dual. Then the irreducible representations (V_{1i}, ρ_{1i}) of G_i are also self-dual. Hence by Lemma 2. 2 there exist involutions ι_{1i} of the algebras $\mathrm{End}(V_{1i})$ given by

$$\varphi^{\iota_{1i}}=\psi_{1i}^{-1}\,{}^t\varphi\psi_{1i}, \tag{2. 18}$$

where ψ_{1i} is a G_i-isomorphism of V_{1i} onto V_{1i}^* such that ${}^t\psi_{1i}=\varepsilon_i\psi_{1i}$. We denote by h_i the corresponding $(\bar{D}_{1i}, \varepsilon_i)$-hermitian forms on $V_{1i}\times V_{1i}$. Identifying W and W^* with $\mathrm{Hom}_{G_1\times G_2}(V_{11}\otimes V_{12}, V_1)$ and $\mathrm{Hom}_{G_1\times G_2}(V_{11}^*\otimes V_{12}^*, V_1^*)$, respectively, we define a linear map $\Psi : W\to W^*$ by

$$\Psi(w)=\psi_1\circ w\circ(\psi_{11}^{-1}\otimes\psi_{12}^{-1})\qquad(w\in W). \tag{2. 19}$$

Then it is easy to check that Ψ is bijective and satisfies the relations

$${}^t\Psi=\varepsilon\varepsilon_1\varepsilon_2\Psi, \tag{2. 20}$$

$$(\alpha\otimes\beta)_{W^*}\circ\Psi=\Psi\circ(\bar{\alpha}^{\iota_{11}}\otimes\bar{\beta}^{\iota_{12}})_W\qquad(\alpha\in D_{11},\ \beta\in D_{12}). \tag{2. 21}$$

Conversely, it is easy to see that, when there are given self-dual irreducible representations (V_{1i}, ρ_{1i}) and a linear map $\Psi : W\to W^*$ satisfying these properties, the irreducible representation (V_1, ρ_1) given by (2. 17) is self-dual and a G-isomorphism $\psi_1 : V_1\to V_1^*$ is determined by (2. 19). In the following lemma, we suppose that self-dual representation-spaces V_1, V_{11}, and V_{12} are set in this manner.

Lemma 2. 4. *The notation being as above, let h_i be a non-degenerate G_i-invariant $(\bar{D}_{1i}, \varepsilon_i)$-hermitian form on $V_{1i}\times V_{1i}$ and h a non-degenerate G-invariant $(\bar{D}_1, \varepsilon)$-hermitian form on $V_1\times V_1$ corresponding to ι_{1i} and ι_1, respectively. Then, after a suitable adjustment of scalars, we have*

$$h(w(v_1\otimes v_2), w'(v_1'\otimes v_2'))=\Psi(w)\circ(h_1(v_1, v_1')\otimes h_2(v_2, v_2'))\circ{}^tw' \tag{2. 22}$$

for all $v_1, v_1'\in V_1$, $v_2, v_2'\in V_2$, $w, w'\in W$.

Proof. The right-hand side of (2. 22) is trilinear in v_1', v_2', w' and invariant under the transformation

$$(v_1', v_2', w') \longmapsto (\alpha^{-1}v_1', \beta^{-1}v_2', (\bar{\alpha}\otimes\bar{\beta})_W w') \qquad (\alpha \in D_{11}^{\times},\ \beta \in D_{12}^{\times}).$$

Hence it depends only on $(w'\otimes_{D_{12}}v_2')\otimes_{D_{11}}v_1' = w'(v_1'\otimes v_2')$. Similarly, by virtue of (2. 21), it depends only on $w(v_1\otimes v_2)$. Hence the right-hand side of (2. 22) may be written as $h'(w(v_1\otimes v_2), w'(v_1'\otimes v_2'))$ with a bilinear map $h' : V_1 \times V_1 \to \bar{D}_1$. Clearly h' is G-invariant, and one has by (2. 19), (2. 21)

$$\begin{aligned}
&h'(w(v_1\otimes v_2), w'(v_1'\otimes v_2'))^{\bar{\iota}_0}\\
&= \psi_1\circ w'\circ{}^t(h_1(v_1, v_1')\otimes h_2(v_2, v_2'))\circ{}^t(\psi_{11}^{-1}\otimes\psi_{12}^{-1})\circ{}^t w\circ{}^t\psi_1\circ\psi_1^{-1}\\
&= \varepsilon\varepsilon_1\varepsilon_2\psi_1\circ w'\circ(\bar{h}_1(v_1, v_1')\otimes\bar{h}_2(v_2, v_2'))\circ(\psi_{11}^{-1}\otimes\psi_{12}^{-1})\circ{}^t w\\
&= \varepsilon\varepsilon_1\varepsilon_2\Psi(w'\circ(\bar{h}_1(v_1, v_1')\otimes\bar{h}_2(v_2, v_2')))\circ{}^t w\\
&= \varepsilon\varepsilon_1\varepsilon_2\Psi(w')\circ(h_1(v_1, v_1')^{\bar{\iota}_{01}}\otimes h_2(v_2, v_2')^{\bar{\iota}_{02}})\circ{}^t w\\
&= \varepsilon\Psi(w')\circ(h_1(v_1', v_1)\otimes h_2(v_2', v_2))\circ{}^t w\\
&= \varepsilon h'(w'(v_1'\otimes v_2'), w(v_1\otimes v_2)),
\end{aligned}$$

where $\iota_{0i} = \iota_{1i}|D_{1i}$. Thus h' is ε-hermitian. Hence by the uniqueness we may assume $h = h'$, q. e. d.

Exercise

1. Let $(e_1, \cdots, e_l)$ and $(e_1^*, \cdots, e_l^*)$ be mutually dual D_1-basis of W and $\bar{D}_1$-basis of W^* such that we have $e_i\circ{}^t e_j^* = \delta_{ij}(\in D_1)$. For $\varphi \in \mathrm{End}_{D_1}(W)$, denote by $\bar{M}(\varphi)$ the matrix of φ with respect to these bases, i. e.,

$$\bar{M}(\varphi) = (e_i^*\circ{}^t(\varphi(e_j))) \in \mathcal{M}_l(\bar{D}_1).$$

Also put

$$\bar{M}(\Psi) = (\Psi(e_i)\circ{}^t e_j) \in \mathcal{M}_l(\bar{D}_1).$$

Then show that

$$\begin{gathered}
{}^t\bar{M}(\Psi)^{\bar{\iota}_0} = \varepsilon\varepsilon_1\varepsilon_2\bar{M}(\Psi),\\
(h(e_i(v_1\otimes v_2), e_j(v_1'\otimes v_2'))) = \bar{M}(\Psi)\cdot\bar{M}((h_1(v_1, v_1')\otimes h_2(v_2, v_2'))_W).
\end{gathered}$$

§ 3. Scalar extensions.

In this section, we examine the effect of a scalar extension to the primary decomposition and invariant hermitian forms. Retaining previous notation, let $V = U_1\otimes_{D_1}V_1$ be a primary representation-space of G over F. Let F' be an extension field of F, and set

$$F_1\otimes_F F' = \bigoplus_{i=1}^{t} F_1^{(i)}, \tag{3. 1}$$

where $F_1^{(i)} (1 \le i \le t)$ are finite extensions of F' (called "composites" of F_1 and F'). If one fixes an algebraic closure $\tilde{F}'$ of F', then, for each i, there exists an F-monomorphism (i. e., an F-linear monomorphism) σ_i of F_1 into $\tilde{F}'$ such that the correspondence

$$(\alpha\otimes\beta)^{(i)} \longmapsto \alpha^{\sigma_i}\beta \qquad (\alpha \in F_1,\ \beta \in F')$$

gives an F'-isomorphism $F_1^{(i)} \cong F_1^{\sigma_i}F' \subset \tilde{F}'$, where $(\ \ )^{(i)}$ denotes the i-th component in the decomposition (3. 1). Such an F-monomorphism σ_i is determined up to a composition with $\tau \in \mathrm{Gal}(\tilde{F}'/F')$, so that t is equal to the number of $\mathrm{Gal}(\tilde{F}'/F')$-orbits in the set of d_1 distinct F-monomorphisms of F_1 into $\tilde{F}'$. From (3. 1) one obtains

$$D_1 \otimes_F F' = \bigoplus_{i=1}^{t} D_1 \otimes_{F_1} F_1^{(i)}, \tag{3. 2}$$

where the tensor product $D_1 \otimes_{F_1} F_1^{(i)}$ is defined by means of the natural injection $F_1 \to F_1^{(i)}$. We denote by $V_{F'}, V_{1F'}, \cdots$ the representation-spaces of G over F' obtained by scalar extension F'/F from $V, V_1, \cdots$. Then it is clear that, after natural identifications, we have

$$\mathrm{End}_G(V_{1F'}) = (\mathrm{End}_G(V_1))_{F'} = D_1 \otimes_F F', \tag{3. 3}$$

so that $F_1^{(i)}$ and $D_1 \otimes_{F_1} F_1^{(i)}$ can be considered as subalgebras of $\mathrm{End}_G(V_{1F'})$. [When G is an F-group, one also has $\mathrm{End}_G(V_{1F'}) = \mathrm{End}_{G_{F'}}(V_{1F'})$.] Hence, from the decompositions (3. 1) and (3. 2), we obtain the corresponding direct sum decomposition

$$V_{1F'} = \bigoplus_{i=1}^{t} F_1^{(i)} V_1 \tag{3. 4}$$

and the natural isomorphisms

$$F_1^{(i)} V_1 \cong V_1 \otimes_{F_1} F_1^{(i)}, \qquad \mathrm{End}_G(F_1^{(i)} V_1) \cong D_1 \otimes_{F_1} F_1^{(i)}. \tag{3. 5}$$

Thus we can conclude that (3. 4) is the primary decomposition of $V_{1F'}$. Similarly,

$$V_{F'} = \bigoplus_{i=1}^{t} F_1^{(i)} V \tag{3. 6}$$

is the primary decomposition of $V_{F'}$.

Now let $(V_1^{(i)}, \rho_1^{(i)})$ be an irreducible representation of G over F' contained in $F_1^{(i)} V$ and set $D_1^{(i)} = \mathrm{End}_G(V_1^{(i)})$. Then one has $D_1 \otimes_{F_1} F_1^{(i)} \sim D_1^{(i)}$ over $F_1^{(i)}$ in the sense of Brauer. We fix an isomorphism

$$M:\ \mathrm{End}_G(F_1^{(i)} V_1) \xrightarrow{\cong} \mathcal{M}_s(D_1^{(i)}) \tag{3. 7}$$

once and for all, where $s = s_i$ is a positive integer depending on i, and let ε_{jk} $(1 \leq j, k \leq s)$ be the corresponding matrix units in $\mathrm{End}_G(F_1^{(i)} V_1)$. Then one has a G-isomorphism

$$V_1^{(i)} \xrightarrow{\cong} F_1^{(i)} \varepsilon_{11} V_1$$

compatible with the inclusion map $D_1^{(i)} \to \mathrm{End}_G(F_1^{(i)} V_1)$ given by $M^{-1}|D_1^{(i)}$. For simplicity, we identify $V_1^{(i)}$ with $F_1^{(i)} \varepsilon_{11} V_1$ through this isomorphism (which is uniquely determined by the isomorphism M up to a scalar multiplication by an element of $F_1^{(i)\times}$). Then one has

$$F_1^{(i)} V_1 = \bigoplus_{j=1}^{s} \varepsilon_{j1} V_1^{(i)}. \tag{3. 8}$$

Moreover, if we put $U_1^{(i)} = \mathrm{Hom}_G(V_1^{(i)}, V_{F'})$, then

$$\begin{aligned} U_1^{(i)} &= \mathrm{Hom}_G(F_1^{(i)} V_1, F_1^{(i)} V) \circ \mathrm{Hom}_G(V_1^{(i)}, F_1^{(i)} V_1) \\ &= F_1^{(i)} U_1 \circ \Big(\bigoplus_{j=1}^{s} D_1^{(i)} \varepsilon_{j1} \Big) \end{aligned}$$

$$= F_1^{(i)} \varepsilon_{11} U_1.$$

Thus one has

(3.9) $$F_1^{(i)} V_{F'} = F_1^{(i)} U_1 \otimes_{\mathcal{M}_s} F_1^{(i)} V_1 = U_1^{(i)} \otimes_{D_1^{(i)}} V_1^{(i)},$$

where $\otimes_{\mathcal{M}_s}$ stands for the tensor product over $\mathcal{M}_s(D_1^{(i)})$.

Now, suppose (V_1, ρ_1) is self-dual, and let $\psi_1, \iota_1, \iota_0, \cdots$ be as in Lemma 2.2. Then, for instance, ψ_1 extends naturally to a G-isomorphism (or $G_{F'}$-isomorphism, if we are dealing with an F-group G) of $V_{1F'}$ onto $V^*_{1F'}$, which will also be denoted by the same letter ψ_1. The same convention will apply to the extensions of other homomorphisms. Then, for each i, there corresponds an index i' such that one has $\psi_1 F_1^{(i)} V_1 = F_1^{(i')} V_1^*$. One then has $\psi_1 F_1^{(i')} V_1 = F_1^{(i)} V_1^*$ and hence $F_1^{(i)\iota_0} = F_1^{(i')}$. In this way, ι_0 induces a permutation of order at most 2 on the set of indices $\{1, \cdots, t\}$. Clearly, an index i is fixed, i. e., $F_1^{(i)\iota_0} = F_1^{(i)}$, if and only if $\rho_1^{(i)}$ is self-dual.

Suppose $\rho_1^{(i)}$ is self-dual. Then, ι_0 induces an involution of the simple component $D_1 \otimes_{F_1} F_1^{(i)}$ of $D_1 \otimes_F F'$, and by Albert's theorem $D_1^{(i)}$ has also an involution $\iota_0^{(i)}$ such that $\iota_0 | F_1^{(i)} = \iota_0^{(i)} | F_1^{(i)}$. As in the proof of Lemma 2.1, one can then find a matrix $L \in GL_s(D_1^{(i)})$ and $\eta = \pm 1$ (both depending on i) such that

(3.10) $$\begin{aligned} M(\alpha^{\iota_0}) &= L^{-1}\, {}^t M(\alpha)^{\iota_0^{(i)}} L \qquad (\alpha \in \mathrm{End}_G(F_1^{(i)} V_1)), \\ {}^t L^{\iota_0^{(i)}} &= \eta L. \end{aligned}$$

As before, we denote by $\bar{D}_1^{(i)}$ a division algebra anti-isomorphic to $D_1^{(i)}$ (identified with $\mathrm{End}_G(V_1^{(i)*})$) and by $\bar{\iota}_0^{(i)}$ the involution of $\bar{D}_1^{(i)}$ corresponding to $\iota_0^{(i)}$.

Now, let h be a non-degenerate G-invariant $(\bar{D}_1, \varepsilon)$-hermitian form on $V_1 \times V_1$. By scalar extension $F_1^{(i)}/F_1$, h is uniquely extended to a $(\bar{D}_1 \otimes_{F_1} F_1^{(i)}, \varepsilon)$-hermitian form on $F_1^{(i)} V \times F_1^{(i)} V$ (with respect to the extension of $\bar{\iota}_0$), which will be denoted by $h_{F_1^{(i)}}$. Using these notations, we have the following lemma.

Lemma 3.1. *Suppose that $(V_1^{(i)}, \rho_1^{(i)})$ is self-dual. Then there exists a uniquely determined non-degenerate G-invariant $(\bar{D}_1^{(i)}, \varepsilon\eta)$-hermitian form $h^{(i)}$ on $V_1^{(i)} \times V_1^{(i)}$ (with respect to $\bar{\iota}_0^{(i)}$) such that one has*

(3.11) $$M(\bar{h}_{F_1^{(i)}}(v, v'))L^{-1} = (\bar{h}^{(i)}(\varepsilon_{1k} v, \varepsilon_{1j} v')) \qquad \textit{for} \quad v, v' \in F_1^{(i)} V.$$

[*This means that the (j, k)-entry of the matrix on the left-hand side is given by $\bar{h}^{(i)}(\varepsilon_{1k} v, \varepsilon_{1j} v')$.*]

Proof. Let $\bar{h}_{jk}^{(i)}(v, v')$ denote the (j, k)-entry of the matrix $M(\bar{h}_{F_1^{(i)}}(v, v'))L^{-1}$. Then $h_{jk}^{(i)}$ is an F'-bilinear map $F_1^{(i)} V_1 \times F_1^{(i)} V_1 \to \bar{D}_1^{(i)}$. First, since h is ε-hermitian, one has from (3.10)

$$M(\bar{h}_{F_1^{(i)}}(v', v))L^{-1} = \varepsilon\eta\, {}^t(M(\bar{h}_{F_1^{(i)}}(v, v'))L^{-1})^{\iota_0^{(i)}},$$

which means

(*) $$h_{jk}^{(i)}(v', v) = \varepsilon\eta\, h_{kj}^{(i)}(v, v')^{\iota_0^{(i)}}.$$

Second, since h is (right) $\overline{\mathcal{M}_s(D_1^{(i)})}$-linear, one has

$$M(\bar{h}_{F_1^{(i)}}(v, \varepsilon_{pq} v'))L^{-1} = M(\varepsilon_{pq} \bar{h}_{F_1^{(i)}}(v, v'))L^{-1},$$

which means

$$(**) \qquad h_{jk}^{(i)}(v, \epsilon_{pq}v') = \delta_{pj} h_{qk}^{(i)}(v, v').$$

From (*), (**) it follows that

$$h_{11}^{(i)}(\epsilon_{kl}v, \epsilon_{pq}v') = \delta_{1k}\delta_{1p} h_{ql}^{(i)}(v, v').$$

Therefore, putting $h^{(i)} = h_{11}^{(i)} | V_1^{(i)} \times V_1^{(i)}$, one obtains the relation

$$h_{jk}^{(i)}(v, v') = h^{(i)}(\epsilon_{1k}v, \epsilon_{1j}v').$$

It is then clear that this $h^{(i)}$ is a (unique) $(\bar{D}_1^{(i)}, \epsilon\eta)$-hermitian form satisfying all the conditions stated in the Lemma, q. e. d.

Similarly, given a non-degenerate (D_1, ϵ')-hermitian form h' on $U_1 \times U_1$, one extends it to a $(D_1 \otimes_{F_1} F_1^{(i)}, \epsilon')$-hermitian form $h'_{F_1^{(i)}}$ on $F_1^{(i)} U_1 \times F_1^{(i)} U_1$, and then obtain a $(D_1^{(i)}, \epsilon'\eta)$-hermitian form $h'^{(i)}$ on $U_1^{(i)} \times U_1^{(i)}$ satisfying the relation

$$(3.12) \qquad L \cdot M(h'_{F_1^{(i)}}(u, u')) = (h'^{(i)}(\bar{\epsilon}_{1j}u, \bar{\epsilon}_{1k}u')) \qquad (u, u' \in F_1^{(i)} U_1).$$

When $\epsilon' = -\epsilon$, let A and $A^{(i)}$ be non-degenerate G-invariant alternating bilinear forms on $V \times V$ and $F_1^{(i)} V \times F_1^{(i)} V$, respectively, given symbolically by

$$A = \mathrm{tr}_{D_1/F}(h' \otimes \bar{h}), \qquad A^{(i)} = \mathrm{tr}_{D_1^{(i)}/F}(h'^{(i)} \otimes \bar{h}^{(i)}).$$

Then, extending A to an alternating bilinear form on $V_{F'} \times V_{F'}$, one obtains the relation

$$(3.13) \qquad A^{(i)} = A | F_1^{(i)} V \times F_1^{(i)} V.$$

In fact, for $u, u' \in F_1^{(i)} U_1$, $v, v' \in F_1^{(i)} V_1$, we have

$$\begin{aligned} A(u \otimes_{\mathcal{M}_s} v, u' \otimes_{\mathcal{M}_s} v') &= \mathrm{tr}_{\mathcal{M}_s(D_1^{(i)})/F}(M(h'_{F_1^{(i)}}(u, u') \bar{h}_{F_1^{(i)}}(v, v'))) \\ &= \mathrm{tr}_{D_1^{(i)}/F}\Big(\sum_{j,k=1}^{s} h'^{(i)}(\bar{\epsilon}_{1j}u, \bar{\epsilon}_{1k}u') \bar{h}^{(i)}(\epsilon_{1j}v, \epsilon_{1k}v')\Big) \\ &= A^{(i)}\Big(\sum_{j=1}^{s} \bar{\epsilon}_{1j}u \otimes_{D_1^{(i)}} \epsilon_{1j}v, \sum_{k=1}^{s} \bar{\epsilon}_{1k}u' \otimes_{D_1^{(i)}} \epsilon_{1k}v'\Big), \end{aligned}$$

where by the identification (3.9) one has

$$u \otimes_{\mathcal{M}_s} v = \sum_{j=1}^{s} \bar{\epsilon}_{1j}u \otimes_{D_1^{(i)}} \epsilon_{1j}v \qquad \text{for} \quad u \in F_1^{(i)} U_1, \ v \in F_1^{(i)} V_1$$

and similarly for $u' \otimes_{\mathcal{M}_s} v'$, which proves our assertion.

Example 1. Let D_1 be a quaternion division algebra over F_1 and ι_0 the "standard" involution (Appendix, § 2). If $D_1 \otimes_{F_1} F_1^{(i)}$ remains division, one may take $\iota_0^{(i)}$ to be the standard involution of $D_1 \otimes_{F_1} F_1^{(i)}$ and $L = 1$. If $D_1 \otimes_{F_1} F_1^{(i)}$ splits, one has $\iota_0^{(i)} = 1_{F^{(i)}}$ and $L = \begin{pmatrix} 0 & -1 \\ 1 & 0 \end{pmatrix}$. Hence, in this case, one has $\eta = -1$.

Example 2. Let $F = \boldsymbol{R}$, $F' = \boldsymbol{C}$. We denote by ρ_{1C} the complex representation of G on V_{1C} obtained from ρ_1 by scalar extension. Then we have the following three cases.

($\boldsymbol{R}$-type) $D_1 = F_1 = \boldsymbol{R}$. In this case, $t = 1$, $V_{1C} = V_1^{(1)}$, and $\rho_{1C} = \rho_1^{(1)}$. Thus ρ_1 is absolutely irreducible.

($\boldsymbol{H}$-type) $D_1 \cong \boldsymbol{H}$, $F_1 = \boldsymbol{R}$. In this case, $t=1$, $D_1 \otimes_{\boldsymbol{R}} \boldsymbol{C} \cong \mathcal{M}_2(\boldsymbol{C})$. To fix an isomorphism $D_1 \cong \boldsymbol{H}$ is equivalent to giving a pair of complex structures (I, J) on V_1 such that $IJ = -JI$. Given such a pair (I, J), one has an isomorphism $D_1 \otimes_{\boldsymbol{R}} \boldsymbol{C} \cong \mathcal{M}_2(\boldsymbol{C})$ defined by $I = i\varepsilon_{11} - i\varepsilon_{22}$, $J = \varepsilon_{12} - \varepsilon_{21}$ (Appendix, § 2). Then, for $V_1^{(1)} = \varepsilon_{11} V_{1C}$, one has

$$V_{1C} = V_1^{(1)} \oplus \varepsilon_{21} V_1^{(1)},$$
$$V_1^{(1)} = V_{1C}(I; i), \qquad \varepsilon_{21} V_1^{(1)} = J V_1^{(1)} = \bar{V}_1^{(1)} = V_{1C}(I; -i).$$

Thus $\rho_{1C} \sim 2\rho_1^{(1)}$, $\overline{\rho_1^{(1)}} \sim \rho_1^{(1)}$. Putting $\varphi(x) = J\bar{x}$ for $x \in V_1^{(1)}$, one obtains an $\boldsymbol{C}$-antilinear automorphism of the representation-space $V_1^{(1)}$ such that $\varphi^2 = -1$.

($\boldsymbol{C}$-type) $D_1 = F_1 \cong \boldsymbol{C}$. In this case, $t=2$, $F_1^{(1)} = F_1^{(2)} = \boldsymbol{C}$, and $F_1^{(1)}$ and $F_1^{(2)}$ correspond to the mutually conjugate isomorphisms σ_1 and σ_2 of F_1 onto $\boldsymbol{C}$. Hence, if we put $I = \sigma_1^{-1}(i)$, I is a complex structure on V_1 and we have

$$V_{1C} = V_1^{(1)} \oplus V_1^{(2)},$$
$$V_1^{(1)} = V_{1C}(I; i), \qquad V_1^{(2)} = \bar{V}_1^{(1)} = V_{1C}(I; -i).$$

Thus $\rho_{1C} \sim \rho_1^{(1)} \oplus \overline{\rho_1^{(1)}}$, $\overline{\rho_1^{(1)}} \nsim \rho_1^{(1)}$.

We see that the $\boldsymbol{R}$-type (resp. $\boldsymbol{H}$-type) is characterized by the self-conjugacy $\overline{\rho_1^{(1)}} \sim \rho_1^{(1)}$ and the existence of a $\boldsymbol{C}$-antilinear G-automorphism φ of $V_1^{(1)}$ such that $\varphi^2 = 1$ (resp. -1).

§ 4. Symplectic representations giving rise to equivariant holomorphic maps.

Let G be a Zariski connected semi-simple $\boldsymbol{R}$-group of hermitian type and $\mathcal{D}$ the associated symmetric domain. We fix an "origin" $o \in \mathcal{D}$ and let $\mathfrak{g} = \mathfrak{k} + \mathfrak{p}$ be the corresponding Cartan decomposition of $\mathfrak{g} = \mathrm{Lie}\, G$. Recall that an "$H$-element" of G (or of $\mathfrak{g}$) at o is an element H_0 in the center of $\mathfrak{k}$ such that $ad_{\mathfrak{p}} H_0$ gives a complex structure of $\mathfrak{p} = T_o(\mathcal{D})$ compatible with the one given on the domain $\mathcal{D}$. Let G' be another (Zariski connected) semi-simple $\boldsymbol{R}$-group of hermitian type and $\mathcal{D}'$ the associated symmetric domain. A pair $(\rho, \rho_{\mathcal{D}})$ of an $\boldsymbol{R}$-homomorphism $\rho: G \to G'$ and a holomorphic map $\rho_{\mathcal{D}}: \mathcal{D} \to \mathcal{D}'$ is "(strongly) equivariant" if the following two conditions are satisfied:

$$\rho_{\mathcal{D}}(gz) = \rho(g)\rho_{\mathcal{D}}(z) \qquad \text{for all } g \in G^\circ,\ z \in \mathcal{D}, \tag{4. 1}$$

$$\rho \circ \theta = \theta' \circ \rho, \tag{4. 2}$$

where θ (resp. θ') is a Cartan involution of G (resp. G') at o (resp. $o' = \rho_{\mathcal{D}}(o)$). Geometrically, the condition (4. 2) implies that the map $\rho_{\mathcal{D}}$ is totally geodesic with respect to the Bergman metrics (II, § 2). For the given pairs $(G, \mathcal{D})$ and $(G', \mathcal{D}')$, one may ask the problem of finding all possible (strongly) equivariant pairs $(\rho, \rho_{\mathcal{D}})$.

A triple (ρ, o, o') formed of an $\boldsymbol{R}$-homomorphism $\rho: G \to G'$ and points $o \in \mathcal{D}$, $o' \in \mathcal{D}'$ will be called *compatible* if the corresponding Lie algebra homomorphism $d\rho: \mathfrak{g} \to \mathfrak{g}'$ satisfies the condition (H_1) with respect to the H-elements H_0 of $\mathfrak{g}$ at o and H_0' of $\mathfrak{g}'$ at o'. We know (II, § 8) that, for any (strongly) equivariant pair $(\rho, \rho_{\mathcal{D}})$ and $o \in \mathcal{D}$, one obtains a compatible triple (ρ, o, o') by setting $o'=\rho_{\mathcal{D}}(o)$ and conversely, if (ρ, o, o') is a compatible triple, then, defining a holomorphic map $\rho_{\mathcal{D}}$ by $\rho_{\mathcal{D}}(go)=\rho(g)o'$ $(g \in G^{\circ})$, one obtains a (strongly) equivariant pair $(\rho, \rho_{\mathcal{D}})$. Thus the above problem is essentially equivalent to finding all compatible triples (ρ, o, o'), where one may choose $o \in \mathcal{D}$ arbitrarily and then fix it once and for all.

Before specializing our problem, we give a general result concerning the freedom of the solution.

Proposition 4. 1. *Let (ρ, o, o') be a compatible triple and set*

$$C_{\rho} = C_{G'}(\rho(G))^{z}, \quad \mathcal{D}_{\rho} = \{z' \in \mathcal{D}' \mid (\rho, o, z') \text{ is compatible}\}. \tag{4. 3}$$

Then C_{ρ} is a reductive subgroup of G' of hermitian type with an H-element $H_0'-d\rho(H_0)$ and $\mathcal{D}_{\rho}$ is the symmetric domain associated with C_{ρ} which is a complex submanifold of $\mathcal{D}$. Moreover, the inclusion maps $C_{\rho}\to G'$ and $\mathcal{D}_{\rho}\to\mathcal{D}'$ are (strongly) equivariant.

This follows immediately from II, § 8, Exerc. 1 and III, § 1, Exerc. 1. When C_{ρ} is compact, i. e., when $\mathcal{D}_{\rho}$ reduces to a point, the (H_1)-homomorphism ρ is called *rigid*.

The special case, where $G=SL_2(\boldsymbol{R})$, was treated in the theory of Wolf-Korányi (Ch. III). In this chapter, we propose to consider another important case, where $G'=Sp(V, A)$ and $\mathcal{D}'=\mathfrak{S}(V, A)$, A denoting a non-degenerate alternating bilinear form on $V\times V$. As explained in II, § 7, an H-element H_0' of G' is then given by $\frac{1}{2}I$, where I is a complex structure on V such that the bilinear form $A(x, Iy)$ is symmetric and positive definite, or symbolically

$$AI \gg 0. \tag{4. 4}$$

Hence for a symplectic representation $\rho: G\to Sp(V, A)$ the condition (H_1) reads

$$\left[d\rho(H_0)-\frac{1}{2}I, d\rho(X)\right] = 0 \qquad \text{for all } X \in \mathfrak{g}. \tag{4. 5}$$

Thus our problem in this special case may be formulated as follows: *For a given (Zariski connected) semi-simple $\boldsymbol{R}$-group of hermitian type (G, H_0), find all quadruples (V, ρ, A, I) such that*

$$\begin{cases} (V, \rho) \text{ is a representation of } G \text{ over } \boldsymbol{R}, \\ A \text{ is a } G\text{-invariant non-degenerate alternating bilinear form on } V\times V, \\ I \text{ is a complex structure on } V \text{ satisfying the conditions (4. 4), (4. 5).} \end{cases}$$

Later on, we shall consider the same problem over the rational number field $\boldsymbol{Q}$, assuming that G is defined over $\boldsymbol{Q}$. In this case, we shall require that V, ρ, A (but not I) are also defined over $\boldsymbol{Q}$. Two solutions (V, ρ, A, I) and (V', ρ', A', I') are said to be *similar* (*$\boldsymbol{Q}$-similar*) if there exists a linear isomorphism $\varphi: V\to V'$ (defined over $\boldsymbol{Q}$) such that

$$(4.6)\qquad \begin{cases} \rho'(g) = \varphi\circ\rho(g)\circ\varphi^{-1} & (g\in G), \\ A'(x',y') = \lambda A(\varphi^{-1}(x'), \varphi^{-1}(y')) & (x', y'\in V'), \\ I' = \varphi\circ I\circ\varphi^{-1}, \end{cases}$$

where λ is a positive (rational) number. When $\lambda=1$, they are said to be *equivalent* (*$\boldsymbol{Q}$-equivalent*). Clearly, over $\boldsymbol{R}$, the similarity implies the equivalence.

First, we shall show that the problem can be reduced to the primary case.

Lemma 4. 2. *Let (V, A, I, ρ) be a solution of our problem and let*

$$(4.7)\qquad V = \bigoplus_i V^{[i]}$$

be the primary decomposition of V over $\boldsymbol{R}$. Then, each $V^{[i]}$ is stable under I, and the decomposition (4. 7) is an orthogonal sum with respect to A.

Proof. By (4. 5) one has $\frac{1}{2}I-d\rho(H_0)\in \mathrm{End}_G(V)$. Since $V^{[i]}$ is stable under both $d\rho(\mathfrak{g})$ and $\mathrm{End}_G(V)$, it is also stable under

$$\frac{1}{2}I = \left(\frac{1}{2}I-d\rho(H_0)\right)+d\rho(H_0).$$

To prove the second assertion, let $V^{[i]\perp}$ denote the orthogonal complement of $V^{[i]}$ with respect to A and let $v\in V^{[i]}\cap V^{[i]\perp}$. Then, by the first assertion of the Lemma, one has $Iv\in V^{[i]}$ and hence $A(v, Iv)=0$. By the condition (4. 4), this implies $v=0$. Thus $V^{[i]}\cap V^{[i]\perp}=\{0\}$. Since $V^{[i]\perp}$ is G-stable, it follows that $V^{[i]\perp}=\bigoplus_{j\neq i} V^{[j]}$, q. e. d.

By this lemma, if we put

$$\rho^{[i]}(g) = \rho(g)|V^{[i]}, \qquad I^{[i]} = I|V^{[i]}, \qquad A^{[i]} = A|V^{[i]}\times V^{[i]},$$

then, for each i, $(V^{[i]}, \rho^{[i]}, A^{[i]}, I^{[i]})$ is a solution of our problem and the solution (V, ρ, A, I) is a direct sum of $(V^{[i]}, \rho^{[i]}, A^{[i]}, I^{[i]})$ in an obvious sense. Thus it is enough to find all "primary solutions". Similarly, when we are dealing with the problem over $\boldsymbol{Q}$, we may restrict ourselves to the $\boldsymbol{Q}$-primary case.

Now, let (V, ρ, A, I) be a primary solution (over $\boldsymbol{R}$). If ρ belongs to the trivial representation, then one has $d\rho=0$ and the condition (4. 5) is void. In this case, one can take as (A, I) an arbitrary pair of a non-degenerate alternating bilinear form A and a complex structure I on V satisfying the condition (4. 4). Such a solution is called a *trivial solution*; the corresponding equivariant map of $\mathcal{D}$ into $\mathcal{D}'=\mathfrak{S}(V, A)$ is a constant map.

So suppose that ρ belongs to a non-trivial irreducible representation (V_1, ρ_1) of G and let the notation $D_1, U_1, \cdots$ be as in § 1. Then we know that ρ_1 is self-dual and D_1 has an involution ι_0. In the case of $\boldsymbol{R}$-type $(D_1=F_1=\boldsymbol{R})$, one has $\iota_0=id$. In the case of $\boldsymbol{H}$-type $(D_1=\boldsymbol{H}, F_1=\boldsymbol{R})$, ι_0 is of the first kind, so that we may choose ι_0 to be the "standard" involution of $\boldsymbol{H}$. In the case of $\boldsymbol{C}$-type $(D_1=F_1=\boldsymbol{C})$, ι_0 is either the identity or the complex conjugation. Actually, as we shall see below, the first case never occurs. Thus, in all cases, we may assume that ι_0 is *the* standard involution of D_1 and $F_{10}=\boldsymbol{R}$. It will then be allowed to identify $\bar{D}_1(={}^tD_1)$ with D_1 by setting $\bar{\alpha}(={}^t\alpha)=\alpha^{\iota_0}$, in accordance with the usual convention of denoting the standard involution by $\alpha\mapsto\bar{\alpha}$. The formula (2. 12) then becomes

$$\psi_1\circ\bar{\alpha}=\bar{\alpha}\circ\psi_1 \qquad (\alpha\in D_1),$$

i. e., the G-isomorphism $\psi_1: V\to V^*$ is also a D_1-isomorphism.

We first consider the case $d\rho_1(H_0)\neq 0$. In this case, by the condition (4. 5) one has

$$\frac{1}{2}I=\varphi\otimes_{F_1}1_{V_1}+1_{U_1}\otimes_{F_1}d\rho_1(H_0) \tag{4. 8}$$

with $\varphi\in\mathrm{End}_{\bar{D}_1}(U_1)$. Taking the square of both sides, one has

$$-\frac{1}{4}1_V=\varphi^2\otimes_{F_1}1_{V_1}+2\varphi\otimes_{F_1}d\rho_1(H_0)+1_{U_1}\otimes_{F_1}d\rho_1(H_0)^2,$$

or

$$-1_{U_1}\otimes_{F_1}d\rho_1(H_0)^2=2\varphi\otimes_{F_1}d\rho_1(H_0)+\left(\varphi^2+\frac{1}{4}1_{U_1}\right)\otimes_{F_1}1_{V_1}.$$

Since $\mathrm{tr}_{V_1/F_1}d\rho_1(H_0)=0$, the linear transformations $d\rho_1(H_0)$ and 1_{V_1} are linearly independent over F_1. Hence, writing

$$-d\rho_1(H_0)^2=\lambda_{V_1}+\mu d\rho_1(H_0)$$

with $\lambda, \mu\in F_1$, one obtains

$$\lambda_{U_1}=\varphi^2+\frac{1}{4}1_{U_1}, \qquad \mu_{U_1}=2\varphi,$$

whence $\lambda=\frac{1}{4}(\mu^2+1)$. Since $\mathrm{tr}_{V/\boldsymbol{R}}I=0$, one has $\mathrm{tr}_{U_1/\boldsymbol{R}}\varphi=0$. This implies

$$\begin{cases}\mu=0 & \text{if } \rho_1 \text{ is of } \boldsymbol{R}\text{-type or } \boldsymbol{H}\text{-type,}\\ \mu \text{ is purely imaginary} & \text{if } \rho_1 \text{ is of } \boldsymbol{C}\text{-type.}\end{cases} \tag{4. 9}$$

In either case, if we put

$$I_1=2d\rho_1(H_0)+\mu_{V_1}, \tag{4. 10}$$

then I_1 is a complex structure on V_1 and we have

$$I=1_{U_1}\otimes_{F_1}I_1. \tag{4. 11}$$

Now by Theorem 2. 3 we have a $(\bar{D}_1, \varepsilon)$-hermitian form h on $V_1\times V_1$ and a $(D_1, -\varepsilon)$-hermitian form h' on $U_1\times U_1$ such that

$$A(u\otimes_{D_1}v, u'\otimes_{D_1}v')=\mathrm{tr}_{D_1/\boldsymbol{R}}(h'(u,u')\overline{h(v,v')}),$$

or symbolically

$$A=\mathrm{tr}_{D_1/\boldsymbol{R}}(h'\otimes\bar{h}). \tag{4. 12}$$

Then by (4. 11) one has

$$A(u\otimes_{D_1}v, I(u'\otimes_{D_1}v'))=\mathrm{tr}_{D_1/\boldsymbol{R}}(h'(u,u')\overline{h(v,I_1v')})$$

or symbolically

$$AI=\mathrm{tr}_{D_1/\boldsymbol{R}}(h'\otimes\overline{hI_1}). \tag{4. 13}$$

Since AI is symmetric and positive definite, we must have $\varepsilon=-1$ and both hI_1 and h' are hermitian and definite in the same sign (i. e., either both positive definite or both negative definite). Replacing simultaneously h and h' by $-h$ and $-h'$ if necessary, we may always assume that they are both positive definite. In particular, in the case of $\boldsymbol{C}$-type, ih and h' are hermitian in the usual sense and $hI_1\gg 0$, $h'\gg 0$. Thus the involution ι_0 must be the complex conjugation, as we have already mentioned. (Hence from now on we can legitimately make the identification $\bar{D}_1=D_1$.) Note also that, in the case of $\boldsymbol{C}$-type, if the signature of ih is (p_1, q_1), then $\mathrm{tr}_{V_1/F_1} I_1 = i(p_1-q_1)$ and hence

$$\mu = i\frac{p_1-q_1}{p_1+q_1}. \tag{4.14}$$

Thus we have obtained the following proposition.

Proposition 4. 3. *Let (V, ρ, A, I) be a primary solution of our problem with $d\rho(H_0)\neq 0$. Then, in terms of the tensor product decomposition $V=U_1\otimes_{D_1}V_1$, A and I are given by* (4. 12) *and* (4. 11), *where I_1 is a complex structure on V_1, uniquely determined by ρ_1 and H_0, h is a D_1-valued, G-invariant, skew-hermitian form on $V_1\times V_1$ such that $hI_1\gg 0$, also unique up to a positive scalar multiple, and h' is an arbitrary D_1-valued positive definite hermitian form on $U_1\times U_1$.*

In this case, in the notation of Proposition 4. 1 with $G'=Sp(V, A)$, one has $C_\rho = U(U_1, h')$, which is compact, so that the solution is "rigid". On the other hand, the special unitary group $G_1'=SU(V_1, h)$ is a "classical" simple $\boldsymbol{R}$-group of hermitian type with an H-element $H_{01}'=\frac{1}{2}(I_1-\mu_{V_1})$, and one has an (H_2)-homomorphism

$$\rho_1 : G \longrightarrow G_1' = SU(V_1, h) \tag{4.15}$$

with respect to the H-elements H_0 and H_{01}'.

In the case $d\rho_1(H_0)=0$, the relation (4. 8) reduces to the form

$$I = I_1'\otimes_{F_1} 1_{V_1}, \tag{4.16}$$

where I_1' is a complex structure on U_1. Hence instead of (4. 13) we obtain

$$AI = \mathrm{tr}_{D_1/\boldsymbol{R}}(h'I_1'\otimes h) \gg 0.$$

This implies that $\varepsilon=1$ and one may assume

$$h\gg 0, \qquad h'I' \gg 0. \tag{4.17}$$

For the given ρ_1, h is uniquely determined up to a positive scalar, but h' and I_1' can be taken arbitrarily subject to the above conditions. The group $C_\rho = U(U_1, h')$ is non-compact except for the case where $D_1=\boldsymbol{C}$ and ih' is definite. The irreducible representation ρ_1 gives an $\boldsymbol{R}$-homomorphism of G into a compact classical group $G_1'=SU(V_1, h)$, which satisfies the condition (H_2) in a trivial sense. Thus in both cases, the determination of primary solutions is reduced to that of irreducible $\boldsymbol{R}$-representations

$$\rho_1 : G \longrightarrow G_1' = SU(V_1, h)$$

satisfying the condition (H_2).

The following proposition shows that, in most cases, the equivalence class of a solution (V, ρ, A, I) is determined only by that of ρ.

Proposition 4. 4. *Two solutions (V, ρ, A, I) and (V', ρ', A', I'), containing no primary component of* **C**-*type with $d\rho_1(H_0)=0$, are equivalent in the sense of* (4. 6) *if and only if $\rho\sim\rho'$.*

Proof. The "only if" part is obvious, and the proof of the "if" part can easily be reduced to the primary case. Hence, suppose that, in the given two solutions, ρ and ρ' are equivalent primary representations of G belonging to the same irreducible representation $(V_., \rho_1)$. Then, we may write

$$V = U_1 \otimes_{D_1} V_1, \qquad V' = U_1' \otimes_{D_1} V_1.$$

Any G-isomorphism φ of V onto V' can be written as $\varphi=\varphi_1\otimes_{F_1}1_{V_1}$ with a D_1-isomorphism φ_1 of U_1 onto U_1'. Let

$$A = \mathrm{tr}(h'\otimes \bar{h}), \qquad A' = \mathrm{tr}(h''\otimes \bar{h})$$

be the expressions of alternating forms similar to (4. 12). Then the D_1-valued forms h' and h'' are both hermitian positive definite if $d\rho_1(H_0)\neq 0$ and both skew-hermitian if $d\rho_1(H_0)=0$. So, in either case, except for the case of **C**-type with $d\rho_1(H_0)=0$, one can find φ_1 such that

$$h'' = h'\circ(\varphi_1^{-1}\times\varphi_1^{-1})$$

(cf. Appendix, § 3, Exerc. 2). Then one has

$$A' = A\circ(\varphi^{-1}\times\varphi^{-1}).$$

In the case $d\rho_1(H_0)\neq 0$, the relation $I'=\varphi\circ I\circ\varphi^{-1}$ is automatically satisfied by (4. 10), (4. 11). In the case $d\rho_1(H_0)=0$, let

$$I = I_1'\otimes_{F_1}1_{V_1}, \qquad I' = I_1''\otimes_{F_1}1_{V_1}$$

be the expressions of complex structures similar to (4. 16). Then the complex structures I_1'' and $\varphi_1 I_1'\varphi_1^{-1}$ of U_1' correspond to two points in the symmetric domain associated with $SU(U_1', h'')$ (Appendix, § 3). Hence there exists $g_1'\in SU(U_1', h'')$ such that $I_1''=g_1'(\varphi_1 I_1'\varphi_1^{-1})g_1'^{-1}$. Replacing φ_1 by $g_1'\varphi_1$, one obtains the relation $I_1''=\varphi_1 I_1'\varphi_1^{-1}$ and hence $I'=\varphi I\varphi^{-1}$. Then φ gives an equivalence of the two solutions, q. e. d.

As an immediate consequence, we obtain

Corollary 4. 5. *Let (V, ρ, A, I) and (V, ρ', A, I') be two solutions with the same (V, A), containing no primary component of* **C**-*type with $d\rho_1(H_0)=0$. If $\rho\sim\rho'$, then ρ and ρ' are equivalent in $Sp(V, A)$, i. e., there exists $g_1'\in Sp(V, A)$ such that $\rho'(g)=g_1'\rho(g)g_1'^{-1}$ for all $g\in G$.*

When (V, ρ) is a primary representation of C-type with $d\rho_1(H_0)=0$, two solutions (V, ρ, A, I) and (V, ρ, A', I') are equivalent if and only if the signatures of ih' in (4. 17) coincide. In particular, the solutions (V, ρ, A, I) and $(V, \rho, -A, -I)$ are *not* equivalent.

Exercise

1. An equivalence φ of a solution (V, ρ, A, I) to itself is called an *automorphism* of the solution. Show that the automorphism group $\mathrm{Aut}(V, \rho, A, I)$ is compact. In the primary case, show that

$$\begin{aligned}\mathrm{Aut}(V, \rho, A, I) &= \{g' \in C_\rho \mid g'I = Ig'\} \\ &= \begin{cases} U(U_1, h') & \text{if } d\rho(H_0) \neq 0, \\ U(U_1, h') \cap U(U_1, h'I_1') & \text{if } d\rho(H_0) = 0. \end{cases}\end{aligned}$$

§ 5. Further reductions and the determination of solutions.

We shall now reduce the problem to the case where $\mathfrak{g}$ is simple and non-compact. Let

$$\mathfrak{g} = \bigoplus_{i=0}^{s} \mathfrak{g}_i \tag{5. 1}$$

be the direct sum decomposition of $\mathfrak{g}$ where $\mathfrak{g}_0$ is compact and $\mathfrak{g}_i\,(1 \leq i \leq s)$ is simple and non-compact. Let

$$H_0 = \sum_{i=1}^{s} H_{0i}, \qquad H_{0i} \in \mathfrak{g}_i$$

be the corresponding decomposition of the H-element H_0. Then each $\mathfrak{g}_i$ is of hermitian type with H-element H_{0i}.

When $d\rho_1(H_0)=0$, the image $d\rho_1(\mathfrak{g})$, being a semi-simple subalgebra of a compact Lie algebra $\mathfrak{su}(V_1, h)$, is compact. Hence one has $d\rho_1(\mathfrak{g}_i)=0$ for all $1 \leq i \leq s$, so that $d\rho_1$ is essentially a representation of the compact part $\mathfrak{g}_0$. When $d\rho_1(H_0) \neq 0$, there exists an index i such that $d\rho_1(H_{0i}) \neq 0$. Rearranging the indices if necessary, we may assume $d\rho_1(H_{01}) \neq 0$. Under this assumption, we claim that $d\rho_1(\mathfrak{g}_i)=0$ for $2 \leq i \leq s$.

To prove this, we put

$$\mathfrak{g}^{(2)} = \mathfrak{g}_0 \oplus \sum_{i=2}^{s} \mathfrak{g}_i, \qquad H_0^{(2)} = \sum_{i=2}^{s} H_{0i}$$

and apply (the Lie algebra version of) the result in § 1 to the direct sum decomposition $\mathfrak{g}=\mathfrak{g}_1 \oplus \mathfrak{g}^{(2)}$. Let $(V_{11}, d\rho_{11})$ be an irreducible representation of $\mathfrak{g}_1$ contained in the primary representation $d\rho_1|\mathfrak{g}_1$ and set

$$\begin{aligned} D_{11} &= \mathrm{End}_{\mathfrak{g}_1}(V_{11}), \qquad F_{11} = \mathrm{Cent}\, D_{11}, \\ U_{11} &= \mathrm{Hom}_{\mathfrak{g}_1}(V_{11}, V_1). \end{aligned}$$

Then one has $V_1 = U_{11} \otimes_{D_{11}} V_{11}$ and

$$d\rho_1(X_1 + X^{(2)}) = 1_{U_{11}} \otimes_{F_{11}} d\rho_{11}(X_1) + d\rho^{(2)}(X^{(2)}) \otimes_{F_{11}} 1_{V_{11}} \qquad (X_1 \in \mathfrak{g}_1,\ X^{(2)} \in \mathfrak{g}^{(2)}),$$

where $d\rho^{(2)}$ is a primary representation of $\mathfrak{g}^{(2)}$ on U_{11}. In particular, one has

(5. 2) $$d\rho_1(H_0) = 1_{U_{11}} \otimes_{F_{11}} d\rho_{11}(H_{01}) + d\rho^{(2)}(H_0^{(2)}) \otimes_{F_{11}} 1_{V_{11}}.$$

Hence, in view of (4. 10), $d\rho_1|\mathfrak{g}_1$ satisfies the condition (4. 5) with respect to H_{01} and I_1. Therefore, by Proposition 4. 3, we obtain

(5. 3) $$\begin{cases} I_1 = 1_{U_{11}} \otimes_{F_{11}} I_{11}, \\ I_{11} = 2\,d\rho_{11}(H_{01}) + \mu'_{V_{11}} \end{cases}$$

with $\mu' \in F_{11}$. From (4. 10), (5. 2), (5. 3) we conclude

(5. 4) $$2\,d\rho^{(2)}(H_0^{(2)}) = \mu'_{U_{11}} - \mu_{U_{11}}.$$

[Recall that μ acts on U_{11} through the isomorphism $D_1 \cong \mathrm{End}_{\mathfrak{g}', D_{11}}(U_{11})$.]

Let $(V_{10}, d\rho_1^{(2)})$ be an irreducible representation of $\mathfrak{g}^{(2)}$ contained in $(U_{11}, d\rho^{(2)})$. Then one has

$$U_{11} = W \otimes_{D_{10}} V_{10},$$

where $D_{10} = \mathrm{End}_{\mathfrak{g}^{(2)}}(V_{10})$ and $W = \mathrm{Hom}_{\mathfrak{g}^{(2)}}(V_{10}, U_{11})$. Since the trace of $d\rho^{(2)}(H_0^{(2)})$ taken over any composite of F_1 and F_{11} (in $\mathrm{End}(U_{11})$) vanishes, the relation (5. 4) implies $\mu_{U_{11}} = \mu'_{U_{11}}$, i. e., $d\rho^{(2)}(H_0^{(2)}) = 0$ and hence $d\rho_1^{(2)}(H_0^{(2)}) = 0$. It follows, as we observed first, that $d\rho_1^{(2)}(\mathfrak{g}_i) = 0$ for all $2 \leq i \leq s$, which proves our claim. Thus, in this case, $d\rho_1$ *is essentially an irreducible representation of* $\mathfrak{g}_0 \oplus \mathfrak{g}_1$.

Let h_0 and h_1 denote the corresponding invariant hermitian forms on $V_{10} \times V_{10}$ and $V_{11} \times V_{11}$. Then, by Lemma 2. 4, we have

(5. 5) $$h(w(v_0 \otimes v_1), w'(v'_0 \otimes v'_1)) = \Psi(w) \circ (h_0(v_0, v'_0) \otimes h_1(v_1, v'_1)) \circ {}^t w'$$
$$(v_0, v'_0 \in V_{10},\ v_1, v'_1 \in V_{11},\ w, w' \in W)$$

and hence by (5. 3)

$$h(w(v_0 \otimes v_1), I_1 w'(v'_0 \otimes v'_1)) = \Psi(w) \circ (h_0(v_0, v'_0) \otimes h_1(v_1, I_{11} v'_1)) \circ {}^t w'.$$

Since $hI_1 \gg 0$, one may assume that

(5. 6) $$\Psi \gg 0, \qquad h_0 \gg 0, \qquad h_1 I_{11} \gg 0,$$

where $\Psi \gg 0$ means that the corresponding bilinear form on $W \times W$ is symmetric and positive definite. Thus the determination of ρ_1 reduces to that of the irreducible representation ρ_{11} of a simple non-compact $\boldsymbol{R}$-group G_1 of hermitian type:

(5. 7) $$\rho_{11}:\ G_1 \longrightarrow SU(V_{11}, h_1)$$

satisfying the condition (H_2). In fact, having such a representation ρ_{11} satisfying (H_2) with respect to H_{01} and $H'_{01} = \frac{1}{2} I_{11} - \mu'_{V_{11}}$, one can construct an irreducible representation ρ_1 of G ((4. 15)) by the following procedure:

1) Take any irreducible representation (V_{10}, ρ_{10}) of the compact group G_0 and a D_{10}-valued G_0-invariant positive definite hermitian form h_0 on $V_{10} \times V_{10}$.

2) Take a simple $D_{10} \otimes D_{11}$-module W satisfying the condition

$$(\rho_{10}(g))_W = (\rho_{11}(g))_W \qquad \text{for} \quad g \in G_0 \cap G_1,$$

and let Ψ be a $D_{10} \otimes D_{11}$-isomorphism of W onto W^* such that $\Psi = {}^t\Psi \gg 0$. (Such Ψ is unique up to a positive scalar multiple.)

3) Let ρ_1 be an irreducible representation of G on $V_1 = (W \otimes_{D_{10}} V_{10}) \otimes_{D_{11}} V_{11}$ obtained

from ρ_{10} and ρ_{11}, and define a D_1-valued skew-hermitian form h on $V_1 \times V_1$ by (5. 5) and a complex structure I_1 on V_1 by (5. 3). Then the quadruple (V_1, ρ_1, h, I_1) satisfies all the requirements in Proposition 4. 3.

Thus we have shown that the whole problem (over $\boldsymbol{R}$) is reduced to the determination of the ($\boldsymbol{R}$-)irreducible representation ρ_{11} satisfying the condition (H_2). Hence, for the sake of simplicity, we shall now assume that G is simple and non-compact (i. e., $G=G_1$ in the above notation, and so $V_1=V_{11}$, $\rho_1=\rho_{11}$, etc.) Let ρ_{1C} denote the complex representation of G_C on V_{1C} obtained from ρ_1 by scalar extension and let $(\tilde{V}_1, \tilde{\rho}_1)$ be an (absolutely) irreducible representation contained in ρ_{1C}. When ρ_1 is of $\boldsymbol{R}$-type, one has $\rho_{1C}=\tilde{\rho}_1$; when ρ_1 is of $\boldsymbol{C}$-type or $\boldsymbol{H}$-type, one has $\rho_{1C}\sim 2\tilde{\rho}_1$ or $\sim\tilde{\rho}_1 \oplus \mathrm{conj}(\tilde{\rho}_1)$ accordingly (§ 3, Ex. 2).

As in II, § 4, take a Cartan subalgebra $\mathfrak{h}$ of $\mathfrak{g}$ contained in $\mathfrak{k}$ and let $\{\alpha_1, \cdots, \alpha_l\}$ be a fundamental system of roots relative to $\mathfrak{h}_C$ such that $\alpha_1(H_0)=\sqrt{-1}$ and $\alpha_j(H_0)=0$ $(2\leq j\leq l)$. A "weight" λ of $\tilde{\rho}_1$ relative to $\mathfrak{h}_C$ is by definition a linear form on $\mathfrak{h}_C$ for which there exists a non-zero vector v in $\tilde{V}_1$ such that

$$d\tilde{\rho}_1(X)v = \lambda(X)v \qquad \text{for all} \quad X \in \mathfrak{h}_C.$$

As is well-known, the (absolutely) irreducible representation $\tilde{\rho}_1$ is uniquely determined (up to equivalence) by the "highest weight" λ_1, which satisfies the condition

$$(5.\ 8) \qquad \frac{2\langle\lambda_1, \alpha_j\rangle}{\langle\alpha_j, \alpha_j\rangle} \text{ is a non-negative integer} \qquad \text{for all } 1\leq j\leq l.$$

Now, if ρ_1 is an (H_2)-homomorphism into $SU(V_1, h)$, then from the relation (4. 10) and (4. 9), (4. 14) one has for any weight λ

$$\lambda(H_0) = \begin{cases} \pm\dfrac{i}{2} & \text{if } \rho_1 \text{ is of } \boldsymbol{R}\text{-type or } \boldsymbol{H}\text{-type,} \\ \dfrac{iq_1}{p_1+q_1} \quad \text{or} \quad \dfrac{-ip_1}{p_1+q_1} & \text{if } \rho_1 \text{ is of } \boldsymbol{C}\text{-type.} \end{cases}$$

In particular, for the highest weight one has

$$(5.\ 9) \qquad -i\lambda_1(H_0) \text{ is a positive (rational) number} < 1.$$

By virtue of the (necessary) conditions (5. 8), (5. 9), it is easy to determine all possible irreducible representations $\tilde{\rho}_1$ for each simple non-compact Lie algebra $\mathfrak{g}$ of hermitian type. We give the result in the following table. For the detail, the reader is referred to Satake [4]. The case $\mathfrak{g}=(IV_p)$ will be treated in Appendix, § 6. When $\mathfrak{g}$ is of exceptional types (V) and (VI), there are *no* solutions (cf. Exerc. 2).

In this table, $\bigwedge^k$ denotes the skew-symmetric tensor representation on the k-fold exterior product $\bigwedge^k(\boldsymbol{C}^{p+q})$ (cf. Exerc. 1). In the cases (II_n), (III_n), $\tilde{\rho}_1=id$ means the natural injections $SU^-(n, \boldsymbol{H})\rightarrow SU(n, n)$, $Sp(2n, \boldsymbol{R})\rightarrow SU(n, n)$ given in Appendix, § 3 and II, § 7.

Type	G		$\tilde{\rho}_1$	$\dim_C \tilde{V}_1$	D_1
$(I_{p,q})$	$SU(p,q)$	$p\geq q\geq 2$	id, $\overline{id}$	$p+q$	$\boldsymbol{C}$
		$p\geq q=1$	$\wedge^k$, $k=\frac{p+1}{2}$	$\binom{p+1}{(p+1)/2}$	$\boldsymbol{R}$ for $p\equiv 1$ (4) $\boldsymbol{H}$ for $p\equiv 3$ (4)
			$\wedge^k$, $k\neq\frac{p+1}{2}$	$\binom{p+1}{k}$	$\boldsymbol{C}$
(II_n)	$SU^-(n,\boldsymbol{H})$	$n\geq 5$	id	$2n$	$\boldsymbol{H}$
(III_n)	$Sp(2n,\boldsymbol{R})$	$n\geq 1$	id	$2n$	$\boldsymbol{R}$
(IV_p)	$Spin(p,2)$	$p\geq 1$ odd	spin rep.	$2^{(p+1)/2}$	$\boldsymbol{R}$ for $p\equiv 1,3$ (8) $\boldsymbol{H}$ for $p\equiv 5,7$ (8)
		$p\geq 4$ even	two spin rep.	$2^{p/2}$	$\boldsymbol{R}$ for $p\equiv 2$ (8) $\boldsymbol{H}$ for $p\equiv 6$ (8) $\boldsymbol{C}$ for $p\equiv 0$ (4)

We note that, as special cases of the above representations $\wedge^k$ and the spin representations, the following well-known isomorphisms between irreducible symmetric domains are obtained:

$$\begin{array}{lll} (I_{1,1}) \cong & (III_1) \cong & (IV_1) \\ & (III_2) \cong & (IV_3) \\ (I_{2,2}) & \cong & (IV_4) \\ (I_{3,1}) \rightleftarrows & (II_3) & \\ & (II_4) \rightleftarrows & (IV_6) \end{array}$$

The symbol $\rightleftarrows$ indicates that, for the corresponding groups in the above list, one has an isogeny of degree 2. By virtue of these isogenies and the trivial isomorphism $(I_{p,q})\cong(I_{q,p})$, we have simplified the above table by excluding $(I_{p,q})$ $(p<q)$ and (II_n) $(n\leq 4)$.

Exercises

1. Let V be a $(p+1)$-dimensional vector space over $\boldsymbol{C}$ with a hermitian form h of signature $(p,1)$ and let $G=SU(V,h)$. Let $\wedge(V)=\bigoplus_{k=0}^{p+1}\wedge^k(V)$ be the exterior algebra of V and define a hermitian form $h^{(k)}$ on $\wedge^k(V)$ by putting

(5.10) $$h^{(k)}(x_1\wedge\cdots\wedge x_k, y_1\wedge\cdots\wedge y_k) = \det(h(x_i,y_j)) \qquad (x_i, y_j\in V).$$

Show that $h^{(k)}$ is an invariant hermitian form on $\wedge^k(V)$ of signature $\left(\binom{p}{k},\binom{p}{k-1}\right)$. Show also that $d(\wedge^k):\mathfrak{g}=\mathfrak{su}(V,h)\to\mathfrak{su}(\wedge^k(V),h^{(k)})$ satisfies the condition (H_2).

2. Let $\mathfrak{g}$ be a simple non-compact Lie algebra of hermitian type, and let (c_{ij}) denote the "inverse Cartan matrix", i.e., $\sum_j c_{ij}\frac{2\langle\alpha_j,\alpha_k\rangle}{\langle\alpha_k,\alpha_k\rangle}=\delta_{ik}$. As in the text, α_1 is the unique non-compact root.

2.1) Putting $r_j=\frac{2\langle\lambda_1,\alpha_j\rangle}{\langle\alpha_j,\alpha_j\rangle}$, show that

$$-i\lambda_1(H_0) = \sum_{j=1}^{l} r_j c_{j1}.$$

Thus the conditions (5. 8), (5. 9) amount to $r_j \in \boldsymbol{Z}$ and the inequalities:

(5. 11) $$0 < \sum_j r_j c_{j1} < 1, \quad r_j \geq 0.$$

2. 2) For the exceptional Lie algebra $\mathfrak{g} = (VI)$, compute the inverse Cartan matrix and show that there are no solutions for (5. 11). (For the location of α_1 in $\tilde{\Delta}$, see the diagram on p. 117 where it is marked n.)

§6. The solutions over $\boldsymbol{Q}$.

Let G be a Zariski connected semi-simple $\boldsymbol{R}$-group of hermitian type defined over $\boldsymbol{Q}$. In this section, we shall consider the problem of finding solutions "over $\boldsymbol{Q}$", i. e., solutions (V, ρ, A, I) in which the triple (V, ρ, A) is defined over $\boldsymbol{Q}$. (This means that V is given a "$\boldsymbol{Q}$-structure", i. e., a $\boldsymbol{Q}$-form $V_{\boldsymbol{Q}}$ with $V = V_{\boldsymbol{Q}} \otimes_{\boldsymbol{Q}} \boldsymbol{R}$, and ρ and A are obtained by scalar extension $\boldsymbol{R}/\boldsymbol{Q}$ from a representation $\rho_{\boldsymbol{Q}}$ of $G_{\boldsymbol{Q}}$ on $V_{\boldsymbol{Q}}$ and an alternating bilinear form $A_{\boldsymbol{Q}}$ on $V_{\boldsymbol{Q}} \times V_{\boldsymbol{Q}}$, respectively.) As we have already mentioned in §4, such a solution is the direct sum of $\boldsymbol{Q}$-primary ones, so that it is sufficient to consider the $\boldsymbol{Q}$-primary case, i. e., the case where $(V_{\boldsymbol{Q}}, \rho_{\boldsymbol{Q}})$ is primary.

So let (V, ρ, A, I) be a (non-trivial) $\boldsymbol{Q}$-primary solution, (V_1, ρ_1) a (non-trivial) $\boldsymbol{Q}$-irreducible representation of $G_{\boldsymbol{Q}}$ contained in $(V_{\boldsymbol{Q}}, \rho_{\boldsymbol{Q}})$, and let

(6. 1) $$V_{\boldsymbol{Q}} = U_1 \otimes_{D_1} V_1$$

be the canonical tensor product decomposition. Then, since $\rho_{\boldsymbol{Q}}$ is self-dual, so is ρ_1, and one has a $G_{\boldsymbol{Q}}$-isomorphism $\psi_1 : V_1 \to V_1^*$ with ${}^t\psi_1 = \varepsilon\psi_1$; let ι_1 be the involution of $\mathrm{End}(V_1)$ defined by ψ_1 and let $\iota_0 = \iota_1 | D_1$. Then, in the notation of §3 ($F = \boldsymbol{Q}$, $F' = \boldsymbol{R}$), one has

(6. 2) $$F_1 \otimes_{\boldsymbol{Q}} \boldsymbol{R} = \bigoplus_{\nu=1}^{t} F_1^{(\nu)},$$

(6. 3) $$D_1 \otimes_{\boldsymbol{Q}} \boldsymbol{R} = \bigoplus_{\nu=1}^{t} D_1 \otimes_{F_1} F_1^{(\nu)} \cong \bigoplus_{\nu=1}^{t} \mathcal{M}_{s_\nu}(D_1^{(\nu)}),$$

where $D_1^{(\nu)}$ is either $\boldsymbol{R}$ or $\boldsymbol{C}$ or $\boldsymbol{H}$. Viewing $F_1^{(\nu)}$ as subalgebras of $\mathrm{End}(V)$, one obtains the primary decomposition of V over $\boldsymbol{R}$:

(6. 4) $$V = \bigoplus_{\nu=1}^{t} V^{[\nu]},$$
$$V^{[\nu]} = F_1^{(\nu)} V = U_1^{(\nu)} \otimes_{D_1^{(\nu)}} V_1^{(\nu)}.$$

By Proposition 4. 1, this gives a decomposition of (V, ρ, A, I) into the direct sum of $\boldsymbol{R}$-primary solutions $(V^{[\nu]}, \rho^{(\nu)}, A^{(\nu)}, I^{(\nu)})$. For every ν, the representation $\rho^{(\nu)}$ and the irreducible representation $\rho_1^{(\nu)}$ contained in $\rho^{(\nu)}$ are self-dual, and $F_1^{(\nu)}$ is stable under (the unique extension of) the involution ι_0. We know that $(F_1^{(\nu)})_0$, the fixed field of $\iota_0 | F_1^{(\nu)}$, coincides with $\boldsymbol{R}$.

First we consider the case where ι_0 is of the first kind. In this case, all the $F_1^{(\nu)}$ are equal to $\boldsymbol{R}$, i. e., F_1 is "totally real". More precisely, in (6. 2) one has $t = d_1 = [F_1 : \boldsymbol{Q}]$, and the components $F_1^{(\nu)} (1 \leq \nu \leq d_1)$ correspond to the d_1 distinct monomorphisms σ_ν of F_1 into $\boldsymbol{R}$ by the relation:

$$(6.5)\qquad (\alpha\otimes 1)_{V^{(\nu)}} = \alpha^{\sigma_\nu} 1_{V^{(\nu)}} \qquad (\alpha\in F_1).$$

Since the degree of D_1 over F_1 is at most 2, there are two possibilities.

Case R 1 : $D_1=F_1$, $\iota_0=id$.

Case R 2 : D_1=a central division quaternion algebra over F_1, ι_0=the standard involution of D_1.

In Case R 1, in view of (6. 5), $V^{(\nu)}$, $V_1^{(\nu)}$, $U_1^{(\nu)}$ may be considered as real vector spaces obtained from $(V_Q)^{\sigma_\nu}$, $V_1^{\sigma_\nu}$, $U_1^{\sigma_\nu}$, respectively, by scalar extension $\boldsymbol{R}/F_1^{\sigma_\nu}$. Thus in (6. 4) one has

$$(6.6)\qquad D_1^{(\nu)} = F_1^{(\nu)} = \boldsymbol{R}, \qquad V_1^{(\nu)} = (V_1^{\sigma_\nu})_{\boldsymbol{R}}, \qquad U_1^{(\nu)} = (U_1^{\sigma_\nu})_{\boldsymbol{R}}.$$

We denote by $\rho_1^{(\nu)}=(\rho_1^{\sigma_\nu})_{\boldsymbol{R}}$ the real representation of G on $V_1^{(\nu)}$ obtained from $\rho_1^{\sigma_\nu}$ by scalar extension $\boldsymbol{R}/F_1^{\sigma_\nu}$; then, since $D_1^{(\nu)}=\boldsymbol{R}$, $\rho_1^{(\nu)}$ is absolutely irreducible. From $\rho_Q=1_{U_1}\otimes\rho_1$, one has

$$(6.7)\qquad \rho^{(\nu)} = 1_{U_1^{(\nu)}}\otimes\rho_1^{(\nu)}.$$

Since the alternating bilinear form A_Q on $V_Q\times V_Q$ is G_Q-invariant, we can write

$$(6.8)\qquad A_Q = \mathrm{tr}_{F_1/Q}(S_1\otimes A_1),$$

where A_1 is a non-degenerate G_Q-invariant alternating (F_1-)bilinear form on $V_1\times V_1$ and S_1 is a non-degenerate symmetric (F_1-)bilinear form on $U_1\times U_1$. Denoting the unique $\boldsymbol{R}$-bilinear extensions of $A_1^{\sigma_\nu}$ and $S_1^{\sigma_\nu}$ by the same letters, we obtain the corresponding decomposition

$$(6.9)\qquad A^{(\nu)} = S_1^{\sigma_\nu}\otimes A_1^{\sigma_\nu}.$$

Finally we have

$$(6.10)\qquad I^{(\nu)} = 1_{U_1^{(\nu)}}\otimes I_1^{(\nu)},$$

where $I_1^{(\nu)}=2d\rho_1^{(\nu)}(H_0)$ is a complex structure on $V_1^{(\nu)}$ satisfying the condition

$$S_1^{\sigma_\nu}\otimes(A_1^{\sigma_\nu}I_1^{(\nu)})\gg 0.$$

Hence, for each ν, $S_1^{\sigma_\nu}$ and $A_1^{\sigma_\nu}I_1^{(\nu)}$ are definite in the same sign. By a well-known theorem in algebraic number theory, one can find an element $\alpha\in F_1^\times$ such that $(\alpha S_1)^{\sigma_\nu}\gg 0$ for all ν. Hence, replacing S_1 and A_1 by αS_1 and $\alpha^{-1}A_1$ if necessary, we may always assume that

$$(6.11)\qquad S_1^{\sigma_\nu}\gg 0, \qquad A_1^{\sigma_\nu}I_1^{(\nu)}\gg 0 \qquad (1\le\nu\le d_1).$$

A symmetric F_1-bilinear form S_1 satisfying (6. 11) is called "totally positive".

Summing up, we obtain the following theorem.

Theorem 6. 1. *In Case* R 1, *a* $\boldsymbol{Q}$*-primary solution* (V, ρ, A, I) *is obtained as follows. Let* F_1 *be a totally real number field of degree* d_1, $\{\sigma_1, \cdots, \sigma_{d_1}\}$ *the set of all monomorphisms of* F_1 *into* $\boldsymbol{R}$, V_1 *a vector space over* F_1, *and* A_1 *a non-degenerate alternating* F_1*-bilinear form on* $V_1\times V_1$. *Let*

$$(6.12)\qquad \rho_1: G_Q \longrightarrow Sp(V_1, A_1)$$

be an absolutely irreducible symplectic representation over F_1 *satisfying the following conditions:*

For every ν $(1 \leq \nu \leq d_1)$, the real representation

$$\rho_1^{(\nu)} = (\rho_1^{\sigma_\nu})_{\boldsymbol{R}} : G \longrightarrow Sp(V_1^{(\nu)}, A_1^{(\nu)}),$$

obtained from the conjugate representation $\rho_1^{\sigma_\nu}$ (of $G_{F_1 \sigma_\nu}$) by the scalar extension $\boldsymbol{R}/F_1^{\sigma_\nu}$, satisfies the condition (H$_2$) *with respect to H_0 and $\frac{1}{2}I_1^{(\nu)}$, where $I_1^{(\nu)}$ is a complex structure on $V_1^{(\nu)} = (V_1^{\sigma_\nu})_{\boldsymbol{R}}$ satisfying* (6. 11). *On the other hand, let U_1 be a vector space over F_1 and S_1 a totally positive symmetric F_1-bilinear form on $U_1 \times U_1$. Then, one has*

$$V_{\boldsymbol{Q}} = U_1 \otimes_{F_1} V_1, \qquad A_{\boldsymbol{Q}} = \mathrm{tr}_{F_1/\boldsymbol{Q}}(S_1 \otimes A_1), \qquad \rho_{\boldsymbol{Q}} = 1_{U_1} \otimes \rho_1,$$

and (V, ρ, A, I) is the direct sum of the $\boldsymbol{R}$-primary solutions $(V^{[\nu]}, \rho^{(\nu)}, A^{(\nu)}, I^{(\nu)})$ given by (6. 4), (6. 6), (6. 7), (6. 9), *and* (6. 10).

In this case, it is clear that the $\boldsymbol{Q}$-similarity class of the solution (V, ρ, A, I) is completely determined by the class of the $\boldsymbol{Q}$-irreducible representation (V_1, ρ_1) and the $\boldsymbol{Q}$-similarity class of (U_1, S_1).

In Case R 2, we may assume that

$$D_1^{(\nu)} \begin{cases} \cong \boldsymbol{H} & \text{for } \ 1 \leq \nu \leq d_1' \\ = \boldsymbol{R} & \text{for } \ d_1'+1 \leq \nu \leq d_1. \end{cases} \tag{6. 13}$$

Then, as explained in § 3, one may put

$$V_1^{(\nu)} = \begin{cases} (V_1^{\sigma_\nu})_{\boldsymbol{R}} \\ \varepsilon_{11}^{(\nu)}(V_1^{\sigma_\nu})_{\boldsymbol{R}}, \end{cases} \qquad U_1^{(\nu)} = \begin{cases} (U_1^{\sigma_\nu})_{\boldsymbol{R}} & (1 \leq \nu \leq d_1') \\ \varepsilon_{11}^{(\nu)}(U_1^{\sigma_\nu})_{\boldsymbol{R}} & (d_1'+1 \leq \nu \leq d_1), \end{cases} \tag{6. 14}$$

where the $\varepsilon_{ij}^{(\nu)}$ are the (fixed) matrix units in $D_1 \otimes_{F_1} F_1^{(\nu)} \cong \mathscr{M}_2(\boldsymbol{R})$, viewed as a subalgebra of $\mathrm{End}(V_1^{(\nu)})$. We again obtain (6. 7), where $\rho_1^{(\nu)} = (\rho_1^{\sigma_\nu})_{\boldsymbol{R}}$ for $1 \leq \nu \leq d_1'$ and $\rho_1^{(\nu)}$ is the restriction of $(\rho_1^{\sigma_\nu})_{\boldsymbol{R}}$ to $V_1^{(\nu)}$ for $d_1'+1 \leq \nu \leq d_1$. Let

$$A_{\boldsymbol{Q}} = \mathrm{tr}_{D_1/\boldsymbol{Q}}(h' \otimes h), \tag{6. 15}$$

where h is a non-degenerate $G_{\boldsymbol{Q}}$-invariant (D_1, ε)-hermitian form on $V_1 \times V_1$ and h' is a non-degenerate $(D_1, -\varepsilon)$-hermitian form on $U_1 \times U_1$. Through the fixed isomorphisms (6. 3) and (6. 13), we obtain from h a quaternion ε-hermitian form $h^{(\nu)}$ on $V_1^{(\nu)} \times V_1^{(\nu)}$ for $1 \leq \nu \leq d_1'$ and a real $(-\varepsilon)$-symmetric bilinear form $B_1^{(\nu)}$ $(= S_1^{(\nu)}$ or $A_1^{(\nu)})$ on $V_1^{(\nu)} \times V_1^{(\nu)}$ for $d_1'+1 \leq \nu \leq d_1$. Similarly, we obtain a quaternion $(-\varepsilon)$-hermitian form $h'^{(\nu)}$ or a real ε-symmetric bilinear form $B_1'^{(\nu)}$ $(= S_1'^{(\nu)}$ or $A_1'^{(\nu)})$ on $U_1^{(\nu)} \times U_1^{(\nu)}$ from h', and we have

$$A^{(\nu)} = \begin{cases} \mathrm{tr}_{\boldsymbol{H}/\boldsymbol{R}}(h_1'^{(\nu)} \otimes h^{(\nu)}) & (1 \leq \nu \leq d_1') \\ B_1'^{(\nu)} \otimes B_1^{(\nu)} & (d_1'+1 \leq \nu \leq d_1). \end{cases} \tag{6. 16}$$

Replacing h and h' by scalar multiples αh and $\alpha^{-1} h'$ $(\alpha \in F_1^\times)$ if necessary, we may again assume that the following conditions are satisfied. Here we distinguish two cases R 2. 1, R 2. 2 according as $\varepsilon = \pm 1$.

Case R 2. 1 : $\varepsilon = 1$,

$$h'^{(\nu)} I_1'^{(\nu)} \gg 0, \qquad h^{(\nu)} \gg 0 \qquad \text{for } \ 1 \leq \nu \leq d_1', \tag{6. 17}$$

$$S_1'^{(\nu)} \gg 0, \qquad A_1^{(\nu)} I_1^{(\nu)} \gg 0 \qquad \text{for } \ d_1'+1 \leq \nu \leq d_1. \tag{6. 18}$$

Case R 2.2: $\varepsilon=-1$,

(6.19) $$h'^{(\nu)} \gg 0, \quad h^{(\nu)} I_1^{(\nu)} \gg 0 \quad \text{for} \quad 1 \leq \nu \leq d_1',$$

(6.20) $$A_1'^{(\nu)} I_1'^{(\nu)} \gg 0, \quad S_1^{(\nu)} \gg 0 \quad \text{for} \quad d_1'+1 \leq \nu \leq d_1,$$

and one has

(6.21) $$I^{(\nu)} = \begin{cases} 1_{U_1^{(\nu)}} \otimes I_1^{(\nu)} & \text{for } d_1'+1 \leq \nu \leq d_1 \text{ in Case R 2.1,} \\ & \text{and for } 1 \leq \nu \leq d_1' \text{ in Case R 2.2,} \\ I_1'^{(\nu)} \otimes 1_{V_1^{(\nu)}} & \text{otherwise.} \end{cases}$$

Therefore, putting $G_1'=SU(V_1, h)$, $G_1''=SU(U_1, h')$ (viewed as F_1-groups), one has

(6.22) $$R_{F_1/\mathbf{Q}}(G_1')_{\mathbf{R}} = \prod_\nu G_1'^{(\nu)},$$
$$G_1'^{(\nu)} = SU(V_1^{(\nu)}, h^{(\nu)}) \quad \text{or} \quad Sp(V_1^{(\nu)}, A_1^{(\nu)}) \quad \text{or} \quad SO(V_1^{(\nu)}, S_1^{(\nu)}),$$

(6.23) $$R_{F_1/\mathbf{Q}}(G_1'')_{\mathbf{R}} = \prod_\nu G_1''^{(\nu)},$$
$$G_1''^{(\nu)} = SU(U_1^{(\nu)}, h'^{(\nu)}) \quad \text{or} \quad Sp(U_1^{(\nu)}, A_1'^{(\nu)}) \quad \text{or} \quad SO(U_1^{(\nu)}, S_1'^{(\nu)}).$$

Thus in both cases $R_{F_1/\mathbf{Q}}(G_1')$ and $R_{F_1/\mathbf{Q}}(G_1'')$ are (simple) $\mathbf{Q}$-groups of hermitian type.
Summing up, we obtain the following theorem.

Theorem 6.2. *In Case* R 2, *a* $\mathbf{Q}$*-primary solution* (V, ρ, A, I) *is obtained as follows. Let* F_1 *be a totally real number field of degree* d_1, D_1 *a central quaternion division algebra over* F_1 *such that* $(D_1^{\sigma_\nu})_{\mathbf{R}} \not\sim 1$ *for* $1 \leq \nu \leq d_1'$ *and* ~ 1 *for* $d_1'+1 \leq \nu \leq d_1$ *(we fix the isomorphisms* (6.3), (6.13)). *Let* V_1 *be a vector space over* F_1 *with a* D_1*-module structure and* h *a non-degenerate* (D_1, ε)*-hermitian form on* $V_1 \times V_1$ *satisfying the conditions* (6.17), (6.20). *Let*

(6.24) $$\rho_1 : G_{\mathbf{Q}} \longrightarrow G_1' = SU(V_1, h)$$

be an irreducible representation over F_1 *with* $\mathrm{End}_{G_{\mathbf{Q}}}(V_1)=D_1$ *and satisfying the following condition : For every* ν, *the corresponding real representation*

$$\rho_1^{(\nu)} : G \longrightarrow G_1'^{(\nu)}$$

obtained from $\rho_1^{\sigma_\nu}$ *by scalar extension satisfies the condition* (H_2) *with respect to* H_0 *and* $\frac{1}{2} I_1^{(\nu)}$ *satisfying* (6.18), (6.19) *when* $G_1'^{(\nu)}$ *is non-compact (i.e., when* $\varepsilon=1$ *and* $d_1'+1 \leq \nu \leq d_1$ *or when* $\varepsilon=-1$ *and* $1 \leq \nu \leq d_1'$). *On the other hand, let* U_1 *be a vector space over* F_1 *with a* D_1*-module structure,* h' *a non-degenerate* $(D_1, -\varepsilon)$*-hermitian form on* $U_1 \times U_1$ *satisfying the conditions* (6.18), (6.19), *and* $I_1'^{(\nu)}$ *is a complex structure on* $U_1^{(\nu)}$ *satisfying* (6.17), (6.20). *Then, one has*

$$V_{\mathbf{Q}} = U_1 \otimes_{D_1} V_1, \qquad A_{\mathbf{Q}} = \mathrm{tr}_{D_1/\mathbf{Q}}(h' \otimes h), \qquad \rho_{\mathbf{Q}} = 1_{U_1} \otimes \rho_1,$$

and (V, ρ, A, I) *is the direct sum of the* $\mathbf{R}$*-primary solutions* $(V^{[\nu]}, \rho^{(\nu)}, A^{(\nu)}, I^{(\nu)})$ *given by* (6.4), (6.13), (6.14), (6.7), (6.16), *and* (6.21).

Thus we see that (the $\mathbf{Q}$-similarity class of) a $\mathbf{Q}$-primary solution (V, ρ, A, I) is determined by (the class of) a $\mathbf{Q}$-irreducible representation (V_1, ρ_1), (the similarity class of) (U_1, h'), and (the class of) a system of complex structures $(I_1'^{(\nu)})$ satisfying the above conditions. Otherwise expressed, for a fixed triple (V, ρ, A), the solution can be parametrized by a point $(I_1'^{(\nu)})$ in the symmetric domain associated with a semi-simple $\boldsymbol{R}$-group of hermitian type, $R_{F_1/Q}(G_1'')_{\boldsymbol{R}}=\prod_\nu G_1''^{(\nu)}(=(C_\rho)^s)$. (For this reason, $I'=\sum I_1'^{(\nu)}\otimes 1_{V_1^{(\nu)}}$ will be called the *parametric part* of I.)

Next we consider the case where ι_0 is of the second kind, which we call the *Case* C. Put $\sigma_0=\iota_0|F_1$. Then, $\mathrm{Gal}(F_1/F_{10})=\{1, \sigma_0\}$, and σ_0 extends uniquely to an algebra automorphism of $F_1\otimes_{\boldsymbol{Q}}\boldsymbol{R}$ over $\boldsymbol{R}$ (denoted also by σ_0), which leaves the decomposition (6. 2) invariant. Therefore, for every ν, we have $F_1^{(\nu)}\cong\boldsymbol{C}$ and $\sigma_0|F_1^{(\nu)}$ is the complex conjugation, i. e., F_1 is a "totally imaginary" quadratic extension of a totally real number field F_{10}. More precisely, one has $d_1=2t$, and the set of all monomorphisms of F_1 into $\boldsymbol{C}$ can be written in the form

$$\{\sigma_1, \cdots, \sigma_t, \sigma_0\sigma_1, \cdots, \sigma_0\sigma_t\},$$

where the pair of mutually conjugate monomorphisms σ_ν, $\sigma_0\sigma_\nu$ correspond to the two mutually conjugate isomorphisms $F_1^{(\nu)}\cong\boldsymbol{C}$ by the relation similar to (6. 5). For each ν, we fix one of such isomorphisms (i. e., a complex structure on $V^{(\nu)}$) corresponding to σ_ν along with the isomorphism (6. 3). Thus we fix isomorphisms:

$$D_1^{(\nu)} = F_1^{(\nu)} \cong \boldsymbol{C}, \qquad D_1\otimes_{F_1}F_1^{(\nu)} \cong \mathcal{M}_{r_1}(\boldsymbol{C}). \tag{6. 25}$$

Then, as before, one may put

$$V_1^{(\nu)} = \varepsilon_{11}^{(\nu)}(V_1^{\sigma_\nu})_{\boldsymbol{C}}, \qquad U_1^{(\nu)} = \bar{\varepsilon}_{11}^{(\nu)}(U_1^{\sigma_\nu})_{\boldsymbol{C}}, \tag{6. 26}$$

where the $\varepsilon_{ij}^{(\nu)}$ are the fixed matrix units in $D_1\otimes_{F_1}F_1^{(\nu)}$, viewed as a subalgebra of $\mathrm{End}(V_1^{(\nu)})$. Then, denoting by $\rho_1^{(\nu)}$ the restriction of $(\rho_1^{\sigma_\nu})_{\boldsymbol{C}}$ to $V_1^{(\nu)}$, one obtains again (6. 7). Let

$$A_{\boldsymbol{Q}} = \mathrm{tr}_{D_1/\boldsymbol{Q}}(h'\otimes h) \tag{6. 27}$$

be an expression of $A_{\boldsymbol{Q}}$ similar to (6. 15). Then, for each ν, we obtain from h and h', through the fixed isomorphism (6. 25), an $\varepsilon\eta_\nu$-hermitian form $h^{(\nu)}$ on $V_1^{(\nu)}\times V_1^{(\nu)}$ and a $(-\varepsilon\eta_\nu)$-hermitian form $h'^{(\nu)}$ on $U_1^{(\nu)}\times U_1^{(\nu)}$, respectively, so that we have

$$A^{(\nu)} = 2\,\mathrm{Re}(h'^{(\nu)}\otimes h^{(\nu)}). \tag{6. 28}$$

Note that in this case, modifying the isomorphism (6. 25), the (ordered) system of signs $\{\eta_\nu\}$ can be chosen arbitrarily. Hence, choosing $\{\eta_\nu\}$ suitably and then replacing h and h' by suitable scalar multiples, one may always assume that for every ν at least one of $h^{(\nu)}$ and $h'^{(\nu)}$ is hermitian positive definite and the other one is skew-hermitian. Thus we may assume

$$h^{(\nu)}I_1^{(\nu)} \gg 0, \qquad h'^{(\nu)} \gg 0 \qquad \text{for} \quad 1\le\nu\le t', \tag{6. 29}$$

$$h^{(\nu)} \gg 0, \qquad h'^{(\nu)}I_1'^{(\nu)} \gg 0 \qquad \text{for} \quad t'+1\le\nu\le t, \tag{6. 30}$$

where

$$I_1^{(\nu)} = 2\,d\rho_1^{(\nu)}(H_0)+\sqrt{-1}\,\frac{p_\nu-q_\nu}{p_\nu+q_\nu}$$

((p_ν, q_ν) denoting the signature of $\sqrt{-1}h^{(\nu)}$) is a complex structure on $V_1^{(\nu)}$ and $I_1'^{(\nu)}$ is one on $U_1^{(\nu)}$, and one has

(6. 31) $$I^{(\nu)} = \begin{cases} 1_{U_1^{(\nu)}} \otimes I_1^{(\nu)} & (1 \le \nu \le t') \\ I_1'^{(\nu)} \otimes 1_{V_1^{(\nu)}} & (t'+1 \le \nu \le t). \end{cases}$$

[One may further assume that the $\sqrt{-1}\, h'^{(\nu)}$ $(t'+1 \le \nu \le t)$ are all indefinite. We then call $I' = \sum_{\nu=t'+1}^{t} I^{(\nu)}$ the "parametric part" of I.]

Summing up, we obtain the following theorem.

Theorem 6. 3. *In Case* C *with involution* ι_0 *of the second kind, a* **Q***-primary solution* (V, ρ, A, I) *is obtained as follows. Let* F_1 *be a totally imaginary quadratic extension of a totally real number field* F_{10} *of degree* t, $\{\sigma_\nu, \sigma_0\sigma_\nu (1 \le \nu \le t)\}$ *the set of all monomorphisms of* F_1 *into* **C**, *and let* D_1 *be a central division algebra over* F_1 *with an involution* ι_0 *of the second kind such that* $\iota_0|F_1 = \sigma_0$. *Let* V_1 *be a vector space over* F_1 *with a* D_1*-structure and* h *a non-degenerate* (D_1, ε)*-hermitian form on* $V_1 \times V_1$ *satisfying* (6. 29), (6. 30), *i. e., such that* $\sqrt{-1}\, h^{(\nu)}$ *is hermitian for* $1 \le \nu \le t'$ *and* $h^{(\nu)}$ *is hermitian, positive definite for* $t'+1 \le \nu \le t$ *(under the fixed isomorphism* (6. 25)*). Let*

(6. 32) $$\rho_1 : G_{\boldsymbol{Q}} \longrightarrow SU(V_1, h)$$

be an irreducible representation over F_1 *with* $\mathrm{End}_{G_{\boldsymbol{Q}}}(V_1) = D_1$ *and satisfying the following condition: For every* ν, *the corresponding representation*

$$\rho_1^{(\nu)} : G \longrightarrow SU(V_1^{(\nu)}, h^{(\nu)})$$

obtained from $\rho_1^{\sigma_\nu}$ *by scalar extension satisfies the condition* (H_2) *with respect to* H_0 *and* $\frac{1}{2}\left(I_1^{(\nu)} - \sqrt{-1}\,\frac{p_\nu - q_\nu}{p_\nu + q_\nu} 1\right)$ *in* (6. 29). *On the other hand, let* U_1 *be a vector space over* F_1 *with* $\bar{D}_1$*-module structure,* h' *a non-degenerate* $(\bar{D}_1, -\varepsilon)$*-hermitian form on* $U_1 \times U_1$ *satisfying* (6. 29), *and* $I_1'^{(\nu)}$ *a complex structure on* $U_1^{(\nu)}$ *satisfying* (6. 30). *Then, one has*

$$V_{\boldsymbol{Q}} = U_1 \otimes_{D_1} V_1, \qquad A_{\boldsymbol{Q}} = \mathrm{tr}_{D_1/\boldsymbol{Q}}(h' \otimes h), \qquad \rho_{\boldsymbol{Q}} = 1_{U_1} \otimes \rho_1,$$

and (V, ρ, A, I) *is the direct sum of the* **R***-primary solutions* $(V^{[\nu]}, \rho^{(\nu)}, A^{(\nu)}, I^{(\nu)})$ *given by* (6. 4), (6. 25), (6. 26), (6. 7), (6. 28), (6. 31).

Thus, quite analogously to Case R 2, we have a pair of groups of hermitian type

$$R_{F_1/\boldsymbol{Q}}(G_1')_{\boldsymbol{R}} = \prod_{\nu=1}^{t} SU(V_1^{(\nu)}, h^{(\nu)}), \qquad R_{F_1/\boldsymbol{Q}}(G_1'')_{\boldsymbol{R}} = \prod_{\nu=1}^{t} SU(U_1^{(\nu)}, h'^{(\nu)}),$$

and the solution with a fixed symplectic representation (V, ρ, A) is parametrized by a point $(I_1'^{(\nu)})$ in the symmetric domain associated with $R_{F_1/\boldsymbol{Q}}(G_1'')_{\boldsymbol{R}}$.

Remark. From the above, we see that a (non-trivial, **Q**-primary) solution (V, ρ, A, I) is rigid (i. e., $R_{F_1/\boldsymbol{Q}}(G_1'')_{\boldsymbol{R}}$ is compact) in the following four cases: (R 1), (R 2. 1) with $d_1' = 0$, (R 2. 2) with $d_1' = d$, and (C) with $t' = t$. In these cases, the simple **Q**-algebra $\mathrm{End}(U_1)$ has a positive involution given by S_1 or h'. It is known (Albert [1], [3], [4]; Siegel [14]) that all simple **Q**-algebras with positive involution are obtained in this manner. Note also that, if G has no compact simple factor, then $R_{F_1/\boldsymbol{Q}}(G_1')_{\boldsymbol{R}}$ has none either, forcing $R_{F_1/\boldsymbol{Q}}(G_1'')_{\boldsymbol{R}}$ to be compact, so that the solution is rigid.

In view of the list in the preceding section, one may say that the possibilities for the representation ρ_1 in (6. 12), (6. 24), (6. 32) are very limited. Actually, under an additional assumption that $\rho_1^{(1)}=\rho_{1\boldsymbol{R}}$ *is essentially a representation of one simple factor of* G (which is automatically satisfied, by § 5, if G has no compact factor), the $\boldsymbol{Q}$-primary solutions are completely classified (Satake [6]). It turns out that, besides the "standard solutions" obtained (for each of the cases (R 1)$\sim$(C)) by setting

$$G = R_{F_1/\boldsymbol{Q}}(G_1')_{\boldsymbol{R}} \qquad (G_1'=Sp(V_1, A_1) \text{ or } SU(V_1, h))$$

and taking as ρ_1 the projection map of $G_{\boldsymbol{Q}}$ onto G_1', there are few non-standard solutions involving skew-symmetric tensor representations and spin representations. However, without the above additional assumption, the general solution seems to become rather complicated, because of the presence of various "mixed types". (For examples of such solutions, see Satake [6], Mumford [4]).

§ 7. Analytic construction of Kuga's fiber varieties.

First we give some general definitions. Two subgroups H_1 and H_2 of a group G are said to be *commensurable* and denoted $H_1\sim H_2$ if $H_1\cap H_2$ is of finite index in both H_1 and H_2. It is easy to see that this is an equivalence relation.

Let V be a (finite-dimensional) real vector space. By a *lattice* in V, we mean a discrete submodule L of V such that V/L is compact. A lattice in V is a free $\boldsymbol{Z}$-module of the rank equal to dim V. A lattice L in V determines a $\boldsymbol{Q}$-structure of V by the relation

$$V_{\boldsymbol{Q}} = L\otimes_{\boldsymbol{Z}}\boldsymbol{Q}. \tag{7. 1}$$

When V is defined over $\boldsymbol{Q}$ and when the relation (7. 1) holds, we say that the lattice L belongs to the given $\boldsymbol{Q}$-structure, or simply L is a $\boldsymbol{Q}$-*lattice*. It is clear that two lattices in V are commensurable if and only if they belong to the same $\boldsymbol{Q}$-structure of V.

Let V be a real vector space defined over $\boldsymbol{Q}$ and G an $\boldsymbol{R}$-group in $GL(V)$ defined over $\boldsymbol{Q}$. For a $\boldsymbol{Q}$-lattice L in V, one puts

$$G_L = \{g\in G\,|\,gL=L\}. \tag{7. 2}$$

Clearly, G_L is a discrete subgroup of G (in the usual topology), contained in $G_{\boldsymbol{Q}}$. For a positive integer N, one also defines the "principal congruence subgroup" (of level N) by

$$G_L(N) = \{g\in G_L\,|\,(g-1_V)L\subset NL\}. \tag{7. 3}$$

Clearly, $G_L(N)$ is a normal subgroup of G_L of finite index.

Lemma 7. 1. *Let G and G' be $\boldsymbol{R}$-groups defined over $\boldsymbol{Q}$ in $GL(V)$ and $GL(V')$, respectively, and let $\rho: G\to G'$ be an $\boldsymbol{R}$-homomorphism defined over $\boldsymbol{Q}$. Let further L and L' be $\boldsymbol{Q}$-lattices in V and V', respectively. Then, for any positive integer N', there exists a positive integer N such that one has $\rho(G_L(N))\subset G'_{L'}(N')$.*

Proof. Let $g \in G$ and write $g = 1_V + x$, $\rho(g) = 1_{V'} + y$, and $x = (x_{ij})$, $y = (y_{kl})$ in terms of any bases of L and L'. Then the y_{kl} are polynomial functions in x_{ij} $(1 \leq i, j \leq \dim V)$ with coefficient in $\boldsymbol{Q}$ without constant terms. Hence, for a given positive integer N', one can find a positive integer N such that, if the x_{ij} are multiples of N, then all the y_{kl} are multiples of N'. This proves our assertion, q. e. d.

Applying this lemma to the case where $V = V'$, $G = G'$, $\rho = id$, $N' = 1$, one has

$$(G_L : G_L \cap G_{L'}) \leq (G_L : G_L(N)) < \infty$$

for some N, and similarly $(G_{L'} : G_L \cap G_{L'}) < \infty$. Thus one has $G_L \sim G_{L'}$. In general, a subgroup Γ of $G_{\boldsymbol{Q}}$ such that $\Gamma \sim G_L$ for some (hence all) $\boldsymbol{Q}$-lattice L is called an *arithmetic subgroup* of G. An arithmetic subgroup is discrete. By Lemma 7. 1, the commensurability class of arithmetic subgroups of G is uniquely determined only by the $\boldsymbol{Q}$-isomorphism class of G.

Lemma 7. 2. *Let Γ be an arithmetic subgroup of an $\boldsymbol{R}$-group G. Then there exists a normal subgroup Γ_1 of Γ of finite index such that Γ_1 is "torsion-free" (i. e., Γ_1 has no element of finite order other than the unit element).*

Proof. Let L_0 be any $\boldsymbol{Q}$-lattice in V. Then one has $\Gamma \sim G_{L_0}$. Let $\Gamma = \bigcup_{i=1}^{s} \gamma_i (\Gamma \cap G_{L_0})$ and put $L_1 = \bigcap_{i=1}^{s} \gamma_i L_0$. Then L_1 is also a $\boldsymbol{Q}$-lattice, and one has $\Gamma \subset G_{L_1}$. Put $\Gamma_1 = \Gamma \cap G_{L_1}(N)$ with a positive integer $N \geq 3$. Then, as is well-known (and easy to prove), $G_{L_1}(N)$ and hence Γ_1 is torsion-free. Therefore Γ_1 satisfies all the requirements stated in the Lemma, q. e. d.

We quote the following theorem due to Borel-Harish-Chandra [1] and Mostow-Tamagawa [1] (cf. also Godement [2]).

Theorem 7. 3. *Let G be a semi-simple $\boldsymbol{R}$-group and Γ an arithmetic subgroup. Then the homogeneous space $\Gamma \backslash G$ is of finite volume with respect to the measure induced from the Haar measure of G. $\Gamma \backslash G$ is compact if and only if G is of $\boldsymbol{Q}$-rank zero.*

Remark. Concerning arithmetic subgroups, there are two famous problems:

1) Let G be a semi-simple Lie group (with center reduced to $\{1\}$) and Γ a discrete subgroup of G such that $\Gamma \backslash G$ is of finite volume. Is it possible to find a compact semi-simple Lie group G_0 (with center reduced to $\{1\}$) and a $\boldsymbol{Q}$-structure on $G_0 \times G$ such that Γ is the projection to G of an arithmetic subgroup of $G_0 \times G$?

2) Let Γ be an arithmetic subgroup of G (defined over $\boldsymbol{Q}$), contained in G_L. Is there a principal congruence subgroup $G_L(N)$ contained in Γ? (In both problems, one has to exclude some cases where G is of low rank.)

For the answers to these problems, see e. g., Tits [6], Raghunathan [4] (Russian edition, Appendix), [6], and Serre [4].

Now, let G_1 be a semi-simple $\boldsymbol{R}$-group of hermitian type and $\mathcal{D}_1$ the associated symmetric domain. Let Γ_1 be a discrete subgroup of G_1 contained in G_1°. Then, since the isotropy subgroup of G_1° at o is compact, it is easy to see that the action of Γ_1 on $\mathcal{D}_1$ is *properly discontinuous*, i. e., the following condition is satisfied: For any compact subset C of $\mathcal{D}_1$, the set

$$\{\gamma \in \Gamma \mid \gamma C \cap C \neq \phi\}$$

is finite. It follows that the quotient space $M_1 = \Gamma_1 \backslash \mathcal{D}_1$ is Hausdorff. By virtue of Lemma 7. 2, if one replaces Γ_1 by a suitable subgroup of finite index, one may assume that Γ_1 has no fixed points. Then M_1 inherits a structure of (non-singular) complex manifold. By Theorem 7. 3, when Γ_1 is arithmetic, M_1 is of finite volume with respect to the invariant measure on $\mathcal{D}_1$ defined by Bergman metric, and M_1 is compact if and only if G_1 is of $\boldsymbol{Q}$-rank zero. As we shall see in the next section, when M_1 is compact, M_1 is "projective", i. e., holomorphically equivalent to a (non-singular) algebraic submanifold of a complex projective space. (In general, when Γ_1 is arithmetic, the quotient space $M_1 = \Gamma_1 \backslash \mathcal{D}_1$ has a standard compactification so that M_1 becomes a Zariski open subset of a projective variety. Cf. Baily-Borel [1], H. Cartan [4], Ash et al. [1].)

Let G_1, $\mathcal{D}_1$, Γ_1 be as above. Fix the origin $o \in \mathcal{D}_1$ and let H_0 be the corresponding H-element in $\mathfrak{g}_1 = \mathrm{Lie}\, G_1$. We consider a 5-tuple (V, ρ, A, I_0, L) satisfying the following conditions:

(K 1) *The quadruple* (V, ρ, A, I_0) *is a solution (over* $\boldsymbol{R}$*) of our problem for* $(G_1, \mathcal{D}_1, o)$. [Thus, $\rho: G_1 \to Sp(V, A)$ is a symplectic $\boldsymbol{R}$-representation satisfying (H_1) with respect to H_0 and $\frac{1}{2} I_0$, and one has $AI_0 = {}^t(AI_0) \gg 0$.]

(K 2) *L is a lattice in V such that one has*

$$A(L, L) \subset \boldsymbol{Z}, \tag{7. 4}$$

$$\rho(\Gamma_1) L \subset L. \tag{7. 5}$$

Under these conditions, we always endow V with the $\boldsymbol{Q}$-structure defined by (7. 1). Then by (7. 4) the bilinear form A is defined over $\boldsymbol{Q}$ and hence so is $G' = Sp(V, A)$. The group $G'_L = Sp(L, A)$ is called a *Siegel paramodular group*. The following lemma is well-known (e. g., Shimura [5]), and can be proved by an easy induction.

Lemma 7. 4 (*Frobenius*). *Let L be a $\boldsymbol{Q}$-lattice in V satisfying* (7. 4). *Then there exists a symplectic basis $(e_1, \cdots, e_{2n})$ of (V, A) such that one has*

$$L = \{e_1, \cdots, e_n, \delta_1 e_{n+1}, \cdots, \delta_n e_{2n}\}_{\boldsymbol{Z}}, \tag{7. 6}$$

where $\delta_1, \cdots, \delta_n$ are positive integers such that $\delta_i \mid \delta_{i+1}$ $(1 \leq i \leq n-1)$. The n-tuple $(\delta_1, \cdots, \delta_n)$ is uniquely determined by the pair (L, A). [The δ_i are called the "elementary divisors" of (L, A).]

In terms of such a basis (e_i), if one writes $g' = \begin{pmatrix} a & b \\ c & d \end{pmatrix} \in Sp(V, A)$, then one has

$$(7.7) \qquad g' \in Sp(L, A) \Longleftrightarrow a,\ b\delta,\ \delta^{-1}c,\ \delta^{-1}d\delta \in \mathcal{M}_n(\boldsymbol{Z}),$$

where δ stands for the diagonal matrix $\mathrm{diag}(\delta_1, \cdots, \delta_n)$. In the case $\delta_1 = \cdots = \delta_n = 1$, $Sp(L, A)$ is the *Siegel modular group.*

Remark 1. The condition (7. 5) means that ρ induces a homomorphism $\Gamma_1 \to Sp(L, A)$. When G_1 is defined over $\boldsymbol{Q}$ and Γ_1 is arithmetic, this implies (by a density theorem of Borel [10]) that ρ is defined over $\boldsymbol{Q}$, so that (V, ρ, A, I_0) becomes a solution over $\boldsymbol{Q}$. For convenience, we are including the case where ρ is trivial, in which case the condition (7. 5) is void. On the contrary, when ρ is faithful, the condition (7. 5) implies that there exists a unique $\boldsymbol{Q}$-structure of G_1 such that Γ_1 is arithmetic and ρ is defined over $\boldsymbol{Q}$.

Remark 2. Suppose that G_1 is defined over $\boldsymbol{Q}$ and (V, ρ, A, I_0) is a solution over $\boldsymbol{Q}$. Then, first, for any $\boldsymbol{Q}$-lattice L, one has $A(L, L) \subset \boldsymbol{Q}$. Hence, replacing A by a suitable scalar multiple λA $(\lambda \in \boldsymbol{Z}, \lambda > 0)$, or replacing L by λL, one can always restore the condition (7. 4). Next, by Lemma 7. 1, there exists a subgroup Γ_1' of Γ_1 of finite index such that the condition (7. 5) is satisfied for Γ_1' and L. It follows that, for a given arithmetic subgroup Γ_1 of G_1 (contained in G_1°), one can always find a $\boldsymbol{Q}$-lattice L in V such that the conditions (7. 4), (7. 5) are satisfied.

Let $(V', \rho', A', I_0', L')$ be another 5-tuple satisfying the conditions (K 1, 2). We say that two 5-tuples (V, ρ, A, I_0, L) and $(V', \rho', A', I_0', L')$ are *equivalent* if there exists an isomorphism $\psi : V \to V'$ such that

$$(7.8) \qquad \begin{cases} \rho(g) = \psi^{-1}\rho'(g)\psi & (g \in G_1), \\ A(x, y) = A'(\psi(x), \psi(y)) & (x, y \in V), \\ I_0 = \psi^{-1}I_0'\psi, \\ L = \psi^{-1}L'. \end{cases}$$

Clearly, this implies that the elementary divisors of (L, A) and (L', A') are the same.

To a given 5-tuple (V, ρ, A, I_0, L), one can associate a holomorphic map $\varphi (= \rho_{\mathcal{D}}) : \mathcal{D}_1 \to \mathfrak{S}(V, A)$ defined by

$$z = go \longmapsto I_{\varphi(z)} = \rho(g) I_0 \rho(g)^{-1} \qquad (g \in G_1^\circ).$$

If $(V', \rho', A', I_0', L')$ is another 5-tuple equivalent to the given one and $\varphi' : \mathcal{D}_1 \to \mathfrak{S}(V', A')$ is the associated holomorphic map, then it follows from (7. 8) that

$$(7.9) \qquad I_{\varphi(z)} = \psi^{-1} I_{\varphi'(z)} \psi \qquad (z \in \mathcal{D}_1).$$

In this sense, two holomorphic maps φ and φ' are equivalent. A converse is given by the following

Lemma 7. 5 (*Kuga*). *Let* (V, ρ, A, I_0, L) *and* $(V', \rho', A', I_0', L')$ *be two 5-tuples satisfying the conditions* (K 1, 2) *and let* φ *and* φ' *be the associated holomorphic maps. Suppose that there exists a linear isomorphism* $\psi : V \to V'$ *satisfying the following conditions :*

$$(7.8') \qquad \begin{cases} \psi(L) = L', \\ A(x, y) = A'(\psi(x), \psi(y)) & (x, y \in V), \\ I_{\varphi(z)} = \psi^{-1} I_{\varphi'(z)} \psi & (z \in \mathcal{D}_1). \end{cases}$$

Then these two 5-tuples are equivalent.

Proof. It is enough to show that the first condition in (7.8) is satisfied. Let Ψ be the set of all linear maps ψ satisfying the condition (7.8′). Then, since Ψ is discrete and compact in the space of all linear maps $V \to V'$, Ψ is finite. The group Γ_1 acts on Ψ by $\psi \mapsto \rho'(\gamma)^{-1} \circ \psi \circ \rho(\gamma)$. Hence there is a subgroup Γ_0 of Γ_1 of finite index such that $\rho'(\gamma)^{-1} \circ \psi \circ \rho(\gamma) = \psi$ for all $\gamma \in \Gamma_0$ and $\psi \in \Psi$. Since Γ_0 is Zariski dense in G_1 (Borel [6]), it follows that $\rho'(g)^{-1} \circ \psi \circ \rho(g) = \psi$ for all $g \in G_1$, which proves our assertion, q. e. d.

For a given 5-tuple (V, ρ, A, I_0, L), we shall now construct a fiber space over $M_1 = \Gamma_1 \backslash \mathcal{D}_1$, whose fibers are abelian varieties (cf. Kuga [1]). First, consider the disjoint union of complex vector spaces $(V, I_{\varphi(z)})$ $(z \in \mathcal{D}_1)$. As in III, § 6, we put $V_\pm = V_C(I_0; \pm i)$ and introduce a V_+-valued function

$$w(z, v) = v_{\varphi(z)} = v_+ - \varphi(z) v_- \qquad (z \in \mathcal{D}_1,\ v \in V), \tag{7.10}$$

where as before $v_\pm$ denote the $V_\pm$-components of $v \in V$ and $\varphi(z) \in \mathfrak{S}(V, A)$ is viewed as a (symmetric) linear map $V_- \to V_+$. Then the correspondence $(z, v) \mapsto (z, w(z, v))$ gives a bijection

$$\coprod_{z \in \mathcal{D}_1} (V, I_{\varphi(z)}) \longrightarrow \mathcal{D}_1 \times V_+, \tag{7.11}$$

which induces a C-linear isomorphism $(V, I_{\varphi(z)}) \cong z \times V_+$ for each $z \in \mathcal{D}_1$. Hence we define a structure of (trivial) holomorphic vector bundle on $\coprod (V, I_{\varphi(z)})$ through the bijection (7.11) and denote the holomorphic vector bundle thus obtained by $\tilde{\mathcal{D}}$. Though this definition is independent of the choice of the origin $o = I_0$, we will often identify $\tilde{\mathcal{D}}$ with $\mathcal{D}_1 \times V_+$ by the isomorphism (7.11).

On the other hand, we define a semi-direct product $\tilde{G} = G_1 \cdot V$ by the linear action of G_1 on V defined by ρ. Then $\tilde{G}^\circ$ acts on $\tilde{\mathcal{D}}$ by

$$g_1 \cdot v_1 :\ (z, v) \longmapsto (g_1(z), \rho(g_1)(v + v_1)). \tag{7.12}$$

If one writes $\rho(g_1) = \begin{pmatrix} a & b \\ \bar{b} & \bar{a} \end{pmatrix}$ according to the decomposition $V_C = V_+ \oplus V_-$, then one has

$$w(g_1(z), \rho(g_1) v) = {}^t(\bar{b}\varphi(z) + \bar{a})^{-1} w(z, v) \tag{7.13}$$

(III, § 5, Exerc. 2). Hence the action of $\tilde{G}^\circ$ on $\tilde{\mathcal{D}} = \mathcal{D}_1 \times V_+$ is given by

$$(z, w) \longmapsto (g_1(z), J(\rho(g_1), \varphi(z))(w + w(z, v_1))). \tag{7.12′}$$

Since $\rho(\Gamma_1)$ leaves L invariant, one has a semi-direct product $\tilde{\Gamma} = \Gamma_1 \cdot L$, which is a discrete subgroup of $\tilde{G}^\circ$. Hence $\tilde{\Gamma}$ acts properly discontinuously on $\tilde{\mathcal{D}} = \mathcal{D}_1 \times V_+$ (as an automorphism group of holomorphic vector bundle). When Γ_1 is assumed to have no fixed point on $\mathcal{D}_1$, $\tilde{\Gamma}$ has none either, and the quotient space $\tilde{M} = \tilde{\Gamma} \backslash \tilde{\mathcal{D}}$ is a (non-singular) complex manifold. $\tilde{M}$ has then a natural structure of holomorphic fiber space over M_1 with the projection map

$$\tilde{M} = \tilde{\Gamma} \backslash \tilde{\mathcal{D}} \longrightarrow M_1 = \Gamma_1 \backslash \mathcal{D}_1$$

given by $\tilde{\Gamma} \cdot (z, w) \mapsto \Gamma_1 \cdot z$. The fiber over the Γ_1-orbit of $z \in \mathcal{D}_1$ is a complex torus $(V/L, I_{\varphi(z)})$. [In general, when $(\Gamma_1)_z$ is not trivial, the fiber is the quotient $(\Gamma_{1z} \cdot L) \backslash$

$(V, I_{\varphi(z)})$.] This torus is actually an abelian variety, since there exists a "Riemann form" A. $\tilde{M}$ is called the *Kuga's fiber variety* constructed by the 5-tuple (V, ρ, A, I_0, L) (and the data $(G_1, \mathcal{D}_1, o, \Gamma_1)$).

Remark. In general, a complex torus $(V/L, I)$ is called an *abelian variety*, if it has a structure of algebraic variety. It is well-known (e. g., Weil [2]) that for a complex torus $(V/L, I)$ to be an abelian variety, it is necessary and sufficient that there exists a non-degenerate alternating bilinear form A on $V\times V$ such that $AI={}^t(AI)\gg 0$ and $A(L, L)\subset \boldsymbol{Z}$. Such an alternating form A, or the corresponding hermitian form $h=AI+iA$, is called a "Riemann form", and an abelian variety endowed with (a similarity class of) a Riemann form A is called a "polarized abelian variety". The above criterion for abelian varieties (due to Lefschetz) may be regarded as a special case of a theorem of Kodaira to be explained in the next section.

Example 1 (The universal family). Let (V, A) be a symplectic space defined over $\boldsymbol{Q}$, and L a $\boldsymbol{Q}$-lattice satisfying the condition (7. 4). Then, for each $z\in\mathfrak{S}(V, A)$, $(V/L, I_z)$ is an abelian variety with a Riemann form A, i. e., $\mathcal{A}_z=(V/L, I_z, A)$ is a polarized abelian variety. Two polarized abelian varieties $\mathcal{A}_z$ and $\mathcal{A}_{z'}$ $(z, z'\in \mathfrak{S}(V, A))$ are "isomorphic", if there exists an analytic isomorphism $(V/L, I_z)\to (V/L, I_{z'})$ preserving the polarization. Clearly this amounts to saying that there exists an $\boldsymbol{R}$-linear automorphism ϕ of V such that

$$\phi(L) = L, \qquad \phi I_z\phi^{-1} = I_{z'}, \qquad A\circ(\phi\times\phi) = \lambda A$$

with $\lambda\in\boldsymbol{Q}^\times$. It then follows immediately that $\lambda=1$ and $\phi\in Sp(L, A)$. Conversely, if one has $z'=\phi(z)$ with $\phi\in Sp(L, A)$, then $\mathcal{A}_z$ and $\mathcal{A}_{z'}$ are isomorphic. Thus the quotient space $Sp(L, A)\backslash\mathfrak{S}(V, A)$ may be viewed as a "space of moduli" for the polarized abelian varieties (with a fixed polarization A). The fiber variety $\tilde{M}= Sp(L, A)\cdot L\backslash(\mathfrak{S}(V, A)\times V_+)$ may be viewed as the universal family of polarized abelian varieties. However, one should note that, in this construction, the fibers are *not* abelian varieties, but the quotients of them by $Sp(L, A)_z$ (which always contains $\{\pm 1_V\}$). This difficulty can be avoided if one replaces $Sp(L, A)$ by a suitable subgroup, e. g., a congruence subgroup $Sp(L, A)(N)$ $(N\geq 3)$. Note also that in this case the $\boldsymbol{Q}$-rank of $Sp(V, A)$ is equal to $n(>0)$, so that both $\tilde{M}$ and M_1 are not compact.

Example 2 (Shimura's families). Let V, A, L be as in Example 1. For each $z\in\mathfrak{S}(V, A)$, the "endomorphism algebra" (over $\boldsymbol{Q}$) of the abelian variety $(V/L, I_z)$ is given by

$$\mathfrak{A}_z = \{\alpha\in \operatorname{End}(V_{\boldsymbol{Q}})\,|\,\alpha I_z=I_z\alpha\}. \tag{7. 14}$$

From the complete reducibility of abelian varieties (cf. Weil [2]), it can be shown that $\mathfrak{A}_z$ is semi-simple ; moreover, $\mathfrak{A}_z$ has a positive involution

$$\alpha\longmapsto\alpha^* = A^{-1}\,{}^t\alpha A \;(=S_z^{-1}\,{}^t\alpha S_z), \tag{7. 15}$$

where $S_z=AI_z$. Now, suppose there is given *a priori* a semi-simple $\boldsymbol{Q}$-algebra $\mathfrak{A}$ with positive involution $*$ and that $\mathfrak{A}$ is realized as a subalgebra of $\mathfrak{A}_0$ (with the same unit element 1_V) as algebra with involution. (For a suitable choice of V, A, L, and o, such a realization is always possible.) Put

$$(7.16)\qquad \begin{aligned} G(\mathfrak{A}) &= \{g \in Sp(V, A) \mid g\alpha=\alpha g \ \text{ for all } \alpha \in \mathfrak{A}\}, \\ \mathcal{D}(\mathfrak{A}) &= \{z \in \mathfrak{S}(V, A) \mid I_z\alpha=\alpha I_z \ \text{ for all } \alpha \in \mathfrak{A}\}. \end{aligned}$$

Then, first, it is clear that $G(\mathfrak{A})$ is a $\boldsymbol{Q}$-subgroup of $Sp(V, A)$ stable under the Cartan involution $g \mapsto S_o^{-1}{}^tg^{-1}S_o (=I_o^{-1}gI_o)$. Hence, by a theorem of Mostow (I, § 4), $G(\mathfrak{A})^z$ is reductive. Moreover, since $I_o \in \operatorname{Lie} G(\mathfrak{A})$, $G(\mathfrak{A})^z$ is of hermitian type and the natural inclusion map

$$\rho: G_1 = G(\mathfrak{A})^z \longrightarrow Sp(V, A)$$

is an (H_2)-homomorphism. The group $G(\mathfrak{A})$ acts on $\mathcal{D}(\mathfrak{A})$ in a natural manner, and by an argument similar to the proof of Proposition III. 6. 3 one can show that $\mathcal{D}_1=\mathcal{D}(\mathfrak{A})$ is a symmetric domain associated with $G_1=G(\mathfrak{A})^z$. Now, let Γ_1 be an arithmetic subgroup of G_1 contained in $G_1^\circ \cap Sp(L, A)$, (e. g., $\Gamma_1=G_1^\circ \cap Sp(L, A)(N)$, $N\geq 3$). Then the conditions (K 1, 2) are clearly satisfied, and one obtains a fiber variety $\tilde{M}=\Gamma_1\cdot L\backslash(\mathcal{D}(\mathfrak{A})\times V_+)$, whose fibers are abelian varieties with a fixed polarization and with endomorphism algebra containing the prescribed algebra $\mathfrak{A}$. This kind of fiber varieties (with congruence subgroup Γ_1) was first introduced by Shimura [6], [11] (where the data determining such a family are called "PEL type"). It is clear (by the theory of semi-simple algebras) that the solution in this case is $\boldsymbol{Q}$-primary if and only if $\mathfrak{A}$ is simple, and when that is so one has $\mathfrak{A}=\operatorname{End}(U_1)$ and $G=G(\mathfrak{A})^s=R_{F_1/\boldsymbol{Q}}(G_1')_{\boldsymbol{R}}$ in the notation of § 6. Moreover, C_ρ is contained in $\mathfrak{A}_o^\times \cap Sp(V, A)$, which is compact. Thus, what we obtain here is a "standard" solution which is rigid. Shimura also studied the standard solutions in non-rigid case (for "special" values of the parametric part I') and, as an important application, constructed the "canonical model" for the base variety $M_1=\Gamma_1\backslash\mathcal{D}_1$. (Cf. Shimura [12]~[18]; also K. Miyake [1], Deligne [2], [3], Shih [4].) Thanks to the works of Shimura and his school, the arithmetic theory of automorphic functions has been enriched enormously in the recent two decades.

Example 3. In Chapter III we have seen that for each boundary component $\mathcal{F}_\kappa$ of a symmetric domain $\mathcal{D}$ there corresponds a quadruple (V, ad_V, A_e, I_0), which is a solution of our problem for $(G_\kappa^{(1)}, \mathcal{D}_\kappa, o_\kappa)$. When G and κ are defined over $\boldsymbol{Q}$ and $e=e_\kappa$ is taken to be $\boldsymbol{Q}$-rational, this is a solution over $\boldsymbol{Q}$. Suppose there is given an arithmetic subgroup Γ of G. Then $\exp(U+V)\cap\Gamma$ is an arithmetic subgroup of the unipotent group $\exp(U+V)$, so that there exist $\boldsymbol{Q}$-lattices M and L in U and V, respectively, such that one has

$$\exp(U+V)\cap\Gamma = \exp(M+L).$$

Then one has $A_e(L, L)\subset\boldsymbol{Q}$. On the other hand, $\Gamma_1=G_\kappa^{(1)}\cap\Gamma$ is an arithmetic subgroup of $G_1=G_\kappa^{(1)}$ and the condition (7. 5) is satisfied. Hence one can construct a Kuga's fiber variety $\Gamma_1\cdot L\backslash(\mathcal{D}_\kappa\times V_+)$, which plays an important role in the theory of compactification.

Remark. Various discussions on Kuga's fiber varieties can be found in *Algebraic Groups and Discontinuous Subgroups* (Amer. Math. Soc., 1966). In particular, Mumford ([1], [4]) has shown (modulo "Hodge conjecture") that, in the rigid case, or in the non-rigid case where the parametric part is "special" (in the sense that the corresponding family contains an abelian variety of *CM*-type), the fibers of Kuga's fiber variety can always be given an algebraic characterization in terms of the "Hodge type". (Cf. also Kuga [2], Weil [10].) Kuga's fiber varieties have been proved to be a very powerful tool in the arithmetic study of automorphic functions (cf. e. g., Kuga-Shimura [1], Satake [14]).

§ 8. The algebraicity of Kuga's fiber varieties.

In this section, we shall prove, by using Kodaira's criterion (Theorem 8. 2 below), that the Kuga's fiber variety $\tilde{M}=\tilde{\Gamma}\backslash\tilde{\mathcal{D}}$ is projective, provided $M_1=\Gamma_1\backslash\mathcal{D}_1$ (hence $\tilde{M}$) is compact. We start with reviewing some basic notions and results relevant to our considerations (cf. Weil [2], Lascoux-Berger [1]).

Let M be a complex manifold and $\{U_\lambda(\lambda\in\Lambda)\}$ an open covering of M. By a 1-*cocycle* of M (relative to $\{U_\lambda\}$), we mean a collection of non-vanishing holomorphic functions $f_{\lambda\mu}$ on $U_\lambda\cap U_\mu$ satisfying the condition

$$(8.1)\qquad \begin{cases} f_{\lambda\mu}\cdot f_{\mu\nu}\cdot f_{\nu\lambda}=1 & \text{on} \quad U_\lambda\cap U_\mu\cap U_\nu, \\ f_{\lambda\lambda}=1 & \text{on} \quad U_\lambda. \end{cases}$$

Two 1-cocycles $(f_{\lambda\mu})$ and $(f'_{\lambda\mu})$ are *cohomologous* if there exists a collection of non-vanishing holomorphic functions g_λ on $U_\lambda(\lambda\in\Lambda)$ such that

$$(8.2)\qquad f'_{\lambda\mu}=g_\lambda f_{\lambda\mu}\, g_\mu^{-1} \qquad \text{on} \quad U_\lambda\cap U_\mu.$$

Given a 1-cocycle $(f_{\lambda\mu})$, one can construct a (holomorphic) complex line bundle E over M as follows: Consider the disjoint union $\coprod_{\lambda\in\Lambda} U_\lambda\times \boldsymbol{C}$ and identify $(z,\zeta^{(\lambda)})\in U_\lambda\times\boldsymbol{C}$ and $(z',\zeta^{(\mu)})\in U_\mu\times\boldsymbol{C}$ if and only if one has $U_\lambda\cap U_\mu\neq\phi$ and

$$(8.3)\qquad z=z', \qquad \zeta^{(\lambda)}=f_{\lambda\mu}(z)\zeta^{(\mu)}.$$

($\zeta^{(\lambda)}$ will be called "fiber coordinate" on U_λ.) Clearly, the isomorphism class of the complex line bundle thus obtained depends only on the cohomology class of $(f_{\lambda\mu})$. In this way, one actually obtains a natural isomorphism between the (additively written) group of complex line bundles over M and the first cohomology group $H^1(M,\mathcal{O}^\times)$ (defined in a standard manner as the direct limit of the cohomology groups relative to coverings), where $\mathcal{O}^\times$ denotes the sheaf of (multiplicative) groups, consisting of germs of non-vanishing holomorphic functions on M.

Example. Suppose that each open set U_λ carries coordinate functions $z_1^{(\lambda)},\cdots,z_n^{(\lambda)}$. Then the jacobians

$$(8.4)\qquad f_{\lambda\mu}=\det\left(\frac{\partial z_\alpha^{(\mu)}}{\partial z_\beta^{(\lambda)}}\right)$$

form a 1-cocycle. The corresponding complex line bundle is called the *canonical bundle* of M and denoted by $K(M)$. The space of C^∞-sections of $K(M)$ is naturally identified with that of C^∞-forms of type $(n,0)$ on M.

Let E be a complex line bundle over M defined by a 1-cocycle $(f_{\lambda\mu})$. Then, putting

$$c_{\lambda\mu\nu} = -\frac{1}{2\pi i}(\log f_{\lambda\mu} + \log f_{\mu\nu} + \log f_{\nu\lambda}), \tag{8.5}$$

one obtains a $\boldsymbol{Z}$-valued 2-cocycle $(c_{\lambda\mu\nu})$, whose cohomology class depends only on that of $(f_{\lambda\mu})$. The cohomology class of $(c_{\lambda\mu\nu})$ in $H^2(M, \boldsymbol{Z})$ is called the (integral) *Chern class* of E and denoted by $c(E)$. The corresponding real cohomology class $c_{\boldsymbol{R}}(E)$, i. e., the image of $c(E)$ in $H^2(M, \boldsymbol{R})$, is the *real Chern class*. The class $c(K(M))$ (or $c_{\boldsymbol{R}}(K(M))$) is called the "first Chern class" of M.

Now suppose there is given a collection of positive C^∞-functions h_λ on U_λ such that

$$h_\lambda(z) = h_\mu(z)|f_{\lambda\mu}(z)|^{-2} \qquad (z \in U_\lambda \cap U_\mu). \tag{8.6}$$

Then the expression $h_\lambda(z)d\zeta^{(\lambda)} \wedge d\overline{\zeta^{(\lambda)}}$ has an invariant meaning and defines a hermitian metric on the fibers of E. For simplicity, we call such a system (h_λ) a *hermitian structure* on the complex line bundle E. Any complex line bundle E admits a hermitian structure.

Lemma 8.1. *Let (h_λ) be a hermitian structure on a complex line bundle E. Then the real Chern class of E is represented (in the de Rham group of M) by the 2-form $(2\pi i)^{-1} d'd''$ $\log h_\lambda$:*

$$c_{\boldsymbol{R}}(E) = \mathrm{Cl}\Big(\frac{1}{2\pi i} d'd'' \log h_\lambda\Big). \tag{8.7}$$

Proof (Weil [2], V, Prop. 2, Cor. 3 ; cf. also [1]). Denote by δ the Čech coboundary operator with respect to the covering $\{U_\lambda\}$. Then, by definition, $-c_{\boldsymbol{R}}(E)$ is represented by $\delta\Big(\frac{1}{2\pi}\mathrm{Im}(\log f_{\lambda\mu})\Big)$. On the other hand, one has

$$\begin{aligned} d\Big(\frac{1}{2\pi}\mathrm{Im}(\log f_{\lambda\mu})\Big) &= \frac{1}{2\pi}\mathrm{Im}(d'\log h_\mu - d'\log h_\lambda) \\ &= \delta\Big(\frac{1}{2\pi}\mathrm{Im}(d'\log h_\lambda)\Big), \\ d\Big(\frac{1}{2\pi}\mathrm{Im}(d'\log h_\lambda)\Big) &= \frac{1}{4\pi i} d(d'\log h_\lambda - d''\log h_\lambda) \\ &= -\frac{1}{2\pi i} d'd''\log h_\lambda. \end{aligned}$$

Hence, under the de Rham isomorphism, $c_{\boldsymbol{R}}(E)$ corresponds to the de Rham class of $(2\pi i)^{-1}d'd''\log h_\lambda$, q. e. d.

We apply this Lemma to the following situation. Let $\mathcal{D}$ be a domain in $\boldsymbol{C}^n$ and Γ a group acting on $\mathcal{D}$ holomorphically, properly discontinuously and without fixed points. Suppose there is given a holomorphic "automorphy factor" $j: \Gamma \times \mathcal{D}$

$\to C^\times$, which means that $j(\gamma, z)$ is holomorphic in z and satisfies the relations

$$j(\gamma\gamma', z) = j(\gamma, \gamma' z)\, j(\gamma', z)$$

for all $\gamma, \gamma' \in \Gamma$, $z \in \mathcal{D}$ (cf. II, § 5). Then one can define a holomorphic action of Γ on $\mathcal{D} \times C$ by

(8. 8) $$\gamma:\ (z, \zeta) \longmapsto (\gamma(z), j(\gamma, z)\zeta).$$

Clearly, this action is also properly discontinuous and fixed point free, so that the quotient space $E(j) = \Gamma \backslash (\mathcal{D} \times C)$ is a complex line bundle over $M = \Gamma \backslash \mathcal{D}$. Let $\pi: \mathcal{D} \to M$ denote the covering map. We may choose an open covering $\{U_\lambda\}$ in such a way that each U_λ is "evenly covered", i. e., if $U_{\lambda,1}$ is one of the connected components of $\pi^{-1}(U_\lambda)$ then π induces a homeomorphism $U_{\lambda,1} \to U_\lambda$ and $\pi^{-1}(U_\lambda)$ is a disjoint union of $\{\gamma U_{\lambda,1} (\gamma \in \Gamma)\}$. One may also assume that $U_\lambda \cap U_\mu$ is connected for all λ, $\mu \in \Lambda$. Then, for each pair (λ, μ) such that $U_\lambda \cap U_\mu \neq \phi$, there exists a unique element $\gamma = \gamma_{\lambda\mu} \in \Gamma$ such that $U_{\lambda,1} \cap \gamma U_{\mu,1} \neq \phi$, and it is easy to see that $E(j)$ is the complex line bundle defined by the 1-cocycle

(8. 9) $$f_{\lambda\mu}(z) = j(\gamma_{\lambda\mu}, z^{(\mu)}),$$

where $z^{(\mu)} = (\pi | U_{\mu,1})^{-1}(z)$. In particular, when j is the ordinary jacobian, one has $E(\text{jac}) = K(M)^{-1}$.

Now suppose further that there exists a kernel function $k: \mathcal{D} \times \mathcal{D} \to C^\times$ associated with j. This means that $k(z, z')$ is holomorphic in z, and satisfies the relations $k(z', z) = \overline{k(z, z')}$ and

(8. 10) $$k(\gamma(z), \gamma(z')) = j(\gamma, z) k(z, z') \overline{j(\gamma, z')}$$

for all $\gamma \in \Gamma$, $z, z' \in \mathcal{D}$. If moreover $k(z, z) > 0$ for all $z \in \mathcal{D}$, then $k(z, z)^{-1}$ is a Γ-invariant hermitian structure on the trivial complex line bundle $\mathcal{D} \times C$. This amounts to saying that in the above notation

(8. 11) $$h_\lambda(z) = k(z^{(\lambda)}, z^{(\lambda)})^{-1}$$

is a hermitian structure on $E(j)$. Hence, by Lemma 8. 1, one has

(8. 12) $$c_R(E(j)) = \mathrm{Cl}\left(-\frac{1}{2\pi i} d'd'' \log k(z, z)\right).$$

Now Kodaira's theorem can be formulated as follows.

Theorem 8. 2 (*Kodaira*). *Let M be a compact complex manifold with a Kählerian metric h and let ω be the associated 2-form. If the cohomology class of ω is "integral", i. e., if it is in the image of the natural homomorphism $H^2(M, Z) \to H^2(M, R)$, then M is algebraic. Moreover, let E be a complex line bundle over M with $c_R(E) = \mathrm{Cl}(\omega)$ and let $(\varphi_0, \cdots, \varphi_N)$ be a basis of the space of holomorphic sections of νE, ν being a positive integer. Then, for a sufficiently large ν, the map*

(8. 13) $$M \ni p \longmapsto (\varphi_0(p), \cdots, \varphi_N(p)) \in P^N(C)$$

is well-defined and gives an analytic isomorphism of M onto a (non-singular) algebraic variety in $P^N(C)$.

For the proof, see Kodaira [2] or Lascoux-Berger [1]. A Kählerian metric h on M satisfying the integrality condition mentioned above is called a *Hodge metric.*

In the situation described above, suppose that the Hessian

$$\left(2\frac{\partial^2}{\partial z^\alpha \partial \bar{z}^\beta}\log k(z, z)\right)$$

is positive (or negative) definite. Then one obtains a hermitian metric

$$ds^2 = (-)\frac{1}{\pi}\sum_{\alpha,\beta}\frac{\partial^2}{\partial z^\alpha \partial \bar{z}^\beta}\log k(z, z)dz^\alpha d\bar{z}^\beta \tag{8. 14}$$

on $\mathcal{D}$, which is clearly Γ-invariant and Kählerian. The induced metric h on $M=\Gamma\backslash\mathcal{D}$ is Hodge, because the class of the associated 2-form $\omega=(-)\frac{i}{2\pi}d'd''\log k(z, z)$ is integral by (8. 12). Therefore M is algebraic. In this case, a holomorphic section of $\nu E(j)(=E(j^\nu))$ is identified with a holomorphic function f on $\mathcal{D}$ satisfying the relation

$$f(\gamma(z)) = j(\gamma, z)^\nu f(z) \qquad \text{for all } \gamma\in\Gamma,\ z\in\mathcal{D}. \tag{8. 15}$$

Such a function is called a (holomorphic) *automorphic form* for $(\Gamma, \mathcal{D})$ with respect to the automorphy factor j^ν, or of "weight" ν when j is fixed. Thus we obtain

Corollary 8. 3. *Let $\mathcal{D}$ be a domain in $\boldsymbol{C}^n$ and Γ a group acting on $\mathcal{D}$ holomorphically, properly discontinuously and without fixed point and such that the quotient space $\Gamma\backslash\mathcal{D}$ is compact. Suppose that there exist a holomorphic automorphy factor $j:\Gamma\times\mathcal{D}\to\boldsymbol{C}^\times$ and the associated kernel function $k:\mathcal{D}\times\mathcal{D}\to\boldsymbol{C}^\times$ such that $k(z, z)>0$ and the Hessian of* $\log k(z, z)$ *is positive (or negative) definite. Then the quotient space $M=\Gamma\backslash\mathcal{D}$ is algebraic and, for a sufficiently large ν, a map defined by a basis of the space of holomorphic automorphic forms for $(\Gamma, \mathcal{D})$ of weight ν gives a projective embedding of M.*

As explained in II, § 6, for a symmetric domain $\mathcal{D}_1$ and any discrete subgroup Γ_1 of G_1 without fixed points such that $\Gamma_1\backslash\mathcal{D}_1$ is compact, the inverse jacobian $j_1=(\text{jac})^{-1}$ and the Bergman kernel function $k_1=k_{\mathcal{D}_1}$ satisfy the conditions stated in Corollary 8. 3, so that one can conclude that the quotient space $M_1=\Gamma_1\backslash\mathcal{D}_1$ is projective. This implies a theorem of Siegel saying that the field of (meromorphic) automorphic functions for $(\Gamma_1, \mathcal{D}_1)$ is an algebraic function field of transcendence degree equal to $\dim\mathcal{D}_1$ (cf. Siegel [8], [11]). At the same time, we see that, if $h_{\mathcal{D}_1}$ denotes the Bergman metric of $\mathcal{D}_1$, then $(2\pi)^{-1}h_{\mathcal{D}_1}$ induces a Hodge metric on M_1 and the associated 2-form $-(2\pi i)^{-1}d'd''\log k_{\mathcal{D}_1}$ belongs to the first Chern class $c_R(K(M_1))$. We will apply the same principle to prove the algebraicity of Kuga's fiber varieties.

The notation being as in § 7, let J_1 and K_1 denote the canonical automorphy factor $G_1\times\mathcal{D}_1\to K_{1C}$ and the canonical kernel function $\mathcal{D}_1\times\mathcal{D}_1\to K_{1C}$ defined in II, § 5 for $(G_1, \mathcal{D}_1)$. For $\gamma=l\cdot\gamma_1\in\tilde{\Gamma}=\Gamma_1\cdot L$, $z=(z_1, w)$, $z'=(z_1', w')\in\tilde{\mathcal{D}}=\mathcal{D}_1\times V_+$, we put

$$\text{(8.16)}\quad \begin{cases} \eta(\gamma, z) = \mathbf{e}\Big(\dfrac{1}{2}A(l, l_{\gamma_1(z_1)}) + A(l, \rho(J_1)w) + \dfrac{1}{2}A(\rho(\gamma_1)w, \rho(J_1)w)\Big), \\ \kappa(z, z') = \mathbf{e}\Big(\dfrac{1}{2}A(\overline{\varphi(z_1')}\rho(K_1)^{-1}w, w) + A(\bar{w}', \rho(K_1)^{-1}w) \\ \qquad\qquad\qquad + \dfrac{1}{2}A(\bar{w}', \rho(K_1)^{-1}\varphi(z_1)\bar{w}')\Big), \end{cases}$$

where J_1, K_1 stand for $J_1(\gamma_1, z_1)$, $K_1(z_1, z_1')$, respectively. Note that, from the definitions, one has the relations

$$\text{(8.17)}\quad \begin{cases} \rho(J_1)|V_+ = J_+(\rho(\gamma_1), \varphi(z_1)) = {}^t(\bar{b}\varphi(z_1)+\bar{a})^{-1}, \\ \rho(K_1)|V_+ = K_+(\varphi(z_1), \varphi(z_1')) = 1_{V_+} - \varphi(z_1)\overline{\varphi(z_1')}, \end{cases}$$

where J_+ and K_+ denote the restrictions on V_+ of the canonical automorphy factor and kernel function of $Sp(V, A)$ (II, § 7). In the notation of III, § 5 with $U=\boldsymbol{R}$, (8.16) can be rewritten as

$$\text{(8.16')}\quad \begin{cases} \eta(\gamma, z) = \mathbf{e}\Big(-\dfrac{1}{4}\mathcal{J}_U((0, l, \gamma_1), (w, z_1))\Big), \\ \kappa(z, z') = \mathbf{e}\Big(-\dfrac{1}{4}\mathcal{K}_U((w, z_1), (w', z_1'))\Big). \end{cases}$$

Recall that one has for $\gamma = l\gamma_1$, $\gamma' = l'\gamma_1'$, $z = (w, z_1)$

$$\begin{aligned} \mathcal{J}_U((0, l, \gamma_1), \gamma'(z)) + \mathcal{J}_U((0, l', \gamma_1'), z) &= \mathcal{J}_U((-2A(l, \gamma_1 l'), l+\gamma_1 l', \gamma_1\gamma_1'), z) \\ &= -2A(l, \gamma_1 l') + \mathcal{J}_U((0, l+\gamma_1 l', \gamma_1\gamma_1'), z), \end{aligned}$$

whence follows that

$$\text{(8.18)}\quad \eta(\gamma, \gamma'(z))\eta(\gamma', z) = \mathbf{e}\Big(\frac{1}{2}A(l, \gamma_1 l')\Big)\eta(\gamma\gamma', z).$$

Similarly, one has

$$\text{(8.19)}\quad \kappa(\gamma(z), \gamma(z')) = \eta(\gamma, z)\kappa(z, z')\overline{\eta(\gamma, z')}.$$

Now, let ψ be a *semi-character* of L relative to A, i. e., a map $L \to \boldsymbol{C}^{(1)}$ such that

$$\text{(8.20)}\quad \psi(l+l') = \mathbf{e}\Big(\frac{1}{2}A(l, l')\Big)\psi(l)\psi(l'),$$

and put

$$Sp(L, A, \psi) = \{\gamma \in Sp(L, A) \mid \psi\circ\gamma = \psi\}.$$

Replacing Γ_1 by a subgroup of finite index if necessary, we may assume that $\rho(\Gamma_1) \subset Sp(L, A, \psi)$. Then from the above formulas, we see that

$$\text{(8.21)}\quad \eta_\psi(\gamma, z) = \psi(l)\cdot\eta(\gamma, z) \qquad (\gamma = l\gamma_1 \in \tilde{\Gamma},\ z \in \tilde{\mathcal{D}})$$

is a holomorphic automorphic factor for $(\tilde{\Gamma}, \tilde{\mathcal{D}})$ and κ is the associated kernel function.

To compute the Hessian of $\log\kappa$, we introduce the following notation. For the function $w = w(z_1, v)$ defined by (7.10), we put

$$\text{(8.22)}\quad d^{z_1}w = dv_+ - \varphi(z_1)d\bar{v}_+ \qquad (\text{and } d^{z_1}\bar{w} = d\bar{v}_+ - \overline{\varphi(z_1)}dv_+),$$

where dv_+ is viewed as a V_+-valued differential form. Then one has

(8. 23) $$dw = d^{z_1}w - d\varphi\cdot\bar{v}_+.$$

($d^{z_1}w$ is the differential of w in the direction of the fiber over z_1.)

Lemma 8. 4. *One has*

(8. 24) $$\begin{cases} d'v_+ = \rho(K_1)^{-1}d^{z_1}w, \\ d''v_+ = \rho(K_1)^{-1}\varphi_1 d^{z_1}\bar{w}, \end{cases}$$

where K_1 and φ_1 stand for $K_1(z_1, z_1)$, $\varphi(z_1)$, respectively.

Proof. From (8. 22) one has

$$d^{z_1}w + \varphi_1 d^{z_1}\bar{w} = (1-\varphi_1\bar{\varphi}_1)dv_+ = \rho(K_1)dv_+.$$

Hence

$$dv_+ = \rho(K_1)^{-1}(d^{z_1}w + \varphi_1 d^{z_1}\bar{w}).$$

From (8. 23) $d^{z_1}w$ is of type (1, 0) and hence $d^{z_1}\bar{w}$ is of type (0, 1). Therefore one obtains (8. 24), q. e. d.

In the following, we fix the basis $(e_1, \cdots, e_{2n})$ of V_C introduced in II, § 7 and consider v_+, w (resp. $\rho(K_1)|V_+$ and φ_1) as expressed by complex column n-vectors (resp. $n\times n$ matrices). Then, for $w, w' \in V_+$, one has $iA(\bar{w}, w') = {}^t\bar{w}\cdot w'$.

Lemma 8. 5. *One has*

(8. 25) $$\frac{i}{2\pi}d'd''\log\kappa(z, z) = i\,{}^t(d^{z_1}w)\wedge\rho(K_1)^{-1}d^{z_1}\bar{w}.$$

Proof. From (8. 16) and (7. 10) one has

$$\begin{aligned} \frac{1}{2\pi}\log\kappa(z, z) &= \frac{1}{2}{}^tw\cdot\bar{\varphi}_1\cdot\rho(K_1)^{-1}w + {}^t\bar{w}\cdot\rho(K_1)^{-1}w + \frac{1}{2}{}^t\bar{w}\cdot\rho(K_1)^{-1}\cdot\varphi_1\cdot\bar{w} \\ &= \frac{1}{2}({}^t\bar{v}_+\cdot\rho(K_1)v_+ + {}^t\bar{w}\cdot w). \end{aligned}$$

Hence one has by Lemma 8. 4 and (8. 17)

$$\begin{aligned} \frac{1}{2\pi}d''\log\kappa &= \frac{1}{2}({}^t(\overline{\rho(K_1)^{-1}d^{z_1}w})\rho(K_1)v_+ + {}^t\bar{v}_+(-\varphi_1 d\bar{\varphi}_1)v_+ \\ &\quad + {}^t\bar{v}_+\cdot\rho(K_1)(\rho(K_1)^{-1}\varphi_1 d^{z_1}\bar{w}) + {}^td\bar{w}\cdot w) \\ &= {}^td\bar{w}\cdot v_+ + \frac{1}{2}{}^tv_+\cdot d\bar{\varphi}_1\cdot v_+ \end{aligned}$$

and

$$\begin{aligned} \frac{1}{2\pi}d'd''\log\kappa &= -{}^td\bar{w}\wedge\rho(K_1)^{-1}d^{z_1}w - {}^tv_+ d\bar{\varphi}_1\wedge\rho(K_1)^{-1}d^{z_1}w \\ &= -{}^t(d^{z_1}\bar{w})\wedge\rho(K_1)^{-1}d^{z_1}w, \end{aligned}$$

which proves (8. 25), q. e. d.

Lemma 8. 5 shows that $(2\pi i)^{-1}d'd''\log \kappa(z, z)$ is the 2-form associated to the hermitian metric on the fiber over z_1 corresponding to the hermitian form

$$2{}^t\bar{w}\cdot\rho(K_1)^{-1}w = 2iA(\bar{w}, \rho(K_1)^{-1}w)\ (=H_{\varphi(z_1)}(v, v)).$$

It follows that, if one defines a hermitian metric

$$ds^2 = \pi^{-1}\sum_{\alpha,\beta}\left(\frac{\partial^2}{\partial\bar{z}_1^\alpha\partial z_1^\beta}\log k_{\mathcal{D}_1}(z_1, z_1)\right)d\bar{z}_1^\alpha dz_1^\beta + 2{}^t d\bar{w}\rho(K_1)^{-1}dw \tag{8. 26}$$

on $\tilde{M}=\tilde{\Gamma}\backslash\tilde{\mathcal{D}}$, then the associated 2-form is given by

$$\omega = \frac{i}{2\pi}d'd''\log k_{\mathcal{D}_1}(z_1, z_1) + i\,{}^t(d^{z_1}w)\wedge\rho(K_1)^{-1}d^{z_1}\bar{w} \tag{8. 27}$$

which represents the real Chern class of the complex line bundle $E(j_1\cdot\eta_\phi)$. Hence, when M_1 and hence $\tilde{M}$ is compact, the hermitian metric (8. 26) is a Hodge metric. Thus by Corollary 8. 3 we obtain the following result (Kuga [1], Satake [14]).

Theorem 8. 6. *When $M_1=\Gamma_1\backslash\mathcal{D}_1$ is compact, the Kuga's fiber variety $\tilde{M}=\Gamma_1\cdot L\backslash(\mathcal{D}_1\times V_+)$ is algebraic. If $(\varphi_0, \cdots, \varphi_N)$ is a basis of the space of holomorphic automorphic forms for $(\Gamma_1\cdot L, \mathcal{D}_1\times V_+)$ with respect to the automorphy factor $(j_1(\gamma_1, z_1)\cdot\eta_\phi(l\gamma_1, (w, z_1)))^\nu$, where $j_1=(\mathrm{jac})^{-1}$ and η_ϕ is given by* (8. 21), (8. 16), *then, for a sufficiently large ν, the map defined by $(\varphi_0, \cdots, \varphi_N)$ gives a projective embedding of $\tilde{M}$.*

It follows also by a theorem of Chow that $\tilde{M}$ is an algebraic fiber variety over M_1; in particular, the fibers are abelian varieties as already remarked. The automorphic forms for $(\Gamma_1\cdot L, \mathcal{D}_1\times V_+)$ mentioned in Theorem 8. 6 are essentially given by "theta-functions" (cf. e. g., Weil [2], Ch. VI, Igusa [8]).

Chapter V
Infinitesimal Automorphisms of Symmetric Siegel Domains

§ 1. Holomorphic vector fields on a Siegel domain.

Let $\mathcal{S}=\mathcal{S}(U, V, \Omega, H)$ be a Siegel domain (of the second kind) defined by the following data:

$$\begin{cases} U = \text{a real vector space of dimension } m, \\ V = \text{a complex vector space of dimension } n, \\ \Omega = \text{a (non-degenerate) open convex cone in } U \text{ (cf. I, § 8)}, \\ H = \text{a "}\Omega\text{-positive" hermitian map, i. e., a sesquilinear map} \\ \qquad V\times V \rightarrow U_c \text{ satisfying the condition} \\ \qquad\qquad v \in V,\ v \neq 0 \Longrightarrow H(v, v) \in \bar{\Omega} - \{0\}. \end{cases}$$

(As before, we assume that H is $\boldsymbol{C}$-linear in the second variable and $\boldsymbol{C}$-antilinear in the first.) Then one has

$$\mathcal{S} = \{(u, v) \in U_c \times V \mid \operatorname{Im} u - H(v, v) \in \Omega\}. \tag{1.1}$$

When V reduces to $\{0\}$, $\mathcal{S}=U+i\Omega$ is a "tube domain" (or a Siegel domain of the first kind). We also use the following notation introduced previously:

$$\begin{aligned} \operatorname{Hol}(\mathcal{S}) &= \text{the Lie group of holomorphic automorphisms of } \mathcal{S}, \\ \operatorname{Aff}(\mathcal{S}) &= \text{the Lie group of affine automorphisms of } \mathcal{S}, \\ G(\Omega) &= \text{the Lie group of linear automorphisms of } \Omega, \\ \mathfrak{g} = \mathfrak{g}(\mathcal{S}) &= \operatorname{Lie} \operatorname{Hol}(\mathcal{S}), \qquad \mathfrak{g}(\Omega) = \operatorname{Lie} G(\Omega). \end{aligned}$$

In this chapter, we identify $X \in \mathfrak{g}(\mathcal{S})$ with the corresponding vector field $\tilde{X}$ on $\mathcal{S}$ defined by

$$(\tilde{X}f)(u, v) = \frac{d}{dt} f((\exp tX)^{-1}(u, v))\Big|_{t=0}, \tag{1.2}$$

where f is any holomorphic (or C^∞) function defined on an open subset of $\mathcal{S}$*). Then $\tilde{X}$ is a holomorphic vector field and the bracket in $\mathfrak{g}(\mathcal{S})$ coincides with the Poisson bracket. When the relation (1. 2) holds, we say that $\tilde{X}$ is "tangential" to the one-parameter subgroup $g(t)=(\exp tX)^{-1}$. A (holomorphic) vector field on $\mathcal{S}$ is called "complete" if it is tangential to a one-parameter subgroup of $\operatorname{Hol}(\mathcal{S})$. Thus $\mathfrak{g}(\mathcal{S})$ is identified with the Lie algebra of all complete holomorphic vector

*) For $X \in \mathfrak{g}$, let $(X)_l$ (resp. $(X)_r$) denote the corresponding left (resp. right) invariant vector field on G tangential to $\exp tX$ at e. In I, § 1, X was identified with $(X)_l$. Here we identify X with the vector field $\tilde{X}$ on $\mathcal{S}$ induced by $-(X)_r$.

fields on $\mathcal{S}$ (with Poisson bracket). Note that, in our notation, the actions of a one-parameter subgroup $\exp tX$ on "points" and "functions" are contragradient (Exerc. 1).

We fix bases of U and V once and for all and introduce complex coordinate functions z^k $(1 \leq k \leq m)$ and w^α $(1 \leq \alpha \leq n)$ in U_c and V. Then a U_c-valued (resp. V-valued) polynomial function on $U_c \times V$ is represented by an m-tuple (resp. n-tuple) of polynomials in the variables $z=(z^k)$, $w=(w^\alpha)$. The space of U_c-valued (resp. V-valued) polynomial functions on $U_c \times V$ whose components are homogeneous of degree μ in z^k's and of degree ν in w^α's will be denoted by $\mathfrak{P}_{\mu,\nu}$ (resp. $\mathfrak{Q}_{\mu,\nu}$). To each U_c-valued (resp. V-valued) polynomial function $p=(p^k)$ (resp. $q=(q^\alpha)$) on $U_c \times V$, we associate a polynomial vector field on $U_c \times V$ defined by

$$p \cdot \frac{\partial}{\partial z} = \sum_{k=1}^{m} p^k(z, w) \frac{\partial}{\partial z^k},$$
$$\left(\text{resp. } q \cdot \frac{\partial}{\partial w} = \sum_{\alpha=1}^{n} q^\alpha(z, w) \frac{\partial}{\partial w^\alpha}\right).$$

Then the direct sum $\bigoplus_{\mu,\nu=0}^{\infty} (\mathfrak{P}_{\mu,\nu} \oplus \mathfrak{Q}_{\mu,\nu})$ with the product defined by (I. 7. 2) is a Lie algebra, isomorphic to the Lie algebra of polynomial vector fields on $U_c \times V$ (with Poisson bracket) by the correspondence $(p, q) \leftrightarrow -p \cdot \partial/\partial z - q \cdot \partial/\partial w$. One has natural identifications

$$\mathfrak{P}_{00} = U_c, \qquad \mathfrak{Q}_{00} = V,$$
$$\mathfrak{P}_{10} = \mathfrak{gl}(U_c), \qquad \mathfrak{Q}_{01} = \mathfrak{gl}(V).$$

To describe $\mathfrak{g} = \mathfrak{g}(\mathcal{S})$, we introduce the following notation. For $a=(a^k) \in U_c$, $b=(b^\alpha) \in V$, we put

(1. 3 a) $$\partial_a = a \cdot \frac{\partial}{\partial z} \left(= \sum_{k=1}^{m} a^k \frac{\partial}{\partial z^k}\right),$$

(1. 3 b) $$\tilde{\partial}_b = 2i\, H(b, w) \cdot \frac{\partial}{\partial z} + b \cdot \frac{\partial}{\partial w} \left(= 2i \sum_{k=1}^{m} H(b, w)^k \frac{\partial}{\partial z^k} + \sum_{\alpha=1}^{n} b^\alpha \frac{\partial}{\partial w^\alpha}\right).$$

Also, for $A=(a^k_j) \in \mathfrak{gl}(U_c)$, $B=(b^\beta_\alpha) \in \mathfrak{gl}(V)$, we put

(1. 3 c) $$X_{A,B} = (Az) \cdot \frac{\partial}{\partial z} + (Bw) \cdot \frac{\partial}{\partial w}$$
$$\left(= \sum_{j,k=1}^{m} a^k_j z^j \frac{\partial}{\partial z^k} + \sum_{\alpha,\beta=1}^{n} b^\beta_\alpha w^\alpha \frac{\partial}{\partial w^\beta}\right).$$

[For the typographical reason, $X_{A,B}$ will sometimes be written as $X(A, B)$.] We say B is *associated with* A (with respect to H) if the relation

(1. 4) $$AH(v, v') = H(Bv, v') + H(v, Bv')$$

holds for all $v, v' \in V$. For instance, $\frac{1}{2} 1_V$ is associated with 1_U, and $i\, 1_V$ is associated with 0. We put

$$\partial = X_{1_U, \frac{1}{2}1_V} = z\cdot\frac{\partial}{\partial z} + \frac{1}{2}w\cdot\frac{\partial}{\partial w}, \tag{1.5}$$
$$\partial' = X_{0, i1_V} = iw\cdot\frac{\partial}{\partial w}.$$

In this chapter, we assume the following results to be known. (Cf. Piatetskii-Shapiro [2], Kaup-Matsushima-Ochiai [1], referred to as K-M-O, Murakami [5], and Nakajima [4]). In particular, all $X\in\mathfrak{g}$ turns out to be a polynomial vector field of degree at most two.

(A) (K-M-O, Th. 4, 5) $\mathfrak{g}$ has the following gradation

$$\mathfrak{g} = \mathfrak{g}_{-1}+\mathfrak{g}_{-1/2}+\mathfrak{g}_0+\mathfrak{g}_{1/2}+\mathfrak{g}_1 \tag{1.6}$$

where $\mathfrak{g}_\nu=\mathfrak{g}(ad(\partial)\,;\nu)$. The non-positive part $\sum_{\nu\leq 0}\mathfrak{g}_\nu$ is the subalgebra corresponding to $\mathrm{Aff}(\mathcal{S})$, and one has

$$\mathfrak{g}_{-1} = \{\partial_u | u\in U\} \cong U, \tag{1.7 a}$$

$$\mathfrak{g}_{-1/2} = \{\tilde{\partial}_v | v\in V\} \cong V, \tag{1.7 b}$$

$$\mathfrak{g}_0 = \{X_{A,B} | A\in\mathfrak{g}(\Omega), B\in\mathfrak{gl}(V), B \text{ is associated with } A\}. \tag{1.7 c}$$

Moreover, the radical of $\mathfrak{g}$ is of the form

$$\mathfrak{r} = \mathfrak{r}_{-1}+\mathfrak{r}_{-1/2}+\mathfrak{r}_0,$$

and one has

$$\dim\mathfrak{g}_\nu = \dim\mathfrak{g}_{-\nu} - \dim\mathfrak{r}_{-\nu} \qquad \left(\nu=\frac{1}{2}, 1\right). \tag{1.8}$$

The bracket products in $\sum_{\nu\leq 0}\mathfrak{g}_\nu$ (which are not identically equal to zero) are given as follows:

$$[X_{A,B}, \partial_u] = -\partial_{Au}, \tag{1.9 a}$$

$$[X_{A,B}, \tilde{\partial}_v] = -\tilde{\partial}_{Bv}, \tag{1.9 b}$$

$$[X_{A,B}, X_{A',B'}] = -X_{[A,A'],[B,B']}, \tag{1.9 c}$$

$$[\tilde{\partial}_v, \tilde{\partial}_{v'}] = 4\partial_{\mathrm{Im}\,H(v,v')}. \tag{1.9 d}$$

(B) (K-M-O, Th. 6) Fix $e\in\Omega$ and denote by $\mathfrak{k}$ (resp. $\mathfrak{k}(\Omega)$) the subalgebra of $\mathfrak{g}(\mathcal{S})$ (resp. $\mathfrak{g}(\Omega)$) corresponding to the isotropy subgroup of $\mathrm{Hol}(\mathcal{S})$ (resp. $G(\Omega)$) at the point $(ie, 0)$ (resp. e). (These isotropy subgroups are known to be compact. They are maximal compact, if $\mathcal{S}$ and Ω are homogeneous.) Then one has

$$\mathfrak{k} = \mathfrak{k}_0 + \{X+\varphi_e(X) | X\in\mathfrak{g}_1\} + \{Y+\psi_e(Y) | Y\in\mathfrak{g}_{-1/2}\}, \tag{1.10}$$

where $\mathfrak{k}_0=\mathfrak{k}\cap\mathfrak{g}_0$ and

$$\varphi_e = \frac{1}{2}ad(\partial_e)^2|\mathfrak{g}_1, \tag{1.11 a}$$

$$\psi_e = ad(\partial')ad(\partial_e)|\mathfrak{g}_{1/2}. \tag{1.11 b}$$

Remark. The maps $\varphi_e:\mathfrak{g}_1\rightarrow\mathfrak{g}_{-1}$, $\psi_e:\mathfrak{g}_{1/2}\rightarrow\mathfrak{g}_{-1/2}$ are both injective. In fact, if $X\in\mathfrak{g}_1$ and $\varphi_e(X)$

$=0$, then by (1. 10) one has $X \in \mathfrak{g}_1 \cap \mathfrak{k}$. $X \in \mathfrak{g}_1$ implies that $ad(X)$ is nilpotent, while $X \in \mathfrak{k}$ implies that $ad(X)$ is semi-simple. Hence one has $ad(X)=0$; in particular, $[X, \partial]=0$, i. e., $X \in \mathfrak{g}_0$. Therefore $X=0$. Similarly, $X \in \mathfrak{g}_{1/2}$ and $\psi_e(X)=0$ imply $X=0$.

(C) (K-M-O, Th. 4) Put

$$\mathcal{S}^{(0)} = \{(u, v) \in \mathcal{S} \mid v=0\}, \qquad (1.12)$$
$$\mathfrak{g}^{(0)} = \mathfrak{g}_{-1}+\mathfrak{g}_0+\mathfrak{g}_1.$$

Then $\mathcal{S}^{(0)}$ is a complex submanifold of $\mathcal{S}$, holomorphically equivalent to the tube domain $U+i\Omega$, and $\mathfrak{g}^{(0)}$ is the subalgebra of $\mathfrak{g}$ corresponding to the subgroup $G^{(0)}=\{g \in \mathrm{Hol}(\mathcal{S}) \mid g\mathcal{S}^{(0)}=\mathcal{S}^{(0)}\}$. The natural restriction map induces a Lie algebra homomorphism $\mathfrak{g}^{(0)} \to \mathfrak{g}(\mathcal{S}^{(0)}) \cong \mathfrak{g}(U+i\Omega)$.

(D) (Murakami [5], Th. 8. 1) $\mathfrak{g}_{1/2}$ consists of all polynomial vector fields Y of the form

$$Y = p_{11}\cdot\frac{\partial}{\partial z}+(q_{10}+q_{02})\frac{\partial}{\partial w} \qquad (p_{11} \in \mathfrak{P}_{11},\ q_{10} \in \mathfrak{Q}_{10},\ q_{02} \in \mathfrak{Q}_{02}) \qquad (1.13)$$

satisfying the following conditions (1. 14 a, b) :

$$[Y, \mathfrak{g}_{-1/2}] \subset \mathfrak{g}_0, \qquad (1.14\text{ a})$$

$$[Y, \mathfrak{g}_{-1}] \subset \mathfrak{g}_{-1/2}. \qquad (1.14\text{ b})$$

(E) (Murakami [5], Th. 8. 2) $\mathfrak{g}_1$ consists of all polynomial vector fields Z of the form

$$Z = p_{20}\cdot\frac{\partial}{\partial z}+q_{11}\cdot\frac{\partial}{\partial w} \qquad (p_{20} \in \mathfrak{P}_{20},\ q_{11} \in \mathfrak{Q}_{11}) \qquad (1.15)$$

satisfying the following conditions (1. 16 a, b) :

$$[Z, \mathfrak{g}_{-1/2}] \subset \mathfrak{g}_{1/2}, \qquad (1.16\text{ a})$$

$$[Z, \mathfrak{g}_{-1}] \subset \mathfrak{g}_0^0, \qquad (1.16\text{ b})$$

where we put

$$\mathfrak{g}_0^0 = \{X_{A,B} \in \mathfrak{g}_0 \mid \mathrm{Im}\,\mathrm{tr}\,B=0\}.$$

Remark. The result of the form (D), (E) was first obtained by Tanaka [2]. It was then reorganized and sharpened (using the result of K-M-O) in the form stated above by Murakami [5] under the assumption that $\mathcal{S}$ is homogeneous (hence complete with respect to the Bergman metric). Nakajima [4] removed the homogeneity assumption by showing that a Siegel domain is always complete. He also noted that the condition (1. 14 b) is superfluous ([4], Prop. 2. 6 (1) ; cf. the proof of Prop. 2. 1 in the next section). For other approaches to the determinations of $\mathfrak{g}_{1/2}$ and $\mathfrak{g}_1$, see Kaup [3], Dorfmeister [4], Rothaus [3], [5].

Exercises

1. Let M be a complex manifold and X a holomorphic vector field on M tangential to a one-parameter group $g(t)=(\exp tX)^{-1}$ in $\mathrm{Hol}(M)$. For a C^∞-function f defined in a neighbourhood of $p \in M$, show that

$$\frac{d^\nu}{dt^\nu} f(g(t)p)\Big|_{t=0} = (X^\nu f)(p) \qquad (\nu=1, 2, \cdots).$$

For an analytic function f, this gives a Maclaurin expansion

$$(1.17) \qquad f((\exp tX)^{-1}p) = \sum_{\nu=0}^{\infty} \frac{t^\nu}{\nu!}(X^\nu f)(p) = ((\exp tX)f)(p).$$

2. Let $A \in \mathfrak{gl}(n, \boldsymbol{C})$ and let U_1 be a real linear subspace of $\boldsymbol{C}^n$ stable under A. Let $\mathcal{D}$ be a domain in $\boldsymbol{C}^n$, stable under affine transformations of the form $z \mapsto (\exp tA)z + b$ $(t \in \boldsymbol{R}, b \in U_1)$. Show that, for any $b \in U_1$, $\mathcal{D}$ admits a (complete) holomorphic vector field $X = (Az+b)\frac{\partial}{\partial z}$ tangential to a one-parameter group

$$(1.18) \qquad (\exp tX)^{-1}(z) = (\exp tA)z + \sum_{\nu=1}^{\infty} \frac{t^\nu}{\nu!} A^{\nu-1} b.$$

[In this formula, and also in (1.19) below, the symbol z is to be understood as representing a "point", not a system of "coordinate functions".]

3. Let Ω be a self-dual homogeneous cone in $U(=\boldsymbol{R}^n)$ and let $\mathcal{S} = U + i\Omega$ be the corresponding (symmetric) tube domain in $U_C (=\boldsymbol{C}^n)$. U is then endowed with a structure of formally real Jordan algebra with unit element $e \in \Omega$ (I, § 8). Show that, for any $a \in U$, $\mathcal{S}$ admits a (complete) holomorphic vector field $X = \{z, a, z\} \frac{\partial}{\partial z}$ tangential to a one-parameter group

$$(1.19) \qquad (\exp tX)^{-1}(z) = (1 - tz\Box a)^{-1} z.$$

Remark. Let $\mathfrak{g}$ be a semi-simple Lie algebra of hermitian type with a canonical decomposition $\mathfrak{g}_C = \mathfrak{p}_+ + \mathfrak{k}_C + \mathfrak{p}_-$ and $\mathcal{D}$ the associated symmetric bounded domain in $\mathfrak{p}_+$. Then the formal polynomial vector fields on $\mathfrak{p}_+$ considered previously (I, § 7, II, § 3) can also be interpreted as actual holomorphic vector fields on the compact dual $M = G_C/K_C P_-$ of $\mathcal{D}$. More precisely, let $z_0 \in \mathfrak{p}_+$, $T \in \mathfrak{k}_C$ and $A = ad_{\mathfrak{p}_+} T$. Then the vector fields $z_0 \frac{\partial}{\partial z}$, $Az\frac{\partial}{\partial z}$, $\{z, z_0, z\}\frac{\partial}{\partial z}$ on $\mathfrak{p}_+$ are tangential to the following one-parameter groups in $\mathrm{Hol}(M)$:

$$\exp tz_0 : z \longmapsto z + tz_0,$$
$$\exp tT : z \longmapsto (\exp tA)z,$$
$$\exp(-t\bar{z}_0) : z \longmapsto \left(1 - \frac{1}{2} t\, ad[z, \bar{z}_0]\right)^{-1} z.$$

(For the last one, compare (II. 5. 23) and Exerc. 3 above. Cf. also I, § 6, Exerc. 6.) It follows that $\mathfrak{g}_C$ can be identified with the Lie algebra of polynomial vector fields on $\mathfrak{p}_+$ by the correspondences

$$z_0 \longleftrightarrow -z_0 \frac{\partial}{\partial z}, \qquad T \longleftrightarrow -Az\frac{\partial}{\partial z},$$

$$\theta \bar{z}_0 = -\bar{z}_0 \longleftrightarrow -\{z, z_0, z\}\frac{\partial}{\partial z}$$

in accordance with our previous result.

§ 2. Explicit determinations of $\mathfrak{g}_{1/2}$ and $\mathfrak{g}_1$.

Using the results (D), (E), we can obtain more explicit expressions of elements in $\mathfrak{g}_{1/2}$ and $\mathfrak{g}_1$.

Proposition 2. 1. *Every element Y in $\mathfrak{g}_{1/2}$ can be written uniquely in the form*

(2. 1) $$Y = 2i\,H(\Phi(\bar{z}), w)\frac{\partial}{\partial z} + (\Phi(z) + c(w, w))\frac{\partial}{\partial w},$$

where Φ is a $\mathbf{C}$-linear map $U_C \to V$ such that for each $v_0 \in V$ one has

(2. 2 a) $$\Phi_{v_0} = [u \mapsto \operatorname{Im} H(v_0, \Phi(u))] \in \mathfrak{g}(\Omega) \qquad (u \in U)$$

and c is a symmetric $\mathbf{C}$-bilinear map $V \times V \to V$ satisfying the condition

(2. 2 b) $$H(v, c(v', v')) = 2i\,H(\Phi(H(v', v)), v')$$

for all $v, v' \in V$. Conversely, for any pair (Φ, c) satisfying these conditions, the vector field Y defined by (2. 1) *belongs to $\mathfrak{g}_{1/2}$.*

Proof. By (D) every element $Y \in \mathfrak{g}_{1/2}$ can be written uniquely in the form

$$Y = \sum_{k,l,\alpha} a^l_{k\alpha} z^k w^\alpha \frac{\partial}{\partial z^l} + \sum_{k,\alpha,\beta,\gamma} (b^\gamma_k z^k + c^\gamma_{\alpha\beta} w^\alpha w^\beta)\frac{\partial}{\partial w^\gamma}$$

where $c^\gamma_{\alpha\beta} = c^\gamma_{\beta\alpha}$. We put

$$H(v, w) = (H^l(v, w)), \qquad H^l(v, w) = \sum_{\gamma,\delta} h^l_{\gamma\delta} \bar{v}^\gamma w^\delta.$$

For any

$$\tilde{\partial}_v = 2i \sum_k H^k(v, w)\frac{\partial}{\partial z^k} + \sum v^\alpha \frac{\partial}{\partial w^\alpha},$$

one has by (1. 14 a)

(2. 3) $$[\tilde{\partial}_v, Y] = X_{A,B}\ (\in \mathfrak{g}_0).$$

A direct computation of the left-hand side gives

$$\begin{aligned}[\tilde{\partial}_v, Y] = {} & \sum (2i a^l_{k\alpha} H^k(v, w) w^\alpha + a^l_{k\alpha} z^k v^\alpha)\frac{\partial}{\partial z^l} \\ & + \sum (2i b^\gamma_k H^k(v, w) + 2c^\gamma_{\alpha\beta} v^\alpha w^\beta)\frac{\partial}{\partial w^\gamma} \\ & - \sum 2i h^l_{\gamma\delta} \bar{v}^\gamma (b^\delta_k z^k + c^\delta_{\alpha\beta} w^\alpha w^\beta)\frac{\partial}{\partial z^l}.\end{aligned}$$

We define $\Phi : U_C \to V$ and $c : V \times V \to V$ by

$$\Phi(u) = (\Phi^\gamma(u)), \qquad \Phi^\gamma(u) = \sum_k b^\gamma_k u^k,$$

$$c(v, v') = (c^\gamma(v, v')), \qquad c^\gamma(v, v') = \sum_{\alpha,\beta} c^\gamma_{\alpha\beta} v^\alpha v'^\beta.$$

Then, comparing the above expression of $[\tilde{\partial}_v, Y]$ with $X_{A,B}$, one obtains the following three equations:

(2. 4 a) $$Au = \Big(\sum_{k,\alpha} a^l_{k\alpha} u^k v^\alpha - 2i H^l(v, \Phi(u))\Big),$$

(2. 4 b) $$Bv' = 2i\,\Phi(H(v, v')) + 2c(v, v'),$$

(2. 4 c) $$\sum_{k,\alpha} a^l_{k\alpha} H^k(v, v') v'^\alpha = H^l(v, c(v', v')).$$

Since $A \in \mathfrak{g}(\Omega)$ is real, it follows from (2. 4 a) that

$$\sum a^{\iota}_{k\alpha} u^k v^\alpha = 2i\overline{H^{\iota}(v, \Phi(\bar{u}))} = 2iH^{\iota}(\Phi(\bar{u}), v)$$

for $u \in U_C$, $v \in V$. Combining this with (2. 4 c), one obtains (2. 2 b). Also we see that Y can be written in the form (2. 1) and one has

(2. 4 a') $$A = [u \mapsto -2i(H(v, \Phi(u)) - H(\Phi(\bar{u}), v))] = 4\Phi_v \qquad (u \in U_C).$$

Since $A \in \mathfrak{g}(\Omega)$, one obtains (2. 2 a). Conversely, given any pair (Φ, c) satisfying (2. 2 a, b), define Y by (2. 1). Then, from the above computation, we see that (2. 3) holds with A, B defined by (2. 4 a'), (2. 4 b). One has $A \in \mathfrak{g}(\Omega)$ by (2. 2 a) and that B is associated with A follows from (2. 2 b) immediately. Thus (1. 14 a) is satisfied. On the other hand, for any $u \in U$, one has

$$[\partial_u, Y] = 2iH(\Phi(u), w)\frac{\partial}{\partial z} + \Phi(u)\frac{\partial}{\partial w}$$
$$= \tilde{\partial}_{\Phi(u)} \in \mathfrak{g}_{-1/2},$$

which proves (1. 14 b). Therefore by (D) one has $Y \in \mathfrak{g}_{1/2}$, q. e. d.

It is clear from (2. 2 b) that the bilinear map c (if it exists) is uniquely determined by Φ. Hence we write $c = c_\Phi$. The vector field Y given by (2. 1) with $c = c_\Phi$ will be written as Y_Φ. Then, from the above proof, one has

(2. 5) $$[\partial_u, Y_\Phi] = \tilde{\partial}_{\Phi(u)} \ (\in \mathfrak{g}_{1/2}),$$

(2. 3) $$[\tilde{\partial}_v, Y_\Phi] = X_{A,B} \ (\in \mathfrak{g}_0),$$

where A, B are given by (2. 4 a') and (2. 4 b), respectively. Moreover, by an easy computation, we get

(2. 6) $$\psi_e(Y_\Phi) = \tilde{\partial}_{-i\Phi(e)}.$$

To eliminate c_Φ from the above formulas, it is convenient to use the "dual expression". We fix a positive definite inner product $\langle\ \rangle$ on U and extend it to a symmetric C-bilinear form on $U_C \times U_C$. We also fix a positive definite hermitian form h on V (which is C-linear in the second variable) and denote the adjoints with respect to $\langle\ \rangle$ and h by t and $*$, respectively. For $u \in U_C$ we define $R_u \in \mathrm{End}(V)$ by

(2. 7) $$\langle u, H(v, v')\rangle = 2h(v, R_u v') \qquad (v, v' \in V).$$

(For the typographical reason, we sometimes write $R(u)$ for R_u.) Then, clearly $R_u^* = R_{\bar{u}}$; in particular, if $u \in U$, then $R_u \in \mathcal{H}(V, h)$ (the space of hermitian transformations of V with respect to h) and, if $u \in \Omega^*$ (the dual cone of Ω), then $R_u \in \mathcal{P}(V, h)$ (the cone of positive definite hermitian transformations of V with respect to h). We assume that h is chosen in such a way that there exists $u_1 \in U$ with $R_{u_1} = 1_V$. It is clear from the definitions that $B \in \mathfrak{gl}(V)$ is associated with $A \in \mathfrak{g}(\Omega)$ if and only if one has

(2. 8) $$R({}^tAu) = B^* R_u + R_u B \qquad \text{for all } u \in U.$$

For a C-linear map $\Phi : U_C \to V$ we define the adjoint $\Phi^* : V \to U_C$ by

(2. 9) $$h(\Phi(u), v) = \langle \bar{u}, \Phi^*(v)\rangle.$$

In these notations we have the following relations

$$'\Phi_v(u) = i(\Phi^* R_u(v) - \overline{\Phi^* R_{\bar{u}}(v)}), \tag{2.10}$$

$$c_\Phi(v, v) = 4iR(\Phi^*(v))v \tag{2.11}$$

for $u \in U_C$, $v \in V$. In fact, (2. 10) follows immediately from the definitions. By (2. 2 b), (2. 7) and (2. 9), one obtains for $u \in U$, $v, v' \in V$

$$\begin{aligned} h(v', R_u c_\Phi(v, v)) &= 2ih(\Phi(H(v, v')), R_u v) \\ &= 2i\langle H(v', v), \Phi^* R_u v\rangle \\ &= 4ih(v', R(\Phi^* R_u v)v). \end{aligned}$$

Thus we see that (2. 2 b) is equivalent to

$$R_u c_\Phi(v, v) = 4iR(\Phi^* R_u v)v \qquad (u \in U, \ v \in V). \tag{2.2 b'}$$

Substituting u_1 for u in (2. 2 b') one obtains (2. 11).

In view of (2. 10) the condition (2. 2 a) can be rewritten as

$$'\Phi_v = [u \mapsto -2 \operatorname{Im} \Phi^* R_u(v)] \in \mathfrak{g}(\Omega^*) \qquad (u \in U). \tag{2.2 a'}$$

Also, making substitution (2. 11) in (2. 2 b) and (2. 2 b'), we see that the existence of the bilinear map c_Φ satisfying (2. 2 b) is equivalent to either one of the following conditions on Φ:

$$H(\Phi(H(v', v)), v') = 2H(v, R(\Phi^*(v'))v'), \tag{2.12}$$

$$R_u R(\Phi^*(v))v = R(\Phi^* R_u v)v \qquad (u \in U, \ v, v' \in V). \tag{2.12'}$$

Thus we can restate Proposition 2. 1 as follows.

Proposition 2. 1'. *$\mathfrak{g}_{1/2}$ consists of polynomial vector fields of the form*

$$Y_\Phi = 2iH(\Phi(z), w)\frac{\partial}{\partial z} + (\Phi(z) + 4iR(\Phi^*(w))w)\frac{\partial}{\partial w}, \tag{2.1'}$$

where Φ is a C-linear map $U_C \to V$ satisfying the conditions (2. 2 a) *and* (2. 12) (*or equivalently,* (2. 2 a') *and* (2. 12')).

In view of (2. 6) and Remark following § 1, (B), we see that the C-linear map Φ in Proposition 2. 1' is uniquely characterized by the conditions (2. 2 a), (2. 12) and the value $\Phi(e)$. It follows, in particular, that the set of all C-linear maps Φ satisfying the conditions (2. 2 a), (2. 12) is a *complex* vector space.

The formulas (2. 3) and (2. 4 b) can also be rewritten as

$$[\partial_v, Y_\Phi] = 4X(\Phi_v, \Psi_v) \ (\in \mathfrak{g}_0), \tag{2.3'}$$

$$\Psi_v(v') = \frac{i}{2}\Phi(H(v, v')) + iR(\Phi^*(v))v' + iR(\Phi^*(v'))v, \tag{2.4 b'}$$

where Φ_v is given by (2. 4 a'). We note that the condition (2. 12) (or (2. 12')) is equivalent to saying that Ψ_v (defined by (2. 4 b')) is associated with Φ_v.

Proposition 2. 2. *Every element in $\mathfrak{g}_1$ can be written uniquely in the form*

(2. 13) $$Z_{a,b} = a(z, z)\frac{\partial}{\partial z} + b(z, w)\frac{\partial}{\partial w},$$

where a is a symmetric **R**-*bilinear map* $U \times U \to U$ *(which we extend to a* **C**-*bilinear map* $U_c \times U_c \to U_c$*) such that for every* $u_0 \in U$ *one has*

(2. 14 a) $$A_{u_0} = [u \mapsto a(u_0, u)] \in \mathfrak{g}(\Omega),$$

and b is a **C**-*bilinear map* $U_c \times V \to V$ *such that, if one puts*

(2. 15) $$B_{u_0} = \left[v \mapsto \frac{1}{2} b(u_0, v)\right],$$

the following conditions are satisfied:

(2. 14 b) B_{u_0} *is associated to* A_{u_0} *and* $\operatorname{Im} \operatorname{tr} B_{u_0} = 0$ *for all* $u_0 \in U$.

(2. 14 c) $[u \mapsto \operatorname{Im} H(v', b(u, v))] \in \mathfrak{g}(\Omega),$

(2. 14 d) $H(v, b(H(v', v''), v'')) = H(b(H(v'', v), v'), v'')$

$(u \in U,\ v, v', v'' \in V).$

Conversely, for any pair (a, b) *satisfying the conditions* (2. 14 a, b, c, d), *the vector field* $Z_{a,b}$ *defined by* (2. 13) *belongs to* $\mathfrak{g}_1$.

Proof. By (E), every $Z \in \mathfrak{g}_1$ can be written uniquely in the form (2. 13). For every $u \in U$ one has

$$[\partial_u, Z] = 2a(u, z)\frac{\partial}{\partial z} + b(u, w)\frac{\partial}{\partial w}.$$

Hence the condition (1. 16 b) is equivalent to (2. 14 a, b). Next, for $v \in V$, one has

$$[\tilde{\partial}_v, Z] = 2i(2a(H(v, w), z) - H(b(w, z), v))\frac{\partial}{\partial z}$$
$$+ (b(z, v) + 2ib(H(v, w), w))\frac{\partial}{\partial w}.$$

Hence, if we set

$$\Phi(u) = b(u, v),$$
$$c(v'', v'') = 2ib(H(v, v''), v''),$$

then, by Proposition 2. 1, the condition (1. 16 a) implies that Φ and c satisfy the conditions (2. 2 a, b), i. e., b satisfies the conditions (2. 14 c, d). [Moreover, in the above expression of $[\tilde{\partial}_v, Z]$ the coefficient of $\frac{\partial}{\partial z}$ should coincide with $2iH(\Phi(\bar{z}), w)$, i. e., one has

$$2a(H(v, w), z) - H(b(w, z), v) = H(b(\bar{z}, v), w).$$

But this follows from the condition that B_u is associated with A_u for all $u \in U$.] Conversely, if (a, b) is a pair satisfying the conditions (2. 14 a∼d), then it is clear from the above that the vector field $Z_{a,b}$ defined by (2. 13) satisfies the conditions (1. 16 a, b), so that one has $Z_{a,b} \in \mathfrak{g}_1$, q. e. d.

From the above proof, we have for $Z_{a,b} \in \mathfrak{g}_1$

$$[\partial_u, Z_{a,b}] = 2X(A_u, B_u), \tag{2.16}$$

$$[\tilde{\partial}_v, Z_{a,b}] = Y_\Phi \qquad (\Phi(u) = b(u, v)). \tag{2.17}$$

Also, by an easy computation, we get

$$\varphi_e(Z_{a,b}) = \partial_{a(e,e)}. \tag{2.18}$$

As a special case of Proposition 2. 2, we obtain

Corollary 2. 3. *For a tube domain* $\mathcal{S} = U + i\Omega$, *one has*

$$\mathfrak{g}_1 = \left\{ Z_a = a(z, z) \frac{\partial}{\partial z} \right\}, \tag{2.19}$$

where a ranges over symmetric bilinear maps $U \times U \to U$ satisfying the condition (2. 14 a) *for all* $u_0 \in U$.

In view of (2. 18) and Remark following § 1, (B), we see that the bilinear map a in Corollary 2. 3 (or the pair (a, b) in Proposition 2. 2) is uniquely characterized by the conditions stated there and the value $a(e, e)$. For the later use, we state this in the form of a lemma.

Lemma 2. 4. *For each $u \in U$, there exists at most one bilinear map $a : U \times U \to U$ satisfying the following conditions:*

$$\begin{cases} a(u', u'') = a(u'', u'), \\ [u'' \mapsto a(u', u'')] \in \mathfrak{g}(\Omega) \qquad (u', u'' \in U), \\ a(e, e) = u. \end{cases} \tag{2.20}$$

It follows, in particular, that the bilinear map a in Proposition 2. 2 coincides with the one in Corollary 2. 3 if the corresponding values $a(e, e)$ are the same. By a result of Kaup [3] (I, § 7, Exerc. 1), it can be seen that any bilinear map a satisfying (2. 20) defines a Jordan algebra structure on U. In the next section, we shall show that, when Ω is homogeneous and self-dual, the bilinear map a satisfying (2. 20) is given by the Jordan triple product $\{u', u, u''\}$. We shall also need the following lemma later.

Lemma 2. 5. *If* $\mathfrak{g}_1 = \{0\}$, *then* $\mathfrak{g}_{1/2} = \{0\}$.

Proof. In general, for two C-bilinear maps $\Phi, \Phi' : U_C \to V$ satisfying (2. 2 a), (2. 12), let

$$[Y_\Phi, Y_{\Phi'}] = Z_{a,b}.$$

Then, by a straightforward computation, one obtains

$$a(u, u) = 4 \operatorname{Im} H(\Phi(u), \Phi'(u)) \qquad (u \in U). \tag{2.21}$$

If $\mathfrak{g}_1 = \{0\}$, the left-hand side of (2. 21) is identically equal to zero. Substituting $i\Phi$

for Φ', one obtains

$$H(\Phi(u), \Phi(u)) = \operatorname{Im} H(\Phi(u), i\Phi(u)) = 0.$$

Since H is Ω-positive, this implies $\Phi(u) \equiv 0$. Thus one has $\mathfrak{g}_{1/2} = \{0\}$, q. e. d.

Exercises

1. Prove the relations (2. 6), (2. 18), (2. 21).

2. For any $Y_\Phi \in \mathfrak{g}_{1/2}$, $Z_{a,b} \in \mathfrak{g}_1$, prove the relations

$$\Phi(a(u, u)) = b(u, \Phi(u)), \tag{2. 22}$$

$$b(a(u, u), v) = b(u, b(u, v)) \qquad (u \in U,\ v \in V). \tag{2. 23}$$

Hint. To obtain (2. 22), compute $[Y_\Phi, Z_{a,b}] = 0$. (2. 23) then follows by substitution $Y_\Phi = [\tilde{\partial}_v, Z_{a,b}]$. (2. 23) implies that $u \mapsto 2B_u$ is a Jordan algebra representation of U with the multiplication defined by a. Cf. Rothaus [5] ; for the symmetric case, see (3. 20), (4. 3).

§ 3. The symmetric case.

We know that a bounded domain $\mathcal{D}$ is symmetric (holomorphically, or isometrically with respect to the Bergman metric), if and only if $\mathcal{D}$ is homogeneous and $\operatorname{Hol}(\mathcal{D})$ is semi-simple (II, §§ 3, 4, 6). In this section, we will give a criterion for a Siegel domain $\mathcal{S}$ to be symmetric in terms of the data (U, V, Ω, H).

Let G_0 be the group of linear automorphisms of $\mathcal{S}$. Then, by Proposition III. 6. 2, G_0 consists of linear transformations of $U_C \oplus V$ given by a pair (g, l) ($g \in G(\Omega)$, $l \in GL(V)$) satisfying the relation

$$gH(v, v') = H(lv, lv') \qquad (v, v' \in V). \tag{3. 1}$$

Hence one has $\operatorname{Lie} G_0 = \mathfrak{g}_0$. If one denotes by π_U and π_V the projections of G_0 into $G(\Omega)$ and $GL(V)$, respectively, as well as the corresponding Lie algebra homomorphisms, then, in view of (1. 9 a, b), one has

$$\pi_U(X_{A,B}) = -A, \qquad \pi_V(X_{A,B}) = -B.$$

Thus the representations π_U and π_V of $\mathfrak{g}_0$ are nothing other than the restrictions of $ad^{\mathfrak{g}}|\mathfrak{g}_0$ on $\mathfrak{g}_{-1} = U$ and $\mathfrak{g}_{-1/2} = V$.

Lemma 3. 1 (K-M-O, Th. 8). *For a Siegel domain $\mathcal{S}$ the following three conditions are equivalent :*

(a) *$\mathcal{S}$ is homogeneous.*

(b) *$\mathcal{S}$ is "affinely homogeneous", i. e.,* $\operatorname{Aff}(\mathcal{S})$ *is transitive on $\mathcal{S}$.*

(c) *$\pi_U(G_0)$ is transitive on Ω.*

Proof. (c)⇒(b)⇒(a) is trivial. To prove (a)⇒(c), suppose $\mathcal{S}$ is homogeneous. Then one has

$$\dim \mathfrak{g} - \dim \mathfrak{k} = \dim_R \mathcal{S} = 2(m+n).$$

By § 1, (B) one has

$$\dim \mathfrak{k} = \dim \mathfrak{k}_0 + \dim \mathfrak{g}_{1/2} + \dim \mathfrak{g}_1.$$

Hence the above equation is equivalent to

$$(*) \qquad \dim \mathfrak{g}_0 - \dim \mathfrak{k}_0 = m.$$

Let K denote the isotropy subgroup of G at $(ie, 0)$. Then $\mathrm{Lie}(K \cap G_0) = \mathfrak{k}_0$. The group G_0 acts (linearly) on $\Omega \subset U$ through π_U, and $\pi_U(g)$ $(g \in G_0)$ fixes e if and only if g fixes $(ie, 0)$, i. e., $g \in K \cap G_0$. Therefore, since Ω is connected, $\pi_U(G_0)$ is transitive on Ω if and only if the relation $(*)$ holds. Thus (a) implies (c), q. e. d.

Proposition 3. 2. *If a Siegel domain $\mathcal{S} = \mathcal{S}(U, V, \Omega, H)$ is symmetric, then Ω is homogeneous, self-dual, and one has $\pi_U(\mathfrak{g}_0) = \mathfrak{g}(\Omega)$.*

Proof. By the assumption, $G^\circ = \mathrm{Hol}(\mathcal{S})^\circ$ is semi-simple and transitive on $\mathcal{S}$. Since the center of G° is trivial, G° may be identified with $\mathrm{Inn}(\mathfrak{g}) = \mathrm{Ad}(\mathfrak{g})^\circ$ (Th. II. 2. 5, Lem. II. 8. 3). Then, from the definition, one has $G_0^\circ = C_{G^\circ}(ad(\partial))^\circ$. Since $ad(\partial)$ is clearly symmetric with respect to the Killing form B of $\mathfrak{g}$, G_0° is self-adjoint (with respect to B) and so, by a theorem of Mostow (Cor. I. 4. 4), "algebraically reductive", i. e., G_0° coincides with the identity connected component of a reductive $\boldsymbol{R}$-group in $GL(\mathfrak{g})$. Therefore $\pi_U(G_0^\circ)$ is also algebraically reductive and hence, for a suitable choice of the inner product $\langle\ \rangle$ on U, we may assume that $\pi_U(G_0^\circ)$ is self-adjoint. Proposition 3. 2 now follows from Lemma I. 8. 3 and Lemma 3. 1, q. e. d.

Remark. If $\mathfrak{g}(\mathcal{S})$ is semi-simple, then $\mathcal{S}$ is necessarily homogeneous, and hence symmetric (Vey [1], Matsushima [9]). In fact, if $\mathfrak{r}_{-1} = 0$, then by § 1, (A) one has $\dim \mathfrak{g}_1 = \dim \mathfrak{g}_{-1}$ and φ_e is surjective. In view of (2. 18), this means that for any $u \in U$ there exists $Z_{a,b} \in \mathfrak{g}_1$ such that $a(e, e) = u$. Since $X = [\partial_e, Z_{a,b}] = 2X(A_e, B_e) \in \mathfrak{g}_0$ and $\pi_U(X)e = -2A_e e = -2a(e, e)$, it follows that $\pi_U(\mathfrak{g}_0)e = U$. Since $e \in \Omega$ can be chosen arbitrarily, this implies that $\pi_U(G_0^\circ)$ is transitive on Ω, and hence (Lem. 3. 1) $\mathrm{Hol}(\mathcal{S})^\circ$ is also transitive on $\mathcal{S}$.

In the symmetric case, we will always assume that the inner product $\langle\ \rangle$ is so chosen that Ω is self-dual with respect to $\langle\ \rangle$ and the Cartan involution of $\mathfrak{g}(\Omega)$ at the "reference point" e is given by $\theta_\Omega : A \mapsto -{}^t A$ $(A \in \mathfrak{g}(\Omega))$. Let

$$\mathfrak{g}(\Omega) = \mathfrak{k}(\Omega) + \mathfrak{p}(\Omega)$$

be the corresponding Cartan decomposition. Then one has

$$(3.2) \qquad A \in \mathfrak{k}(\Omega) \Longleftrightarrow Ae = 0 \Longleftrightarrow {}^t A = -A$$

for $A \in \mathfrak{g}(\Omega)$ (I, § 8).

Proposition 3. 3. *Suppose that $\mathcal{S}$ is symmetric and let θ be the Cartan involution of $\mathfrak{g} = \mathfrak{g}(\mathcal{S})$ at $(ie, 0)$. Then θ reverses the gradation (1. 6), i. e., $\theta(\mathfrak{g}_\nu) = \mathfrak{g}_{-\nu}$ and one has*

$$(3.3) \qquad \theta|\mathfrak{g}_1 = \varphi_e, \qquad \theta|\mathfrak{g}_{1/2} = \psi_e,$$

$$(3.4) \qquad \pi_U \circ (\theta|\mathfrak{g}_0) = \theta_\Omega \circ \pi_U.$$

Proof. Let $X \in \mathfrak{g}_1$. First we shall show that $X - \varphi_e(X) \in \mathfrak{p} = \mathfrak{k}^{\perp}$. Then, since we know that $X + \varphi_e(X) \in \mathfrak{k}$, it will follow that $\theta(X) = \varphi_e(X)$ $(\in \mathfrak{g}_1)$. Let $X' \in \mathfrak{g}_1$. Then, in view of the orthogonality $B(\mathfrak{g}_\nu, \mathfrak{g}_\mu) = 0$ for $\nu \neq -\mu$, one has

$$\begin{aligned} &B(X - \varphi_e(X),\ X' + \varphi_e(X')) \\ &= \frac{1}{2} B(X, ad(\partial_e)^2 X') - \frac{1}{2} B(ad(\partial_e)^2 X, X') \\ &= -\frac{1}{2} B(ad(\partial_e) X, ad(\partial_e) X') + \frac{1}{2} B(ad(\partial_e) X, ad(\partial_e) X') \\ &= 0, \end{aligned}$$

which proves that $X - \varphi_e(X) \in \mathfrak{p}$. Similarly, noting that $[\partial', \partial_e] = 0$, we can prove that $\theta Y = \psi_e Y \in \mathfrak{g}_{-1/2}$ for $Y \in \mathfrak{g}_{1/2}$. As one has $\dim \mathfrak{g}_\nu = \dim \mathfrak{g}_{-\nu}$ in the semi-simple case (§ 1, (A)), it follows that $\theta(\mathfrak{g}_\nu) = \mathfrak{g}_{-\nu}$ $(\nu > 0)$ and hence $\theta(\mathfrak{g}_0) = \mathfrak{g}_0$. It is clear that $\{(iu, 0) \mid u \in \Omega\}$ is a totally geodesic real submanifold of $\mathcal{S}$ and the vector field $X_{A,B}$ $(\in \mathfrak{g}_0)$ induces a (linear) vector field $Au \cdot \frac{\partial}{\partial u}$ on Ω (cf. § 1, (C)). Hence $\theta(X_{A,B})$ induces $\theta_\Omega(A) u \cdot \frac{\partial}{\partial u}$ on Ω, which proves (3. 4), q. e. d.

From now on we assume that $\mathcal{S}$ is symmetric. Let $K^{(1)}$ be the kernel of π_U in G_0. Then, one has

$$K^{(1)} = \{(1_U, B) \mid H(Bv, Bv') = H(v, v')\} \cong U(V, H),$$

and so $K^{(1)}$ is compact. Since G_0 is algebraically reductive, there exists an (algebraic) ideal $\mathfrak{g}^{(2)}$ in $\mathfrak{g}_0$ such that

$$\mathfrak{g}_0 = \mathfrak{k}^{(1)} \oplus \mathfrak{g}^{(2)}, \qquad \mathfrak{g}^{(2)} \cong \mathfrak{g}(\Omega),$$

where $\mathfrak{k}^{(1)} = \mathrm{Lie}\, K^{(1)}$. Hence the restriction $\theta_0 = \theta|\mathfrak{g}_0$ is uniquely determined by the relation (3. 4). (Note that the ideal $\mathfrak{g}^{(2)}$ is also uniquely determined.)

We define a positive definite hermitian form h on V by

(3. 5) $$h(v, v') = \langle e, H(v, v') \rangle,$$

and accordingly $R_u \in \mathfrak{gl}(V)$ by (2. 7). Then clearly $R_e = \frac{1}{2} 1_V$, and so the assumption on h mentioned in § 2 is certainly satisfied. We shall see later on that θ_0 is given by

$$\theta_0 : X(A, B) \longmapsto X(-{}^tA, -B^*),$$

where $*$ denotes the adjoint with respect to h.

Now, for $u \in U$, $v \in V$, we put

(3. 6) $$\partial_u^* = \theta \partial_u, \qquad \tilde{\partial}_v^* = \theta \tilde{\partial}_v.$$

Then by Proposition 3. 3 one has

$$\mathfrak{g}_1 = \{\partial_u^* \mid u \in U\}, \qquad \mathfrak{g}_{1/2} = \{\tilde{\partial}_v^* \mid v \in V\}.$$

In the notation of Proposition 2. 2, when $\partial_u^* = Z_{a,b}$, we write $a = a^u$, $b = b^u$ and denote the corresponding $A_{u'}$, $B_{u'}$ by $A_{u',u}$, $B_{u',u}$, i. e.,

$$(3.7)\qquad \begin{cases} A_{u',u}(u'') = a^u(u', u''), \\ B_{u',u}(v) = \frac{1}{2} b^u(u', v) \qquad (u, u', u'' \in U,\ v \in V). \end{cases}$$

Then the formula (2. 16) can be rewritten as

$$(3.8)\qquad [\partial_{u'}, \partial^*_u] = 2X(A_{u',u}, B_{u',u}).$$

As explained in I, § 8, there exists for each $u \in U$ a unique element T_u in $\mathfrak{p}(\Omega)$ such that $T_u e = u$; in particular, $T_e = 1_U$. The space U endowed with the product $uu' = T_u u'$ becomes a formally real Jordan algebra with the unit element e. In terms of this Jordan algebra structure, the bilinear map a^u can be described as follows.

Lemma 3. 4. *One has*

$$(3.9)\qquad a^u(u', u'') = \{u', u, u''\} = (u'u)u'' + u'(uu'') - u(u'u''),$$

or equivalently,

$$(3.9')\qquad A_{u',u} = T_{u'u} + [T_{u'}, T_u].$$

Proof. If we define the bilinear map a^u by (3. 9), or equivalently, $A_{u',u}$ by (3. 9′), then the condition (2. 20) in Lemma 2. 4 is clearly satisfied. Therefore, by the uniqueness of a^u, one obtains the relations (3. 9), (3. 9′), q. e. d.

On the other hand, in the notation of Proposition 2. 1′, when $\tilde{\partial}^*_v = Y_\Phi$, we write $\Phi = \Phi^v$, and denote the corresponding $\Phi_{v'}$, $\Psi_{v'}$ by $\Phi_{v',v}$, $\Psi_{v',v}$, i. e.,

$$(3.10\,\mathrm{a})\qquad \Phi_{v',v}(u) = \operatorname{Im} H(v', \Phi^v(u)),$$

$$(3.10\,\mathrm{b})\qquad \Psi_{v',v}(v'') = \frac{i}{2}\Phi^v(H(v', v'')) + iR(\Phi^{v*}(v'))v'' + iR(\Phi^{v*}(v''))v' \qquad (u \in U,\ v, v', v'' \in V).$$

Then (2. 5), (2. 3′), (2. 6) can be rewritten as

$$(3.11)\qquad [\partial_u, \tilde{\partial}^*_v] = \tilde{\partial}_{\Phi^v(u)},$$

$$(3.12)\qquad [\tilde{\partial}_{v'}, \tilde{\partial}^*_v] = 4X(\Phi_{v',v}, \Psi_{v',v}),$$

$$(3.13)\qquad \Phi^v(e) = iv.$$

To determine $B_{u',u}$, Φ^v, etc., we first apply θ on both sides of (3. 11). Comparing the resulting equation with (2. 17), we obtain

$$(3.14)\qquad \Phi^{v'}(u') = -b^u(u', v) = -2B_{u',u}(v),$$

where $v' = \Phi^v(u)$. In view of (3. 13), this implies

$$B_{u,e}(v) = -\frac{1}{2}\Phi^{iv}(u),$$

$$B_{e,u}(v) = -\frac{i}{2}\Phi^v(u).$$

As we remarked in § 2 (or as is seen from either one of these equations), Φ^v is $\boldsymbol{C}$-linear in v. Hence, the right-hand sides of the above two equations are equal, and we get

(3. 15) $$B_{u,e} = B_{e,u} = \left[v \mapsto -\frac{i}{2}\Phi^v(u)\right].$$

Next, from (3. 10 a) and (3. 15) one has

(3. 16) $$\Phi_{v',v}(u) = 2\,\mathrm{Re}\,H(v', B_{u,e}(v)),$$

or equivalently,

(3. 16′) $$\langle\Phi_{v',v}(u), u'\rangle = 4\,\mathrm{Re}\,h(v', R_{u'}B_{u,e}(v))$$
$$(u, u' \in U,\ v, v' \in V).$$

Applying θ on both sides of (3. 12) and using (3. 4), one has

(3. 17) $${}^t\Phi_{v',v} = \Phi_{v,v'}.$$

Hence, by (3. 16′), we get

$$R_{u'}B_{u,e} = B^*_{u',e}R_u.$$

Since $R_e = B_{e,e} = \frac{1}{2}1_V$, this implies

(3. 18) $$B_{u,e} = R_u.$$

Therefore, by (3. 15) and (3. 14) we obtain

(3. 19) $$\Phi^v(u) = 2iR_uv,$$

(3. 20) $$B_{u',u} = 2R_{u'}R_u.$$

Also, since $B_{u,e}$ is associated with $A_{u,e}$ and since $A_{u,e} = T_u$ by (3. 9′), we see that R_u is associated with T_u. Actually, as is easily seen, R_u is the unique element in $\mathscr{H}(V, h)$ associated with T_u. (See the proof of Proposition 4. 1.)

Finally, from (2. 7), (2. 9) and (3. 19), one obtains

(3. 21) $$(\Phi^v)^*(v') = -iH(v, v').$$

Hence the formulas (3. 10 a, b) become

(3. 10′ a) $$\Phi_{v',v}(u) = 2\,\mathrm{Re}\,H(v', R_uv),$$

(3. 10′ b) $$\Psi_{v',v}(v'') = R(H(v, v'))v'' + R(H(v, v''))v' - R(H(v', v''))v$$
$$(u \in U,\ v, v', v'' \in V).$$

Also, the conditions (2. 12), (2, 12′) for $\Phi = \Phi^v$ read

(3. 22) $$H(R(H(v'', v'))v, v'') = H(v', R(H(v, v''))v''),$$

(3. 22′) $$R_uR(H(v, v'))v' = R(H(v, R_uv'))v'.$$

Summing up, we obtain the following theorems except for the "if" part of Theorem 3. 5, which will be proved in the next section.

Theorem 3. 5. *A Siegel domain $\mathscr{S} = \mathscr{S}(U, V, \Omega, H)$ is symmetric if and only if the following three conditions are satisfied.*

(i) *Ω is homogeneous and self-dual.* [*Hence we take $\langle\ \rangle$ and h in such a way that Ω is self-dual with respect to $\langle\ \rangle$ and* (3. 2) *and* (3. 5) *are satisfied. Then we define R_u by* (2. 7).]

(ii) R_u *is associated with* T_u *for all* $u \in U$.
(iii) R_u *satisfies the relation* (3. 22) *for all* $v, v', v'' \in V$.

Theorem 3. 6. *For a symmetric Siegel domain* $\mathcal{S}=\mathcal{S}(U, V, \Omega, H)$, $\mathfrak{g}_1$ *(resp.* $\mathfrak{g}_{1/2}$*) consists of vector fields* ∂_u^* *(resp.* $\tilde{\partial}_v^*$*) given as follows.*

$$\partial_u^* = \{z, u, z\}\frac{\partial}{\partial z}+4R_zR_u(w)\frac{\partial}{\partial w}, \tag{3. 23 a}$$

$$\tilde{\partial}_v^* = 4H(R_{\bar{z}}v, w)\frac{\partial}{\partial z}+(2iR_zv+4R(H(v, w))w)\frac{\partial}{\partial w}, \tag{3. 23 b}$$

where { } *is the Jordan triple product given by* (3. 9) *and* R_u $(u \in U)$ *is the unique element in* $\mathcal{H}(V, h)$ *which is associated with* T_u. *Moreover, the linear transformation* $A_{w,u}$, $B_{w,u}$, $\Phi_{v',v}$, $\Psi_{v',v}$ *in the formulas* (3. 8), (3. 12) *are given by* (3. 9′), (3. 20), (3. 10′ a, b), *respectively.*

In the tube domain case, the conditions (ii), (iii) are trivially satisfied. Hence we obtain

Corollary 3. 7 (*Rothaus*). *A tube domain* $\mathcal{S}=U+i\Omega$ *corresponding to a (non-degenerate) open convex cone* Ω *is symmetric if and only if* Ω *is homogeneous and self-dual. In that case, one has*

$$\partial_u^* = \{z, u, z\}\frac{\partial}{\partial z} \qquad (u \in U). \tag{3. 24}$$

This means that $\mathfrak{g}(U+i\Omega)$ can naturally be identified with the "symmetric Lie algebra" associated with the JTS $(U, \{\ \})$ (I, § 7).

Remark 1. Let $\Gamma=\Gamma(U, \{\ \})$ be the structure group corresponding to a self-dual homogeneous cone Ω (I, §§ 6, 8). For simplicity, assume that Ω is irreducible; then one has $\Gamma=G(\Omega)\cup(-G(\Omega))$ (Prop. I. 9. 4). For $-g \in -G(\Omega)$, one defines its "antiholomorphic action" on $\mathcal{S}=U+i\Omega$ by $z \mapsto -g\bar{z}$. Then one has a natural injection $\Gamma \rightarrow I(\mathcal{S})$, which is compatible with the isomorphisms given by Proposition I. 7. 2 and Theorem II. 2. 5. It can be seen that each coset in $I(\mathcal{S})/I(\mathcal{S})^\circ$ has a representative in Γ (cf. Appendix, §§ 3, 6, and Kaup [3]). Since the semi-direct product $\Gamma^s \cdot U$ is a parabolic subgroup of $I(\mathcal{S})^s$, one has $\Gamma^s=\Gamma\cap I(\mathcal{S})^s$. Therefore one obtains the relation $\Gamma/\Gamma^s \cong I(\mathcal{S})/I(\mathcal{S})^s$ (cf. I, § 9, II, § 8).

Remark 2. For a symmetric Siegel domain $\mathcal{S}$, it is easy to see that the H_0-element at $(ie, 0)$ is given by

$$H_0 = -\frac{1}{2}(\partial_e+\partial'+\partial_e^*) \tag{3. 25}$$

(§ 4, Exerc. 4). It follows that the linear map $\kappa_0 : \mathfrak{sl}_2(\boldsymbol{R}) \rightarrow \mathfrak{g}$ defined by

$$\kappa_0(e_+) = -\partial_e, \quad \kappa_0(e_-) = \partial_e^*, \quad \kappa_0(h) = -2\partial \tag{3. 26}$$

is an (H_1)-homomorphism with respect to $H_0^1=\frac{1}{2}(e_+-e_-)$ and the above H_0, (and $\mathfrak{g}_{\kappa_0}^{(1)}=\mathfrak{k}^{(1)}$ is compact). The corresponding Siegel domain realization $\mathcal{S}_{\kappa_0}\subset U_C+V_+$ (III, § 7) coincides with the given one $\mathcal{S}$ through the natural isomorphism $V\cong V_+$. Thus, in the symmetric case, our result implies the "uniqueness" of the Siegel domain expression (up to a linear isomorphism). For convenience of the reader, we give a list of the corresponding symbols in Chapters III and V:

$$\begin{aligned}
\kappa_0(e_+) &= e_{\kappa_0} \longleftrightarrow -\partial_e = -e\cdot\frac{\partial}{\partial z},\\
-\kappa_0(e_-) &= \theta e_{\kappa_0} \longleftrightarrow -\partial_e^* = -z^2\cdot\frac{\partial}{\partial z} - 2R_z w\cdot\frac{\partial}{\partial w},\\
\kappa_0(h'') &= H_{\kappa_0} \longleftrightarrow -(\partial_e + \partial_e^*),\\
\kappa_0(h) &= X_{\kappa_0} \longleftrightarrow -2\partial = -X_{2\cdot 1_U, 1_V},\\
H_0 - \frac{1}{2}H_{\kappa_0} &= H_0^{(1)} \longleftrightarrow -\frac{1}{2}\partial' = -X_{0,(i/2)\cdot 1_V}.
\end{aligned}$$

§ 4. The sufficiency of the conditions (i) ~ (iii).

In this section, we prove the "if" part of Theorem 3. 5, i. e., that the conditions (i) ~ (iii) are also sufficient for $\mathcal{S}$ to be symmetric. First we assume only the condition (i). Fix $e \in \Omega$ and the inner product $\langle\ \rangle$ so that Ω is self-dual with respect to $\langle\ \rangle$ and (3. 2) holds. Then (U, e) is endowed with a Jordan algebra structure. As before, h and R_u are defined by (3. 5) and (2. 7).

Proposition 4. 1. *Under the condition* (i), *the following conditions are equivalent.*

(ii_1) (= (ii)) *R_u is associated with T_u for all $u \in U$.*

(ii_1') *For each $u \in U$, there exists $R_u' \in \mathcal{H}(V, h)$ which is associated with T_u.*

(ii_2) *The map $u \mapsto 2R_u$ is a (unital) Jordan algebra representation of (U, e) into $\mathcal{H}(V, h)$.*

(ii_3) *There exists a Lie algebra representation $\beta: \mathfrak{g}(\Omega) \to \mathfrak{gl}(V)$ such that, for every $A \in \mathfrak{g}(\Omega)$, $\beta(A)$ is associated with A, i. e.,*

$$R({}^tAu) = \beta(A)^* R_u + R_u \beta(A), \tag{4. 1}$$

and that

$$\beta({}^tA) = \beta(A)^*. \tag{4. 2}$$

(ii_4) *The Lie algebra $\mathfrak{g}_0$, embedded in $\mathfrak{g}(\Omega) \oplus \mathfrak{gl}(V)$ by the natural injection $X_{A,B} \mapsto (-A, -B)$, is stable under the standard involution $(A, B) \mapsto (-{}^tA, -B^*)$, and the natural projection $\pi_U: \mathfrak{g}_0 \to \mathfrak{g}(\Omega)$ is surjective.*

Proof. (ii_1)⇔(ii_1') Suppose there exists $R_u' \in \mathcal{H}(V, h)$ associated with T_u. Then by (2. 8) one has

$$R_{uu'} = R_u' R_{u'} + R_{u'} R_u' \qquad \text{for all } u' \in U.$$

Putting $u' = e$ and using $R_e = \frac{1}{2} 1_V$, one obtains $R_u = R_u'$. Hence (ii_1') implies (ii_1). The converse is trivial.

(ii_1)⇔(ii_2) (ii_1) is equivalent to saying that

$$R_{uu'} = R_u R_{u'} + R_{u'} R_u \tag{4. 3}$$

holds for all $u, u' \in U$. Since $R_e = \frac{1}{2} 1_V$, this means that $u \mapsto 2R_u$ is a (unital) Jordan algebra homomorphism $(U, e) \to \mathcal{H}(V, h)$, i. e., ($\mathrm{ii}_2$). The converse is clear.

(ii_1)⇒(ii_4) (ii_1) implies that for every $u\in U$ one has $X(T_u, R_u)\in\mathfrak{g}_0$. Since $\mathfrak{g}(\Omega)$ is generated by $\mathfrak{p}(\Omega)=\{T_u(u\in U)\}$, it follows that $\pi_U:\mathfrak{g}_0\to\mathfrak{g}(\Omega)$ is surjective. Put $X_u=X(T_u, R_u)$ and $\mathfrak{k}^{(1)}=\mathrm{Ker}\,\pi_U$. If we denote by $\mathfrak{g}^{(2)\prime}$ the subalgebra of $\mathfrak{g}_0$ generated by $\{X_u\,(u\in U)\}$, then one has $\mathfrak{g}_0=\mathfrak{k}^{(1)}+\mathfrak{g}^{(2)\prime}$. To show that $\mathfrak{g}_0$ is stable under the Cartan involution $(A, B)\mapsto(-{}^tA, -B^*)$ of $\mathfrak{g}(\Omega)\oplus\mathfrak{gl}(V)$, it is enough to show that both $\mathfrak{k}^{(1)}$ and $\mathfrak{g}^{(2)\prime}$ are stable. For $\mathfrak{g}^{(2)\prime}$, this follows from the fact that T_u and R_u are self-adjoint. For $\mathfrak{k}^{(1)}$, this is clear from the following equivalence.

$$(4.4)\qquad X_{0,B}\in\mathfrak{k}^{(1)}\Longleftrightarrow B^*R_u+R_uB=0\quad\text{for all } u\in U$$
$$\Longleftrightarrow B^*=-B \text{ and } R_uB=BR_u\quad\text{for all } u\in U.$$

(ii_4)⇒(ii_3) Since $\mathfrak{g}_0$, viewed as an (algebraic) Lie subalgebra of $\mathfrak{gl}(U\oplus V)$, is stable under the standard Cartan involution, it is (algebraically) reductive and

$$(4.5)\qquad \theta_0:\ X(A, B)\longmapsto -X({}^tA, B^*)$$

is a Cartan involution of $\mathfrak{g}_0$. Let $\mathfrak{g}_0=\mathfrak{k}_0+\mathfrak{p}_0$ be the corresponding Cartan decomposition. Since, by assumption, $\pi_U:\mathfrak{g}_0\to\mathfrak{g}(\Omega)$ is surjective, there exists an (algebraic) ideal $\mathfrak{g}^{(2)}$ of $\mathfrak{g}_0$ such that

$$(4.6)\qquad \mathfrak{g}_0=\mathfrak{k}^{(1)}\oplus\mathfrak{g}^{(2)},\qquad \mathfrak{g}^{(2)}\cong\mathfrak{g}(\Omega).$$

Here both $\mathfrak{k}^{(1)}$ and $\mathfrak{g}^{(2)}$ are stable under θ_0. [We note that, since $\mathfrak{k}^{(1)}\subset\mathfrak{k}_0$, one has $\mathfrak{g}^{(2)}\supset\mathfrak{p}_0$, so that the Cartan decomposition of $\mathfrak{g}^{(2)}$ corresponding to $\theta_0|\mathfrak{g}^{(2)}$ is given by $\mathfrak{g}^{(2)}=(\mathfrak{k}_0\cap\mathfrak{g}^{(2)})+\mathfrak{p}_0$. Therefore $\mathfrak{g}^{(2)}$ coincides with the subalgebra generated by $\mathfrak{p}_0$.] If we write

$$(4.7)\qquad \mathfrak{g}^{(2)}=\{X(A, \beta(A))\,|\,A\in\mathfrak{g}(\Omega)\},$$

then $\beta:\mathfrak{g}(\Omega)\to\mathfrak{gl}(V)$ is a Lie algebra homomorphism satisfying the conditions (4.1), (4.2).

(ii_3)⇒(ii_1') Clearly $R_u'=\beta(T_u)$ meets the requirements in (ii_1'), q. e. d.

Corollary 4.2. *Under the assumptions* (i), (ii), *if $B\in\mathfrak{gl}(V)$ is associated with $A\in\mathfrak{g}(\Omega)$, then B^* is associated with ${}^tA$.*

Remark 1. The equivalence of (ii_2) and (ii_3) is a special case of Proposition I. 9. 2. We see from the above proof that, under the assumptions (i), (ii), one has $\mathfrak{g}^{(2)}=\mathfrak{g}^{(2)\prime}$, $\mathfrak{p}_0=\{X_u\,(u\in U)\}$, and hence the representation β in (ii_3) is uniquely determined by the relation $\beta(T_u)=R_u\,(u\in U)$ or by (4.7).

Remark 2. In Proposition 4.1, the condition (ii_3) can be weakened by dropping (4.2), and also the condition (ii_4) can be replaced by the following:

(ii_4') The natural projection $\pi_U:\mathfrak{g}_0\to\mathfrak{g}(\Omega)$ is surjective. (Cf. Takeuchi [9], Lem. 1, and Satake [17], Prop. 1.) Actually, it can be shown that, if $\rho:\mathfrak{g}(\Omega)\to\mathfrak{gl}(V)$ is a Lie algebra representation satisfying only (4.1), then ρ is a commutative sum of two representations ρ_0 and ρ_1 such that $\rho_0(\mathfrak{g}(\Omega))\subset\mathfrak{u}(V, H)$ and ρ_1 satisfies the conditions (4.1), (4.2) (Satake [15], p. 127).

A Siegel domain $\mathcal{S}$ satisfying the conditions (i), (ii) will be called *quasi-symmetric.* For a quasi-symmetric Siegel domain $\mathcal{S}$, we define the symbols ∂_u^*, $\bar\partial_v^*$, $A_{u',u}$, $B_{u',u}$,

$\Phi_{v',v}, \Psi_{v',v}$ by (3. 23 a, b), (3. 9′), (3. 20), (3. 10′ a, b), respectively.

Lemma 4. 3. *Under the assumptions* (i), (ii), $B_{u',u}$ *is associated with* $A_{u',u}$ *for all* u, $u' \in U$. *More precisely, one has*

$$\beta(A_{u',u}) = B_{u',u} \qquad (u, u' \in U), \tag{4. 8}$$

where β *is the unique representation of* $\mathfrak{g}(\Omega)$ *in* V *determined by* (4. 7).

Proof. Since $\beta(T_u)=R_u$, one has by the definitions and (4. 3)

$$\begin{aligned}\beta(A_{u',u}) &= \beta(T_{u'u}+[T_{u'}, T_u])\\ &= \beta(T_{u'u})+[\beta(T_{u'}), \beta(T_u)]\\ &= R_{u'u}+(R_{u'}R_u-R_uR_{u'})\\ &= 2R_{u'}R_u = B_{u',u},\end{aligned}$$

q. e. d.

Lemma 4. 4. *Under the assumptions* (i), (ii), *one has* ${}^t\Phi_{v',v}=\Phi_{v,v'}$ *and* $\Phi_{v',v} \in \mathfrak{g}(\Omega)$ *for all* $v, v' \in V$.

Proof. From the definition (3. 10′ a) one has

$$\langle u', \Phi_{v',v}(u)\rangle = 4 \operatorname{Re} h(v', R_{u'}R_u v) \qquad (u, u' \in U,\ v, v' \in V). \tag{4. 9}$$

Since the right-hand side is invariant under the permutation $(u, u', v, v')\mapsto(u', u, v', v)$, it follows that ${}^t\Phi_{v',v}=\Phi_{v,v'}$. We put

$$\Phi^{(+)}_{v',v} = \frac{1}{2}(\Phi_{v',v}+\Phi_{v,v'}),$$

$$\Phi^{(-)}_{v',v} = \frac{1}{2}(\Phi_{v',v}-\Phi_{v,v'}).$$

Then from (3. 10′ a) one has

$$\Phi^{(+)}_{v',v} = T(\operatorname{Re} H(v, v')) \in \mathfrak{p}(\Omega) \tag{4. 10}$$

(where we write $T(u)$ for T_u). Hence it is sufficient to show that $\Phi^{(-)}_{v',v}$ belongs to $\mathfrak{k}(\Omega)$, or what amounts to the same thing, that $\Phi^{(-)}_{v',v}$ is a derivation of the Jordan algebra U ((I. 8. 8)). From (4. 9) one has

$$\langle u', \Phi^{(-)}_{v',v}(u)\rangle = 2 \operatorname{Re} h(v', [R_{u'}, R_u]v).$$

Hence the relation

$$\Phi^{(-)}_{v',v}(uu') = (\Phi^{(-)}_{v',v}u)u'+u(\Phi^{(-)}_{v',v}u')$$

follows from the identity

$$[R_{uu'}, R_{u''}] = [R_u, R_{u'u''}]+[R_{u'}, R_{uu''}],$$

which can be verified by a simple computation (or from the identity (I. 6. 2)), q. e. d.

Remark (Nakajima). Contrary to Lemma 4. 3, there are always $v, v' \in V$ such that $\beta(\Phi_{v',v}) \neq$

$\Psi_{v',v}$, if $\mathscr{S}$ is symmetric and not of tube type. In fact, one has

$$\begin{aligned}\mathfrak{g}_0 &= [\mathfrak{g}_0, \mathfrak{g}_0]+[V, \theta V]+[U, \theta U] \\ &= \mathfrak{k}^{(1)s}+\mathfrak{g}^{(2)}+[V, \theta V]\end{aligned}$$

by Lemma 4. 3. On the other hand, $V \neq \{0\}$ implies $\mathfrak{k}^{(1)a} \neq \{0\}$, because one has $H_0^{(1)} = -\frac{1}{2}\partial' \in \mathfrak{k}^{(1)a}$. Hence one has $[V, \theta V] \not\subset \mathfrak{g}^{(2)}$.

Lemma 4. 5. *Under the assumptions* (i), (ii), *one has*

$$\Psi^*_{v',v} = \Psi_{v,v'} \qquad (v, v' \in V). \tag{4. 11}$$

The proof is straightforward, and left for an exercise of the reader.

Lemma 4. 6. *Under the assumptions* (i), (ii), *the following conditions are equivalent.*

(iii$_1$) $\quad R_u R(H(v, v'))v' = R(H(v, R_u v'))v'$,

(iii$_1'$) $\quad R(H(v, v'))R_u v' = R(H(R_{\bar{u}} v, v'))v'$,

(iii$_2$) (= (iii)) $\quad H(v', R(H(v, v''))v'') = H(R(H(v'', v'))v, v'')$,

(iii$_3$) $\quad \langle H(v, v''), H(R_{\bar{u}} v', v'')\rangle = \langle H(v, R_u v''), H(v', v'')\rangle$

$\qquad (u \in U_C,\ v, v', v'' \in V)$.

(iii$_4$) $\Psi_{v',v}$ *is associated with* $\Phi_{v',v}$ *for all* $v, v' \in V$.

Proof. (iii$_1$)$\Leftrightarrow$(iii$_3'$) (iii$_1'$) means

$$h(v, R(H(v', v''))R_u v'') = h(v, R(H(R_{\bar{u}} v', v''))v'')$$

for all $u \in U_C$, $v, v', v'' \in V$. By (2. 7) this is equivalent to (iii$_3$).

(iii$_1$)$\Leftrightarrow$(iii$_3$) can be proved similarly. [(iii$_1$)$\Leftrightarrow$(iii$_1'$) follows also from (ii$_1$) and (4. 3). (iii$_1$)$\Leftrightarrow$(iii$_2$) was already shown in § 3.]

(iii$_2$)$\Leftrightarrow$(iii$_3$) (iii$_2$) means that one has

$$\langle u, H(v', R(H(v, v''))v'')\rangle = \langle u, H(R(H(v'', v'))v, v'')\rangle$$

for all $u \in U_C$, $v, v', v'' \in V$. The left-hand side is equal to

$$2h(R_{\bar{u}} v', R(H(v, v''))v'') = \langle H(v, v''), H(R_{\bar{u}} v', v'')\rangle,$$

while the right-hand side is equal to

$$2h(R(H(v'', v'))v, R_u v'') = \langle H(v', v''), H(v, R_u v'')\rangle.$$

Hence (iii$_2$) is equivalent to (iii$_3$).

(iii$_2$)$\Leftrightarrow$(iii$_4$) Put

$$\tilde{H}(v, v', w, w') = H(R(H(v', v))w, w') - H(v, R(H(w, w'))v') \qquad (v, v', w, w' \in V).$$

Then from the definitions (3. 10$'$ a, b), we see that the relation

$$\Phi_{v',v} H(w, w') = H(\Psi_{v',v} w, w') + H(w, \Psi_{v',v} w')$$

is equivalent to

$$\tilde{H}(w, w', v, v') + \tilde{H}(w, v', v, w') = \tilde{H}(v', v, w, w') + \tilde{H}(w, v, v', w').$$

But the left-hand side of this equation is $\boldsymbol{C}$-antilinear in v, while the right-hand side is $\boldsymbol{C}$-linear in v. Hence (iii_4) is equivalent to the following systems of two equations:

$$(*) \qquad \tilde{H}(w, w', v, v') + \tilde{H}(w, v', v, w') = 0,$$

$$(**) \qquad \tilde{H}(v', v, w, w') + \tilde{H}(w, v, v', w') = 0$$

$$\text{for all} \quad v, v', w, w' \in V.$$

On the other hand, the condition (iii_2) is equivalent to saying that $\tilde{H}(v, v', w, w')$ is alternating in v', w', i. e., (*). Since $\tilde{H}(v, v', w, w') = -\tilde{H}(w', w, v', v)$, this is also equivalent to saying that $\tilde{H}(v, v', w, w')$ is alternating in v, w, i. e., (**). Thus the conditions (iii_2), (*), (**), (iii_4) are all equivalent, q. e. d.

Proposition 4. 7. *Let $\mathcal{S}$ be a quasi-symmetric Siegel domain. Then $\tilde{\partial}_v^*$ defined by* (3. 23 b) *belongs to $\mathfrak{g}_{1/2}$ if and only if the relation* (3. 22′) *holds for all $u \in U$, $v' \in V$, and in that case one has $\varphi_e(\tilde{\partial}_v^*) = \tilde{\partial}_v$. ∂_u^* defined by* (3. 23 a) *belongs to $\mathfrak{g}_1$ if and only if the relation*

$$(4.\ 12) \qquad R_{u'} R(H(v, v')) R_u v' = R(H(R_u v, R_{u'} v')) v'$$

holds for all $u' \in U$, $v, v' \in V$, and in that case one has $\varphi_e(\partial_u^*) = \partial_u$.

Proof. From the definitions (3. 23 b) and (2. 1′) one has

$$\tilde{\partial}_v^* = Y_{\Phi^v} \qquad \text{with} \quad \Phi^v = [u \mapsto 2iR_u v]$$

(and hence $(\Phi^v)^* = [v' \mapsto -iH(v', v)]$). Therefore, by Proposition 2. 1′, one has $\tilde{\partial}_v^* \in \mathfrak{g}_{1/2}$ if and only if $\Phi = \Phi^v$ satisfies (2. 2 a) and (2. 12). By Lemma 4. 4, the condition (2. 2 a) is satisfied (under the assumptions (i), (ii)). The condition (2. 12), or equivalently, (2. 12′) for $\Phi = \Phi^v$ amounts to the relation (3. 22) or (3. 22′). Hence, when this condition is satisfied, one has $\tilde{\partial}_v^* \in \mathfrak{g}_{1/2}$, and by (2. 6) and (3. 19) one has $\varphi_e(\tilde{\partial}_v^*) = \tilde{\partial}_v$. Similarly, one has

$$\partial_u^* = Z_{a^u, b^u},$$

where a^u, b^u are given by (3. 7), (3. 9′), (3. 20). By (3. 9′) one has $A_{u',u} \in \mathfrak{g}(\Omega)$ and by Lemma 4. 3 $B_{u',u}$ is associated with $A_{u',u}$. Moreover by (4. 3) one has

$$\mathrm{tr}(B_{u',u}) = 2\,\mathrm{tr}(R_{u'} R_u) = \mathrm{tr}(R_{u'u}) \in \boldsymbol{R}.$$

Thus the conditions (2. 14 a, b) are satisfied. Lemma 4. 4, applied to $\Phi_{v', iR_u v}$, shows that the condition (2. 14 c) is also satisfied. The condition (2. 14 d) amounts to the equation

$$(4.\ 12') \qquad H(v, R(H(v', v'')) R_u v'') = H(R(H(v'', v)) R_u v', v'')$$

$$\text{for all} \quad v, v', v'' \in V.$$

By the same argument as in the proof of Lemma 4. 6, one can see that this is equivalent to saying that (4. 12) holds for all $u' \in U$, v, $v' \in V$. Hence, by Proposition 2. 2, when this condition is satisfied, one has $\partial_u^* \in \mathfrak{g}_1$. Moreover, by (2. 18) and (3. 9) one has $\varphi_e(\partial_u^*) = \partial_u$, q. e. d.

Proof of the "if" part of Theorem 3. 5. Assume the conditions (i), (ii), (iii). Then, by Lemma 4. 6 and Proposition 4. 7, one has $\partial_u^* \in \mathfrak{g}_1$, $\tilde{\partial}_v^* \in \mathfrak{g}_{1/2}$ for all $u \in U$, $v \in V$, and

φ_e and ψ_e are surjective. Hence by § 1, (A) the radical $\mathfrak{r}$ of $\mathfrak{g}$ is contained in $\mathfrak{g}_0$. Then one has $[\mathfrak{r}, \mathfrak{g}_{-\nu}] \subset \mathfrak{r} \cap \mathfrak{g}_{-\nu} = \{0\}$ for $\nu = 1/2, 1$. Therefore $\mathfrak{r} = \{0\}$ and $\mathfrak{g}$ is semi-simple. Moreover, since Ω is homogeneous, so is $\mathcal{S}$. Thus $\mathcal{S}$ is symmetric, q. e. d.

Remark. My original proof of Theorem 3. 5 has been simplified here by Proposition 3. 3, due to Nakajima [5]. A similar characterization of symmetric Siegel domains was obtained independently by Dorfmeister [3].

We add here some properties of quasi-symmetric Siegel domains, of which the proofs depend on the "complete reducibility" to be given in the next section (Th. 5. 3).

Proposition 4. 8. *Let $\mathcal{S}$ be an irreducible quasi-symmetric Siegel domain which is not symmetric. Then one has $\mathfrak{g}_1 = \mathfrak{g}_{1/2} = \{0\}$.*

Proof. Let $\mathfrak{r} = \mathfrak{r}_0 + \mathfrak{r}_{1/2} + \mathfrak{r}_{-1}$ be the radical of $\mathfrak{g}$. Since $\mathcal{S}$ is irreducible, so is Ω, and hence the identical representation $\mathfrak{g}(\Omega) \to \mathfrak{gl}(U)$ is irreducible. By the assumption (ii), this implies that the space $\mathfrak{g}_{-1}$, viewed as a $\mathfrak{g}_0$-module, is also irreducible. Therefore one has either $\mathfrak{r}_{-1} = \{0\}$ or $\mathfrak{g}_{-1}$. If $\mathfrak{r}_{-1} = \{0\}$, one has

$$[\mathfrak{g}_{-1/2}, \mathfrak{r}_{-1/2}] \subset \mathfrak{r}_{-1} = \{0\},$$

which implies $\mathfrak{r}_{-1/2} = \{0\}$, since the bilinear map $\operatorname{Im} H: V \times V \to U$ is non-degenerate. Then, by the same argument as above, one has $\mathfrak{r} = \mathfrak{r}_0 = \{0\}$ and $\mathfrak{g}$ is semi-simple. But this is the case which is excluded. If $\mathfrak{r}_{-1} = \mathfrak{g}_{-1}$, one has $\mathfrak{g}_1 = \{0\}$ by § 1, (A). Then it follows by Lemma 2. 5 that $\mathfrak{g}_{1/2}$ is also $= \{0\}$, q. e. d.

Remark 1. Proposition 4. 8 is essentially due to Tsuji [3], Th. 2. 1. Combining this with a result of Kaup [3], Satz 6, one can conclude that, under the same assumption, one has $\operatorname{Hol}(\mathcal{S}) = \operatorname{Aff}(\mathcal{S})$.

Remark 2. There are Siegel domains for which Ω is irreducible, homogeneous and self-dual, but one has $\mathfrak{g}_1 \neq \{0\}$, $\mathfrak{r}_{-1} \neq \{0\}$ (see Nakajima [3], Th. 7. 6, 7. 8, 7. 12). These are examples of Siegel domains satisfying the condition (i) but not (ii). Siegel domains satisfying only the condition (i) have been studied by Piatetskii-Shapiro [2], Takeuchi [8], Tsuji [3] and others.

Proposition 4. 9. *Let $\mathcal{S}$ be a quasi-symmetric Siegel domain. Then, if $Z \in \mathfrak{g}_1$ and $\varphi_e(Z) = \partial_u$, one has $Z = \partial_u^*$. Similarly, if $Y \in \mathfrak{g}_{1/2}$ and $\psi_e(Y) = \tilde{\partial}_v$, one has $Y = \tilde{\partial}_v^*$.*

By Theorem 5. 3, it is sufficient to prove this for irreducible quasi-symmetric Siegel domains, in which case the assertion follows immediately from Theorem 3. 6 and Proposition 4. 8. From Propositions 4. 7 and 4. 9, one can conclude that, for a quasi-symmetric Siegel domain $\mathcal{S}$, $\mathfrak{g}_1$ (resp. $\mathfrak{g}_{1/2}$) consists of all ∂_u^* (resp. $\tilde{\partial}_v^*$) for which $u \in U$ (resp. $v \in V$) satisfies the condition mentioned in Proposition 4. 7. (Similar results for arbitrary homogeneous Siegel domains have been obtained by Dorfmeister.)

Exercises

1. Show that (under the conditions (i), (ii)) the condition (iii) can also be written as

(iii′) $$H(v, v')^2 = 2H(v, R(H(v, v'))v') \qquad (v, v' \in V).$$

2. Consider the condition

(iii*) $$H(w, R(H(v, v'))w') = H(R(H(w', w))v, v') \qquad (v, v', w, w' \in V),$$

which is stronger than (iii). Show that under the condition (iii*) $R(H(v, v'))$ is associated with $\Phi_{v,v'}$ for all $v, v' \in V$. [The condition (iii*) is satisfied, e. g., for $(I_{p,q})$.]

3. Under the assumptions (i), (ii), (iii), prove the following formulas by a direct computation from the definitions (1. 3 a, b, c), (3. 23 a, b), (3. 9′), (3. 20), (3. 10′ a, b).

(4. 13 a) $$[\partial_{u'}, \partial^*_u] = 2X(A_{u',u}, B_{u',u}),$$

(4. 13 b*) $$[\tilde{\partial}_v, \partial^*_u] = -2i\tilde{\partial}^*_{R_u v},$$

(4. 13 b) $$[\partial_u, \tilde{\partial}^*_v] = 2i\tilde{\partial}_{R_u v},$$

(4. 13 c) $$[\tilde{\partial}_{v'}, \tilde{\partial}^*_v] = 4X(\Phi_{v',v}, \Psi_{v',v}),$$

(4. 13 d) $$[X_{A,B}, \partial^*_u] = \partial^*_{{}^tAu},$$

(4. 13 e) $$[X_{A,B}, \tilde{\partial}^*_v] = \tilde{\partial}^*_{B^*v}.$$

Hint. The formulas (4. 13 a, b, d) can be proved without using (iii). The relation (4. 13 e) can be reduced (by (i), (ii)) to

$$\langle H(v, v''), H(v', Bv'')\rangle + \langle H(v, Bv''), H(v', v'')\rangle$$
$$= \langle H(v, v''), H(B^*v', v'')\rangle + \langle H(B^*v, v''), H(v', v'')\rangle,$$

which can be proved by separating the cases where $B^* = -B$ (associated with ${}^tA = -A$) and where $B^* = B\,(=R_u)$ and using (iii$_3$).

4. Let $\mathscr{S}$ be a symmetric Siegel domain and define H_0 by (3. 25). For

$$X = \partial_{u_1} + \tilde{\partial}_{v_1} + X_{A,B} + \tilde{\partial}^*_{v_2} + \partial^*_{u_2} \in \mathfrak{g} \qquad (u_1, u_2 \in U,\ v_1, v_2 \in V)$$

prove the following formulas:

(4. 14) $$\frac{d}{dt}(\exp tX)^{-1}(ie, 0)\Big|_{t=0} = (u_1 - u_2 + iAe,\ v_1 - v_2),$$

(4. 15) $$2[H_0, X] = -\partial_{Ae} + \tilde{\partial}_{i(v_1-v_2)} + 2X_{u_1-u_2} - \tilde{\partial}^*_{i(v_1-v_2)} + \partial^*_{Ae}.$$

(4. 14) shows that $\mathfrak{p}$ can be identified with the tangent space $T_{(ie,0)}(\mathscr{S}) = U_C \oplus V$ by the correspondence

$$X = \partial_{u_1} + \tilde{\partial}_{v_1} + 2X_{u_2} - \tilde{\partial}^*_{v_1} - \partial^*_{u_1} \longleftrightarrow 2(u_1 + iu_2, v_1).$$

(4. 15) shows that $[H_0, \mathfrak{k}] = \{0\}$ and $ad\, H_0|\mathfrak{p}$ coincides with the scalar multiplication by i through the above identification. Thus H_0 is the H-element of $\mathfrak{g}$ at $(ie, 0)$ compatible with the given complex structure of $\mathscr{S}$. (This follows also from III, § 7, Exerc. 3. 2).)

5. Given a quasi-symmetric Siegel domain $\mathscr{S} = \mathscr{S}(U, V, \Omega, H)$, let $\tilde{V} = U_C \oplus V$ and define a triple product $\{\ \}$ on $\tilde{V}$ by

(4. 16) $$\left\{\begin{pmatrix} u \\ v \end{pmatrix}, \begin{pmatrix} u' \\ v' \end{pmatrix}, \begin{pmatrix} u'' \\ v'' \end{pmatrix}\right\} = \begin{pmatrix} \{u, \bar{u}', u''\} + 2H(R_{\bar{u}''}v', v) + 2H(R_{\bar{u}}v', v'') \\ 2R_{u''}R_{\bar{u}'}v + 2R_u R_{\bar{u}'}v'' + 2R(H(v', v))v'' + 2R(H(v', v''))v \end{pmatrix},$$

or equivalently by

(4. 16′) $$\begin{pmatrix} u \\ v \end{pmatrix} \square \begin{pmatrix} u' \\ v' \end{pmatrix} = \begin{pmatrix} A_{u,\bar{u}'} + 2\Phi^0_{v,v'} & i(\Phi^{R_{\bar{u}}v'})^* \\ -i\Phi^{R_{\bar{u}'}v} & B_{u,\bar{u}'} + 2\Psi^0_{v,v'} \end{pmatrix},$$

where

(4.17 a) $$\Phi^0_{v,v'} = \frac{1}{2}(\Phi_{v,v'} - i\Phi_{iv,v'}) = [u \mapsto H(R_{\bar{u}}v', v)],$$

(4.17 b) $$\Psi^0_{v,v'} = \frac{1}{2}(\Psi_{v,v'} - i\Psi_{iv,v'}) = [v'' \mapsto R(H(v', v))v'' + R(H(v', v''))v],$$

($\Phi^0_{v,v'}$ and $\Psi^0_{v,v'}$ are the $\boldsymbol{C}$-linear parts of $\Phi_{v,v'}$ and $\Psi_{v,v'}$ with respect to the variable v). Then it is clear that { } is $\boldsymbol{C}$-bilinear and symmetric in the first and third variables and $\boldsymbol{C}$-antilinear in the second one. Define the "trace form" by

$$\tau\left(\begin{pmatrix} u' \\ v' \end{pmatrix}, \begin{pmatrix} u \\ v \end{pmatrix}\right) = \mathrm{tr}\left(\begin{pmatrix} u \\ v \end{pmatrix} \square \begin{pmatrix} u' \\ v' \end{pmatrix}\right).$$

Prove the following.

(a) One has the relation

(4.18) $$\tau\left(\begin{pmatrix} u \\ v \end{pmatrix}, \begin{pmatrix} u' \\ v' \end{pmatrix}\right) = \left(1 + \frac{n}{2m}\right)(\langle \bar{u}, u' \rangle + 2h(v, v')),$$

where $\langle\ \rangle$ is the "normalized" inner product of U_C, i. e., $\langle u, u' \rangle = \mathrm{tr}(T_{uu'})$ $(u, u' \in U_C)$, h is defined by (3.5), and $m = \dim_{\boldsymbol{R}} U$, $n = \dim_{\boldsymbol{C}} V$. Thus τ is a positive definite hermitian form on $\tilde{V}$. (*Hint.* Use the result of I, § 9, Exerc. 2.)

(b) Denoting by * the adjoint with respect to τ, one has

$$\left(\begin{pmatrix} u \\ v \end{pmatrix} \square \begin{pmatrix} u' \\ v' \end{pmatrix}\right)^* = \begin{pmatrix} u' \\ v' \end{pmatrix} \square \begin{pmatrix} u \\ v \end{pmatrix}.$$

(c) The condition (JT 2) for { } (I, § 6) is equivalent to (iii). Thus, in the symmetric case, ($\tilde{V}$, { }) becomes a positive definite hermitian JTS.

Remark. We treat the problem 5, (c) in a more general context in the next section. Let us here explain the meaning of the triple product (4.16). First, to simplify the notation, we identify $\mathfrak{g}_{-1}, \mathfrak{g}_{-1/2}, \mathfrak{g}_0$ with U, V, and a subalgebra of $\mathfrak{g}(\Omega) \oplus \mathfrak{gl}(V)$, respectively, by the correspondences

(4.19) $$u \longleftrightarrow -\partial_u, \qquad v \longleftrightarrow -\tilde{\partial}_v,$$
$$(A, B) \longleftrightarrow -X_{A,B}$$

(see § 1, (A)). This identification is compatible with the action of $\mathfrak{g}(\Omega) \oplus \mathfrak{gl}(V)$ on $U \oplus V$, i. e., one has $[(A, B), u] = Au$, etc., and by (1.9 d)

(4.20) $$[v, v'] = -4 \,\mathrm{Im}\, H(v, v').$$

In the symmetric case, we have

(4.21) $$\mathfrak{g} = U + V + \mathfrak{g}_0 + \theta V + \theta U,$$

and by the result in § 3 ((3.8), (3.12), (3.11), (3.19))

(4.22 a) $$[u, \theta u'] = -2(A_{u,u'}, B_{u,u'}),$$

(4.22 b) $$[v, \theta v'] = -4(\Phi_{v,v'}, \Psi_{v,v'}),$$

(4.22 c) $$[u, \theta v] = -2IR_u v \qquad \text{for all } u, u' \in U,\ v, v' \in V,$$

where I denotes the complex structure on V (cf. III, § 3, Exerc. 3, 4). Putting $V_\pm = V_C(I; \pm i)$, one has

(4.23) $$\mathfrak{g}_C = (U_C + V_+) + (V_- + \mathfrak{g}_{0C} + \theta V_+) + (\theta V_- + \theta U_C).$$

For $v, v', \cdots \in V$, we denote by $v_+, v'_+, \cdots$ (resp. $v_-, v'_-, \cdots$) their V_+-parts (resp. V_--parts). Then, first from (4.20) one has

(4.20′) $$H(v, v') = -\frac{i}{2}[v_-, v'_+]$$

and by the correspondence (4.19)

(4. 24) $$v_+ \longleftrightarrow -v\frac{\partial}{\partial w}, \qquad v_- \longleftrightarrow -2iH(v, w)\frac{\partial}{\partial z}.$$

Next, from (4. 22 c) one has

(4. 22 c′) $$[u, \theta v_+] = -2iR_u v_+, \qquad [u, \theta v_-] = 2iR_u v_-.$$

In view of the relation

$$\Phi_{Iv,Iv'} = \Phi_{v,v'}, \qquad \Psi_{Iv,Iv'} = \Psi_{v,v'},$$

it follows from (4. 22 b) that

(4. 25) $$[V_+, \theta V_+] = [V_-, \theta V_-] = 0,$$

and hence, taking the C-linear parts of both sides of (4. 22 b), one obtains

(4. 22 b′) $$[v_+, \theta v'_-] = -4(\Phi^0_{v,v'}, \Psi^0_{v,v'}).$$

Thus the decomposition (4. 23), along with the Cartan involution $X \mapsto \theta \bar{X}$, defines a structure of real symmetric Lie algebra on $\mathfrak{g}_C$, so that $U_C \oplus V_+$ becomes a positive definite hermitian JTS with the triple product

$$\{u+v_+, u'+v'_+, u''+v''_+\} = -\frac{1}{2}[[u+v_+, \theta(\bar{u}'+v'_-)], u''+v''_+].$$

By (4. 22 a, b′, c′) one has

(4. 16″) $$(u+v_+)\square(\bar{u}'+v'_-) = -iR_u v'_- + (A_{u,\bar{u}'} + 2\Phi^0_{v,v'}, B_{u,\bar{u}'} + 2\Psi^0_{v,v'}) - i\theta(R_{\bar{u}'}v_+).$$

In particular, we notice that $U_C \square V_+ = V_-$ and $2ie\square v_+ = v_-$. Transforming (4. 16″) by the natural isomorphism $U_C + V_+ \cong \tilde{V}$, one obtains (4. 16′). In terms of the triple product (4. 16), the formulas (3. 23 a, b) can be simplified as

(4. 26 a) $$\theta u \longleftrightarrow -\partial^*_u = -P(\mathfrak{z})u \cdot \frac{\partial}{\partial \mathfrak{z}},$$

(4. 26 b) $$\theta v_+ \longleftrightarrow -2iR_z v \cdot \frac{\partial}{\partial w}, \qquad \theta v_- \longleftrightarrow -P(\mathfrak{z})v \cdot \frac{\partial}{\partial \mathfrak{z}},$$

where $\mathfrak{z}$ and $\frac{\partial}{\partial \mathfrak{z}}$ stand for $\begin{pmatrix} z \\ w \end{pmatrix}$ and $\begin{pmatrix} \partial/\partial z \\ \partial/\partial w \end{pmatrix}$, respectively, and $P(\mathfrak{z})t = \{\mathfrak{z}, t, \mathfrak{z}\}$.

§ 5. Classification of quasi-symmetric Siegel domains.

In the preceding section, we have seen that, for a quasi-symmetric Siegel domain $\mathcal{S} = \mathcal{S}(U, V, \Omega, H)$ (with a fixed reference point $e \in \Omega$ and inner products $\langle\ \rangle$ and h on U and V), one has a unique Lie algebra representation $\beta : \mathfrak{g}(\Omega) \to \mathfrak{gl}(V)$ satisfying the following relations:

(5. 1) $$\langle u, H(v_1, v_2)\rangle = 2h(v_1, \beta(T_u)v_2),$$

(5. 2 a) $$\beta(T_{Au}) = \beta(A)\beta(T_u) + \beta(T_u)\beta(A)^*,$$

(5. 2 b) $$\beta({}^tA) = \beta(A)^*,$$

(5. 2 c) $$\beta(1_U) = \frac{1}{2}1_V$$

for all $u \in U$, $v_1, v_2 \in V$, and $A \in \mathfrak{g}(\Omega)$. Suppose, conversely, that there are given a self-dual homogeneous cone Ω in U (with $\langle\ \rangle$ and $e \in \Omega$ satisfying (3. 2)) and a representation β of $\mathfrak{g}(\Omega)$ in V (with a hermitian inner product h) satisfying the conditions (5. 2 a, b, c). Then by (5. 2 b) $R_u = \beta(T_u)$ is hermitian and (5. 2 a) implies

(5. 2 a′) $$R_{u_1 u_2} = R_{u_1}R_{u_2} + R_{u_2}R_{u_1} \qquad (u_1, u_2 \in U).$$

Thus $u \mapsto 2R_u = 2\beta(T_u)$ is a unital Jordan algebra representation $U \to \mathcal{H}(V, h)$, and one has $\{R_u (u \in \Omega)\} \subset \mathcal{P}(V, h)$ (Prop. I. 9. 2). Therefore, defining an Ω-positive hermitian map $H: V \times V \to U_c$ by (5. 1), one obtains a quasi-symmetric Siegel domain $\mathcal{S}(U, V, \Omega, H)$. Thus, for a fixed self-dual homogeneous cone Ω in U (with $\langle \ \rangle$ and $e \in \Omega$), there exists a one-to-one correspondence between quasi-symmetric Siegel domains $\mathcal{S}(U, V, \Omega, H)$ and the Lie algebra representations β of $\mathfrak{g}(\Omega)$ in V (with h) satisfying (5. 2 a, b, c), or equivalently, the unital Jordan algebra representations $2R: (U, e) \to \mathcal{H}(V, h)$. For simplicity, we denote by $\mathcal{S}(\Omega, e, \beta)$ the quasi-symmetric Siegel domain defined by the data Ω, e, and β (depending also on $\langle \ \rangle$ and h). We note that, under the assumptions (5. 2 a, b), the condition (5. 2 c) is equivalent to a weaker condition

(5. 2 c') $$\beta(1_U) \text{ is non-singular.}$$

In fact, putting $u_1 = u_2 = e$ in (5. 2 a'), one has $R_e = 2R_e^2$. Hence, if $R_e = \beta(1_U)$ is non-singular, one has $R_e = \frac{1}{2} 1_V$.

Now, let $\mathcal{S}' = \mathcal{S}(U', V', \Omega', H') = \mathcal{S}(\Omega', e', \beta')$ be another quasi-symmetric Siegel domain, where the objects relative to $\mathcal{S}'$ are denoted by the corresponding symbols with a prime. Using the uniqueness of Siegel domain expression (K-M-O, Th. 11, or Kaneyuki [1]), we will give a condition for $\mathcal{S}$ and $\mathcal{S}'$ to be holomorphically equivalent. The "uniqueness theorem" says that two (arbitrary) Siegel domains $\mathcal{S}$ and $\mathcal{S}'$ are holomorphically equivalent, if and only if they are *linearly equivalent*, i. e., if and only if there exists a pair of linear isomorphisms

$$\varphi: U \longrightarrow U', \qquad \psi: V \longrightarrow V'$$

satisfying the following relations (5. 3), (5. 4):

(5. 3) $$\varphi(\Omega) = \Omega',$$

(5. 4) $$\varphi(H(v_1, v_2)) = H'(\psi(v_1), \psi(v_2)) \qquad (v_1, v_2 \in V).$$

When $\mathcal{S}$ and $\mathcal{S}'$ are quasi-symmetric, we may further assume, by the homogeneity of Ω and the condition (ii), that

(5. 5) $$\varphi(e) = e'.$$

Then, $\varphi: (U, e) \to (U', e')$ is a unital Jordan algebra isomorphism (Cor. I. 9. 3) and hence

(5. 6) $$\varphi \circ T_u \circ \varphi^{-1} = T'_{\varphi(u)}.$$

We may also assume that the inner products $\langle \ \rangle$ and $\langle \ \rangle'$ on U and U' are so chosen that

$$\langle \varphi(u_1), \varphi(u_2) \rangle' = \langle u_1, u_2 \rangle \qquad (u_1, u_2 \in U),$$

i. e., denoting by ${}^t\varphi$ the adjoint of φ with respect to $\langle \ \rangle$ and $\langle \ \rangle'$, one has

(5. 7) $${}^t\varphi = \varphi^{-1}.$$

[This assumption follows automatically from (5. 5) if $\langle \ \rangle$ and $\langle \ \rangle'$ are normalized to be the trace forms of (U, e) and (U', e').] From (5. 1) and (5. 4) we have for

$u' \in U'$, $v_1, v_2 \in V$

$$\begin{aligned} 2h(v_1, \beta(T_{{}^t\varphi(u')})v_2) &= \langle {}^t\varphi(u'), H(v_1, v_2)\rangle \\ &= \langle u', \varphi(H(v_1, v_2))\rangle' \\ &= \langle u', H'(\psi(v_1), \psi(v_2))\rangle' \\ &= 2h'(\psi(v_1), \beta'(T'_{u'})\psi(v_2)). \end{aligned}$$

Hence, denoting by ψ^* the adjoint of ψ with respect to h and h', one has

(5. 4′) $$\beta(T_{{}^t\varphi(u')}) = \psi^* \circ \beta'(T'_{u'}) \circ \psi \qquad (u' \in U').$$

In view of (5. 6), (5. 7), this is equivalent to

$$\beta(T_u) = \psi^* \circ \beta'(T'_{\varphi(u)}) \circ \psi = \psi^* \circ \beta'(\varphi T_u \varphi^{-1}) \circ \psi \qquad (u \in U).$$

Putting $u=e$, one has

(5. 8) $$\psi^* = \psi^{-1}.$$

Then, since $\mathfrak{g}(\Omega)$ is generated by $\mathfrak{p}(\Omega)$, one obtains

(5. 9) $$\beta'(\varphi A \varphi^{-1}) = \psi \circ \beta(A) \circ \psi^{-1} \qquad \text{for all } A \in \mathfrak{g}(\Omega).$$

In this (weaker) sense, two representations β and β' are "equivalent". Conversely, if β and β' are "equivalent", i. e., if there exist linear isomorphisms φ and ψ satisfying (5. 3), (5. 5) and (5. 9), then we can first modify $\langle\ \rangle'$ so that (5. 7) is satisfied. Then, writing

$$h'(\psi(v_1), \psi(v_2)) = h(v_1, Pv_2) \qquad (v_1, v_2 \in V)$$

with $P \in \mathcal{P}(v, h)$, we see by (5. 9) and (5. 6) (or (5. 2 b), (5. 7)) that P commutes with all $\beta(A)$ ($A \in \mathfrak{g}(\Omega)$). Hence, replacing ψ by $\psi \circ P^{-1/2}$, one may assume that $P = 1_V$, i. e., (5. 8) holds. Then, (5. 4′) and hence (5. 4) is also satisfied, and $\mathcal{S}$ and $\mathcal{S}'$ are linearly equivalent.

In particular, in the case $U=U'$, $\Omega=\Omega'$, $e=e'$, $\langle\ \rangle = \langle\ \rangle'$, the conditions on φ amount to saying that φ belongs to the isotropy subgroup $G(\Omega)_e$ of $G(\Omega)$ at e. Two representations β and β' of $\mathfrak{g}(\Omega)$ will be called *automorphically equivalent* (at e) if there exist $\varphi \in G(\Omega)_e$ and a $\boldsymbol{C}$-linear isomorphism $\psi : V \to V'$ satisfying the condition (5. 9). (Notation : $\beta \approx \beta'$.)

Summing up, we obtain the following

Theorem 5. 1. *Let $\mathcal{S} = \mathcal{S}(\Omega, e, \beta)$ and $\mathcal{S}' = \mathcal{S}(\Omega', e', \beta')$ be two quasi-symmetric Siegel domains. Then $\mathcal{S}$ and $\mathcal{S}'$ are holomorphically equivalent if and only if there exists a pair of linear isomorphisms $\varphi : U \to U'$ and $\psi : V \to V'$ satisfying the conditions* (5. 3), (5. 5), (5. 9). *In particular, when $(\Omega, e) = (\Omega', e')$, $\mathcal{S}$ and $\mathcal{S}'$ are holomorphically equivalent if and only if two representations β and β' of $\mathfrak{g}(\Omega)$ are automorphically equivalent (at e).*

Thus the classification of quasi-symmetric Siegel domains is completely reduced to that of the representations β of $\mathfrak{g}(\Omega)$ satisfying the conditions (5. 2 a, b, c) up to an automorphic equivalence.

Now, in general, let (U, V, Ω, H) be data defining a Siegel domain $\mathcal{S}$ and suppose that we have direct sum decompositions

$$U = U' \oplus U'' \quad \text{and} \quad V = V' \oplus V''$$

such that

$$\Omega = \Omega' \times \Omega'', \qquad \Omega' \subset U', \qquad \Omega'' \subset U''$$

and

$$H(V', V') \subset U'_{C}, \qquad H(V'', V'') \subset U''_{C},$$
$$H(V', V'') = 0.$$

Then it is clear that $\mathcal{S} = \mathcal{S}(U, V, \Omega, H)$ decomposes into the direct product of two Siegel domains

$$\mathcal{S}' = \mathcal{S}(U', V', \Omega', H') \quad \text{and} \quad \mathcal{S}'' = \mathcal{S}(U'', V'', \Omega'', H''),$$

where $H' = H|V' \times V'$ and $H'' = H|V'' \times V''$. It is known that, conversely, any direct product decomposition of a Siegel domain (as a Kähler manifold) is obtained in this manner (cf. Kaneyuki [1], Nakajima [4]). A Siegel domain $\mathcal{S}$ which can not be decomposed non-trivially (i. e., with $U' \neq \{0\}$, $U'' \neq \{0\}$) is called *irreducible*. In the following, we consider the irreducible decomposition of a quasi-symmetric Siegel domain $\mathcal{S}$.

Going back to the previous notation, we first note that the representation $\beta : \mathfrak{g}(\Omega) \to \mathfrak{gl}(V)$ satisfying (5. 2 b) is fully reducible. In fact, if V_1 is an invariant subspace of V (under β), then by (5. 2 b) the orthogonal complement $V_1^{\perp}$ of V_1 in V (with respect to h) is also invariant. It is clear that, if a representation β satisfying the conditions (5. 2 a, b, c) is decomposed into an orthogonal sum of irreducible representations, then each irreducible component also satisfies the same conditions.

Now, let

$$(5.10) \qquad U = \bigoplus_{i=1}^{r} U_i,$$
$$\Omega = \Omega_1 \times \cdots \times \Omega_r, \qquad \Omega_i \subset U_i$$

be the (unique) irreducible decomposition of Ω, where Ω_i is an irreducible self-dual homogeneous cone in U_i (I, § 8). Then one has the corresponding decomposition of the Lie algebra

$$\mathfrak{g}(\Omega) = \bigoplus_{i=1}^{r} \mathfrak{g}(\Omega_i), \qquad \mathfrak{g}(\Omega_i) \subset \mathfrak{gl}(U_i).$$

Moreover, if

$$e = \sum_{i=1}^{r} e_i, \qquad e_i \in U_i,$$

then $(U, e) = \bigoplus_{i=1}^{r} (U_i, e_i)$ is the direct sum decomposition of Jordan algebra ; in particular, one has

$$(5.11) \qquad e_i e_j = \delta_{ij} e_i \qquad (1 \leq i, j \leq r).$$

Note that the above decomposition of U is an orthogonal decomposition with respect to $\langle\ \rangle$, and T_{e_i} is the orthogonal projection $U \to U_i$.

Proposition 5. 2*). *Let β be a representation of $\mathfrak{g}(\Omega)$ in V satisfying the conditions* (5. 2 a, b), *and put*

$$V^{(i)} = R_{e_i}V, \qquad V^{(0)} = \left(\sum_{i=1}^{r} V^{(i)}\right)^{\perp}. \tag{5. 12}$$

Then one has the orthogonal decomposition (with respect to h)

$$V = \bigoplus_{i=0}^{r} V^{(i)} \tag{5. 13}$$

satisfying the relation

$$\beta(\mathfrak{g}(\Omega_i))V^{(j)} = \delta_{ij}V^{(i)} \qquad (1 \leq i \leq r,\ 0 \leq j \leq r). \tag{5. 14}$$

Proof. For $u \in U_i$, $u' \in U_j$ $(i \neq j)$, one has $[T_u, T_{u'}]=0$, and so $[R_u, R_{u'}]=0$; in particular, $[R_{e_i}, R_{e_j}]=0$. Hence, from (5. 2 a') and (5. 11) one has

$$R_{e_ie_j} = \delta_{ij}R_{e_i} = 2R_{e_i}R_{e_j}.$$

Therefore $\{2R_{e_i}\ (1 \leq i \leq r)\}$ is a set of mutually orthogonal projection operators on V and the corresponding orthogonal decomposition of V is given by (5. 13). Since T_{e_i} is in the center of $\mathfrak{g}(\Omega)$, each $V^{(i)}=R_{e_i}V$, and hence also $V^{(0)}$, is stable under $\beta(\mathfrak{g}(\Omega))$. For $u \in U_i$, one has $R_u = R_{e_iu} = 2R_{e_i}R_u$, and so $R_uV \subset V^{(i)}$. Since $\mathfrak{g}(\Omega_i)$ is generated by $\{T_u\ (u \in U_i)\}$, one has

$$\beta(\mathfrak{g}(\Omega_i))V \subset V^{(i)},$$

whence follows (5. 14), q. e. d.

Proposition 5. 2 shows that $V^{(i)}$ is essentially a representation-space of the factor $\mathfrak{g}(\Omega_i)$ (or the corresponding Jordan algebra (U_i, e_i)). In the quasi-symmetric case, one has $V^{(0)} = \{0\}$ by (5. 2 c). Hence, denoting by $\beta^{(i)}$ the representation of $\mathfrak{g}(\Omega_i)$ on $V^{(i)}$ induced by β, one has

$$\beta\left(\sum_{i=1}^{r} X_i\right) = \bigoplus_{i=1}^{r} \beta^{(i)}(X_i) \qquad (X_i \in \mathfrak{g}(\Omega_i)).$$

It follows that the Siegel domain $\mathcal{S} = \mathcal{S}(\Omega, e, \beta)$ is decomposed into the direct product

$$\mathcal{S} = \prod_{i=1}^{r} \mathcal{S}(\Omega_i, e_i, \beta^{(i)}). \tag{5. 15}$$

By what we mentioned above, this is the (unique) irreducible decomposition of $\mathcal{S}$. Thus we obtain

Theorem 5. 3. *A quasi-symmetric Siegel domain $\mathcal{S}=\mathcal{S}(\Omega, e, \beta)$ is irreducible if and only if the cone Ω is irreducible. Any quasi-symmetric Siegel domain is decomposed uniquely into the direct product of irreducible quasi-symmetric Siegel domains.*

---

*) This is a special case of the full reducibility of representations of a semi-simple Jordan algebra. Cf. Lemma 6. 4.

Our problem is thus reduced to the determination of irreducible representations β of $\mathfrak{g}(\Omega)$, for Ω irreducible, satisfying the conditions (5. 2 a, b, c).

The problem still involves implicitly the determination of the hermitian form h. However, by virtue of Proposition I. 3. 6, for any representation β of $\mathfrak{g}(\Omega)$ (with Ω irreducible) satisfying (5. 2 c), the image $\beta(\mathfrak{g}(\Omega))$ is algebraic and the corresponding $\boldsymbol{R}$-group is (algebraically) reductive. Hence, by Theorem I. 4. 2, one can always find, for the given $\langle\ \rangle$ on U, a hermitian form h on V, for which (5. 2 b) is satisfied. If β is irreducible, such an h is uniquely determined up to a positive multiple. In the general case, all possible h can easily be determined by a method similar to the one given in IV, § 2. Moreover, since $\mathfrak{g}(\Omega)$ is generated by $\mathfrak{p}(\Omega)$, the condition (5. 2 a) is equivalent to (5. 2 a'), under the assumption (5. 2 b). Therefore, we may disregard the hermitian form h, reducing our problem to the determination of all irreducible representations β of $\mathfrak{g}(\Omega)$, for Ω irreducible, satisfying the conditions (5. 2 a', c).

It should be noted, furthermore, that the condition (5. 2 a') is independent of the choice of the reference point e, i. e., if it is satisfied for e, then it is also satisfied for any $e'\in\Omega$. In fact, let $e'=\varphi e$ with $\varphi=\exp X_1$, $X_1\in\mathfrak{g}(\Omega)$. Then, putting $\psi=\exp\beta(X_1)$, one has

$$(*)\qquad \begin{aligned}\beta(\varphi A\varphi^{-1}) &= \beta(\exp(ad\,X_1)A) = \exp(ad\,\beta(X_1))\beta(A)\\ &= \psi\circ\beta(A)\circ\psi^{-1}\end{aligned}$$

for $A\in\mathfrak{g}(\Omega)$, (i. e., β is "equivalent" to itself with respect to (φ,ψ)). Denoting by T'_u and R'_u the operators T_u and R_u relative to e', one has clearly

$$T'_u = \varphi\circ T_{\varphi^{-1}u}\circ\varphi^{-1}.$$

Hence by (*) one has

$$R'_u = \psi\circ R_{\varphi^{-1}u}\circ\psi^{-1}$$

and so for $u_1, u_2\in U$

$$\begin{aligned}R'(T'_{u_1}u_2) &= \psi\circ R(T_{\varphi^{-1}u_1}\varphi^{-1}u_2)\circ\psi^{-1}\\ &= \psi\circ(R_{\varphi^{-1}u_1}R_{\varphi^{-1}u_2}+R_{\varphi^{-1}u_2}R_{\varphi^{-1}u_1})\circ\psi^{-1}\\ &= R'_{u_1}R'_{u_2}+R'_{u_2}R'_{u_1},\end{aligned}$$

which proves our assertion. It is also easy to see that the notion of "automorphic equivalence" is independent of the choice of e.

Now, to determine the representation β satisfying (5. 2 a, b, c) explicitly, take a maximal abelian subalgebra $\mathfrak{a}$ of $\mathfrak{g}(\Omega)$ contained in $\mathfrak{p}(\Omega)$. For a weight μ (resp. λ) relative to $\mathfrak{a}$ of the identical representation of $\mathfrak{g}(\Omega)$ on U (resp. the representation β on V), let U_μ (resp. V_λ) denote the corresponding eigenspace, i. e.,

$$\begin{aligned}U_\mu &= \{u\in U\,|\,Xu=\mu(X)u \quad \text{for all } X\in\mathfrak{a}\},\\ V_\lambda &= \{v\in V\,|\,\beta(X)v=\lambda(X)v \quad \text{for all } X\in\mathfrak{a}\}.\end{aligned}$$

Then one has the orthogonal decompositions

$$(5.\ 16)\qquad U=\bigoplus_\mu U_\mu \quad\text{and}\quad V=\bigoplus_\lambda V_\lambda.$$

Note that all weights λ, μ are real.

Lemma 5. 4. *If one has* $h(v, \beta(T_u)v') \neq 0$ *for some* $u \in U_\mu$, $v \in V_\lambda$, $v' \in V_{\lambda'}$, *then one has* $\mu = \lambda + \lambda'$.

Proof. For any $X \in \mathfrak{a}$, one has by (5. 2 a, b)

$$\begin{aligned}\lambda(X)h(v, \beta(T_u)v') &= h(\beta(X)v, \beta(T_u)v') \\ &= h(v, \beta(X)\beta(T_u)v') \\ &= h(v, \beta(T_{Xu})v' - \beta(T_u)\beta(X)v') \\ &= (\mu(X) - \lambda'(X))h(v, \beta(T_u)v'),\end{aligned}$$

whence follows our assertion, q. e. d.

Proposition 5. 5. *Let* β *be a (non-trivial) representation of* $\mathfrak{g}(\Omega)$ *satisfying the conditions* (5. 2 a, b, c). *Then, for every weight* λ *of* β *(relative to* $\mathfrak{a}$*),* $\mu = \frac{1}{2}\lambda$ *is a weight of the identical representation of* $\mathfrak{g}(\Omega)$ *on* U.

Proof. Write $e = \sum_\mu u_\mu$ with $u_\mu \in U_\mu$. Then, for any $v \in V$, $v \neq 0$, one has

$$\begin{aligned}h(v, v) &= 2h(v, \beta(T_e)v) \\ &= 2\sum_\mu h(v, \beta(T_{u_\mu})v) \neq 0.\end{aligned}$$

Hence there exists a μ such that $h(v, \beta(T_{u_\mu})v) \neq 0$. Our assertion then follows from Lemma 5. 4, q. e. d.

By virtue of Proposition 5. 5, it is easy to determine, for each irreducible self-dual homogeneous cone Ω, all irreducible representations β of $\mathfrak{g}(\Omega)$ satisfying the conditions (5. 2 a, b, c) (cf. Satake [15]). The result is similar to the case of symplectic representations discussed in IV, § 5. Denoting by N the number of inequivalent (non-trivial) irreducible representations β satisfying our conditions, we obtain $N=2$ for $\Omega = \mathcal{P}_\nu(\boldsymbol{C})$ ($\nu \geq 2$) and $\mathcal{P}(1, \nu-1)$ (ν even, ≥ 6), $N=0$ for $\mathcal{P}_3(\boldsymbol{O})$, and $N=1$ for all other cases. For the cases $\Omega = \mathcal{P}_\nu(\boldsymbol{R})$ ($\nu \geq 1$) and $\mathcal{P}_\nu(\boldsymbol{H})$ ($\nu \geq 3$), the basic (irreducible) representation is given by the natural injections $\mathfrak{gl}_n(\boldsymbol{R}) \to \mathfrak{gl}_n(\boldsymbol{C})$ and $\mathfrak{gl}_n(\boldsymbol{H}) \to \mathfrak{gl}_{2n}(\boldsymbol{C})$, respectively, which we denote by (id). For $\Omega = \mathcal{P}_\nu(\boldsymbol{C})$ ($\nu \geq 2$), the two basic representations are given by (id) and $(\overline{id})$. For $\Omega = \mathcal{P}(1, \nu-1)$ ($\nu \geq 3$, $\nu \neq 4$), the basic representations are the spin representations, which we denote by (sp) or (sp_i) ($i=1, 2$). In general, we write $\beta_r = r\rho_1$ or $\beta_{r,s} = r\rho_1 + s\rho_2$, according as $N=1, 2$, ρ_i ($1 \leq i \leq N$) denoting the basic (irreducible) representations. Then it can be shown that one has $\beta_{r,s} \approx \beta_{r',s'}$ if and only if $(r', s') = (r, s)$ or (s, r). Hence, as the representatives of automorphic equivalence classes of the representations β satisfying (5. 2 a, b, c), one can take β_r ($r \geq 0$) or $\beta_{r,s}$ ($r \geq s \geq 0$).

In the following table, we list all (non-isomorphic) irreducible quasi-symmetric Siegel domains without overlappings.

Type	Ω	β	dim $\mathcal{S}$	Symmetric cases
$(III_{\nu;r})$ $(\nu\geq 1, r\geq 0)$	$\mathcal{P}_\nu(\boldsymbol{R})$	$\beta_r=r(id)$	$\frac{1}{2}\nu(\nu+1)+r\nu$	$(III_{1;r})=(I_{r+1,1})$ $(r\geq 0)$ $(III_{\nu;0})=(III_\nu)$ $(\nu\geq 1)$
$(I_{\nu;r,s})$ $(\nu\geq 2, r\geq s\geq 0)$	$\mathcal{P}_\nu(\boldsymbol{C})$	$\beta_{r,s}=r(id)+s(\overline{id})$ $(\approx\beta_{s,r})$	$\nu^2+(r+s)\nu$	$(I_{\nu;r,0})=(I_{\nu+r,\nu})$ $(\nu\geq 2, r\geq 0)$
$(II_{\nu;r})$ $(\nu\geq 3, r\geq 0)$	$\mathcal{P}_\nu(\boldsymbol{H})$	$\beta_r=r(id)$	$2\nu^2-\nu+2r\nu$	$(II_{\nu;r})=(II_{2\nu+r})$ $(\nu\geq 3, r=0, 1)$
$(IV_{\nu;r,s})$ $\begin{pmatrix}\nu \text{ even}, \geq 6\\ r\geq s\geq 0\end{pmatrix}$	$\mathcal{P}(1, \nu-1)$	$\beta_{r,s}=r(sp_1)+s(sp_2)$ $(\approx\beta_{s,r})$	$\nu+(r+s)2^{\frac{\nu}{2}-1}$	$(IV_{\nu;0,0})=(IV_\nu)$ $(\nu \text{ even}, \geq 6)$ $(IV_{6;1,0})=(II_5)$ $(IV_{8;1,0})=(V)$
$(IV_{\nu;r})$ $\begin{pmatrix}\nu \text{ odd}, \geq 5\\ r\geq 0\end{pmatrix}$	$\mathcal{P}(1, \nu-1)$	$\beta_r=r(sp)$	$\nu+r2^{\frac{\nu-1}{2}}$	$(IV_{\nu;0})=(IV_\nu)$ $(\nu \text{ odd}, \geq 5)$
(VI_0)	$\mathcal{P}_3(\boldsymbol{O})$	$\beta_0=(triv)$	27	$(VI_0)=(VI)$

For the case $\Omega=\mathcal{P}_\nu(\mathcal{K})$, the basic representations (id) (and $(\overline{id})$) are of $\boldsymbol{R}$-type, $\boldsymbol{C}$-type, or $\boldsymbol{H}$-type, according as $\mathcal{K}=\boldsymbol{R}$, $\boldsymbol{C}$, or $\boldsymbol{H}$. For $\Omega=\mathcal{P}(1, \nu-1)$, the spin representations are of $\boldsymbol{R}$-type if $\nu\equiv 1, 2, 3 \pmod 8$, of $\boldsymbol{C}$-type if $\nu\equiv 0 \pmod 4$, and of $\boldsymbol{H}$-type if $\nu\equiv 5, 6, 7 \pmod 8$ (see Appendix (4. 6)). In particular, one obtains the following isomorphisms given by spin representations

$$\mathcal{P}(1, 2)\cong\mathcal{P}_2(\boldsymbol{R}), \qquad \mathcal{P}(1, 5)\cong\mathcal{P}_2(\boldsymbol{H}).$$

Therefore, to eliminate overlappings, we have excluded $\mathcal{P}_2(\boldsymbol{H})$ and $\mathcal{P}(1, 2)$ from the above table.

Remark 1. The classification of quasi-symmetric Siegel domains was first given by Takeuchi [9] (by making use of "S-algebras"). Here we followed essentially the method in Satake [17], [18]. Historically, it was the dawning of the theory of Jordan algebras that Albert [2] (1934) proved the non-existence of faithful representation (the so-called "non-speciality") of the exceptional Jordan algebra $\mathcal{H}_3(\boldsymbol{O})$, which amounts, in our formulation, to the non-existence of the non-trivial representation β, and implies that the only quasi-symmetric Siegel domain with $\Omega=\mathcal{P}_3(\boldsymbol{O})$ is the symmetric tube domain (VI).

Remark 2. The non-symmetric quasi-symmetric Siegel domain of the lowest dimension is $(III_{2;1})$ which is of dimension 5. Piatetskii-Shapiro [1], has given an example of non-quasi-symmetric homogeneous Siegel domain of dimension 4.

From the result of Chapter III we can see that, in the realization of a symmetric domain as a Siegel domain of the third kind, the fibers are always quasi-symmetric Siegel domains. To see this, let $\mathcal{S}_\kappa$ be a Siegel domain of the third kind corresponding to an (H_1)-homomorphism $\kappa: \mathfrak{sl}_2(\boldsymbol{R}) \to \mathfrak{g}$. Clearly it is enough to examine the fiber $\mathcal{S}_{\kappa,o}$ of $\mathcal{S}_\kappa$ over o. By III, §§ 6, 7 (especially, p. 133) one has

$$\mathcal{S}_{\kappa,o} = \mathcal{S}(U, V_+, \Omega, H_o^+) \approx \mathcal{S}(U, V, I_0, \Omega, H),$$

where Ω is homogeneous and self-dual (Th. III. 2. 3). Since B_κ acts on $\mathcal{S}_\kappa$ by "quasi-linear transformations" (III. 5. 34), one has

$$\text{Lie Aff}(\mathcal{S}_{\kappa,o}) \supset U + V + (\mathfrak{k}_\kappa^{(1)} + \mathfrak{g}_\kappa^{(2)}).$$

As we mentioned in III, § 3 (p. 103), there exists, for each $u \in U$, a unique element $X_u \in \mathfrak{p}_\kappa^{(2)}$ such that $ad_U X_u = T_u$; then one has $R_u = ad_V X_u \in \mathcal{H}(V, I_0, h)$ and R_u is associated with T_u. Thus the condition (ii) is satisfied (Prop. 4. 1). Note that, in the notation of § 4, one has

$$(5.17) \qquad \mathfrak{k}_\kappa^{(1)} \subset \mathfrak{k}^{(1)}, \qquad \mathfrak{g}_\kappa^{(2)} = \mathfrak{g}^{(2)}.$$

[Note also that in III the inner products $\langle\ \rangle$ and h on U and V are defined by (III. 2. 10), (III. 3. 8), (III. 3. 12).]

We now introduce the symbol $(X)_k$ to denote the type of the fiber over the k-th boundary component of an irreducible symmetric domain of type (X). Then, from the list in III, § 4, we obtain the following result.

$$\underset{(p \geq q \geq 1)}{(I_{p,q})_k} = \begin{cases} (III_{1;p+q-2}) & (k=1), \\ (I_{k;p-k,q-k}) & (2 \leq k \leq q), \end{cases}$$

$$\underset{(n \geq 3)}{(II_n)_k} = \begin{cases} (III_{1;2(n-2)}) & (k=1), \\ (IV_{6;n-4,0}) & (k=2), \\ (II_{k;n-2k}) & \left(3 \leq k \leq \left[\frac{n}{2}\right]\right), \end{cases}$$

$$\underset{(n \geq 1)}{(III_n)_k} = (III_{k;n-k}) \qquad (1 \leq k \leq n),$$

$$\underset{(n \geq 5)}{(IV_n)_k} = \begin{cases} (III_{1;n-2}) & (k=1), \\ (IV_{n;0}) \text{ or } (IV_{n;0,0}) & (k=2), \end{cases}$$

$$(V)_k = \begin{cases} (III_{1;10}) & (k=1), \\ (IV_{8;1,0}) & (k=2), \end{cases}$$

$$(VI)_k = \begin{cases} (III_{1;16}) & (k=1), \\ (IV_{10;1,0}) & (k=2), \\ (VI_0) & (k=3). \end{cases}$$

Exercises

1. Consider the case $\Omega = \mathcal{P}_\nu(\boldsymbol{C})$ $(\nu \geq 2)$.

1. 1) Using Proposition 5. 5, show that the only irreducible representations β of $\mathfrak{gl}_\nu^0(\boldsymbol{C})$ satisfying (5. 2 a, b, c) are given by $\beta = (id)$, $(\overline{id})$.

1. 2) For $(I_{\nu;r,s})$, give R_u, $H(v, v')$ explicitly. Then show that the condition (iii) is satisfied if and only if $s=0$ (cf. III, § 3, Exerc. 1).

2. Prove the similar results for $\Omega = \mathcal{P}_\nu(\boldsymbol{R})$ (cf. III, § 3, Exerc. 2).

§ 6. Peirce decompositions of a JTS.

The algebraic meaning of the "strange" condition (iii) in Theorem 3. 5 will probably be best understood from the view point of JTS's (cf. § 4, Exerc. 5). To clarify this aspect, we shall here explain the Peirce decompositions of a JTS in general.

Returning to the notation in I, § 6, let $(V, \{\ \})$ be a (not necessarily non-degenerate) JTS over a field F of characteristic zero. Let e be an "idempotent" of V, i. e., an element such that $\{e, e, e\} = e$. Then by the rules (JT 1, 2) one has

$$\{e, e, \{e, x, e\}\} = 2\{\{e, e, e\}, x, e\} - \{e, \{e, e, x\}, e\},$$
$$\{e, x, \{e, e, e\}\} = 2\{\{e, x, e\}, e, e\} - \{e, \{x, e, e\}, e\},$$

which give (after a division by 3)

$$\{e, e, \{e, x, e\}\} = \{e, \{e, e, x\}, e\} = \{e, x, e\},$$

or

(6. 1 a) $$(e\square e)P(e) = P(e)(e\square e) = P(e)$$

(cf. (I. 6. 17′)). Similarly, computing $\{x, e, \{e, e, e\}\}$, one obtains

(6. 1 b) $$e\square e = 2(e\square e)^2 - P(e)^2.$$

Combining these two relations, one has

$$(e\square e-1)(2e\square e-1)(e\square e) = (e\square e-1)P(e)^2 = 0$$

and $P(e)^3 = P(e)$. Thus the linear transformations $e\square e$ and $P(e)$ are commutative, both semi-simple, and the possible combinations of the eigenvalues are as follows:

$e\square e$	0	$\frac{1}{2}$	1
$P(e)$	0	0	± 1

We put $V_\nu = V(e\square e; \nu)$ ($\nu = 0, 1/2, 1$). Then we have the direct sum decomposition

(6. 2) $$V = V_1 \oplus V_{1/2} \oplus V_0,$$

called the *Peirce decomposition* of V with respect to e. By (JT 2) one has clearly

(6. 3) $$\{V_\lambda, V_\mu, V_\nu\} \subset V_{\lambda-\mu+\nu}.$$

In particular, one has

(6. 3 a) $$\{V_1, V_1, V_1\} = V_1,$$

(6. 3 b) $$\{V_1, V_1, V_{1/2}\} = V_{1/2}, \qquad \{V_1, V_{1/2}, V_1\} = 0,$$

(6. 3 c) $$\{V_1, V_{1/2}, V_{1/2}\} \subset V_1, \qquad \{V_{1/2}, V_1, V_{1/2}\} \subset V_0,$$

(6. 3 d) $$\{V_{1/2}, V_{1/2}, V_{1/2}\} \subset V_{1/2}.$$

In this section, we will denote the elements of V_1 and $V_{1/2}$ by $a, b, c, \cdots$ and $x, y, z, \cdots$, respectively.

First, for $a, b \in V_1$, define

(6. 4) $$a \cdot b = \{a, e, b\},$$

(6. 5) $$a^* = \{e, a, e\} = P(e)a.$$

Clearly one has $e \in V_1$ and $e^2 = e^* = e$.

Proposition 6. 1. *$V_1 = V(e \square e; 1)$ becomes a Jordan algebra with unit element e, and $P(e)|V_1 : a \mapsto a^*$ is an involution of the Jordan algebra V_1.*

Proof. The first assertion follows from I, § 6, Exerc. 4. Since $P(e)^2 = id$ on V_1, one has $a^{**} = a$. By the "fundamental formula" (I. 6. 20) one has

$$\begin{aligned} a^* b^* &= \{P(e)a, e, P(e)b\} = P(e)\{a, P(e)e, b\} \\ &= (ab)^*, \end{aligned}$$ q. e. d.

By (I. 6. 23), it follows that

(6. 6) $$\{a, b, c\} = (ab^*)c + a(b^*c) - b^*(ac).$$

Next, we define a linear map $R : V_1 \to \operatorname{End} V_{1/2}$ by

(6. 7) $$R(a)x = \{a, e, x\} \qquad (a \in V_1,\ x \in V_{1/2})\,;$$

in particular, $R(e) = \frac{1}{2} 1_{V_{1/2}}$. Then, for $a, b \in V_1$, $x \in V_{1/2}$, one has

$$\{a, b, x\} = \{\{e, a^*, e\}, b, x\} = 2\{e, a^*, \{e, b, x\}\} - \{e, \{a^*, x, b\}, e\}.$$

In view of (6. 3 b), the last term vanishes. Hence, putting $b = e$, one obtains

$$\{a, e, x\} = \{e, a^*, x\}$$

and in general

(6. 8) $$\{a, b, x\} = 2R(a)R(b^*)x.$$

We note that by a similar argument, but replacing x by an element in V_0, one obtains

(6. 3 e) $$\{V_1, V_1, V_0\} = 0.$$

Proposition 6. 2. *The map $2R : V_1 \to \operatorname{End} V_{1/2}$ is a unital Jordan algebra representation, i. e., one has*

(6. 9) $$R(ab) = R(a)R(b) + R(b)R(a)$$

and $R(e) = \frac{1}{2} 1_{V_{1/2}}$.

Proof. From

$$\{a, e, \{b, e, x\}\} = \{\{a, e, b\}, e, x\} - \{b, \{e, a, e\}, x\} + \{b, e, \{a, e, x\}\}$$

and (6. 8) one obtains

$$R(a)R(b) = R(ab) - 2R(b)R(a) + R(b)R(a),$$

whence follows (6. 9), q. e. d.

Next we define a bilinear map $H: V_{1/2}\times V_{1/2}\to V_1$ by

(6. 10) $$H(x,y)=\{e, x, y\}.$$

Then one has the following relations

(6. 11) $$H(x,y)=H(y,x)^*,$$

(6. 12) $$a\cdot H(x,y)=H(R(a^*)x,y)+H(x,R(a)y),$$

(6. 13) $$\{a, x, y\}=2H(R(a^*)x,y).$$

In fact, from

$$\{a, x, y\}=\{\{e, a^*, e\}, x, y\}=2\{\{e, x, y\}, a^*, e\}-\{e, \{a^*, y, x\}, e\}$$

one has

(*) $$\{a, x, y\}+\{a^*, y, x\}^*=2a\cdot H(x,y).$$

On the other hand, from

$$0=\{x, a, \{e, y, e\}\}=2\{\{x, a, e\}, y, e\}-\{e, \{a, x, y\}, e\}$$

one has

(**) $$\{a, x, y\}^*=2H(y, R(a^*)x).$$

Putting $a=e$ in either (*) or (**), one obtains (6. 11). Then, from (6. 11), (**), and (*), one obtains (6. 13) and (6. 12). We note that by a similar argument, but replacing x, y by elements in V_0 and using (6. 3 e), one obtains $\{V_1, V_0, V_0\}=0$. Since obviously $\{V_1, V_0, V_\nu\}=0$ for $\nu>0$ and $\{V_0, V_1, V_\nu\}=0$ for $\nu<1$ by (6. 3), one obtains

(6. 3 e') $$V_1\square V_0=V_0\square V_1=0.$$

From now on, in the rest of this section, we will further assume that $V_0=\{0\}$; an idempotent e satisfying this condition is called *principal**). (The general case will be treated in Exerc. 5.) Under this assumption, one has

(6. 3 c') $$\{V_{1/2}, V_1, V_{1/2}\}=0.$$

Hence, for $x, y, z\in V_{1/2}$, one has

$$0=\{e, y, \{x, e, z\}\}=\{\{e, y, x\}, e, z\}-\{x, \{y, e, e\}, z\}+\{x, e, \{e, y, z\}\}$$

which gives

(6. 14) $$\{x, y, z\}=2R(H(y,x))z+2R(H(y,z))x.$$

Moreover, from

$$0=\{e, x, \{y, a^*, y\}\}=2\{\{e, x, y\}, a^*, y\}-\{y, \{x, e, a^*\}, y\}$$

one obtains by (6. 8), (6. 14)

(6. 15) $$R(H(x,y))R(a)y=R(H(R(a^*)x,y))y.$$

This is an analogue of the condition (iii$'_1$). In view of (6. 9) and (6. 12), this is equivalent to

*) In general, principal idempotents may not exist. But, for instance, when $F=\mathbf{R}$ and τ is positive definite, one can prove the existence of principal idempotent (Exerc. 6).

$$R(a)R(H(x,y))y = R(H(x, R(a)y))y, \tag{6.15'}$$

which is an analogue of (iii_1). Also, computing $\{e, x, \{z, y, z\}\}$, one obtains

$$H(x, R(H(y, z))z) = H(R(H(z, x))y, z), \tag{6.15''}$$

which is an analogue of (iii_2).

Now, suppose conversely that there are given a Jordan algebra V_1 with unit element e, an involution $*$ of V_1, a unital Jordan algebra representation $2R: V_1 \to \mathrm{End}\, V_{1/2}$, and a bilinear map $H: V_{1/2} \times V_{1/2} \to V_1$ satisfying the relations (6. 11), (6. 12). Then, on $V = V_1 \oplus V_{1/2}$, we can define a triple product $\{\ \}$ satisfying (JT 1) by (6. 3 a~d), (6. 3 c'), (6. 6), (6. 8), (6. 13), (6. 14), i. e., by putting

$$\begin{aligned}\{a+x, b+y, c+z\} = {} & (\{a, b^*, c\} + 2H(R(c^*)y, x) + 2H(R(a^*)y, z)) \\ & + (2R(a)R(b^*)z + 2R(c)R(b^*)x \\ & + 2R(H(y, x))z + 2R(H(y, z))x),\end{aligned}$$

or in a matrix expression,

$$\begin{pmatrix} a \\ x \end{pmatrix} \square \begin{pmatrix} b \\ y \end{pmatrix} = \begin{pmatrix} T(ab^*) + [T(a), T(b^*)] + 2\Phi^0_{x,y} & [z \mapsto 2H(R(a^*)y, z)] \\ [c \mapsto 2R(c)R(b^*)x] & 2R(a)R(b^*) + 2\Psi^0_{x,y} \end{pmatrix}, \tag{6.16}$$

where $T(a) = [c \mapsto ac]$ and

$$\begin{aligned}\Phi^0_{x,y} &= [c \mapsto H(R(c^*)y, x)], \\ \Psi^0_{x,y} &= [z \mapsto R(H(y, x))z + R(H(y, z))x].\end{aligned}$$

Proposition 6. 3. *In the above setting, the triple product* $\{\ \}$ *defined by* (6. 16) *satisfies the condition* (JT 2) *if and only if the conditions* (6. 15) *and* (6. 15'') *are satisfied.*

Proof. The "only if" part is proved above. To prove the "if" part, it is enough to verify the condition (JT 2) for the following twelve cases :

$$\begin{array}{llll} \{a, b, \{c, d, c\}\}, & \{a, b, \{c, d, x\}\}, & \{x, a, \{b, c, b\}\}, & \{a, b, \{c, x, y\}\}, \\ \{a, x, \{c, d, y\}\}, & \{x, y, \{a, b, a\}\}, & \{a, b, \{x, y, x\}\}, & \{x, a, \{b, y, z\}\}, \\ \{x, y, \{a, b, z\}\}, & \{a, x, \{y, z, y\}\}, & \{x, y, \{a, t, z\}\}, & \{x, y, \{z, t, z\}\}, \end{array}$$

where $a, b, c, d \in V_1$ and $x, y, z, t \in V_{1/2}$. In the first six cases, the condition (JT 2) can be verified without using (6. 15) and (6. 15''). But we need (6. 15) (or (6. 15')) for the seventh, eighth, and nineth cases, and (6. 15'') for the tenth. We need furthermore the following formulas, which follow immediately from the definitions, (6. 9), (6. 11), and (6. 12) :

$$R(\{a, b, c\}) = 2R(a)R(b^*)R(c) + 2R(c)R(b^*)R(a), \tag{6.17}$$

$$\{a, b, H(x, y)\} = 2H(R(a^*)R(b)x, y) + 2H(x, R(a)R(b^*)y), \tag{6.18}$$

$$\{a, H(x, y), b\} = 2H(R(a^*)y, R(b)x) + 2H(R(b^*)y, R(a)x). \tag{6.19}$$

Here we give proofs only for the eleventh and twelfth cases. Proofs for other cases are left for an exercise of the reader.

The eleventh case: We want to prove the equality

$$\{x, y, \{a, t, z\}\} = \{\{x, y, a\}, t, z\} - \{a, \{y, x, t\}, z\} + \{a, t, \{x, y, z\}\}.$$

The left-hand side is equal to

$$4H(R(H(z, R(a^*)t))y, x)$$

and the right-hand side is equal to

$$\begin{aligned}&4H(R(H(x, R(a^*)y))t, z)-4H(R(a^*)(R(H(x,y))t+R(H(x, t))y), z)\\&+4H(R(a^*)t, R(H(y, x))z+R(H(y, z))x)\\&= 4H(R(a^*)t, R(H(y, x))z+R(H(y, z))x)-4H(R(H(x, R(a^*)t))y, z)\end{aligned}$$

by (6. 15′). Thus the equality to be proved is symmetric in x and z. Hence, putting $x=z$, $t'=R(a^*)t$, one can reduce it to

$$H(t', R(H(y, x))x) = H(R(H(x, t'))y, x),$$

which is nothing but (6. 15″).

The twelfth case: We want to prove the equality

$$\{x, y, \{z, t, z\}\} = 2\{\{x, y, z\}, t, z\} - \{z, \{y, x, t\}, z\}.$$

The left-hand side is equal to

$$8R(H(y, x))R(H(t, z))z+8R(H(y, R(H(t, z))z))x$$

and the right-hand side is equal to

$$\begin{aligned}&8R(H(t, R(H(y, x))z+R(H(y, z))x))z\\&+8R(H(t, z))(R(H(y, x))z+R(H(y, z))x)\\&-8R(H(R(H(x,y))t+R(H(x, t))y, z))z.\end{aligned}$$

Hence, by (6. 15), (6. 15′), the equality to be proved is reduced to

$$\begin{aligned}&R(H(y, R(H(t, z))z))x-R(H(t, z))R(H(y, z))x\\&= R(H(t, R(H(y, z))x))z-R(H(y, z))R(H(t, x))z.\end{aligned}$$

In view of (6. 9), (6. 12) and (6. 15″), this is equivalent to

$$\begin{aligned}&R(H(R(H(z,y))t, z))x-R(H(t, z))R(H(y, z))x\\&= -R(H(R(H(z,y))t, x))z+R(H(t, x))R(H(y, z))z.\end{aligned}$$

Fixing t and y, we put

$$F(x, z, z') = R(H(R(H(z',y))t, z))x-R(H(t, z))R(H(y, z'))x.$$

Then we have to show that

$$F(x, z, z) = -F(z, x, z).$$

But by (6. 15) one has $F(z, z, x)=0$. Hence one has

$$\begin{aligned}&F(x, z, z)+F(z, x, z)\\&= F(x+z, x+z, z)-F(x, x, z)-F(z, z, z)\\&= 0,\end{aligned}$$

as desired, q. e. d.

Now, returning to the JTS $(V, \{\ \})$ with a principal idempotent e, suppose that the condition (JT 3) is satisfied, i. e., the trace form τ is non-degenerate (and hence symmetric). Then it is clear that the subspaces V_1 and $V_{1/2}$ are orthogonal with respect to τ. By (I. 6. 7) one has

(6. 20) $$\tau(ab, c) = \tau(b, a^*c) \qquad (a, b, c \in V_1),$$

(6. 21) $$\tau(R(a)x, y) = \tau(x, R(a^*)y) \qquad (x, y \in V_{1/2}),$$

or

$$T(a)^* = T(a^*), \qquad R(a)^* = R(a^*),$$

where the * on the left-hand sides denote the adjoint with respect to τ. From (6. 20) one also has

$$\tau(a, b) = \tau(b^*, a^*).$$

Hence, putting

(6. 22) $$\langle a, b\rangle = \tau(a^*, b),$$

one obtains an inner product (i. e., a non-degenerate symmetric bilinear form) on V_1 satisfying the conditions

(6. 23) $$\langle ab, c\rangle = \langle b, ac\rangle,$$

(6. 24) $$\langle a, b\rangle = \langle a^*, b^*\rangle.$$

In general, an inner product $\langle\ \rangle$ satisfying (6. 23) (resp. (6. 24)) will be called *invariant* (resp. **-invariant*). The existence of an invariant inner product $\langle\ \rangle$ on V_1 implies that the Jordan algebra V_1 is semi-simple (I, § 6). Moreover, by (6. 21) the representation $2R: V_1 \to \mathrm{End}\, V_{1/2}$ commutes with the *-operator. Hence, if we put

$$V_1^+ = \{a \in V_1 \mid a^* = a\},$$
$$\mathrm{Sym}(V_{1/2}, \tau) = \{S \in \mathrm{End}\, V_{1/2} \mid S^* = S\},$$

then they are Jordan subalgebras of V_1 and $\mathrm{End}\, V_{1/2}$, respectively, and $2R$ induces a Jordan algebra homomorphism

$$V_1^+ \longrightarrow \mathrm{Sym}(V_{1/2}, \tau).$$

From the relation (I. 6. 7), one obtains

(6. 25) $$\langle a, H(x, y)\rangle = \tau(x, R(a)y).$$

Thus, under the assumption (JT 3), the maps R and H determine each other uniquely and the conditions (6. 9) and (6. 12) are equivalent (cf. Prop. 4. 1). We also note that the conditions (6. 15) and (6. 15″) are also equivalent, as is seen from (the proof of) Lemma 4. 6.

Conversely, suppose there are given a semi-simple Jordan algebra V_1 with an involution *, a unital Jordan algebra representation $2R: V_1 \to \mathrm{End}\, V_{1/2}$, and an inner product f on $V_{1/2}$ such that

(6. 21′) $$f(R(a)x, y) = f(x, R(a^*)y).$$

We fix an invariant and *-invariant inner product $\langle\ \rangle$ on V_1 (e. g., the trace form of V_1) and define a bilinear map $H: V_{1/2} \times V_{1/2} \to V_1$ by the relation

(6. 25′) $$\langle a, H(x, y)\rangle = 2f(x, R(a)y).$$

Then from (6. 23), (6. 24), (6. 21′) and (6. 25′), it is easy to see that H satisfies the conditions (6. 11), (6. 12). Hence we can define a triple product $\{\ \}$ by (6. 16), which satisfies (JT 2) if the equivalent conditions (6. 15) and (6. 15″) are satisfied.

In this setting, We shall now show, without assuming (6. 15), i. e., (JT 2), that the trace form τ of $(V, \{\ \})$ defined by

$$\tau(a+x, b+y) = \mathrm{tr}((a+x)\,\square\,(b+y))$$

is non-degenerate, i. e., (JT 3) is satisfied. For that purpose, let

(6. 26) $$V_1 = \bigoplus_{i=1}^{s} V_1^{(i)}$$

be the decomposition of V_1 into the direct sum of simple ideals. Then the involution * induces a permutation of order at most 2, $i \mapsto i^*$, of the set of indices defined by $V_1^{(i)*} = V_1^{(i^*)}$. In case $i^* = i$, the involution * induces an involution of $V_1^{(i)}$. Let

$$e = \sum e_i, \qquad e_i \in V_1^{(i)}.$$

Then clearly one has

(6. 27) $$e_i e_j = \delta_{ij} e_i, \qquad e_i^* = e_{i^*}.$$

Lemma 6. 4. *For $a \in V_1^{(i)}$, one has*

(6. 28) $$R(a)R(e_j) = R(e_j)R(a) = \frac{1}{2}\delta_{ij}R(a).$$

Proof. Let $j \neq i$. Then, from $ae_j = 0$, $e_j^2 = e_j$, and (6. 9), one has

$$R(a)R(e_j) + R(e_j)R(a) = 0, \qquad 2R(e_j)^2 = R(e_j).$$

Hence

$$\begin{aligned} R(a)R(e_j) &= 2R(a)R(e_j)^2 = -2R(e_j)R(a)R(e_j) \\ &= 2R(e_j)^2 R(a) = R(e_j)R(a) \end{aligned}$$

and so

$$R(a)R(e_j) = R(e_j)R(a) = 0.$$

From $R(e_i) = R(e) - \sum_{j \neq i} R(e_j)$ and $R(e) = \frac{1}{2} 1_{V_{1/2}}$, one has

$$R(a)R(e_i) = R(e_i)R(a) = \frac{1}{2}R(a), \qquad \text{q. e. d.}$$

It follows that $\{2R(e_i)\ (1 \leq i \leq s)\}$ is an orthogonal system of idempotents in $\mathrm{End}\, V_{1/2}$, i. e., one has

$$\sum_{i=1}^{s} R(e_i) = \frac{1}{2} 1_{V_{1/2}}, \qquad R(e_i)R(e_j) = \frac{1}{2}\delta_{ij}R(e_i).$$

Hence, putting

$$V_{1/2}^{(i)} = R(e_i)\, V_{1/2} \qquad (1 \leq i \leq s),$$

we obtain a direct sum decomposition

(6. 29) $$V_{1/2} = \bigoplus_{i=1}^{s} V_{1/2}^{(i)},$$

and by Lemma 6. 4

$$R(V_1^{(i)})V_{1/2}^{(j)} = \delta_{ij}V_{1/2}^{(i)} \qquad (1\le i, j\le s).$$

Thus $V_{1/2}^{(i)}$ is essentially a representation-space of $V_1^{(i)}$. Denoting by $2R^{(i)}$ the representation of $V_1^{(i)}$ on $V_{1/2}^{(i)}$ (obtained by restricting $2R$) and writing $a=\sum a_i$ $(a_i\in V_1^{(i)})$, $x=\sum x_i$ $(x_i\in V_{1/2}^{(i)})$, one has

(6. 30)
$$R(a)x = \sum_{i=1}^{s} R^{(i)}(a_i)x_i.$$

Moreover, from (6. 29) and (6. 21′), we see that

$$f(V_{1/2}^{(i)}, V_{1/2}^{(j)}) = 0 \qquad \text{if } j \neq i^*.$$

Hence $f|V_{1/2}^{(i)}\times V_{1/2}^{(i^*)}$ is non-degenerate, and $V_{1/2}^{(i^*)}$ may be identified with the dual space $(V_{1/2}^{(i)})^*$. One has

$$f(R^{(i)}(a)x, y) = f(x, R^{(i^*)}(a^*)y)$$

for $a\in V_1^{(i)}$, $x\in V_{1/2}^{(i)}$, $y\in V_{1/2}^{(i^*)}$. In particular, in case $i^*=i$, $2R^{(i)}$ induces a Jordan algebra homomorphism $V_1^{(i)+}\to \mathrm{Sym}(V_{1/2}^{(i)}, f)$.

Now by (6. 16) one has

$$\tau(a+x, b+y) = \tau_1(a, b)+\tau_1(x, y) + \tau_{1/2}(a, b)+\tau_{1/2}(x, y),$$

where

(6. 31)
$$\begin{aligned} \tau_1(a, b) &= \mathrm{tr}(T(ab^*)), & \tau_1(x, y) &= 2\,\mathrm{tr}(\Phi^0_{x,y}),\\ \tau_{1/2}(a, b) &= 2\,\mathrm{tr}(R(a)R(b^*)), & \tau_{1/2}(x, y) &= 2\,\mathrm{tr}(\Psi^0_{x,y}). \end{aligned}$$

In order to compute these traces, we need some more lemmas. In what follows, the components of $a\in V_1$ (resp. $x\in V_{1/2}$) in $V_1^{(i)}$ (resp. $V_{1/2}^{(i)}$) will always be denoted by a_i (resp. x_i).

Lemma 6. 5. *Let $\langle\ \rangle$ be an invariant inner product on V_1. Then there exists a uniquely determined invertible element γ in the center of V_1 such that*

(6. 32)
$$\langle a, b\rangle = \mathrm{tr}\ T(\gamma\cdot ab).$$

Moreover $\langle\ \rangle$ is $*$-invariant if and only if $\gamma^*=\gamma$.

Proof. Let $F^{(i)}$ be the center of $V_1^{(i)}$. Then $F^{(i)}$ is a finite extension of F and (6. 32) is equivalent to

$$\langle a_i, b_i\rangle = \mathrm{tr}\ T^{(i)}(\gamma_i\cdot a_ib_i) \qquad (1\le i\le s),$$

$T^{(i)}$ denoting the Jordan multiplication in $V_1^{(i)}$. Hence, in proving the first assertion, we may (hence shall) assume that V_1 is simple. As the trace form of V_1 is non-degenerate, one has

$$\langle a, b\rangle = \mathrm{tr}\ T((Pa)b)$$

with a uniquely determined $P\in GL(V_1)$. From the invariance of $\langle\ \rangle$ and the trace form of V, one has

$$PT(c) = T(c)P \qquad \text{for all } c \in V_1.$$

Since V_1, viewed as a Jordan algebra over its center $F^{(1)}$, is absolutely simple, the set of $T(c)$ $(c \in V_1)$, viewed as $F^{(1)}$-linear transformations, is absolutely irreducible. Hence, by Schur's lemma, one has $P = T(\gamma)$ with $\gamma \in F^{(1)\times}$, which proves (6. 32). The last assertion follows immediately from the uniqueness of γ, q. e. d.

From (6. 32) one has

(6. 32′) $$\operatorname{tr} T(a) = \langle a, \gamma^{-1} \rangle.$$

This relation can also be expressed in a different form as follows. Let (c_k), (c'_k) be mutually dual bases of V_1 with respect to $\langle\ \rangle$, i. e., $\langle c_k, c'_l \rangle = \delta_{kl}$. Then

(6. 32″) $$\sum_{k=1}^{m} c'_k c_k = \gamma^{-1} \qquad (m = \dim V_1).$$

In fact, for $a \in V_1$, one has

$$\operatorname{tr} T(a) = \sum_k \langle c'_k, a c_k \rangle = \langle \sum_k c'_k c_k, a \rangle.$$

Hence (6. 32′) and (6. 32″) are equivalent (cf. Lem. I. 6. 1).

Lemma 6. 6. *Let* $\dim V_1^{(i)} = m_i$, $\dim V_{1/2}^{(i)} = n_i$ $(1 \leq i \leq s)$. *Then*

(6. 33) $$\operatorname{tr} R(a) = \sum_{i=1}^{s} \frac{n_i}{2m_i} \operatorname{tr} T^{(i)}(a_i).$$

Proof. We define a bilinear form r on $V_1 \times V_1$ by

$$r(a, b) = \operatorname{tr}(R(a)R(b)) \qquad (a, b \in V_1).$$

Then r is symmetric, and by (6. 9) satisfies the relation

$$r(ab, c) = r(b, ac) \qquad (a, b, c \in V_1).$$

Hence by Lemma 6. 5 (which is also applicable to the case where $\langle\ \rangle$ is degenerate), there exists an element γ' in the center of V_1 such that

$$r(a, b) = \operatorname{tr} T(\gamma' ab).$$

Putting $b = e$, one has

(*) $$\frac{1}{2} \operatorname{tr} R(a) = \operatorname{tr} T(\gamma' a).$$

Let $d_i = [F_i : F]$. Then d_i is a divisor of m_i and n_i. Putting $a = e_i$ in (*), one has

(**) $$\frac{1}{4} n_i = \operatorname{tr}_{F^{(i)}/F}(\gamma'_i) \cdot \frac{m_i}{d_i}.$$

Let $\bar{F}$ be the algebraic closure of F, and let

$$V_1^{(i)} \otimes_F \bar{F} = \bigoplus_{j=1}^{d_i} V_1^{(i,j)}, \qquad V_{1/2}^{(i)} \otimes_F \bar{F} = \bigoplus_{j=1}^{d_i} V_{1/2}^{(i,j)}$$

be the corresponding direct sum decompositions similar to (6. 26) and (6. 29), where $\dim_{\bar{F}} V_1^{(i,j)} = m_i/d_i$ and $\dim_{\bar{F}} V_{1/2}^{(i,j)} = n_i/d_i$. Then, applying (**) to each component of

γ'_i in $V_1^{(i,j)}$, we see that the (i, j)-component of γ' is given by $\frac{n_i}{4m_i}e_{ij}$, e_{ij} denoting the (i, j)-component of e. Thus one has

$$\gamma' = \sum_{i=1}^{s} \frac{n_i}{4m_i} e_i.$$

Hence (6. 33) follows from (*), q. e. d.

Combining (6. 32') and (6. 33), one has

$$\text{(6. 34)} \qquad \operatorname{tr} R(a) = \left\langle a, \sum_i \frac{n_i}{2m_i}\gamma_i^{-1}\right\rangle$$

(I, § 9, Exerc. 2). Let (z_ν), (z'_ν) be mutually dual bases of $V_{1/2}$ with respect to f, i. e., $f(z'_\mu, z_\nu) = \delta_{\mu\nu}$. Then, for $a \in V_1$, one has

$$\operatorname{tr} R(a) = \sum_\nu f(z'_\nu, R(a)z_\nu) = \frac{1}{2}\sum_\nu \langle a, H(z'_\nu, z_\nu)\rangle.$$

Hence (6. 34) is equivalent to

$$\text{(6. 34')} \qquad \sum_{\nu=1}^{n} H(z'_\nu, z_\nu) = \sum_i \frac{n_i}{m_i}\gamma_i^{-1} \qquad (n = \dim V_{1/2}).$$

Lemma 6. 7. *For $x \in V_{1/2}$, one has*

$$\text{(6. 35)} \qquad \sum_{\nu=1}^{n} R(H(z'_\nu, x))z_\nu = R(\gamma^{-1})x.$$

Proof. Put $H(z'_\lambda, z_\mu) = \sum_{k=1}^{m} h_{\lambda\mu}^k c_k$. Then

$$h_{\lambda\mu}^k = \langle c'_k, H(z'_\lambda, z_\mu)\rangle = 2f(z'_\lambda, R(c'_k)z_\mu).$$

Hence

$$2R(c'_k)z_\mu = \sum_{\lambda=1}^{n} h_{\lambda\mu}^k z_\lambda.$$

It follows by (6. 32'')

$$\begin{aligned} \sum_\nu R(H(z'_\nu, z_\mu))z_\nu &= \sum_{\nu, k} h_{\nu\mu}^k R(c_k)z_\nu \\ &= 2\sum_k R(c_k)R(c'_k)z_\mu \\ &= R(\sum_k c_k c'_k)z_\mu = R(\gamma^{-1})z_\mu, \end{aligned}$$

which implies (6. 35), q. e. d.

Now we are ready to compute the traces in (6. 31).

$$\begin{aligned} \tau_1(a, b) &= \operatorname{tr} T(ab^*) = \langle \gamma^{-1}a, b^*\rangle \qquad \text{(by (6. 32))}, \\ \tau_{1/2}(a, b) &= 2\operatorname{tr}(R(a)R(b^*)) = \operatorname{tr} R(ab^*) \\ &= \sum_i \frac{n_i}{2m_i}\langle \gamma_i^{-1}a, b^*\rangle \qquad \text{(by (6. 34))}, \end{aligned}$$

$$\begin{aligned}\tau_1(x,y) &= 2\operatorname{tr}_{V_1}[c\mapsto H(R(c^*)y,x)]\\ &= 2\sum_k \langle c'_k, H(R(c_k^*)y,x)\rangle\\ &= 4\sum_k f(R(c_k^*)y, R(c'_k)x)\\ &= 4f(y, \sum_k R(c_k)R(c'_k)x)\\ &= 2f(y, R(\gamma^{-1})x) \qquad (\text{by } (6.\,32'')),\\ \tau_{1/2}(x,y) &= 2\operatorname{tr} R(H(y,x)) + 2\operatorname{tr}_{V_{1/2}}[z\mapsto R(H(y,z))x],\end{aligned}$$

where

$$\begin{aligned}\operatorname{tr} R(H(y,x)) &= \sum_i \frac{n_i}{2m_i}\langle \gamma_i^{-1}, H(y,x)\rangle \qquad (\text{by } (6.\,34))\\ &= \sum_i \frac{n_i}{m_i} f(y, R(\gamma_i^{-1})x),\\ \operatorname{tr}[z\mapsto R(H(y,z))x] &= \sum_\nu f(z'_\nu, R(H(y,z_\nu))x)\\ &= f(\sum_\nu R(H(z_\nu,y))z'_\nu, x)\\ &= f(R(\gamma^{-1})y, x) \qquad (\text{by } (6.\,35)).\end{aligned}$$

Hence

$$\tau_{1/2}(x,y) = 2\sum_i \left(1+\frac{n_i}{m_i}\right) f(y, R(\gamma_i^{-1})x).$$

Summing up, one obtains

$$(6.\,36) \qquad \tau(a+x, b+y) = \sum_{i=1}^{s}\left(1+\frac{n_i}{2m_i}\right)(\langle b^*, \gamma_i^{-1}a\rangle + 4f(y, R(\gamma_i^{-1})x)),$$

which shows that τ is non-degenerate. Note that in these formulas, one has $m_{i^*}=m_i$, $n_{i^*}=n_i$ and $\gamma_{i^*}=\gamma_i^*$. We have thus proved the following

Theorem 6. 8. *Let V_1 be a semi-simple Jordan algebra with involution $*$ over F, $V_{1/2}$ a (finite-dimensional) vector space over F with an inner product f, and $2R: V_1 \to \operatorname{End} V_{1/2}$ a unital Jordan algebra representation satisfying (6. 21′). Fix an invariant and $*$-invariant inner product $\langle\ \rangle$ on V_1 and define a bilinear map $H: V_{1/2}\times V_{1/2}\to V_1$ by (6. 25). Then the triple product $\{\ \}$ defined by (6. 16) on $V=V_1+V_{1/2}$ satisfies the conditions (JT 1), (JT 3). The condition (JT 2) is satisfied if and only if one (hence all) of the equivalent conditions (6. 15), (6. 15′), (6. 15″) is satisfied. Every non-degenerate JTS $(V, \{\ \})$, along with a Peirce decomposition $V=V_1+V_{1/2}$ with respect to a principal idempotent e, is obtained in this manner.*

By virtue of this theorem, it is not difficult to obtain the classification of JTS's with a principal idempotent (e. g., that of positive definite JTS's over $\boldsymbol{R}$).

By a slight modification, we can obtain similar results concerning hermitian JTS's. Let $(V, \{\ \})$ be a JTS over F (satisfying only (JT 1, 2)) and let F'/F be a quadratic extension with Galois group $\{1, \sigma_0\}$. When V is given an "F'-structure" (i. e., a structure of vector space over F') extending the given F-structure such that the triple product $\{u, v, w\}$ is F'-linear in u, w and F'/F-antilinear in v, the pair $(V, \{\ \})$ is called a *hermitian JTS over* F'/F. In this case, the map $u\square v$ is F'-linear and the *trace form over* F' is defined by

(6. 37) $$\tau'(u, v) = \mathrm{tr}_{V/F'}(v\square u).$$

Then τ' is "sesquilinear" (F'-linear in v and F'/F-antilinear in u) and one has $\tau(u, v) = \mathrm{tr}_{F'/F}\tau'(u, v)$. The condition (JT 3) is equivalent to

(JT 3′) *the sesquilinear form* τ' *is non-degenerate.*

The condition (JT 3′) implies that τ' is hermitian (with respect to F'/F).

Let e be a (principal) idempotent of V. Then, since $e\square e$ is F'-linear, the eigenspaces V_ν are F'-spaces, and so $m = \dim_F V_1$ and $n = \dim_F V_{1/2}$ are even. On the other hand, $P(e)$ is F'-antilinear. Hence, in Proposition 6. 1, one has that V_1 is a Jordan algebra over F' with unit element e and the involution $a \mapsto a^*$ is of "the second kind", i. e., F'/F-antilinear. Therefore, V_1^+ is an "F-form" of the F'-algebra V_1 (for which the conjugation is given by $a \mapsto a^*$) and one has $V_1 = V_1^+ \otimes_F F'$. Furthermore, in Proposition 6. 2, the map $2R$ is an F'-linear representation of V_1 into $\mathrm{End}_{F'} V_{1/2}$.

Now, assuming the condition (JT 3′), we put

(6. 22′) $$\langle a, b\rangle' = \tau'(a^*, b) \qquad (a, b \in V_1).$$

Then $\langle\ \rangle'$ is an (F'-bilinear) invariant inner product in V_1 and one has $\langle a, b\rangle = \mathrm{tr}_{F'/F}\langle a, b\rangle'$. As an analogue of (6. 24), one has

(6. 24′) $$\langle a^*, b^*\rangle' = \langle a, b\rangle'^{\sigma_0} \qquad (a, b \in V_1).$$

In particular, for $a, b \in V_1^+$, one has $\langle a, b\rangle' = \frac{1}{2}\langle a, b\rangle \in F$. Thus $\langle\ \rangle'$ induces an (F-bilinear) invariant inner product in V_1^+; therefore V_1^+ is semi-simple. On the other hand, if we put

$$h(x, y) = \frac{1}{2}\tau'(x, y) \qquad (x, y \in V_{1/2}),$$

h is a non-degenerate hermitian form on $V_{1/2}$ and the representation $2R$ induces an (F-linear) representation of V_1^+ into $\mathcal{H}(V_{1/2}, h)$. The map $H\colon V_{1/2} \times V_{1/2} \to V_1$ is also hermitian (i. e., F'-linear in the second variable and satisfies (6. 11)), and one has the relation

(6. 25″) $$\langle a, H(x, y)\rangle' = h(x, R(a)y).$$

As for the converse construction, we can, in virtue of Proposition 6. 3 and Theorem 6. 8, obtain the following

Theorem 6. 9. *Let* V_1^+ *be a semi-simple Jordan algebra over* F *with unit element* e *and let* $\langle\ \rangle'$ *be an invariant inner product on* V_1^+. *Let* F' *be a quadratic extension of* F, $V_{1/2}$ *a vector space over* F', h *a non-degenerate hermitian form on* $V_{1/2}$ *(with respect to* F'/F*), and* $2R$ *a unital Jordan algebra representation over* F *of* V_1^+ *into* $\mathcal{H}(V_{1/2}, h)$. *Let further* $V_1 = V_1^+ \otimes_F F'$, *and extend* $\langle\ \rangle'$ *and* $2R$ *to an* F'*-bilinear form on* $V_1 \times V_1$ *and an* F'*-linear representation* $V_1 \to \mathrm{End}_{F'} V_{1/2}$, *respectively. Define a hermitian map* $H\colon V_{1/2} \times V_{1/2} \to V_1$ *by* (6.

25"). *Then the triple product* $\{\ \}$ *defined by* (6. 16) *on* $V=V_1+V_{1/2}$ (*where* $*$ *denotes the conjugation of* V_1 *over* V_1^+) *satisfies the conditions* (JT 1), (JT 3'). *The condition* (JT 2) *is satisfied if and only if one* (*hence all*) *of the equivalent conditions* (6. 15), (6. 15'), (6. 15") *is satisfied. Every non-degenerate hermitian JTS* $(V, \{\ \})$, *with a principal idempotent* e *and the corresponding Peirce decomposition* $V=V_1+V_{1/2}$, *is obtained in this manner.*

From (6. 32'), (6. 33), (6. 36), one obtains

$$(6.38)\qquad \begin{cases} \mathrm{tr}_{V_1/F'} T(a) = \langle a, \gamma^{-1}\rangle', \\ \mathrm{tr}_{V_{1/2}/F'} R(a) = \langle a, \delta-\gamma^{-1}\rangle', \\ \tau'(a+x, b+y) = \langle a^*, \delta b\rangle' + 4h(x, R(\delta)y) \\ \qquad\qquad (a, b \in V_1,\ x, y \in V_{1/2}), \end{cases}$$

where $\gamma=\sum\gamma_i$ is an invertible element in the center of V_1^+ (uniquely determined by $\langle\ \rangle'$) and $\delta=\sum_{i=1}^{s}\left(1+\frac{n_i}{2m_i}\right)\gamma_i^{-1}$. For convenience, let us assume that the indices i are so arranged that $i^*=i$ for $1\le i\le s_1$ and $i^*=i+s_2$ for $s_1+1\le i\le s_1+s_2$, where $s=s_1+2s_2$. Then, for $1\le i\le s_1$, $R^{(i)}=R|V_1^{(i)}$ induces a representation $V_1^{(i)+}\to\mathcal{H}(V_{1/2}^{(i)}, h)$. For $s_1+1\le i\le s_1+s_2$, one has

$$(V_1^{(i)}+V_1^{(i^*)})^+ \cong R_{F'/F}(V_1^{(i)})$$

by the projection $a+a^*\mapsto a$ $(a\in V_1^{(i)})$, $R^{(i)}$ is an F'-linear representation of $V_1^{(i)}$ into $\mathrm{End}_{F'} V_{1/2}^{(i)}$, and $R^{(i^*)}(a^*)$ is the "conjugate dual" of $R^{(i)}(a)$ with respect to $h|V_{1/2}^{(i)}\times V_{1/2}^{(i^*)}$.

In particular, in the case $F=\boldsymbol{R}$, $F'=\boldsymbol{C}$, a hermitian JTS $(V, \{\ \})$ is by definition "positive definite" if the trace form τ' (or τ) is positive definite. Under this assumption, if $a, b\in V_1^+$ and $a^2+b^2=0$, then one has

$$0=\tau'(e, a^2+b^2)=\tau'(a, a)+\tau'(b, b),$$

whence $a=b=0$. Thus V_1^+ is formally real. It follows that all simple components of V_1^+ are formally real (hence central). Therefore, in the above notation, one has $s=s_1$ and $\gamma_i=\langle e_i, e_i\rangle/m_i$ in (6. 32). Hence the invariant inner product $\langle\ \rangle'$ (or $\langle\ \rangle$) is positive definite if and only if all γ_i $(1\le i\le s)$ are positive. Conversely, in view of (6. 38), we see that τ' is positive definite if both $\langle\ \rangle'$ and h are taken to be positive.

Assuming the last condition, let Ω be the self-dual homogeneous cone corresponding to V_1^+. Then the Jordan algebra homomorphism $2R: V_1^+\to\mathcal{H}(V_{1/2}, h)$ induces an equivariant map

$$\Omega \longrightarrow \mathcal{P}(V_{1/2}, h).$$

The condition $R(\Omega)\subset\mathcal{P}(V_{1/2}, h)$ is equivalent to saying that the corresponding hermitian map H is "Ω-positive". In fact, this condition amounts to saying that for all $a\in\Omega$ and $x\in V_{1/2}$, $x\neq 0$, one has

$$\langle a, H(x, x)\rangle' = 2h(x, R(a)x) > 0,$$

which implies $H(x, x)\in\bar{\Omega}-\{0\}$. Thus, in this case, the data $(V_1^+, V_{1/2}, \Omega, H)$ define a quasi-symmetric Siegel domain $\mathcal{S}$. By Theorem 6. 9, we can then conclude that the condition (iii) in Theorem 3. 5 is equivalent to the condition (JT 2) for the triple product { } defined by (6. 16) on $V=V_1\oplus V_{1/2}$ (§ 4, Exerc. 5).

Exercises

1*). Let $(V, \{\ \})$ be a (not necessarily non-degenerate) JTS and $x\in V$. For an odd positive integer r, we define the power x^r inductively by

(6. 39) $$x^r = P(x)x^{r-2} \qquad (r=3, 5, \cdots).$$

1. 1) Prove

(6. 40) $$x^r\Box x = 2(x\Box x^{r-2})(x\Box x)-P(x)P(x^{r-2}, x),$$

(6. 41) $$P(x^r, x) = P(x)(x^{r-2}\Box x) = (x\Box x^{r-2})P(x),$$

where $P(a, b)$ denotes the linear transformation $x\mapsto\{a, x, b\}$ (cf. (I. 6. 18), (I. 6. 17′)).

1. 2) Show that

(6. 42) $$x^r = \{x, x, x^{r-2}\} \qquad (r=3, 5, \cdots).$$

Hint. Induction on r, making use of $[P(x), x\Box x]=0$.

1. 3) Show that, for any odd positive integers p, q, the linear transformation $x^p\Box x^q$ depends only on $p+q=m$, i. e., one has

(6. 43) $$x^{m-1}\Box x = x^{m-3}\Box x^3 = \cdots = x\Box x^{m-1}.$$

Hint. Prove (6. 43) by induction on m, using the relations

$$[x\Box x, x^p\Box x^{m-p}] = x^{p+2}\Box x^{m-p}-x^p\Box x^{m-p+2} \qquad (1\leq p\leq m-1),$$
$$[x^{m-1}\Box x, x\Box x] = x^{m+1}\Box x-x\Box x^{m+1}.$$

1. 4) Show that, for any positive odd integers p, q, r, s, one has

(6. 44) $$\{x^p, x^q, x^r\} = x^{p+q+r},$$

(6. 45) $$[x^p\Box x^q, x^r\Box x^s] = 0,$$

(6. 46) $$[x^p\Box x^q, P(x^r, x)] = 0.$$

2. Let $x\in V$ be such that $\{x, x, x\}=\lambda x$ with $\lambda\in F$. Show that, if $\lambda\neq 0$, $x\Box x$ and $P(x)$ are semi-simple and the possible combinations of eigenvalues are as follows:

$x\Box x$	0	$\frac{1}{2}\lambda$	λ
$P(x)$	0	0	$\pm\lambda$

When τ is non-degenerate and "anisotropic" (i. e., $\tau(x, x)=0$ implies $x=0$), show that $\{x, x, x\}=0$ implies $x=0$.

3. Assume that τ is non-degenerate and anisotropic and that $x\in V$ is such that all eigenvalues of $x\Box x$ are in F. Put $V_\lambda=V(x\Box x\,;\,\lambda)$. Prove the following.

3. 1) One has an orthogonal decomposition $V=\bigoplus V_\lambda$. Write $x=\sum x_\lambda$ with $x_\lambda\in V_\lambda$. Then for any positive odd integer $2k+1$ one has

(6. 47) $$x^{2k+1} = \sum_\lambda \lambda^k x_\lambda.$$

3. 2) For $f(X)\in F[X]$, put $\tilde{f}(X)=X\cdot f(X^2)$. Let $\Lambda=\{\lambda_1, \cdots, \lambda_s\}$ be the totality of (distinct) eigenvalues of $x\Box x$ for which $x_\lambda\neq 0$ and set

$$f_i(X) = (\prod_{j\neq i}(\lambda_i-\lambda_j))^{-1}\prod_{j\neq i}(X-\lambda_j).$$

*) The results in Exerc. 1~4 are taken essentially from a talk of R. Kunze at Kleebach in 1976, where he developed a spectral theory of $x\Box x$. Cf. also Loos [9], § 3.

Then one has

(6. 48) $$x_{\lambda_i} = \tilde{f}_i(x).$$

3. 3) One has

(6. 49) $$x_{\lambda_i}\square x_{\lambda_j} = 0 \qquad \text{for} \quad i \neq j.$$

[*Hint.* Prove first that $x_{\lambda_i}\square x_{\lambda_j} = x_{\lambda_j}\square x_{\lambda_i}$ by 1. 3) and 3. 2), and then note that $(x_{\lambda_i}\square x_{\lambda_j})V_{\lambda_k} \subset V_{\lambda_j-\lambda_i+\lambda_k}$.] It follows that

(6. 50) $$x\square x = \sum_{i=1}^{s} x_{\lambda_i}\square x_{\lambda_i},$$

(6. 51) $$x_\lambda^3 = \lambda x_\lambda.$$

4. Suppose that $F=\boldsymbol{R}$ and τ is positive definite. Then for any $x \in V$, $x\square x$ is hermitian with respect to τ, so that all assumptions of 3 are satisfied. Show that, in the notation of 3, one has $\lambda_i > 0$ $(1 \leq i \leq s)$. It follows that $x\square x$ is positive semi-definite and one has $x\square x = 0 \Leftrightarrow x = 0$. (In this case, $e_i = \sqrt{\lambda_i}^{-1} x_{\lambda_i}$ is an idempotent.)

5. Let $(V, \{\ \})$ be a non-degenerate JTS over F, and e an idempotent, which is not principal. Retaining the notation in the text, we denote elements of $V_0 = V(e\square e; 0)$ by $u, v, w, \cdots$. Define maps $R_0 : V_0 \to \operatorname{End} V_{1/2}$ and $S_0 : V_{1/2} \times V_{1/2} \to V_0$ by

(6. 52) $$R_0(u)x = \{u, x, e\}, \qquad S_0(x, y) = \{x, e, y\} \qquad (x, y \in V_{1/2},\ u \in V_0).$$

Prove the following.

5. 1) $2R_0$ is a JTS homomorphism of V_0 into $\operatorname{Sym}(V_{1/2}, \tau)$, i. e., one has

(6. 53) $$R_0(\{u, v, w\}) = 2(R_0(u)R_0(v)R_0(w) + R_0(w)R_0(v)R_0(u)).$$

5. 2) R_0 and S_0 determine each other uniquely by the relation

(6. 54) $$\tau(u, S_0(x, y)) = \tau(R_0(u)x, y).$$

5. 3) One has the following formulas

(6. 55) $$\{u, v, x\} = 2R_0(u)R_0(v)x,$$

(6. 56) $$\{a, x, u\} = 2R(a)R_0(u)x = 2R_0(u)R(a^*)x,$$

(6. 57) $$\{x, y, u\} = 2S_0(x, R_0(u)y),$$

(6. 58) $$\{x, u, y\} = 2H(R_0(u)y, x),$$

(6. 59) $$\{x, a, y\} = 2S_0(R(a^*)x, y) = 2S_0(x, R(a^*)y),$$

(6. 60) $$\{x, y, z\} = 2R(H(y, x))z + 2R(H(y, z))x - 2R_0(S_0(x, z))y.$$

Hint. To prove (6. 55), first prove (6. 57) by computing $\{\{e, e, x\}, y, u\}$, and then show that

$$\tau(\{u, v, x\}, y) = \tau(R_0(u)R_0(v)x, y)$$

for all $y \in V_{1/2}$.

5. 4) One has the following relation

(6. 61) $$\begin{aligned} &2R(a)R(H(y, x))x - R(a)R_0(S_0(x, x))y \\ &= 2R(H(y, R(a)x))x - R_0(S_0(x, R(a)x))y \end{aligned}$$

for all $a \in V_1$, $x, y \in V_{1/2}$.

6. Let e and e' be two non-zero idempotents in a JTS V such that $\{e, e, e'\} = 0$. Show that one has $e\square e' = e'\square e = 0$, $e + e'$ is an idempotent, and $V_0(e+e') \subsetneqq V_0(e)$. Combining this with the result of 3, 4, prove the existence of principal idempotent in a positive definite JTS over $\boldsymbol{R}$.

7. Let F' be a quadratic extension of F with Galois group $\{1, \sigma_0\}$ and let $V = \mathcal{M}_{p,q}(F')$ endowed with triple product

$$\{x, y, z\} = \frac{1}{2}(x^t y^{\sigma_0} z + z^t y^{\sigma_0} x).$$

Give the Peirce decomposition with respect to the idempotent $e=\begin{pmatrix}1_k & 0\\ 0 & 0\end{pmatrix}$ and describe the corresponding H, R, S_0, R_0.

8. Let A be a Jordan algebra and e an "idempotent" in A (i. e., $e^2=e$). Then, one has $e\square e=T_e$, so that e is an idempotent of the JTS $(A, \{\ \})$. By (6. 2) one has $A=A_1+A_{1/2}+A_0$ with $A_\nu=A(T_e;\nu)$.

8. 1) Prove the relation

$$A_\mu \cdot A_\nu \subset A_{\mu+\nu}+A_{\mu+\nu-1}, \tag{6. 62}$$

(in particular, A_1 and A_0 are subalgebras).

8. 2) Show that $A_1\cdot A_0=0$ and that A_1 and A_0 are "strongly commutative", i. e., one has $[T_a, T_u]=0$ for all $a\in A_1$, $u\in A_0$. (Use (6. 3 e').)

8. 3) Check that the Jordan algebra structure on A_1 defined by (6. 4) coincides with the one induced from A and that the involution (6. 5) is trivial. Then show that, for $a\in A_1$, $x, y\in A_{1/2}$, $u\in A_0$, one has

$$R(a)x = ax, \qquad R_0(u)x = ux,$$
$$H(x, y)+S_0(x, y) = xy\,;$$

moreover, $2R_0$ is a Jordan algebra representation $A_0\to \mathrm{End}\, A_{1/2}$.

8. 4) When e is principal (i. e., $A_0=0$), prove the following identities:

$$ax^2 = 2(ax)x, \qquad x^3 = 0,$$
$$(xy)(ay) = ((ax)y)y, \qquad a((xy)y) = (x(ay))y,$$
$$x(z(yz)) = ((xz)y)z \qquad (a\in A_1,\ x, y, z\in A_{1/2}).$$

9. Let A be a formally real Jordan algebra (with unit element 1) and $\{e_1, \cdots, e_r\}$ a maximal system of orthogonal idempotents. (This means that $e_ie_j=\delta_{ij}e_i$, $\sum_{i=1}^{r} e_i=1$, and each e_i is "primitive", i. e., can not be decomposed into the sum of two non-zero orthogonal idempotents.) Put

$$A_{ii} = A(T_{e_i}; 1) \quad \text{and} \quad A_{ij} = A\left(T_{e_i}; \frac{1}{2}\right)\cap A\left(T_{e_j}; \frac{1}{2}\right) \qquad \text{for} \quad i\neq j.$$

9. 1) Show that $A=\bigoplus_{i\leq j} A_{ij}$, $A_{ii}=\{e_i\}_{\boldsymbol{R}}$, and $A_{ij}\subset A(T_{e_k}; 0)$ for $k\neq i, j$.

9. 2) Let $\mathfrak{g}(A)=A+\mathrm{Der}(A)$ be the "structure algebra" of A (I, § 7). Show that $\mathfrak{a}=\{e_1, \cdots, e_r\}_{\boldsymbol{R}}$ is an abelian Lie subalgebra of $\mathfrak{g}(A)$, contained in A and is maximal with respect to this property. [This implies that rank $\mathfrak{g}(A)=r$; r is also called the "degree" of the Jordan algebra A. In case A is simple, one can show that dim A_{ij} $(i\neq j)$ are all equal (this is the number d in the table on p. 37). From these, it is easy to determine the root structure of $\mathfrak{g}(A)$. Cf. e. g., Ash et al. [1], pp. 92–97.]

Remark. From the results of 6 and 9, one can obtain an alternate proof for the existence of a system of orthogonal (H_1)-homomorphisms κ_i mentioned in III, § 4. Let $\mathcal{D}$ be a bounded symmetric domain (in $\mathfrak{p}_+$) and $\mathfrak{g}=\mathrm{Lie\ Hol}(\mathcal{D})$. By 6, there exists a principal idempotent o_1 in the JTS $\mathfrak{p}_+$. Let $\kappa: \mathfrak{sl}_2(\boldsymbol{R})\to\mathfrak{g}$ be the corresponding (H_1)-homomorphism, determined by the relation $o_1=\frac{1}{2}(X_\kappa+iY_\kappa)$. Then, by Prop. III. 3. 4, one has a Peirce decomposition $c_\kappa^{-1}\mathfrak{p}_+=U_{\boldsymbol{C}}+V_+$ (with $\mathfrak{p}_{\kappa,+}=0$) and hence, by Th. III. 7. 1, $\mathcal{D}$ is realized as a Siegel domain of the second kind: $\mathcal{D}\approx\mathcal{S}_\kappa$. In the notation of §§ 3, 4, one then has a canonical decomposition

$$\mathfrak{g} = U+V+\mathfrak{g}_0+\theta V+\theta U,$$

where $\mathfrak{g}_0=\mathfrak{k}^{(1)}\oplus\mathfrak{g}(\Omega)$ and U may be viewed as a formally real Jordan algebra with unit element e_κ. By I, § 9, Exerc. 5, it follows that rank $\mathfrak{g}$=rank $\mathfrak{g}(\Omega)=r$. Hence, in view of the categorical equivalence (II, § 8), the decomposition of the unit element e_κ in 9 gives rise to a decomposition of κ into a sum of r (H_1)-homomorphisms $\kappa_i:\mathfrak{sl}_2(\boldsymbol{R})\to\mathfrak{g}$ which are mutually commutative.

§ 7. Morphisms of symmetric and quasi-symmetric Siegel domains.

Let $\mathcal{S}=\mathcal{S}(U, V, \Omega, H)$ and $\mathcal{S}'=\mathcal{S}(U', V', \Omega', H')$ be two Siegel domains. A pair of linear maps $\varphi: U\to U'$ and $\psi: V\to V'$ is called a *Siegel domain map* of $\mathcal{S}$ into $\mathcal{S}'$ if the following three conditions are satisfied:

(S 1) $\varphi(\Omega)\subset\Omega'$,

(S 2) ψ *is* $\boldsymbol{C}$*-linear,*

(S 3) $\varphi_C(H(v_1, v_2)) = H'(\psi(v_1), \psi(v_2)) \qquad (v_1, v_2\in V)$,

where φ_C denotes the $\boldsymbol{C}$-linear extension of φ to $U_C\to U'_C$. It is clear that under these conditions $\tilde{\varphi}=\varphi_C\times\psi: U_C\times V\to U'_C\times V'$ induces a holomorphic map $\mathcal{S}\to\mathcal{S}'$, which we also call a "Siegel domain map".

Let

$$\mathfrak{g} = \Sigma\mathfrak{g}_\nu \quad\text{and}\quad \mathfrak{g}' = \Sigma\mathfrak{g}'_\nu \tag{7. 1}$$

be the canonical gradations of $\mathfrak{g}$=Lie Hol($\mathcal{S}$) and $\mathfrak{g}'$=Lie Hol($\mathcal{S}'$) defined in § 1. We recall that, if we put

$$\mathfrak{g}_- = \sum_{\nu\le 0}\mathfrak{g}_\nu \quad\text{and}\quad \mathfrak{g}'_- = \sum_{\nu\le 0}\mathfrak{g}'_\nu, \tag{7. 1'}$$

then $\mathfrak{g}_-$ and $\mathfrak{g}'_-$ (resp. $\mathfrak{g}_0$ and $\mathfrak{g}'_0$) are identified with the Lie algebras of the groups of affine (resp. linear) automorphisms of $\mathcal{S}$ and $\mathcal{S}'$. In this section, we assume $\mathcal{S}$ and $\mathcal{S}'$ to be quasi-symmetric. As before, we fix a reference point $e\in\Omega$, an inner product $\langle\ \rangle$ on U, a hermitian form h on V satisfying the relations (3. 2), (3. 5), and similarly e', $\langle\ \rangle'$, h' for $\mathcal{S}'$. θ_0 and θ'_0 are the Cartan involutions of $\mathfrak{g}_0$ and $\mathfrak{g}'_0$ at ie and ie', respectively. In general, the objects relative to $\mathcal{S}'$ are denoted by the corresponding symbols with a prime.

Now let $\varphi: U\to U'$, $\psi: V\to V'$ be linear maps and $\rho_0:\mathfrak{g}_0\to\mathfrak{g}'_0$ a Lie algebra homomorphism. A triple (φ, ψ, ρ_0) is called an *equivariant triple,* if the following conditions (a), (b 1, 2), (c 1, 2, 3) are satisfied:

(a) $\qquad \varphi(\operatorname{Im} H(v_1, v_2)) = \operatorname{Im} H'(\psi(v_1), \psi(v_2)) \qquad (v_1, v_2\in V)$,

(b 1) $\qquad \varphi(\pi_U(X)u) = \pi_{U'}(\rho_0(X))\varphi(u)$,

(b 2) $\qquad \psi(\pi_V(X)v) = \pi_{V'}(\rho_0(X))\psi(v) \qquad (u\in U,\ v\in V,\ X\in\mathfrak{g}_0)$,

(c 1) $\qquad \varphi(e) = e'$,

(c 2)(=(S 2)) ψ is $\boldsymbol{C}$-linear,

(c 3) $\qquad \rho_0\circ\theta_0 = \theta'_0\circ\rho_0$.

Clearly (a) is equivalent to (S 3) under the assumption (c 2). Let

$$\mathfrak{g}_0 = \mathfrak{k}_0 + \mathfrak{p}^{(2)} \quad \text{and} \quad \mathfrak{g}_0' = \mathfrak{k}_0' + \mathfrak{p}^{(2)\prime}$$

be the Cartan decompositions corresponding to θ_0 and θ_0'. Then, given an equivariant triple (φ, ψ, ρ_0), one has from (c 3)

$$\rho_0(\mathfrak{k}_0) \subset \mathfrak{k}_0' \quad \text{and} \quad \rho_0(\mathfrak{p}^{(2)}) \subset \mathfrak{p}^{(2)\prime}.$$

Hence, denoting by $\mathfrak{g}^{(2)}$ and $\mathfrak{g}^{(2)\prime}$ the subalgebras of $\mathfrak{g}_0$ and $\mathfrak{g}_0'$ generated by $\mathfrak{p}^{(2)}$ and $\mathfrak{p}^{(2)\prime}$, respectively, one has

(7. 2) $$\rho_0(\mathfrak{g}^{(2)}) \subset \mathfrak{g}^{(2)\prime}.$$

To simplify the notation, we set

$$X(A) = X_{A,\beta(A)}, \qquad X'(A') = X'_{A',\beta'(A')}$$

for $A \in \mathfrak{g}(\Omega)$, $A' \in \mathfrak{g}(\Omega')$. Then we know that the correspondences $A \mapsto -X(A)$ and $A' \mapsto -X'(A')$ give isomorphisms $\mathfrak{g}(\Omega) \cong \mathfrak{g}^{(2)}$, $\mathfrak{g}(\Omega') \cong \mathfrak{g}^{(2)\prime}$. Hence it follows from (7. 2) that there exists a unique Lie algebra homomorphism $\rho_\Omega : \mathfrak{g}(\Omega) \to \mathfrak{g}(\Omega')$ such that one has

(7. 3) $$\rho_0(X(A)) = X'(\rho_\Omega(A)) \qquad (A \in \mathfrak{g}(\Omega)).$$

In view of (b 1, 2), this implies

(7. 4) $$\varphi(Au) = \rho_\Omega(A)\varphi(u),$$

(7. 5) $$\psi(\beta(A)v) = \beta'(\rho_\Omega(A))\psi(v)$$

for $A \in \mathfrak{g}(\Omega)$. From (7. 4) and (c 1) one obtains (S 1). Thus $\tilde{\varphi} = \varphi_C \times \psi$ is a Siegel domain map. Moreover, in view of (4. 5), (c 3) and (7. 3), one has

(7. 6) $$\rho_\Omega({}^tA) = {}^t\rho_\Omega(A),$$

which, together with (c 1) and (7. 4), shows that (ρ_Ω, φ) is an equivariant pair in the sense of I, § 9. In view of Proposition I. 9. 2, this is equivalent to saying that

(7. 7) $\varphi : (U, e) \to (U', e')$ is a unital Jordan algebra homomorphism *and ρ_Ω is determined by the relation*

(7. 7′) $$\rho_\Omega(T_u) = T'_{\varphi(u)} \qquad (u \in U).$$

Since $\beta(T_u) = R_u$, $\beta'(T'_{u'}) = R'_{u'}$, one has by (7. 5), (7. 7′)

(7. 8) $$\psi \circ R(u) = R'(\varphi(u)) \circ \psi \qquad (u \in U).$$

Remark 1. As is easily seen, the condition (S 3) (under (S 2)) is equivalent to

(7. 9) $$\psi^* \circ R'(u') \circ \psi = R({}^t\varphi(u')) \qquad (u' \in U').$$

When φ is "isometric", i. e., ${}^t\varphi \circ \varphi = 1_U$, one has ${}^t\varphi(e') = e$ and hence

$$\begin{aligned} h(v_1, v_2) &= \langle {}^t\varphi(e'), H(v_1, v_2) \rangle \\ &= \langle e', H(\psi(v_1), \psi(v_2)) \rangle \\ &= h'(\psi(v_1), \psi(v_2)), \end{aligned}$$

i. e., ψ is also isometric : $\psi^* \circ \psi = 1_V$. Hence we see that, if φ is isometric and surjective (hence bijective), then (7. 8) implies (7. 9). On the other hand, if φ is isometric and ψ is surjective (hence bijective), then (7. 9) implies (7. 8).

Remark 2. We put $\mathfrak{u}(V, H)=\{B\in\mathfrak{gl}(V)\,|\,H(Bv_1, v_2)+H(v_1, Bv_2)=0\}$, $Y(B)=X_{0,B}$ for $B\in\mathfrak{u}(V, H)$, and $\mathfrak{k}^{(1)}=\{Y(B)\,|\,B\in\mathfrak{u}(V, H)\}$. Similarly we define $\mathfrak{k}^{(1)\prime}=\{Y'(B')\,|\,B'\in\mathfrak{u}(V', H')\}$. Then we have

$$\mathfrak{g}_0 = \mathfrak{k}^{(1)}\oplus\mathfrak{g}^{(2)}, \qquad \mathfrak{g}_0' = \mathfrak{k}^{(1)\prime}\oplus\mathfrak{g}^{(2)\prime}.$$

For $B\in\mathfrak{u}(V, H)$, let $\rho_0(Y(B))=X'_{A',B'}$, and put

$$\rho^{(2)}(B) = A', \qquad \rho^{(1)}(B) = B'-\beta'(A').$$

Then $\rho^{(2)}$ and $\rho^{(1)}$ are homomorphisms of $\mathfrak{u}(V, H)$ into $\mathfrak{g}(\Omega')$ and $\mathfrak{u}(V', H')$, respectively, and one has

$$\rho_0(Y(B)) = X'(\rho^{(2)}(B))+Y'(\rho^{(1)}(B)). \tag{7.10}$$

From (b 1) one has

$$\rho^{(2)}(B)\varphi(u) = 0 \qquad \text{for all } B\in\mathfrak{u}(V, H),\ u\in U. \tag{7.11}$$

In particular, $\rho^{(2)}(B)e'=0$, which means ${}^t\rho^{(2)}(B)=-\rho^{(2)}(B)$. On the other hand, from (b 2) one has

$$\psi(Bv) = (\rho^{(1)}(B)+\beta'(\rho^{(2)}(B)))\psi(v) \tag{7.12}$$

for all $B\in\mathfrak{u}(V, H)$, $v\in V$. If φ is surjective, (7. 11) implies $\rho^{(2)}=0$. Hence one has $\rho_0(\mathfrak{k}^{(1)})\subset\mathfrak{k}^{(1)\prime}$ and by (7. 12) $(\rho^{(1)}, \psi)$ is an equivariant pair. In general, the homomorphism ρ_0 is uniquely determined by three homomorphisms ρ_Ω, $\rho^{(1)}$, $\rho^{(2)}$.

Now, for a given equivariant triple (φ, ψ, ρ_0), we define a linear map $\rho_- : \mathfrak{g}_-\to\mathfrak{g}_-'$ by

$$\rho_-(\partial_u+\tilde{\partial}_v+X) = \partial_{\varphi(u)}+\tilde{\partial}_{\psi(v)}+\rho_0(X) \qquad (u\in U,\ v\in V,\ X\in\mathfrak{g}_0). \tag{7.13}$$

Then, in view of (1. 9 a~d) and (a), (b 1, 2), ρ_- is a Lie algebra homomorphism. Moreover, for $a\in U$, $b\in V$ and $X_{A,B}\in\mathfrak{g}_0$, one has

$$\begin{aligned}\exp(\partial_a+\tilde{\partial}_b)^{-1}(u, v) &= (u+2iH(b, v)+a+iH(b, b),\ v+b),\\ \exp(X_{A,B})^{-1}(u, v) &= ((\exp A)u,\ (\exp B)v)\end{aligned} \tag{7.14}$$

(§ 1, Exerc. 2). Hence it is clear that $(\rho_-, \tilde{\varphi})$ is an equivariant pair, i. e., one has

$$\tilde{\varphi}((\exp X)(u, v)) = (\exp \rho_-(X))\tilde{\varphi}(u, v) \tag{7.15}$$

for all $X\in\mathfrak{g}_-$, $(u, v)\in\mathcal{S}$. In this sense, $\tilde{\varphi}$ is an "affinely equivariant" Siegel domain map.

Conversely, suppose there is given a Lie algebra homomorphism $\rho_- : \mathfrak{g}_-\to\mathfrak{g}_-'$ preserving the gradation (7. 1′). Then the triple (φ, ψ, ρ_0) defined by (7. 13) satisfies clearly (a), (b 1, 2). If moreover there exists a Siegel domain map $\tilde{\varphi}=\varphi_C'\times\psi' : \mathcal{S}\to\mathcal{S}'$ equivariant with ρ_-, then by (7. 15) applied to $X=\partial_a+\tilde{\partial}_b$ one has $\varphi=\varphi'$, $\psi=\psi'$, which implies (c 2). Therefore (φ, ψ, ρ_0) is an equivariant triple if the conditions (c 1, 3) are satisfied. Summing up, we obtain the following

Proposition 7. 1. *Let $\mathcal{S}=\mathcal{S}(U, V, \Omega, H)$, $\mathcal{S}'=\mathcal{S}(U', V', \Omega', H')$ be two quasi-symmetric Siegel domains and let (φ, ψ, ρ_0) be an equivariant triple formed of linear maps $\varphi : U\to U'$, $\psi : V\to V'$ and a Lie algebra homomorphism $\rho_0 : \mathfrak{g}\to\mathfrak{g}'$ satisfying the conditions (a), (b 1, 2), (c 1, 2, 3). Then $\tilde{\varphi}=\varphi_C\times\psi : U_C\times V\to U_C'\times V'$ induces a Siegel domain map $\mathcal{S}\to\mathcal{S}'$, affinely equivariant with respect to the Lie algebra homomorphism $\rho_- : \mathfrak{g}_-\to\mathfrak{g}_-'$*

defined by (7. 13). *Conversely, for any Lie algebra homomorphism* $\rho_-:\mathfrak{g}_-\to\mathfrak{g}'_-$ *preserving the gradations* (7. 1′), *the triple* (φ, ψ, ρ_0) *defined by* (7. 13) *is an equivariant triple, if the conditions* (c 1, 2, 3) *are satisfied.*

In the rest of this section, we assume that $\mathcal{S}$ and $\mathcal{S}'$ are symmetric, and let $\mathcal{D}\subset\mathfrak{p}_+$ and $\mathcal{D}'\subset\mathfrak{p}'_+$ be the corresponding Harish-Chandra realizations. Let $\kappa_0:\mathfrak{g}^1\to\mathfrak{g}$ be an (H_1)-homomorphism given by $\kappa_0(e_+)=-\partial_e$, $\kappa_0(e_-)=\partial_e^*$ with respect to the H-element $H_0=-\frac{1}{2}(\partial_e+X_{0,i1_V}+\partial_e^*)$. Then, as we remarked at the end of § 3, the corresponding boundary component F_{κ_0} reduces to a point o_{κ_0}, and $(\mathcal{S}, ie)$ may be identified with the Siegel domain realization $(\mathcal{S}_{\kappa_0}, ie_{\kappa_0})$ of $\mathcal{D}$ in U_c+V_+ through the natural isomorphism $V\cong V_+$. Similarly, $(\mathcal{S}', ie')$ is identified with $(\mathcal{S}'_{\kappa_0'}, ie'_{\kappa_0'})$, where $\kappa_0':\mathfrak{g}^1\to(\mathfrak{g}', H_0')$ is a similar (H_1)-homomorphism for $\mathcal{S}'$.

Theorem 7. 2. *Let* $\mathcal{S}=\mathcal{S}(U, V, \Omega, H)$ *and* $\mathcal{S}'=\mathcal{S}(U', V', \Omega', H')$ *be two symmetric Siegel domains and* $\tilde{\varphi}:\tilde{V}=U_c\oplus V\to\tilde{V}'=U'_c\oplus V'$ *a* $\boldsymbol{C}$*-linear map with* $\tilde{\varphi}(e)=e'$. *Then the following three conditions are equivalent:*

1) *There exist an* $\boldsymbol{R}$*-linear map* $\varphi: U\to U'$ *and a* $\boldsymbol{C}$*-linear map* $\psi: V\to V'$ *such that* $\tilde{\varphi}=\varphi_C\oplus\psi$, *and the conditions* (a), (7. 7), (7. 8) *are satisfied.*

2) $\tilde{\varphi}$ *is a JTS homomorphism.*

3) $\tilde{\varphi}$ *induces a holomorphic map* $\mathcal{S}\to\mathcal{S}'$, *(strongly) equivariant at ie, i. e., there exists a Lie algebra homomorphism* $\rho:\mathfrak{g}\to\mathfrak{g}'$ *such that* $\rho\circ\theta=\theta'\circ\rho$ *and one has*

$$\tilde{\varphi}((\exp X)z) = (\exp\rho(X))\tilde{\varphi}(z)$$

for all $X\in\mathfrak{g}$, $z\in\mathcal{S}$.

Under these conditions, $\tilde{\varphi}$ *is a Siegel domain map (i. e., satisfies the conditions* (S 1∼3)), ρ *satisfies the condition* (H_2') *with respect to* (H_0, κ_0) *and* (H_0', κ_0'), *and* $\tilde{\varphi}$ *can be identified with the restriction of* ρ_C *to* $U_c\oplus V_+$ *through the natural isomorphisms* $V\cong V_+$, $V'\cong V'_+$.

Proof. 1)⇔2) First assume 1). Then, in view of (6. 6), (6. 8), (6. 13), (6. 14), it is clear that $\tilde{\varphi}$ is a JTS homomorphism. Conversely, since $U_c\oplus V$ and $U'_c\oplus V'$ are Peirce decompositions of $\tilde{V}$ and $\tilde{V}'$ with respect to e and e', respectively, and we have (6. 4), (6. 5), (6. 7), (6. 10), any JTS homomorphism $\tilde{\varphi}$ with $\tilde{\varphi}(e)=e'$ can be written in the form $\varphi_C\oplus\psi$ with φ, ψ satisfying (a), (7. 7), (7. 8). Also, as we have seen in the proof of Proposition 7. 1, these conditions on φ, ψ imply that $\tilde{\varphi}$ is a Siegel domain map.

2)⇒3) Assume that $\tilde{\varphi}$ is a JTS homomorphism. Then by Proposition III. 3. 4 the map

$$(ad\, c'_{\kappa_0'})\tilde{\varphi}(ad\, c_{\kappa_0})^{-1}:\ \mathfrak{p}_+\longrightarrow\mathfrak{p}'_+$$

is also a ($\boldsymbol{C}$-linear) JTS homomorphism, which maps o_{κ_0} to $o'_{\kappa_0'}$. Hence, by Propositions II. 8. 1, II. 8. 2, there exists a uniquely determined (H_1)-homomorphism ρ: $(\mathfrak{g}, H_0)\to(\mathfrak{g}', H_0')$ such that one has

(7. 16) $$\rho_C|\mathfrak{p}_+ = (ad\, c'_{\kappa_0'})\tilde{\varphi}(ad\, c_{\kappa_0})^{-1}$$

and the restriction $\rho_{\mathcal{D}}=\rho_C|\mathcal{D}$ is a (strongly) equivariant holomorphic map of $\mathcal{D}$ into

$\mathcal{D}'$ with respect to ρ. From $\rho_c(o_{\kappa_0})=o'_{\kappa_0'}$ one has $\kappa_0'=\rho\circ\kappa_0$, i. e., ρ satisfies (H_2') with respect to (H_0, κ_0) and (H_0', κ_0') (cf. III, § 10). It follows that $\rho(X_{\kappa_0})=X'_{\kappa_0'}$, which implies that ρ preserves the canonical gradation. Moreover, from (7. 16) and $\rho_{G_c}(c_{\kappa_0})=c'_{\kappa_0'}$, one obtains

$$\tilde{\varphi} = \rho_c|(U_c \oplus V_+) \tag{7. 17}$$

and

$$\tilde{\varphi}((ad\, c_{\kappa_0})^{-1}c_{\kappa_0}(z)) = (ad\, c'_{\kappa_0'})^{-1}c'_{\kappa_0'}\rho_c(z) \tag{7. 18}$$

for all $z\in\mathcal{D}$. In view of the result in III, § 7, (7. 18) shows that $\tilde{\varphi}$ induces a (strongly) equivariant map $\mathcal{S}\to\mathcal{S}'$ at ie with respect to ρ.

3)⇒2) We now assume that $\tilde{\varphi}: \tilde{V}\to\tilde{V}'$ is a $\boldsymbol{C}$-linear map with $\tilde{\varphi}(e)=e'$ inducing a (strongly) equivariant map $\rho_{\mathcal{S}}: \mathcal{S}\to\mathcal{S}'$ with respect to a homomorphism $\rho: \mathfrak{g}\to\mathfrak{g}'$. Then, by Proposition II. 8. 1, ρ satisfies the condition (H_1) with respect to H_0 and H_0'. Put

$$\rho_{\mathcal{D}} = c'^{-1}_{\kappa_0'}(ad\, c'_{\kappa_0'})\rho_{\mathcal{S}}(ad\, c_{\kappa_0})^{-1}c_{\kappa_0}.$$

Then, by the assumption, the pair $(\rho, \rho_{\mathcal{D}})$ is (strongly) equivariant at o. Hence, again by Proposition II. 8. 1, one has $\rho_{\mathcal{D}}=\rho_c|\mathcal{D}$, i. e., (7. 18) holds for all $z\in\mathcal{D}$. Then the same is true for any $z\in\mathfrak{p}_+$ as long as the both sides of (7. 18) are defined. Since one has $c_{\kappa_0}(-o_{\kappa_0})=o$, one has $c'_{\kappa_0'}\rho_c(-o_{\kappa_0})=o'$ and hence $\rho_c(o_{\kappa_0})=o'_{\kappa_0'}$. This proves that ρ satisfies (H_2') with respect to κ_0 and κ_0'. It follows that one has

$$(ad\, c'_{\kappa_0'})^{-1}c'_{\kappa_0'}\rho_c = \rho_c(ad\, c_{\kappa_0})^{-1}c_{\kappa_0}$$

on $\mathcal{D}$, and hence $\rho_{\mathcal{S}}=\rho_c|\mathcal{S}$, i. e., (7. 17). Hence one has also (7. 16), which shows that $\tilde{\varphi}$ is a JTS homomorphism by Proposition III. 3. 4, q. e. d.

We recall (Th. III. 10. 5) that, conversely, for any Lie algebra homomorphism $\rho: \mathfrak{g}\to\mathfrak{g}'$ satisfying (H_2') with respect to (H_0, κ_0) and (H_0', κ_0'), the restriction $\rho_{\mathcal{S}}=\rho_c|\mathcal{S}_{\kappa_0}$ is a (strongly) equivariant Siegel domain map $\mathcal{S}_{\kappa_0}\to\mathcal{S}'_{\kappa_0'}$.

Applying Theorem 7. 2 to the tube domain case, we obtain

Corollary 7. 3. *Let $\mathcal{S}=U+i\Omega$ and $\mathcal{S}'=U'+i\Omega'$ be two symmetric tube domains and let $\tilde{\varphi}: U_c\to U_c'$ be a $\boldsymbol{C}$-linear map with $\tilde{\varphi}(e)=e'$. Then the following three conditions are equivalent.*

1) *There exist an $\boldsymbol{R}$-linear map $\varphi: U\to U'$ such that $\tilde{\varphi}=\varphi_c$ and a Lie algebra homomorphism $\rho_\Omega: \mathfrak{g}(\Omega)\to\mathfrak{g}(\Omega')$ such that (ρ_Ω, φ) is an equivariant pair at e.*

2) *$\tilde{\varphi}$ is a Jordan algebra homomorphism defined over $\boldsymbol{R}$.*

3) *There exists a homomorphism $\rho: \mathfrak{g}\to\mathfrak{g}'$ such that $(\rho, \tilde{\varphi})$ is a (strongly) equivariant pair at ie.*

Under these conditions, ρ satisfies the condition (H_2) and $\tilde{\varphi}$ coincides with the restriction $\rho_c|U_c$.

The last assertion follows from the facts that in the tube domain case one has

$H_0=-\frac{1}{2}(\partial_e+\partial_e^*)$, $H'_0=-\frac{1}{2}(\partial_{e'}+\partial_{e'}^*)$ so that the conditions (H_2) and (H'_2) are equivalent (Prop. III. 10. 12) and that the homomorphism ρ_Ω is uniquely determined by φ.

From the above observations, we can also conclude that there exist natural equivalences between the following four categories.

$(\mathcal{SDP})$ The category of symmetric bounded domains $\mathcal{D}$ with origin o and point boundary component o_{κ_0}, a morphism $\rho_{\mathcal{D}}:(\mathcal{D}, o, o_{\kappa_0})\to(\mathcal{D}', o', o'_{\kappa_0'})$ being a holomorphic map $\rho_{\mathcal{D}}:\mathcal{D}\to\mathcal{D}'$ with $\rho_{\mathcal{D}}(o)=o'$, $\rho_+(o_{\kappa_0})=o'_{\kappa_0'}$, (strongly) equivariant at o.

$(\mathcal{SS})$ The category of symmetric Siegel domains $\mathcal{S}$ with reference point ie, a morphism $\rho_{\mathcal{S}}:(\mathcal{S}, ie)\to(\mathcal{S}', ie')$ being a Siegel domain map $\mathcal{S}\to\mathcal{S}'$ with $\rho_{\mathcal{S}}(ie)=ie'$, (strongly) equivariant at ie.

$(\mathcal{HLP})$ The category of semi-simple Lie algebras of hermitian type $(\mathfrak{g}, H_0)$ with an (H_1)-homomorphism $\kappa_0:\mathfrak{sl}(2, \boldsymbol{R})\to(\mathfrak{g}, H_0)$ corresponding to a point boundary component, a morphism $\rho:(\mathfrak{g}, H_0, \kappa_0)\to(\mathfrak{g}', H'_0, \kappa'_0)$ being an (H'_2)-homomorphism.

$(\mathcal{HTJ})$ The category of positive definite hermitian JTS's $\tilde{V}$ with principal idempotent e, a morphism $\tilde{\varphi}:(\tilde{V}, e)\to(\tilde{V}', e')$ being a C-linear JTS homomorphism with $\tilde{\varphi}(e)=e'$.

It is also possible to describe the category $(\mathcal{HTJ})$ in terms of formally real Jordan algebras plus representations.

In the situation of Theorem 7. 2, the homomorphism ρ preserves the canonical gradation and, if we set $\rho_-=\rho|\mathfrak{g}_-$, $\rho_0=\rho|\mathfrak{g}_0$, then by Proposition 7. 1 (φ, ψ, ρ_0) is an equivariant triple. However, in general, an equivariant triple need not be extendible to a (strongly) equivariant homomorphism. If an equivariant triple (φ, ψ, ρ_0) is extendible to a homomorphism $\rho:\mathfrak{g}\to\mathfrak{g}'$ satisfying the condition $\rho\circ\theta=\theta'\circ\rho$, then one has

$$\begin{aligned}\rho(\partial_{u_1}+\tilde{\partial}_{v_1}+X_{A,B}+\tilde{\partial}^*_{v_2}+\partial^*_{u_2})\\ =\partial_{\varphi(u_1)}+\tilde{\partial}_{\psi(v_1)}+\rho_0(X_{A,B})+\tilde{\partial}^*_{\psi(v_2)}+\partial^*_{\varphi(u_2)}.\end{aligned}\tag{7. 19}$$

Hence, in view of (4. 13 c), one has

$$\rho_0(X(\Phi_{v_1,v_2}, \Psi_{v_1,v_2}))=X'(\Phi'_{\psi(v_1),\psi(v_2)}, \Psi'_{\psi(v_1),\psi(v_2)})\tag{7. 20}$$

for all $v_1, v_2\in V$.

Conversely, suppose that for a given equivariant triple (φ, ψ, ρ_0) the condition (7. 20) is satisfied. Define map $\rho:\mathfrak{g}\to\mathfrak{g}'$ by (7. 19). Then the condition $\rho\circ\theta=\theta'\circ\rho$ is clearly satisfied. By (3. 9′), (4. 8), (7. 3), (7. 7), (7. 7′), we obtain

$$\rho_0(X(A_{u_1,u_2}))=X'(A'_{\varphi(u_1),\varphi(u_2)})\qquad(u_1, u_2\in U).$$

Hence, in view of (4. 13 a∼e), the condition (7. 20) assures that ρ is a Lie algebra homomorphism. To prove that $(\rho, \tilde{\varphi})$ is equivariant, we need the following

Lemma 7. 4. *Let $\mathcal{D}\subset C^N$ and $\mathcal{D}'\subset C^{N'}$ be two domains and let*

$$X=\sum_{k=1}^{N}p^k(z)\frac{\partial}{\partial z^k},\qquad X'=\sum_{k'=1}^{N'}p'^{k'}(z')\frac{\partial}{\partial z'^{k'}}$$

be complete polynomial vector fields on $\mathcal{D}$ *and* $\mathcal{D}'$. *Then, for a holomorphic map* $\varphi=(\varphi^{k'})$: $\mathcal{D}\to\mathcal{D}'$, *the following three conditions are equivalent.*

(E 1) $$\varphi((\exp tX)^{-1}z) = (\exp tX')^{-1}\varphi(z) \qquad (t\in \boldsymbol{R},\ z\in\mathcal{D}).$$

(E 2) *For any* C^∞*-function* f' *on* $\mathcal{D}'$, *one has*

$$X(f'\circ\varphi) = (X'f')\circ\varphi.$$

(E 3) $$p'^{k'}(\varphi(z)) = \sum_{k=1}^{N} p^k(z)\frac{\partial\varphi^{k'}}{\partial z^k} \qquad (z\in\mathcal{D},\ 1\leq k'\leq N').$$

Proof. (E 1)$\Rightarrow$(E 2) follows directly from the definitions:

$$(X(f'\circ\varphi))(z) = \frac{d}{dt}(f'\circ\varphi)((\exp tX)^{-1}z)\Big|_{t=0},$$

$$(X'f')\varphi(z) = \frac{d}{dt}f'((\exp tX')^{-1}\varphi(z))\Big|_{t=0}.$$

(E 3) follows from (E 2) applied to the coordinate functions $f'=z'^{k'}$.

(E 3)$\Rightarrow$(E 1) By (1. 17) (§ 1, Exerc. 1) one has

$$f'\circ\varphi((\exp tX)^{-1}z) = \sum_{\nu=0}^{\infty}\frac{t^\nu}{\nu!}(X^\nu(f'\circ\varphi))(z),$$

$$f'((\exp tX')^{-1}\varphi(z)) = \sum_{\nu=0}^{\infty}\frac{t^\nu}{\nu!}(X'^\nu f')(\varphi(z)).$$

Hence it is enough to show

$$X^\nu(f'\circ\varphi) = (X'^\nu f')\circ\varphi \qquad \text{for}\quad \nu=1,2,\cdots.$$

The case $\nu=1$ (i. e., (E 2)) follows immediately from (E 3). The general case can easily be verified by induction on ν, q. e. d.

We note that, when φ is linear, the condition (E 3) reduces to

(E 3′) $$p'(\varphi(z)) = \varphi(p(z)) \qquad (z\in\mathcal{D}).$$

Now, in our case, applying Lemma 7. 4 to the linear map $\tilde{\varphi}:\mathcal{S}\to\mathcal{S}'$, $X=\partial_u$, $\tilde{\partial}_v$, $X_{A,B}$, $\tilde{\partial}_v^*$, ∂_u^*, and $X'=\rho(X)$, it is easy to check the condition (E 3′) by virtue of (1. 3 a$\sim$c), (3. 23 a, b) and (S 3), (7. 7), (7. 8). Therefore we can conclude that $(\rho,\tilde{\varphi})$ is equivariant. It follows by Theorem 7. 2 (or by a direct verification) that ρ satisfies (H_2'), and ρ is uniquely determined by (φ,ψ).

We have thus proved the following

Proposition 7. 5. *Let* $\mathcal{S}$ *and* $\mathcal{S}'$ *be two symmetric Siegel domains. Then an equivariant triple* (φ,ψ,ρ_0) *(satisfying* (a), (b 1, 2), (c 1, 2, 3)) *can be extended to a Lie algebra homomorphism* $\rho:\mathfrak{g}\to\mathfrak{g}'$ *satisfying* $\rho\circ\theta=\theta'\circ\rho$, *if and only if the condition* (7. 20) *is satisfied. When this condition is satisfied, the extension* ρ *is uniquely determined by* (φ,ψ), *satisfies* (H_2'), *and the pair* $(\rho,\tilde{\varphi})$ *is (strongly) equivariant.*

When $\mathcal{S}$ and $\mathcal{S}'$ are both tube domains, the condition (7. 20) is trivially satisfied, and we obtain Corollary 7. 3 once again. Another special case, where the condition (7. 20) is automatically satisfied, is given by

Corollary 7. 6. *Let $\mathcal{S}$ and $\mathcal{S}'$ be two symmetric Siegel domains and (φ, ψ, ρ_0) an equivariant triple such that both φ and ψ are surjective. Then the triple can (uniquely) be extended to a homomorphism $\rho: \mathfrak{g}\to\mathfrak{g}'$ (satisfying (H_2')), with which $\tilde{\varphi}=\varphi_C\times\psi$ is equivariant.*

Proof. It is enough to verify the condition (7. 20). Let $u'\in U'$ and write $u'=\varphi(u)$ with $u\in U$. Then by (b 1), (3. 10′ a), (S 3), (7. 8) one has

$$\begin{aligned}\pi_{U'}(\rho_0(X(\Phi_{v_1,v_2}, \Psi_{v_1,v_2})))\varphi(u) &= \varphi(\Phi_{v_1,v_2}(u)) \\ &= \varphi(2\,\mathrm{Re}\,H(v_1, R_u v_2)) \\ &= 2\,\mathrm{Re}\,H'(\psi(v_1), \psi(R_u v_2)) \\ &= 2\,\mathrm{Re}\,H'(\psi(v_1), R_{\varphi(u)}\psi(v_2)) \\ &= \Phi'_{\psi(v_1),\psi(v_2)}(\varphi(u)).\end{aligned}$$

Similarly one obtains

$$\pi_{V'}(\rho_0(X(\Phi_{v_1,v_2}, \Psi_{v_1,v_2})))\psi(v) = \Psi'_{\psi(v_1),\psi(v_2)}(\psi(v))$$

for $v\in V$, which proves (7. 20), q. e. d.

Remark. Corollary 7. 6 remains valid for quasi-symmetric Siegel domains $\mathcal{S}$, $\mathcal{S}'$ (except for the uniqueness and the condition (H_2') on ρ). In fact, when φ and ψ are surjective, we can show, by Proposition 4. 7, that $\partial_u^*\in\mathfrak{g}_1$ (resp. $\tilde{\partial}_v^*\in\mathfrak{g}_{1/2}$) implies $\partial_{\varphi(u)}^*\in\mathfrak{g}_1'$ (resp. $\tilde{\partial}_{\psi(v)}^*\in\mathfrak{g}_{1/2}'$). Hence the same argument as above shows that (7. 19) defines a homomorphism $\mathfrak{g}\to\mathfrak{g}'$ with which $\tilde{\varphi}$ is equivariant.

Example (Nakajima). Let $\mathfrak{g}=\mathfrak{su}(p, q)$, $\mathfrak{g}'=\mathfrak{su}(p', q)$ $(p'>p>q)$. Taking κ and κ' as in III, § 2, Exerc. 1 (with $k=q$), we identify $U=U'$, V, V' with $\mathcal{H}_q(\boldsymbol{C})$, $\mathcal{M}_{p-q,q}(\boldsymbol{C})$, $\mathcal{M}_{p'-q,q}(\boldsymbol{C})$, respectively. Let

$$\varphi = id, \qquad \psi: v\longmapsto \begin{pmatrix}0\\ v\end{pmatrix}\begin{matrix}\}\,p'-p\\ \}\,p-q\end{matrix}.$$

Then it is easy to see that (φ, ψ, ρ_0) is an equivariant triple if and only if $\rho_0: \mathfrak{g}_0\to\mathfrak{g}_0'$ is of the form

$$\rho_0: \begin{cases} a\longmapsto a & \text{on}\quad \mathfrak{g}^{(2)} = \mathfrak{gl}^0(q, \boldsymbol{C}),\\ b\longmapsto \begin{pmatrix}\lambda_1 1_{p'-p} & 0\\ 0 & b\end{pmatrix}\ (\in\mathfrak{k}^{(1)\prime}) & \text{on}\quad \mathfrak{k}^{(1)} = \mathfrak{u}(p-q, 0),\end{cases}$$

where $\lambda_1=\gamma\,\mathrm{tr}\,b$ (γ a real constant). In this case, one has

$$\Phi_{v_1,v_2}(u) = \frac{1}{2}({}^t\bar{v}_2 v_1 u + u\,{}^t\bar{v}_1 v_2),$$

$$\Psi_{v_1,v_2}(v) = \frac{1}{2}(v\,{}^t\bar{v}_1 v_2 + v_2\,{}^t\bar{v}_1 v - v_1\,{}^t\bar{v}_2 v).$$

Hence the $\mathfrak{k}^{(1)}$- (resp. $\mathfrak{g}^{(2)}$-)component of $X(\Phi_{v_1,v_2}, \Psi_{v_1,v_2})$ is identified with

$$b=\frac{1}{2}(v_2{}^t\bar{v}_1-v_1{}^t\bar{v}_2)+\nu 1_{p-q}$$

$$\left(\text{resp. } a=\frac{1}{2}{}^t\bar{v}_2 v_1+\nu 1_q\right),$$

where $\nu=\frac{i}{2q}\operatorname{Im}\operatorname{tr}(v_2{}^t\bar{v}_1)=\frac{1}{p+q}\operatorname{tr} b$. Therefore the condition (7. 20) is satisfied if and only if $\gamma=\frac{1}{p+q}$.

Appendix : Classical Domains

§ 1. Classical groups.

In this section, we summarize the definitions of classical groups and their connection to algebras with involution. One of our aims is to fix some basic notations which are used throughout the book. Standard references will be Bourbaki [1] and Dieudonné [1].

Let F be a field of characteristic zero and D a division algebra over F. Denoting by F_1 the center of D, we set

$$(1.1)\qquad [F_1 : F] = d, \qquad [D : F_1] = r^2.$$

Let V be a finite-dimensional vector space over F with a structure of a *right* D-module (or, what amounts to the same thing, a left $\bar{D}$-module, $\bar{D}$ denoting an algebra anti-isomorphic to D). Conventionally, we fix a basis $(e_1, \cdots, e_n)$ of V over D and often identify vectors, D-linear transformations, etc. with the corresponding matrix expressions ; for instance, $v \in V$ is identified with its "coordinate vector" (α_i) (which is to be understood as a "column vector", i. e., an $n \times 1$ matrix) defined by $v = \sum_{i=1}^{n} e_i \alpha_i$.

Let $\mathfrak{A}$ denote the algebra of D-linear transformations of V. Then, for the fixed basis (e_i), one has $\mathfrak{A} \cong \mathcal{M}_n(D)$. We introduce the following notation.

$$\begin{aligned} \mathfrak{A}^\times &= (\text{the multiplicative group of units of } \mathfrak{A}) \\ &= GL(V/D), \\ \mathfrak{A}^{(1)} &= \{a \in \mathfrak{A}^\times \mid N(a) = 1\} = SL(V/D), \end{aligned}$$

where "units" mean invertible elements and N denotes the reduced norm of $\mathfrak{A}$ relative to its center F_1. The corresponding matrix groups are denoted by $GL_n(D)$ and $SL_n(D)$, respectively. They have a natural structure of (algebraic) "F-group", i. e., group of F-rational points of an affine algebraic group defined over F, which is connected in Zariski topology (I, § 1). The special linear group $SL_n(D)\,(n \geq 2)$ is "F-simple", i. e., there are no proper normal F-subgroups with positive dimension. It is also "simply connected" in the sense that there are no proper covering F-groups. The "F-rank" (I, § 5) of $SL_n(D)$ is $n-1$. The Lie algebra of $SL_n(D)$ is given by $\mathfrak{sl}_n(D) = \{x \in \mathcal{M}_n(D) \mid \mathrm{tr}(x) = 0\}$, tr denoting the (reduced) trace of $\mathfrak{A}$ relative to F_1.

Next, suppose D has an F-linear involution ι_0 and let $\varepsilon = \pm 1$. A D-valued, ε-hermitian form, or *(D, ε)-hermitian form,* with respect to ι_0 is by definition an F-bilinear map $h : V \times V \to D$ satisfying the following conditions :

$$(1.2)\qquad \begin{aligned} h(v, v'\alpha) &= h(v, v')\alpha, \\ h(v', v) &= \varepsilon h(v, v')^{\iota_0} \qquad \text{for all } v, v' \in V,\ \alpha \in D. \end{aligned}$$

For the fixed basis (e_i), h may be expressed by an $n\times n$ hermitian matrix $(h(e_i, e_j))$. Identifying h with this matrix, one can write $h(v, v')={}^t v^{\iota_0} h v'$. A (D, ε)-hermitian form h is "non-degenerate", if the matrix h is invertible. To any non-degenerate (D, ε)-hermitian form h, we can associate an involution ι of $\mathfrak{A}$ by the relation

$$(1.3)\qquad h(xv, v') = h(v, x^{\iota}v'),$$

or in matrix expression,

$$x^{\iota} = h^{-1}\,{}^t x^{\iota_0} h,$$

where for $x=(\xi_{ij})$ we set ${}^t x^{\iota_0}=(\xi_{ji}^{\iota_0})$ (Lem. IV. 2. 1).

In general, an involution is called *of the first kind* if it fixes all elements in the center of the algebra, and *of the second kind* otherwise. If we set

$$(1.4)\qquad \begin{aligned} D^{\pm} &= \{\alpha \in D \mid \alpha^{\iota_0}=\pm\alpha\}, \\ F_1^{\pm} &= D^{\pm} \cap F_1, \end{aligned}$$

then we have

$$(1.5)\qquad \dim_{F_1^+} D^+ = \begin{cases} \dfrac{1}{2}(r^2+\eta r) & \text{if } \iota_0 \text{ is of the first kind,} \\ r^2 & \text{if } \iota_0 \text{ is of the second kind,} \end{cases}$$

where $\eta=\pm 1$ is called the *sign* of the involution ι_0. The sign of an involution ι of $\mathfrak{A}$ is defined similarly. (Sometimes, we say the sign of ι is $=0$, if ι is of the second kind.) It is then easy to see that if ι_0 is of the first kind with sign η and if ι is defined by a (D, ε)-hermitian form with respect to ι_0, then ι is of the first kind with sign $\varepsilon\eta$.

We define the *unitary group* and the *special unitary group* by

$$(1.6)\qquad \begin{aligned} U(V, h) &= \{g \in \mathfrak{A}^{\times} \mid h(gv, gv')=h(v, v')\ (v, v' \in V)\} \\ &= \{g \in \mathfrak{A}^{\times} \mid g^{\iota}g=1\}, \\ SU(V, h) &= U(V, h) \cap SL(V/D)\,; \end{aligned}$$

the corresponding matrix groups are denoted by $U_n(D, h)$, $SU_n(D, h)$, respectively. Thus, for instance, one has

$$(1.7)\qquad SU_n(D, h) = \{g \in SL_n(D) \mid {}^t g^{\iota_0} h g=h\}.$$

The corresponding Lie algebra is given by

$$\mathfrak{su}_n(D, h) = \{x \in \mathfrak{sl}_n(D) \mid {}^t x^{\iota_0}+x=0\}.$$

When $D=F$, an ε-hermitian form is also called *ε-symmetric*, that is to say, symmetric or alternating, according as the sign $\varepsilon=1$ or -1, and the corresponding unitary group is called an *orthogonal group* or *symplectic group*. In this case, the letter U is replaced by O or Sp.

The special unitary group $SU_n(D, h)$ is a Zariski connected F-group, and so is $U_n(D, h)$ if ι_0 is of the second kind. $SU_n(D, h)$ is F-simple except for the following cases:

(The orthogonal case) : ι_0 is of the first kind, $\varepsilon\eta=1$, $nr=1$, 2, or 4.
(The unitary case) : ι_0 is of the second kind, $nr=1$.

The F-rank of $SU_n(D, h)$ is given by the "Witt index" of h, i. e., the common dimension of maximal totally isotropic subspaces of V with respect to h. ("Totally isotropic" means that the restriction of h to that space is identically equal to zero.) Moreover, $SU_n(D, h)$ is simply connected except for the case where ι_0 is of the first kind and $\varepsilon\eta=1$. These facts (F-simplicity, simply connectedness, etc.) can be checked by the following observations.

Let $\boldsymbol{F}$ be an algebraically closed extension of F. Then, we have

$$V_{\boldsymbol{F}} = V\otimes_F \boldsymbol{F} = \bigoplus_{\nu=1}^{d} \tilde{V}^{(\nu)}, \quad \tilde{V}^{(\nu)} = V^{\sigma_\nu}\otimes_{F_1^{\sigma_\nu}} \boldsymbol{F}, \tag{1. 8}$$

σ_ν being the d distinct monomorphisms of F_1 into $\boldsymbol{F}$ over F, and V^{σ_ν} the "conjugates" of V, viewed as a vector space over $F_1^{\sigma_\nu}$. Similarly, we have

$$D_{\boldsymbol{F}} = D\otimes_F \boldsymbol{F} = \bigoplus_{\nu=1}^{d} \tilde{D}^{(\nu)}, \quad \tilde{D}^{(\nu)} = D^{\sigma_\nu}\otimes_{F_1^{\sigma_\nu}} \boldsymbol{F} \cong \mathcal{M}_r(\boldsymbol{F}). \tag{1. 9}$$

Let $\varepsilon_{ij}^{(\nu)}$ denote the matrix units in $\tilde{D}^{(\nu)}$ giving the above isomorphism. Then, for each ν, the algebra $\tilde{D}^{(\nu)}$ acts on $\tilde{V}^{(\nu)}$ on the right and a basis of $\tilde{V}^{(\nu)}$ over $\boldsymbol{F}$ is given by $(e_k\varepsilon_{ij}^{(\nu)})$ $(1\leq i, j\leq r,\ 1\leq k\leq n)$. Hence we have a natural isomorphism

$$GL(V/D)_{\boldsymbol{F}} \cong \prod_{\nu=1}^{d} GL(\tilde{V}^{(\nu)}\varepsilon_{11}^{(\nu)}) \cong (GL_{nr}(\boldsymbol{F}))^d. \tag{1. 10}$$

Thus $GL_n(D)$ and $SL_n(D)$ are "F-forms" of $(GL_{nr}(\boldsymbol{F}))^d$, $(SL_{nr}(\boldsymbol{F}))^d$, respectively. More precisely, $GL_n(D)$, viewed as an F_1-group, is an F_1-form of $GL_{nr}(\boldsymbol{F})$ and $GL_n(D)$, viewed as an F-group, is obtained from it by scalar restriction F_1/F. The situations are quite similar for all other classical groups to be discussed below.

Now, suppose an involution ι_0 of D and a (D, ε)-form h are given. We extend them in a natural manner to an involution of $D_{\boldsymbol{F}}$ and a $(D_{\boldsymbol{F}}, \varepsilon)$-hermitian form on $V_{\boldsymbol{F}}$. If ι_0 is of the second kind, the extension of ι_0 induces a permutation of order two without fixed element on the set of simple components $\{\tilde{D}^{(\nu)}\}$. From this one has that d is even and

$$U_n(D, h)_{\boldsymbol{F}} \cong (GL_{nr}(\boldsymbol{F}))^{d/2}, \tag{1. 11}$$

where $d/2=[F_1^{\dagger} : F]$. Thus $U_n(D, h)$ and $SU_n(D, h)$ are F-forms of $(GL_{nr}(\boldsymbol{F}))^{d/2}$ and $(SL_{nr}(\boldsymbol{F}))^{d/2}$, respectively. If ι_0 is of the first kind, the extension of ι_0 leaves each $\tilde{D}^{(\nu)}$ invariant, and we have an $\varepsilon\eta$-symmetric bilinear form $B^{(\nu)}$ on $\tilde{V}^{(\nu)}\varepsilon_{11}^{(\nu)}$ such that one has

$$L^{(\nu)}M^{(\nu)}(h(v, v')) = (B^{(\nu)}(v\varepsilon_{i1}, v'\varepsilon_{j1})) \qquad (v, v'\in \tilde{V}^{(\nu)}), \tag{1. 12}$$

where $M^{(\nu)}$ denotes the fixed isomorphism $\tilde{D}^{(\nu)}\rightarrow\mathcal{M}_r(\boldsymbol{F})$ and $L^{(\nu)}$ is an η-symmetric matrix determined by the relation

(1. 13) $$M^{(\nu)}(\alpha^{\iota_0}) = L^{(\nu)-1\,t}M^{(\nu)}(\alpha)L^{(\nu)} \qquad (\alpha \in \tilde{D}^{(\nu)})$$

(Lem. IV. 3. 1, (IV. 3. 12)). As is well-known, the matrix of $B^{(\nu)}$ can be brought to the canonical form 1_{nr} if $\varepsilon\eta=1$. If $\varepsilon\eta=-1$, nr is necessarily even, and a canonical form for $B^{(\nu)}$ is given by

$$J_\mu = \begin{pmatrix} 0 & -1_\mu \\ 1_\mu & 0 \end{pmatrix}$$

where $\mu=nr/2$. By definition, the corresponding unitary groups are the orthogonal group $O_{nr}(\boldsymbol{F})$ and the symplectic group $Sp_{nr}(\boldsymbol{F})$. Thus one obtains

(1. 14) $$\begin{aligned} &U_n(D, h)_{\boldsymbol{F}} \cong (O_{nr}(\boldsymbol{F}))^d, \quad SU_n(D, h)_{\boldsymbol{F}} \cong (SO_{nr}(\boldsymbol{F}))^d && \text{if} \quad \varepsilon\eta = 1, \\ &U_n(D, h)_{\boldsymbol{F}} = SU_n(D, h)_{\boldsymbol{F}} \cong (Sp_{nr}(\boldsymbol{F}))^d && \text{if} \quad \varepsilon\eta = -1. \end{aligned}$$

Hence, in the second case, one has $U_n(D, h)=SU_n(D, h)$. As is well-known, $Sp_{nr}(\boldsymbol{F})$ is simply connected, while $SO_{nr}(\boldsymbol{F})$ has a universal covering group of order 2 (the "spin group"), which will be discussed in § 4.

In general, the groups of the form $SL_n(D)$, $SU_n(D)$ are called *classical groups*. It is known that all F-simple groups of type A, B, C, D, i. e., those which are F-forms of some power of SL_m, SO_m, Sp_m are obtained in the manner explained above except for the case SO_8 (cf. Weil [3]).

§ 2. Quaternion algebras.

Let F_1 be a field of characteristic different from 2. A *quaternion algebra* $\mathfrak{A}$ over F_1 is by definition a central simple (associative) algebra over F_1 with $[\mathfrak{A} : F_1]=4$. If $\mathfrak{A}$ is not division, one has $\mathfrak{A}\cong \mathcal{M}_2(F_1)$, in which case $\mathfrak{A}$ is called a "split" quaternion algebra.

Given $\alpha_1, \alpha_2 \in F_1^\times$, one can define a quaternion algebra $\mathfrak{A}=(\alpha_1, \alpha_2)$ as an algebra with unit element 1 over F_1 generated by two elements $\varepsilon_1, \varepsilon_2$ satisfying the relations

(2. 1) $$\varepsilon_1^2 = \alpha_1, \qquad \varepsilon_2^2 = \alpha_2, \qquad \varepsilon_1\varepsilon_2 = -\varepsilon_2\varepsilon_1.$$

If one sets $\varepsilon_0=1$, $\varepsilon_3=\varepsilon_1\varepsilon_2$, $\alpha_3=-\alpha_1\alpha_2$, then $(\varepsilon_0, \varepsilon_1, \varepsilon_2, \varepsilon_3)$ is a basis of (α_1, α_2) over F_1 and one has

(2. 2) $$\begin{aligned} \varepsilon_i^2 &= \alpha_i\varepsilon_0 \qquad (1 \leq i \leq 3), \\ \varepsilon_i\varepsilon_j &= -\varepsilon_j\varepsilon_i = \varepsilon_k, \end{aligned}$$

where (i, j, k) is any cyclic permutation of $(1, 2, 3)$. This is a special case of the "crossed product", and it is known that any quaternion algebra can be expressed in the form (α_1, α_2) (but not uniquely).

Proposition 2. 1. *Let $\mathfrak{A}$ be a quaternion algebra over F_1. Then there exists a unique involution $\iota_0 : \xi \mapsto \bar{\xi}$ of $\mathfrak{A}$ of the first kind satisfying the following mutually equivalent conditions:*

(a) $\mathfrak{A}^+ = \{\xi \in \mathfrak{A} \mid \bar{\xi}=\xi\} = F_1$.

(b) *The sign of ι_0 is -1.*

(c) *The reduced trace of $\xi \in \mathfrak{A}$ is given by $\xi+\bar{\xi}$.*

(d) *The reduced norm $N(\xi)$ of $\xi \in \mathfrak{A}$ is given by $\xi\bar{\xi}$.*

The involution ι_0 is called the *standard involution* of $\mathfrak{A}$.

For convenience of the reader, we sketch a proof. First, for an involution ι_0 of the first kind, it is clear that

$$(\mathrm{c}) \Longrightarrow (\mathrm{a}) \Longleftrightarrow (\mathrm{b}).$$

If we assume (d), we have $N(1+\xi)=1+(\xi+\bar{\xi})+N(\xi)$, whence follows (c) ; thus (d)$\Rightarrow$(c). If we assume (a), we have $\xi\bar{\xi}\in F_1$. Since the reduced norm N is invariant under any automorphism or anti-automorphism of $\mathfrak{A}$, one has $N(\xi)=N(\bar{\xi})$ and so $N(\xi)^2=N(\xi)N(\bar{\xi})=(\xi\bar{\xi})^2$. Since $N(\xi)$ is an irreducible polynomial function on $\mathfrak{A}$ with $N(1)=1$, it follows that $N(\xi)=\xi\bar{\xi}$. (If F_1 is finite, $\mathfrak{A}$ is necessarily split, and we obtain the same result by a direct computation. See below.) Thus one has (a)$\Rightarrow$(d). The uniqueness of the standard involution ι_0 is clear from (c) or (d). To prove the existence, let $\boldsymbol{F}$ be an algebraically closed extension of F_1. Then $\tilde{\mathfrak{A}}=\mathfrak{A}_{\boldsymbol{F}}$ may be identified with $\mathcal{M}_2(\boldsymbol{F})$, and in view of (b) a standard involution of $\tilde{\mathfrak{A}}$ is given by

$$(2.3) \qquad \xi = \begin{pmatrix} \xi_{11} & \xi_{12} \\ \xi_{21} & \xi_{22} \end{pmatrix} \longmapsto \bar{\xi} = J_1^{-1}\,{}^t\xi J_1 = \begin{pmatrix} \xi_{22} & -\xi_{12} \\ -\xi_{21} & \xi_{11} \end{pmatrix},$$

where $J_1=\begin{pmatrix} 0 & -1 \\ 1 & 0 \end{pmatrix}$. Since this involution of $\tilde{\mathfrak{A}}$ satisfies the condition (c), it leaves $\mathfrak{A}$ stable and induces a standard involution of $\mathfrak{A}$.

In the case of $\mathfrak{A}=(\alpha_1, \alpha_2)$, it is clear that for $\xi=\sum_{i=0}^{3}\xi_i e_i$ one has

$$(2.4) \qquad \bar{\xi} = \xi_0 e_0 - \sum_{i=1}^{3}\xi_i e_i, \qquad N(\xi) = \xi\bar{\xi} = \xi_0^2 - \sum_{i=1}^{3}\alpha_i\xi_i^2.$$

If we denote by $\mathrm{Cl}(\mathfrak{A})$ the Brauer class of $\mathfrak{A}$ and by ${}_2\mathcal{B}(F_1)$ the subgroup of the Brauer group $\mathcal{B}(F_1)$ consisting of all elements of order at most two, then it is clear that $\mathrm{Cl}((\alpha_1, \alpha_2))$ gives a bilinear pairing

$$(2.5) \qquad F_1^\times/(F_1^\times)^2 \times F_1^\times/(F_1^\times)^2 \longrightarrow {}_2\mathcal{B}(F_1).$$

It is known (and easy to verify) that $\mathrm{Cl}((\alpha_1, \alpha_2))=1$, i. e., (α_1, α_2) splits, if and only if $\alpha_1x_1^2+\alpha_2x_2^2=1$ has a solution in F_1. Thus, for instance, we see that the Hamilton quaternion algebra $\boldsymbol{H}=(-1, -1)$ is the unique division quaternion algebra over the real field $\boldsymbol{R}$. More generally, it is known that when F_1 is a local field or an algebraic number field (of finite degree), the above pairing is surjective and complete. It follows, in particular, that a central division algebra D over such a field F_1 with an involution of the first kind is necessarily a quaternion algebra and any involution of D of the first kind with sign η can be written as $\xi\mapsto\alpha^{-1}\bar{\xi}\alpha$ with $\alpha\in D^\times$, $\bar{\alpha}=-\eta\alpha$.

Let $D=(\alpha, \beta)$ be a quaternion division algebra over F_1 and let $F_1'=F_1(\sqrt{\alpha})$. Then one has an isomorphism

$$D\otimes_{F_1}F_1' \cong \mathcal{M}_2(F_1'),$$

determined by

(2.6) $$M(\varepsilon_1)=\begin{pmatrix}\sqrt{\alpha} & 0\\ 0 & -\sqrt{\alpha}\end{pmatrix},\qquad M(\varepsilon_2)=\begin{pmatrix}0 & \beta\\ 1 & 0\end{pmatrix}.$$

Denoting by ε_{ij} the corresponding matrix units in $D\otimes_{F_1}F_1'$, one has

(2.7) $$\begin{aligned}\varepsilon_{11}&=\frac{1}{2}\left(1+\frac{1}{\sqrt{\alpha}}\varepsilon_1\right), & \varepsilon_{12}&=\frac{1}{2\beta}\left(\varepsilon_2+\frac{1}{\sqrt{\alpha}}\varepsilon_1\varepsilon_2\right),\\ \varepsilon_{21}&=\frac{1}{2}\left(\varepsilon_2-\frac{1}{\sqrt{\alpha}}\varepsilon_1\varepsilon_2\right), & \varepsilon_{22}&=\frac{1}{2}\left(1-\frac{1}{\sqrt{\alpha}}\varepsilon_1\right).\end{aligned}$$

Let $\mathrm{Gal}(F_1'/F_1)=\{1,\sigma_0\}$, where σ_0 is the non-trivial automorphism of F_1' over F_1 determined by $\sqrt{\alpha}^{\sigma_0}=-\sqrt{\alpha}$. Then from (2. 7) one obtains the relations

(2.8) $$\varepsilon_{11}^{\sigma_0}=\varepsilon_{22},\qquad \varepsilon_{12}^{\sigma_0}=\beta^{-1}\varepsilon_{21},$$

and hence

(2.9) $$M(D)=\left\{\begin{pmatrix}\xi & \beta\eta\\ -\eta^{\sigma_0} & \xi^{\sigma_0}\end{pmatrix}\,\middle|\,\xi,\eta\in F_1'\right\}.$$

Now let V be a finite-dimensional vector space over F_1 having a structure of a (right) D-module. Let $V_{F_1'}=V\otimes_{F_1}F_1'$ and put $V_i=V_{F_1'}\varepsilon_{ii}\,(i=1,2)$. We denote by ψ the linear isomorphism $V_1\to V_2$ defined by

(2.10) $$\psi(v_1)=v_1\varepsilon_{12}\qquad\text{for}\quad v_1\in V_1.$$

Then from (2. 7) we see that ψ satisfies the "cocycle" condition

(2.11) $$\psi^{\sigma_0}\circ\psi=\beta^{-1}1_{V_1}.$$

Let (V,ε_1) denote the space V viewed as a vector space over F_1' by $\sqrt{\alpha}\,v=v\varepsilon_1$. Then the correspondence

(2.12) $$V\ni v\longmapsto v_1=v\varepsilon_{11}\in V_1$$

gives an isomorphism $(V,\varepsilon_1)\cong V_1$, under which one has

$$v\varepsilon_2\longmapsto v\varepsilon_{21}=v_1^{\sigma_0}\varepsilon_{21}=\psi^{-1}(v_1^{\sigma_0}).$$

Let further h be a non-degenerate (D,ε)-hermitian form on V with respect to the standard involution. By Lemma IV. 3. 1, there corresponds a non-degenerate $(-\varepsilon)$-symmetric bilinear form B_1 on V_1 satisfying the relation

$$\begin{pmatrix}0 & -1\\ 1 & 0\end{pmatrix}M(h(v,v'))=(B_1(v\varepsilon_{i1},v'\varepsilon_{j1})),$$

or

(2.13) $$M(h(v,v'))=\begin{pmatrix}B_1(v\varepsilon_{21},v'\varepsilon_{11}) & B_1(v\varepsilon_{21},v'\varepsilon_{21})\\ -B_1(v\varepsilon_{11},v'\varepsilon_{11}) & -B_1(v\varepsilon_{11},v'\varepsilon_{21})\end{pmatrix}.$$

In view of (2. 9) and (2. 13), we see that B_1 satisfies the condition

(2.14) $$B_1(\psi^{-1}(v^{\sigma_0}),\psi^{-1}(v'^{\sigma_0}))=\beta B_1(v_1,v_1')^{\sigma_0}.$$

Clearly the relation (2. 13) gives a one-to-one correspondence between non-degenerate (D,ε)-hermitian forms h on V and non-degenerate $(-\varepsilon)$-symmetric bilinear forms B_1 on V_1 satisfying the condition (2. 14).

For $g\in SU(V, h)$, let g_1 denote the restriction to V_1 of the F_1'-linear extension of g. Then it is clear that $g_1\in SU(V_1, B_1)$ and from the relation $g(v\varepsilon_2)=(gv)\varepsilon_2$ one has $g_1\psi^{-1}v_1^{\sigma_0}=\psi^{-1}(g_1v_1)^{\sigma_0}$, or

$$g_1 = \psi^{-1}\circ g_1^{\sigma_0}\circ\psi. \tag{2.15}$$

The correspondence $g\mapsto g_1$ gives an isomorphism of $SU(V, h)$ onto an F_1-form of $SU(V_1, B_1)$ defined by the equation (2.15).

§ 3. Classical domains of types $(I)\sim(III)$.

In this section, we consider classical groups over $\boldsymbol{R}$ and single out a certain class of groups corresponding to symmetric domains. Let D be either $\boldsymbol{R}$ or $\boldsymbol{C}$ or $\boldsymbol{H}$. We always denote by $\iota_0:\xi\mapsto\bar{\xi}$ the standard involution of D, i. e., the identity for $D=\boldsymbol{R}$, the complex conjugation for $D=\boldsymbol{C}$, and the "standard" involution for $D=\boldsymbol{H}$.

Let V be a vector space over $\boldsymbol{R}$ with a (right) D-module structure, and we use the same notation as in § 1. For a given non-degenerate D-valued hermitian form h on $V\times V$, we can find an orthogonal basis (e_i) of V over D such that

$$h(e_i, e_i) = \begin{cases} 1 & (1\leq i\leq p) \\ -1 & (p+1\leq i\leq p+q), \end{cases} \tag{3.1}$$

where the pair (p, q) $(p+q=n)$ is the "signature" of h. [We note that Min $\{p, q\}$ is the "Witt index" of the hermitian space (V, h), so that the signature is uniquely determined by h independently of the choice of orthogonal basis.] In the matrix notation, the above result says that any hermitian matrix h is "equivalent" to a canonical form $1_{p,q}=\mathrm{diag}(1_p, -1_q)$, i. e., ${}^t\bar{g}hg=1_{p,q}$ for some $g\in GL_{p+q}(D)$. h is called *positive* (resp. *negative*) *definite* and written as $h\gg 0$ (resp. $h\ll 0$) if $q=0$ (resp. $p=0$) ; and *indefinite* if $p>0$, $q>0$. We put

$$U(p, q, D) = U_n(D, 1_{p,q}), \qquad U_n(D) = U_n(D, 1_n),$$
$$SU(p, q, D) = U(p, q, D)\cap SL_n(D), \qquad SU_n(D) = U_n(D)\cap SL_n(D).$$

For $D=\boldsymbol{R}$ and $\boldsymbol{C}$, these groups are also denoted by $O(p, q)$, $O(n)$, $SO(p, q)$, $SO(n)$, and by $U(p, q)$, $U(n)$, $SU(p, q)$, $SU(n)$, respectively. We set

$$\begin{aligned} \mathcal{H}_n(D) &= \{h\in \mathcal{M}_n(D)\,|\,{}^t\bar{h}=h\}, \\ \mathcal{P}_n(D) &= \{h\in \mathcal{H}_n(D)\,|\,h\gg 0\}, \\ \mathcal{P}_n^{(1)}(D) &= \{h\in \mathcal{P}_n(D)\,|\,N(h)=1\}. \end{aligned} \tag{3.2}$$

Then, from what we mentioned above, the group $GL_n(D)$ acts transitively on $\mathcal{P}_n(D)$ by

$$(g, p)\longmapsto {}^t\bar{g}^{-1}pg^{-1} \qquad (g\in GL_n(D),\ p\in\mathcal{P}_n(D)), \tag{3.3}$$

and the stabilizer of $p\in\mathcal{P}_n(D)$ is given by $U_n(D, p)$. Thus $\mathcal{P}_n(D)$ can be identified with the coset space $GL_n(D)/U_n(D)$. Similarly $\mathcal{P}_n^{(1)}(D)$ is identified with $SL_n(D)/SU_n(D)$. Note that both $SL_n(D)$ and $SU_n(D)$ are connected in the usual topology.

An involution α of $\mathfrak{A}=\mathcal{M}_n(D)$ is called *positive* if it corresponds to a (D-valued) positive definite hermitian form p, i. e., if in matrix expression we have

(3. 4) $$x^\alpha = p^{-1}\,{}^t\bar{x}p \qquad \text{for some} \quad p \in \mathcal{P}_n(D).$$

(This condition is also equivalent to saying that the real quadratic form $\mathrm{tr}(x^\alpha x)$ on $\mathfrak{A}$ is positive definite. Cf. Weil [3].) It is clear that for a positive involution α

(3. 5) $$\theta_\alpha\colon g \longmapsto (g^\alpha)^{-1} = p^{-1}\,{}^t\bar{g}^{-1}p$$

is a (global) Cartan involution of $GL_n(D)$ and $SL_n(D)$ (I, § 4). Thus $U_n(D)$ (resp. $SU_n(D)$) is a maximal compact subgroup of $GL_n(D)$ (resp. $SL_n(D)$) and one has (global) Cartan decompositions

(3. 6) $$GL_n(D) = \mathcal{P}_n(D)\cdot U_n(D), \qquad SL_n(D) = \mathcal{P}_n^{(1)}(D)\cdot SU_n(D).$$

The symmetric space associated with $SL_n(D)$ is given by $\mathcal{P}_n^{(1)}(D)$, which can also be identified with the space of all positive involutions of $\mathfrak{A}$. The action of the Cartan involution θ_α on $\mathfrak{sl}_n(D)$ is given by

(3. 5′) $$\theta_\alpha\colon x \longmapsto -x^\alpha = -p^{-1}\,{}^t\bar{x}p.$$

Now let h be a D-valued skew-hermitian form on V (with respect to the standard involution) and ι the corresponding involution of $\mathfrak{A}$, i. e., $x^\iota = h^{-1}\,{}^t\bar{x}h$, ${}^t\bar{h} = -h$. Then, the special unitary group $SU_n(D, h)$ is connected in the usual sense as well as in Zariski topology. The Cartan involution θ_α of $\mathfrak{sl}_n(D)$ leaves the subalgebra $\mathfrak{g} = \mathfrak{su}_n(D, h)$ stable, if and only if one has $\alpha\iota = \iota\alpha$ and, in that case, its restriction to $\mathfrak{g}$ gives a Cartan involution of $\mathfrak{g}$. Putting $c = h^{-1}p$, one has

(3. 7) $$x^\alpha = c^{-1}x^\iota c, \qquad c^\alpha = c^\iota = -c.$$

Hence the condition $\alpha\circ\iota = \iota\circ\alpha$ is satisfied if and only if

(3. 8) $$c^2 \in F_1 (= \mathrm{Cent}\ D),$$

which also implies that

$$\gamma = c^2 = -cc^\alpha \in F_1^+ = \boldsymbol{R}$$

and so $\gamma < 0$. Hence, putting $I = |\gamma|^{-1/2}c$, one obtains a complex structures on V satisfying the condition

(3. 9) $$hI \in \mathcal{P}_n(D).$$

Conversely, if I is a complex structure on V satisfying this condition, then one has clearly $I^\iota = -I$ and the involution defined by (3. 7) with $c = I$ is a positive involution which commutes with ι. Thus we obtain the following

Proposition 3. 1. *Let $\mathfrak{g} = \mathfrak{su}_n(D, h)$, where h is a (D-valued) skew-hermitian form on V. Then there exists a one-to-one correspondence between Cartan involutions θ of $\mathfrak{g}$ and complex structures I of V satisfying* (3. 9) *by the relation*

(3. 10) $$\theta(x) = -I^{-1}x^\iota I = I^{-1}xI \qquad (x \in \mathfrak{g}).$$

Thus the symmetric space $\mathcal{D} = \mathcal{D}(V, h)$ associated with $G = SU(V, h)$ can be identified with the space of all complex structures I on V satisfying the condition (3. 9).

Now let θ be a Cartan involution of $\mathfrak{g}$ given by (3. 10) and let $\mathfrak{g} = \mathfrak{k} + \mathfrak{p}$ be the corresponding Cartan decomposition of $\mathfrak{g}$. Then, for $x \in \mathfrak{g}$, one has

$$x \in \mathfrak{k} \Longleftrightarrow [I, x] = 0.$$

When $F_1 = \boldsymbol{R}$ (i. e., $D = \boldsymbol{R}$ or $\boldsymbol{H}$), the condition (3. 9) together with $\mathrm{tr}_V(I) = 0$ implies that $I \in \mathfrak{k}$. Putting

(3. 11) $$H_0 = \frac{1}{2} I,$$

we have

(3. 12) $$[H_0, x] = \begin{cases} 0 & \text{if } x \in \mathfrak{k} \\ I \cdot x & \text{if } x \in \mathfrak{p}. \end{cases}$$

Thus $(\mathfrak{g}, H_0)$ is "of hermitian type" in the sense of II, § 3. When $F_1 = \boldsymbol{C}$, $\mathrm{tr}_V(I) = 0$ means that the reduced trace of I is purely imaginary. If the signature of ih is (p, q), one can take a basis (e_k) of V over $\boldsymbol{C}$ in such a way that

(3. 13) $$h = \begin{pmatrix} -i 1_p & 0 \\ 0 & i 1_q \end{pmatrix}, \quad I = \begin{pmatrix} i 1_p & 0 \\ 0 & -i 1_q \end{pmatrix}.$$

Hence the reduced trace of I is $(p-q)i$. Therefore, putting

(3. 14) $$H_0 = \frac{1}{2}\left(I - i\frac{p-q}{n} 1_V\right),$$

one has $H_0 \in \mathfrak{k}$ and the relation (3. 12) remains true. Thus, $\mathfrak{g}$ is again of hermitian type. We call the symmetric domain $\mathcal{D} = \mathcal{D}(V, h)$ thus obtained a symmetric domain *of classical type* or a *classical domain.* In Siegel's notation, they are of type $(III_{n/2})$, $(I_{p,q})$, (II_n) according as $D = \boldsymbol{R}$, $\boldsymbol{C}$, $\boldsymbol{H}$ (Siegel [8]). For simplicity, we sometimes write $\mathcal{D} = (X)$ or $\mathfrak{g} = (X)$ if $\mathcal{D}$ is of type (X).

The case $D = \boldsymbol{R}$ is treated in II, § 7. In this case n is even and $G = Sp_n(\boldsymbol{R})$.

$(I_{p,q})$ (the case $D = \boldsymbol{C}$). Let (V, h) be a skew-hermitian space over $\boldsymbol{C}$. For a complex structure $I \in \mathcal{D}(V, h)$, put $V_\pm = V(I; \pm i)$. Then it is clear that

(3. 15) $$ih | V_+ \times V_+ \gg 0, \qquad ih | V_- \times V_- \ll 0.$$

Also, for $v_\pm \in V_\pm$, one has

$$\begin{aligned} ih(v_+, v_-) &= -h(iv_+, v_-) = h(-Iv_+, v_-) = h(v_+, Iv_-) \\ &= h(v_+, -iv_-) = -ih(v_+, v_-), \end{aligned}$$

so that $ih(v_+, v_-) = 0$. Thus one has

(3. 16) $$V = V_+ \oplus V_-, \qquad V_+ \perp V_-.$$

Conversely, given a pair of complex subspaces (V_+, V_-) satisfying the conditions (3. 15), (3. 16), one obtains $I \in \mathcal{D}(V, h)$ by putting

$$I = (i 1_{V_+}) \oplus (-i 1_{V_-}).$$

Thus we have a one-to-one correspondence between $I \in \mathcal{D}(V, h)$ and a pair (V_+, V_-) satisfying (3. 15), (3. 16). Since such a pair (V_+, V_-) is uniquely determined by one of $V_\pm$, say V_-, $\mathcal{D}$ can be identified with an open subset of the Grassmannian $\mathcal{G}_q(V/\boldsymbol{C})$ formed of q-dimensional complex subspaces of V. That the complex structures on these spaces agree under this identification is seen from the following observation.

We introduce complex coordinates in $\mathcal{D}=\mathcal{D}(V, h)$ as follows. Fix a complex structure I_0 and the corresponding pair (V^0_+, V^0_-) satisfying the above conditions. Take a basis (e_i) of V over $\boldsymbol{C}$ such that (3. 13) holds for I_0; then $(e_1, \cdots, e_p)$ and $(e_{p+1}, \cdots, e_{p+q})$ are bases of V^0_+ and V^0_-, respectively. To each $V_- \in \mathcal{G}_q(V/\boldsymbol{C})$, we assign an $n\times q$ complex matrix $\tilde{Z}=(z_{ij})$ such that V_- is spanned by $\sum_{i=1}^{n} e_i z_{ij}$ $(1\leq j \leq q)$; $\tilde{Z}$ is then uniquely determined up to a right multiplication by a non-singular $q\times q$ matrix. Writing $\tilde{Z}$ in blocks as $\tilde{Z}=\begin{pmatrix} Z_1 \\ Z_2 \end{pmatrix}$ with $Z_1 \in \mathcal{M}_{p,q}(\boldsymbol{C})$, $Z_2 \in \mathcal{M}_q(\boldsymbol{C})$, one sees that the second condition in (3. 15) is equivalent to

$$ {}^t\bar{\tilde{Z}}\begin{pmatrix} 1_p & 0 \\ 0 & -1_q \end{pmatrix}\tilde{Z} = {}^t\bar{Z}_1 Z_1 - {}^t\bar{Z}_2 Z_2 \ll 0. $$

This implies that ${}^t\bar{Z}_2 Z_2 \gg 0$ and so Z_2 is non-singular. Hence, replacing $\tilde{Z}$ by $\tilde{Z}Z_2^{-1}$, one can normalize $\tilde{Z}$ to be of the form $\begin{pmatrix} Z \\ 1_q \end{pmatrix}$, where Z is a $p\times q$ matrix satisfying the condition

$$ 1_q - {}^t\bar{Z}Z \gg 0. \tag{3. 17} $$

The matrix Z is uniquely determined by V_-. Conversely, for any $p\times q$ matrix $Z=(z_{ij})$ satisfying this condition, the q-dimensional complex subspace

$$ V_-(Z) = \left\{ \sum_{i=1}^{p} e_i z_{ij} + e_{p+j} \ (1\leq j\leq q) \right\}_{\boldsymbol{C}} $$

satisfies the condition (3. 15). Thus $\mathcal{D}$ is realized as a (circular) bounded domain in $\mathcal{M}_{p,q}(\boldsymbol{C})$ defined by (3. 17). Most often $\mathcal{D}=(I_{p,q})$ is identified with this domain of matrices.

Now we write $g \in SL_n(\boldsymbol{C})$ in blocks as

$$ g = \begin{pmatrix} \overset{p}{\overline{A}} & \overset{q}{\overline{B}} \\ C & D \end{pmatrix} \begin{matrix} \}p \\ \}q \end{matrix}. $$

Then one has

$$ g \in SU(p, q) \Longleftrightarrow g^{-1} = 1_{p,q}^{-1}\,{}^t\bar{g}\,1_{p,q} = \begin{pmatrix} {}^t\bar{A} & -{}^t\bar{C} \\ -{}^t\bar{B} & {}^t\bar{D} \end{pmatrix}. \tag{3. 18} $$

$g \in SU(p, q)$ acts on $\mathcal{D}$ by $I \mapsto gIg^{-1}$ or $V_- \mapsto gV_-$. By the definition $gV_-(Z)$ is spanned over $\boldsymbol{C}$ by n vectors

$$ (e_1, \cdots, e_n)\begin{pmatrix} A & B \\ C & D \end{pmatrix}\begin{pmatrix} Z \\ 1_q \end{pmatrix} = (e_1, \cdots, e_n)\begin{pmatrix} AZ+B \\ CZ+D \end{pmatrix}. $$

Therefore, putting

$$ g(Z) = (AZ+B)(CZ+D)^{-1}, \tag{3. 19} $$

one obtains

$$gV_-(Z) = V_-(g(Z)).$$

Thus $G=SU(p, q)$ acts on $\mathcal{D}=(I_{p,q})$ as a group of linear fractional transformations. In the notation of II, § 4, the subgroup P_+ of G_C consists of matrices of the form $\begin{pmatrix} 1_p & Z \\ 0 & 1_q \end{pmatrix}$. It follows that the above parametrization of $\mathcal{D}$ coincides essentially with the Harish-Chandra realization. If we denote by Hol$(\mathcal{D})$ the group of all holomorphic automorphisms of $\mathcal{D}$, then, since $SU(p, q)$ is connected, the identity connected component Hol$(\mathcal{D})^\circ$ is equal to $SU(p, q)/$Center. It is known (II, § 8) that

$$(3.20) \qquad (\mathrm{Hol}(\mathcal{D}) : \mathrm{Hol}(\mathcal{D})^\circ) = \begin{cases} 2 & \text{if } p = q > 1, \\ 1 & \text{otherwise.} \end{cases}$$

In the first case, a holomorphic automorphism of $\mathcal{D}$ not belonging to Hol$(\mathcal{D})^\circ$ is given by $Z \mapsto {}^tZ$.

(II_n) (the case $D=\boldsymbol{H}$). Let (V, h) be an n-dimensional skew-hermitian space over $\boldsymbol{H}$. We fix an isomorphism $M: \boldsymbol{H}\otimes_{\boldsymbol{R}}\boldsymbol{C} \to M_2(\boldsymbol{C})$ given by

$$M(\boldsymbol{i}) = \begin{pmatrix} i & 0 \\ 0 & -i \end{pmatrix}, \qquad M(\boldsymbol{j}) = \begin{pmatrix} 0 & -1 \\ 1 & 0 \end{pmatrix}$$

and use the notation of § 2. Put $V_1 = V_C\varepsilon_{11}$. Then the correspondence $v \mapsto v_1 = v\varepsilon_{11}$ gives a $\boldsymbol{C}$-linear isomorphism $(V, \boldsymbol{i}) \cong V_1$. To the given skew-hermitian form h on V, there corresponds a symmetric bilinear form S_1 on V_1 by the relation

$$(3.21) \qquad \begin{pmatrix} 0 & -1 \\ 1 & 0 \end{pmatrix} M(h(v, v')) = (S_1(v\varepsilon_{i1}, v'\varepsilon_{j1})).$$

By (2. 14), S_1 satisfies the relation

$$(3.22) \qquad S_1(\bar{v}_1\varepsilon_{21}, \bar{v}'_1\varepsilon_{21}) = \overline{S_1(v_1, v'_1)},$$

where bars denote the complex conjugation relative to the scalar extension $\boldsymbol{C}/\boldsymbol{R}$. We define the corresponding symmetric $\boldsymbol{C}$-bilinear form S on $(V, \boldsymbol{i})$ by

$$S(v, v') = \mathrm{Re}\, S_1(v_1, v'_1) + \boldsymbol{i}\, \mathrm{Im}\, S_1(v_1, v'_1)\,;$$

then S satisfies the relation

$$(3.23) \qquad S(v\boldsymbol{j}, v'\boldsymbol{j}) = \overline{S(v, v')},$$

or equivalently, $h_1(v, v') = S(v\boldsymbol{j}, v')$ is a skew-hermitian form on $(V, \boldsymbol{i})$. By an easy computation from (3. 21), one obtains a relation

$$(3.24) \qquad h(v, v') = S(v\boldsymbol{j}, v') - \boldsymbol{j}S(v, v').$$

Now, for $I \in \mathcal{D}(V, h)$, let

$$(3.25) \qquad V_\pm = \{v \in V \mid Iv = \pm v\boldsymbol{i}\}.$$

Then one has $V_+\boldsymbol{j} = V_-$ and it is easy to check the following properties:

$$(3.26) \qquad S|V_-\times V_- = 0,$$

$$(3.27) \qquad \boldsymbol{i}h_1|V_-\times V_- \ll 0.$$

It follows, in particular, that the hermitian form $\boldsymbol{i}h_1$ is of signature (n, n). Conversely, given an n-dimensional *complex* subspace V_- satisfying (3. 26), (3. 27), one can define V_+ by $V_+=V_-\boldsymbol{j}$ and then a complex structure I by (3. 25) to obtain $I\in\mathcal{D}(V, h)$. Thus we obtain an embedding of $\mathcal{D}(V, h)$ into $\mathcal{D}((V, \boldsymbol{i}), h_1)$.

To obtain matrix expressions, fix $I_0\in\mathcal{D}(V, h)$ and the corresponding pair (V^0_+, V^0_-). Let (e_i) be an orthonormal basis of V^0_+ with respect to $\boldsymbol{i}h_1$ and put $e_{n+i}=e_i\boldsymbol{j}$; then (e_{n+i}) is an orthonormal basis of V^0_- with respect to $-\boldsymbol{i}h_1$. Using this basis, we introduce the complex coordinates $Z=(z_{ij})$ for $V_-\in\mathcal{D}((V, \boldsymbol{i}), h_1)$. Then, from (3. 26), it is easy to see that V_- belongs to $\mathcal{D}(V, h)$ if and only if

$$(3.28)\qquad {}^tZ = -Z.$$

On the other hand, by the natural injection $SU(V, h)\rightarrow SU((V, \boldsymbol{i}), h_1)=SU(n, n)$, $SU(V, h)$ is identified with $SU(n, n)\cap O_n(\boldsymbol{C}, S)$. Hence, writing $g\in SU(n, n)$ in the form $g=\begin{pmatrix}A & B\\ C & D\end{pmatrix}$, we have by (3. 18)

$$(3.29)\qquad g\in SU(V, h)\Longleftrightarrow g^{-1}=\begin{pmatrix}0 & 1_n\\ 1_n & 0\end{pmatrix}{}^t\begin{pmatrix}A & B\\ C & D\end{pmatrix}\begin{pmatrix}0 & 1_n\\ 1_n & 0\end{pmatrix}=\begin{pmatrix}{}^tD & {}^tB\\ {}^tC & {}^tA\end{pmatrix}$$
$$\Longleftrightarrow C=-\bar{B},\quad D=\bar{A}.$$

The action of g on $\mathcal{D}$ is again given by (3. 19). It is known that, except for the case $n=4$, $SU(V, h)$ and $\mathrm{Hol}(\mathcal{D})$ are both connected and so $\mathrm{Hol}(\mathcal{D})$ coincides with $SU(V, h)/\mathrm{Center}$. (For the case $n=4$, cf. § 6, Exerc. 1).

Exercises

1. Let (V, h) be a skew-hermitian space over $\boldsymbol{C}$ such that ih is of signature (p, q). Show that the correspondence $I\mapsto -I$ (or $V_-\mapsto V_+$) gives an anti-holomorphic isomorphism $\mathcal{D}(V, h)\rightarrow\mathcal{D}(V, -h)$. In a suitable choice of bases, this isomorphism is given by $(I_{p,q})\ni Z\mapsto{}^t\bar{Z}\in(I_{q,p})$. (The holomorphic isomorphism $Z\mapsto{}^tZ$ cannot be defined canonically.)

2. In the case $D=\boldsymbol{H}$, in the notation of the text, show that the subspace over $\boldsymbol{H}$ spanned by $e_k+e_{n-k}\boldsymbol{j}$ $(1\leq k\leq[n/2])$ is (maximal) totally isotropic. Thus the Witt index of (V, h) is $[n/2]$, and the isomorphism class of skew-hermitian space over $\boldsymbol{H}$ is uniquely determined by its dimension $n=\dim_{\boldsymbol{H}} V$. For this reason the matrix group corresponding to $SU(V, h)$ with respect to a canonical basis will be denoted by $SU^-_n(\boldsymbol{H})$ or $SU^-(n, \boldsymbol{H})$.

3. *Holomorphic embeddings* $(I_{p,q})\rightarrow(III_{p+q})$, $(I_{p,q})\rightarrow(II_{p+q})$. Let (V, h) be an n-dimensional skew-hermitian space over $\boldsymbol{C}$. Prove the following.

3. 1) Let ${}_{\boldsymbol{R}}V$ be a $2n$-dimensional real vector space obtained from V by scalar restriction, and let $A=\mathrm{Re}\, h$; then A is a non-degenerate alternating bilinear form on ${}_{\boldsymbol{R}}V$. We have a natural injection $SU(V, h)\rightarrow Sp({}_{\boldsymbol{R}}V, A)$, which gives rise to an equivariant holomorphic embedding of $\mathcal{D}(V, h)$ into $\mathcal{D}({}_{\boldsymbol{R}}V, A)$.

3. 2) We have a natural isomorphism

$$\tilde{V} = ({}_{\boldsymbol{R}}V)\otimes_{\boldsymbol{R}}\boldsymbol{C}\cong V\oplus\bar{V}.$$

We can define a quaternion structure on $\tilde{V}$ by setting

$$(v, \bar{v}')\boldsymbol{j} = (-v', \bar{v})\qquad (v, v'\in V).$$

Then the symmetric bilinear form $\tilde{S}$ on $\tilde{V}$ defined by

$$\check{S}(v_1+\bar{v}_2, v_1'+\bar{v}_2') = h(v_2, v_1')+h(v_2', v_1) \qquad (v_i, v_i' \in V)$$

satisfies the condition (3. 23) so that we can define a quaternion skew-hermitian form $\check{h}$ on $\check{V}$ by (3. 24). Then, we have a natural injection $SU(V, h) \to SU(\check{V}, \check{h})$, which gives rise to an equivariant holomorphic embedding of $\mathcal{D}(V, h)$ into $\mathcal{D}(\check{V}, \check{h})$. [Note that $\check{S}$ is the $\boldsymbol{C}$-bilinear extension of a (purely imaginary) symmetric bilinear form $2i \operatorname{Im} h$ on ${}_{\boldsymbol{R}}V$.]

In a suitable choice of bases, these holomorphic embeddings can be expressed as

$$(I_{p,q}) \ni Z \longmapsto \begin{pmatrix} 0 & Z \\ {}^tZ & 0 \end{pmatrix} \in (III_{p+q}), \quad \begin{pmatrix} 0 & Z \\ -{}^tZ & 0 \end{pmatrix} \in (II_{p+q}).$$

4. Combining the holomorphic embeddings in the text and in Exerc. 3, one obtains equivariant holomorphic embeddings $(II_n) \to (III_{2n})$ and $(III_n) \to (II_{2n})$. Describe these embeddings.

§4. Clifford algebras and spin groups.

Let V be a vector space of dimension n over F. (In this section, F can be any field of characteristic different from 2.) Suppose there is given a non-degenerate symmetric bilinear form S on V. We denote the corresponding quadratic form on V by the same letter S, i. e.,

(4. 1) $$S(x) = S(x, x) = {}^txSx \qquad (x \in V).$$

As is well-known, there exists an orthogonal basis (e_i) of V such that

$$S(x) = \sum_{i=1}^{n} \alpha_i \xi_i^2 \qquad \text{for} \quad x = \sum_{i=1}^{n} \xi_i e_i \in V,$$

or equivalently

(4. 2) $$S(e_i, e_j) = \delta_{ij}\alpha_i \qquad (1 \leq i, j \leq n).$$

Then, by definition, the "discriminant" of S is given by

(4. 3) $$\Delta(S) = (-1)^{\frac{1}{2}n(n-1)} \prod_{i=1}^{n} \alpha_i \ (\mathrm{mod}\ (F^\times)^2).$$

Let $\mathcal{T}(V)$ denote the "tensor algebra" of V, i. e.,

$$\mathcal{T}(V) = \bigoplus_{r=0}^{\infty} V^{\otimes r},$$

where $V^{\otimes r}$ $(r>0)$ denotes the r-fold tensor product $V \otimes \cdots \otimes V$ over F, and $V^{\otimes 0}$ stands for F. $\mathcal{T}(V)$ is an (infinite-dimensional) associative algebra over F with unit element 1. The *Clifford algebra* $C = C(V, S)$ of the quadratic space (V, S) is defined by

(4. 4) $$C(V, S) = \mathcal{T}(V)/\mathcal{J}_S,$$

where $\mathcal{J}_S$ is a two-sided ideal of $\mathcal{T}(V)$ generated by

$$\{x \otimes x - S(x) \ (x \in V)\}.$$

To simplify the notation, we identify V with its image in C by the canonical map $x \mapsto (x \bmod \mathcal{J}_S)$. Then it is known that C is of dimension 2^n and a basis of C over F is given by 1 and

$$e_{i_1} \cdots e_{i_r} \qquad (i_1 < \cdots < i_r, \ 1 \leq r \leq n).$$

Thus C is an (associative) algebra of dimension 2^n with unit element 1 and is generated by $e_1, \cdots, e_n$ satisfying the relations

$$(4.5)\qquad \begin{cases} e_i^2 = \alpha_i & (1 \leq i \leq n), \\ e_i e_j + e_j e_i = 0 & (i \neq j). \end{cases}$$

We put

$$C^+ = \{e_{i_1} \cdots e_{i_r}\ (i_1 < \cdots < i_r,\ r \text{ even})\}_F,$$
$$C^- = \{e_{i_1} \cdots e_{i_r}\ (i_1 < \cdots < i_r,\ r \text{ odd})\}_F.$$

Since $\mathcal{J}_S$ is generated by "even" elements, the definition of $C^\pm$ is independent of the choice of basis. One has clearly

$$C = C^+ \oplus C^- \qquad \text{(as vector space)},$$
$$(C^+)^2 = (C^-)^2 = C^+, \qquad C^+C^- = C^-C^+ = C^-.$$

In particular, C^+ is a subalgebra of C of dimension 2^{n-1} with the same unit element 1.

It is known (e. g., Bourbaki [1]) that C and C^+ are semi-simple algebra over F and the centers of C and C^+ are given as follows:

	Cent C	Cent C^+
n even	F	$\{1, \tilde{e}\}_F$
n odd	$\{1, \tilde{e}\}_F$	F

where $\tilde{e} = e_1 \cdots e_n$. Thus C (n even) and C^+ (n odd) are central simple algebras over F. Moreover, they are both in the Brauer class of $\bigotimes_{i<j}((-1)^{i+1}\alpha_i, (-1)^j\alpha_j)$. On the other hand, it is clear that $\{1, \tilde{e}\}_F$ is a field if and only if $\tilde{e}^2 = \Delta(S) \notin (F^\times)^2$. Hence, if $\Delta(S) \notin (F^\times)^2$, C (n odd) and C^+ (n even) are simple; otherwise, they are the direct sum of two (isomorphic) central simple algebras. In either case, it is known that

$$C \sim C^+ \qquad \text{over} \quad F(\sqrt{\Delta(S)}),$$

where $\sim$ means that all simple components of both sides belong to the same Brauer class (over $F(\sqrt{\Delta(S)})$). In particular, if $F = \boldsymbol{R}$ and the signature of S is (p, q), one has

$$(4.6)\qquad C^+ \sim \begin{cases} \boldsymbol{R} \\ \boldsymbol{C} \\ \boldsymbol{H} \end{cases} \text{if } p - q \equiv \begin{cases} 0, \pm 1 \\ \pm 2 \\ \pm 3, 4 \end{cases} \pmod 8.$$

We denote by ι the *canonical involution*, i. e., the unique involution of C (or its restriction to C^+) defined by $e_i^\iota = e_i$ $(1 \leq i \leq n)$, or equivalently by

$$(x_1 \cdots x_r)^\iota = x_r \cdots x_1 \qquad (x_i \in V).$$

Then one has

$$(4.7)\qquad \tilde{e}^\iota = (-1)^{\frac{1}{2}n(n-1)}\tilde{e}.$$

Thus ι is of the second kind, for C if and only if $n \equiv 3 \pmod 4$, and for C^+ if and only if $n \equiv 2 \pmod 4$. When ι is of the first kind, the sign $\varepsilon\eta$ (cf. § 1) can easily be determined by just computing the dimension of the subspace of ι-stable elements. For instance, for C^+ one has the following result:

$$\varepsilon\eta = \begin{cases} 1 \\ 0 \\ -1 \end{cases} \text{if } n \equiv \begin{cases} 0, \pm 1 \\ \pm 2 \\ \pm 3, 4 \end{cases} \pmod 8, \tag{4.8}$$

where $\varepsilon\eta=0$ means that ι is of the second kind. When C^+ is not simple, i. e., when n is even and $\Delta(S) \in (F^\times)^2$, (4. 8) means that either (in the case $n\equiv 2$ (4)) ι interchanges the two simple components of C^+, or (in the case $n\equiv 0$ (4)) ι leaves the simple components fixed and induces on each of them an involution of the first kind with the same sign as indicated.

We define the *spin group* by

$$Spin(V, S) = \{g \in C^+ \mid g^\iota g=1,\ gVg^{-1}=V\}. \tag{4.9}$$

It is known that, for $n\geq 3$, $Spin(V, S)$ is a Zariski connected, simply connected semi-simple F-group. For $g \in Spin(V, S)$, put

$$\varphi(g)x = gxg^{-1} \qquad (x \in V). \tag{4.10}$$

Then one has $\varphi(g) \in SO(V, S)$ and

$$\varphi:\ Spin(V, S) \longrightarrow SO(V, S)$$

is an "isogeny" of order 2, i. e., if we denote by $\mathcal{G}$ and $\mathcal{G}'$ the algebraic groups (over an algebraically closed field) associated with $Spin(V, S)$ and $SO(V, S)$, then φ is the restriction to $\mathcal{G}(F)=Spin(V, S)$ of a surjective homomorphism of algebraic group $\mathcal{G}\to\mathcal{G}'$ with a kernel of order 2 (i. e., $\{\pm 1\}$).

It is also known that C^+ is the linear closure of $Spin(V, S)$, so that the restriction of the (left) regular representation of C^+ to an (absolutely) simple left ideal of C^+ gives an (absolutely) simple irreducible representation of $Spin(V, S)$, called a *spin representation*. Let e be a primitive idempotent element of C^+. Then

$$V_1 = C^+e, \qquad U_1 = eC^+ \tag{4.11}$$

are simple left and right ideals of C^+ and (the dual of) $D_1=\mathrm{End}_G(V_1)$ can naturally be identified with eC^+e; the center of D_1 is given by

$$F_1 = \begin{cases} F(\sqrt{\Delta(S)}) & \text{if } n \text{ is even and } \Delta(S) \notin (F^\times)^2, \\ F & \text{otherwise.} \end{cases} \tag{4.12}$$

When C^+ is simple over F, we have only one spin representation over F given by V_1 and, in the notation of IV, § 1, the representation-space C^+ can be decomposed as

$$C^+ = U_1 \otimes_{D_1} V_1\ (\cong \mathcal{M}_{n_1}(D_1)), \tag{4.13}$$

where $n_1=\dim_{D_1} V_1=\dim_{D_1} U_1=2^{[(n-1)/2]}/r_1$, $[D_1 : F_1]=r_1^2$ ($r_1=1$ or 2), and $u\otimes_{D_1}v=v\cdot u$ for $u \in U_1$, $v \in V_1$. When n is even and $\Delta(S) \in (F^\times)^2$, C^+ decomposes into the direct sum

$$C^+ = C_1^+ \oplus C_2^+, \tag{4.14}$$

so that one has two (inequivalent) spin representations over F given by the simple left ideals of C_1^+ and C_2^+. In this case, in the notation similar to the above, one has $C_i^+\cong\mathcal{M}_{n_1}(D_1)$ with $n_1=2^{(n/2)-1}/r_1$.

§ 5. The spin representations satisfying (H_2).

From now on, we assume $F=\boldsymbol{R}$. (V, S) is a (non-degenerate) quadratic space over $\boldsymbol{R}$ and (p, q) is the signature of S. We fix an orthogonal basis (e_i) of V such that $(S(e_i, e_j))=\operatorname{diag}(1_p, -1_q)$, and put

$$(5.1)\qquad \begin{aligned} V^0_+ &= \{e_i(1\leq i\leq p)\}_{\boldsymbol{R}}, & V^0_- &= \{e_i(p+1\leq i\leq n)\}_{\boldsymbol{R}},\\ e_+ &= e_1\cdots e_p, & e_- &= e_{p+1}\cdots e_n. \end{aligned}$$

Proposition 5.1. *The correspondence*

$$(5.2)\qquad \alpha : x \longmapsto e_+^{-1}x^\iota e_+ \ (=e_-^{-1}x^\iota e_-)$$

is a positive involution of C^+ which commutes with ι.

Proof. Since $e_+^\iota=e_+^{-1}=(-1)^{p(p-1)/2}e_+$, it is clear that α is an involution commutative with ι. Since $V^0_\pm\subset V$, one has a natural inclusion $C(V^0_\pm)\subset C(V)$. Then it is clear that one has the following direct sum decomposition (as vector space)

$$(5.3)\qquad C^+ = C^+_+ \oplus C^+_-,$$

where

$$C^+_+ = C^+(V^0_+)\cdot C^+(V^0_-),\qquad C^+_- = C^-(V^0_+)\cdot C^-(V^0_-).$$

Clearly one has

$$e_+^{-1}e_ie_+ = \begin{cases} (-1)^{p-1}e_i & \text{for } 1\leq i\leq p,\\ (-1)^p e_i & \text{for } p+1\leq i\leq n.\end{cases}$$

It follows that

$$\begin{aligned} C^+_+ &= \{x\in C^+ \,|\, e_+^{-1}xe_+=x\},\\ C^+_- &= \{x\in C^+ \,|\, e_+^{-1}xe_+=-x\}. \end{aligned}$$

Hence, writing $x\in C^+$ in the form $x=x_++x_-$ with $x_\pm\in C^+_\pm$, one has

$$\operatorname{tr}(x^\alpha x) = \operatorname{tr}(x_+^\iota x_+ - x_-^\iota x_-).$$

It is clear that, with respect to the inner product on C^+ associated with this quadratic form, $\{e_{i_1}\cdots e_{i_r}\ (i_1<\cdots<i_r,\ r\text{ even})\}$ is an orthogonal basis and for each $x=e_{i_1}\cdots e_{i_r}$ one has $x^\alpha x=1$. Therefore this quadratic form is positive definite, q. e. d.

Proposition 5.2. *The following two conditions are equivalent:*

(a) *The standard involution ι of C^+ leaves simple components of C^+ stable (when C^+ is not simple) and induces on each of them an involution corresponding to a skew-hermitian form (with respect to the standard involution ι_0).*

(b) $p\equiv 2$ *or* $q\equiv 2 \pmod 4$.

In fact, the condition (a) is equivalent to saying that (a′) either ι is of the first kind and $\varepsilon=-1$ on each simple component, or ι is of the second kind and $C^+\sim\boldsymbol{C}$. The equivalence of (a′) and (b) follows immediately from (4.6) and (4.8).

Let

(5. 4) $$\tilde{G} = \{g \in C^+ \mid N(g)=1,\ g^t g=1\}.$$

Then, for $n \geq 3$, $\tilde{G}$ is simple except for the case n is even and $n/2 \equiv q \pmod 2$, in which case $\tilde{G}$ is the direct product of two (isomorphic) simple groups $\tilde{G}_i$ coming from the simple components $C_i^+ (i=1, 2)$. By Proposition 5. 2, $\tilde{G}$ is of classical hermitian type (i. e., one of the types (*I*), (*II*), (*III*)) if and only if $p \equiv 2$ or $q \equiv 2 \pmod 4$ and, in that case, by Proposition 5. 1 the complex structure mentioned in Proposition 3. 1 (in each component) is given by (the projection of) $\pm e_+$ or $\pm e_-$.

Let $G = Spin(V, S)$ be the spin group. For $n \geq 3$, G is a connected semi-simple Lie group, which is simply connected (in the ordinary sense) if S is definite, but not so if S is indefinite. The Lie algebra of G is given by

(5. 5) $$\mathfrak{g} = \{x \in C^+ \mid x^t + x = 0,\ [x, V] \subset V\},$$

which is clearly stable under the positive involution in Proposition 5. 1. Hence a Cartan involution of $\mathfrak{g}$ is given by

(5. 6) $$\theta : x \longmapsto -e_+^{-1} x^t e_+ = e_+^{-1} x e_+$$

and the corresponding Cartan decomposition of $\mathfrak{g}$ is given by

(5. 7) $$\mathfrak{g} = \mathfrak{k} + \mathfrak{p} \qquad \text{with} \quad \mathfrak{k} = \mathfrak{g} \cap C_+^+, \quad \mathfrak{p} = \mathfrak{g} \cap C_-^+.$$

The maximal compact subgroup of G corresponding to $\mathfrak{k}$ is given by $K = Spin(V_+^0) \cdot Spin(V_-^0)$.

Let $G' = O(V, S)$. Then, when S is indefinite, one has $(G' : G'^0) = 4$ and G is a covering group of G'^0 of order 2 in the ordinary sense.

When $q \equiv 2 \pmod 4$, one has $e_-^t = e_-^{-1} = -e_-$ and

$$[e_-, x] = 0 \qquad \text{for all } x \in V_+^0,$$
$$[e_-, e_i] = (-1)^{n-i+1} 2 e_{p+1} \cdots \hat{e}_i \cdots e_n \qquad \text{for } p+1 \leq i \leq n,$$

where $\hat{e}_i$ means that the factor e_i is omitted. Hence one has $e_- \in \mathfrak{g}$ if and only if $q=2$. In that case, e_- belongs to the center of $\mathfrak{k}$ and $\mathfrak{g}$ is of hermitian type with an H-element

(5. 8) $$H_0 = \frac{1}{2} e_-.$$

From now on, we assume that $q=2$ and consider the problem in IV, § 4 for $G = Spin(V, S)$. In this case, it is easy to see that

(5. 9) $$\mathfrak{h} = \left\{\frac{1}{2} e_{2k-1} e_{2k}\ (1 \leq k \leq l-1),\ \frac{1}{2} e_-\right\} \qquad \left(l = \left[\frac{p}{2}\right] + 1\right)$$

is a Cartan subalgebra of $\mathfrak{g}$ contained in $\mathfrak{k}$. Moreover, denoting by $\{\xi_k (1 \leq k \leq l)\}$ a basis of $\mathfrak{h}^*$ dual to the basis of $\mathfrak{h}$ given above, we see that a fundamental system $\Delta = \{\alpha_1, \cdots, \alpha_l\}$ is given by

(5. 10) $$\begin{cases} \alpha_1 = \sqrt{-1}(\xi_l - \xi_1), \\ \alpha_k = \sqrt{-1}(\xi_{k-1} - \xi_k) \qquad (2 \leq k \leq l-1), \\ \alpha_l = \begin{cases} \sqrt{-1}\,\xi_{l-1} & \text{if } p \text{ is odd}, \\ \sqrt{-1}(\xi_{l-2} + \xi_{l-1}) & \text{if } p \text{ is even}, \end{cases} \end{cases}$$

where α_1 is the unique non-compact root with $\alpha_1(H_0)=\sqrt{-1}$. Let λ_1 be the highest weight of an absolutely irreducible representation ρ_1 of G and write

$$\lambda_1 = \sqrt{-1}\sum_{k=1}^{l} m_k \xi_k. \tag{5. 11}$$

Then, in view of (5. 10), the condition (IV. 5. 8) reads

$$\begin{aligned} &m_l \geq m_1 \geq \cdots \geq m_{l-2} \geq \begin{cases} m_{l-1} \geq 0 & (p \text{ odd}), \\ |m_{l-1}| & (p \text{ even}), \end{cases} \\ &m_1 \equiv \cdots \equiv m_l \equiv 0 \text{ or } \frac{1}{2} \pmod 1. \end{aligned} \tag{5. 12}$$

On the other hand, the condition (IV. 5. 9) gives

$$-\sqrt{-1}\lambda_1(H_0) = m_l < 1.$$

From these one has

$$m_l = m_1 = \cdots = m_{l-2} = \frac{1}{2}, \qquad m_{l-1} = \begin{cases} \dfrac{1}{2} & \text{if } p \text{ is odd}, \\ \pm\dfrac{1}{2} & \text{if } p \text{ is even}. \end{cases} \tag{5. 13}$$

Thus ρ_1 is a spin representation.

To prove that the spin representations give actually a solution of our problem, it is enough to show that the "regular" representation ρ of G on C^+, $\rho(g)v=gv$ ($g\in G$, $v\in C^+$), is a solution. Let V_1 be a simple left ideal of C^+, contained in a simple component C_1^+, and put $\rho_1(g)=\rho(g)|V_1$. Fix a basis of V_1 over D_1 and make the identifications $C_1^+=\mathcal{M}_{n_1}(D_1)$, $V_1=C_1^+\varepsilon_{11}$. Then the simple factor of $\tilde{G}$ corresponding to C_1^+ is given by $\tilde{G}_1=SU(V_1, h)$ with ${}^th=-h$ and an H-element of $\tilde{G}_1$ is given by $\frac{1}{2}I_1$, where $I_1=\rho_1(e_-)$. Thus ρ_1 satisfies the condition (H_2).

In this case, the involution ι_1 in Lemma IV. 2. 2 is given by $\iota|C_1^+$, i. e.,

$$\iota_1 : x \longmapsto h^{-1}{}^t\bar{x}h.$$

Hence the corresponding hermitian form on V_1 is given by

$$h(v, v') = {}^t\bar{v}hv' = hv^{\iota}v' \qquad (v, v' \in V_1). \tag{5. 14}$$

On the other hand, for the regular representation ρ on C^+, the space $U_1=\mathrm{Hom}_G(V_1, C^+)$ may be identified with the simple right ideal $\varepsilon_{11}C^+$ by the correspondence

$$\varepsilon_{11}C^+ \ni x \longleftrightarrow \mu_x \in U_1,$$

where $\mu_x(y)=yx$ for $y\in V_1$. One considers U_1 as a right D_1-space by representing $u\in U_1$ by ${}^t\bar{u}$. Then the hermitian form on U_1 corresponding to $h'\in\mathcal{P}_{n_1}(D_1)$ is given by $h'(u, u')=uh'{}^t\bar{u}'$. Putting $a=h'h$, one can write

$$h'(u, u') = uah^{-1}{}^t\bar{u}' = uau'^{\iota}h^{-1} \qquad (u, u' \in U_1). \tag{5. 15}$$

Clearly one has $a^{\iota}=-a$ and $x\mapsto a^{-1}x^{\iota}a$ is a positive involution of C_1^+.

When C^+ is simple, let A_a denote the corresponding invariant alternating form on $C^+=V_1\cdot U_1$. Then by (IV. 4. 12) one has

$$A_a(v\cdot u,\, v'\cdot u') = \mathrm{tr}_{D_1/\boldsymbol{R}}(\overline{h'(u,\,u')}h(v,\,v'))$$
$$= \mathrm{tr}_{D_1/\boldsymbol{R}}(u'au'h^{-1}\cdot hv'v')$$
$$= \mathrm{tr}_{C^+/\boldsymbol{R}}(a(v\cdot u)^t(v'\cdot u')).$$

Thus, for $x, y \in C^+$, one has

(5. 16) $$A_a(x, y) = \mathrm{tr}_{C^+/\boldsymbol{R}}(ax^ty).$$

Moreover, for any solution (C^+, ρ, A_a, I), one has by (IV. 4. 11) $I = 1_{U_1}\otimes\rho_1(e_-) = \rho(e_-)$ and so

(5. 17) $$A_a(x, Iy) = \mathrm{tr}(ax^te_-y).$$

Since $A_aI \gg 0$, ae_- should be "positive" with respect to the positive involution $x\mapsto e_-^{-1}x^te_-$. We express this property by $ae_-\gg 0$ and put

(5. 18) $$\mathcal{P}(C^+) = \{a\in C^+ \mid a^t = -a,\ ae_-\gg 0\}.$$

Then $\mathcal{P}(C^+)$ is a self-dual, homogeneous open convex cone in C^+ (I, § 8). By Proposition IV. 4. 3 we can conclude that all solutions for (G, H_0) in which ρ is the regular representation of G on C^+ are obtained in this manner. When C^+ is not simple, we define $\mathcal{P}(C_i^+)$ $(i=1, 2)$ in an analogous way and put $\mathcal{P}(C^+) = \mathcal{P}(C_1^+)\times\mathcal{P}(C_2^+)$. Then we obtain the same result. Thus we have

Theorem 5. 3. *Let $G = Spin(V, S)$ where S is of signature $(p, 2)$ and let ρ be the regular representation of G on the even Clifford algebra C^+. Define the self-dual cone $\mathcal{P}(C^+)$ by* (5. 18). *Then, for a given H-element $H_0 = \frac{1}{2}e_-$ of G, the quadruple $(C^+, \rho, A_a, \rho(e_-))$ $(a\in\mathcal{P}(C^+))$ is a solution of the problem in* IV, § 4, *where A_a is an alternating form on $C^+\times C^+$ defined by* (5. 16). *Moreover, all solutions, in which ρ is the regular representation of G on C^+, are obtained in this manner.*

When (V, S) is defined over $\boldsymbol{Q}$, the regular representation (C^+, ρ) is also defined over $\boldsymbol{Q}$. Clearly, a quadruple $(C^+, \rho, A_a, \rho(e_-))$ is then a solution over $\boldsymbol{Q}$ if and only if $a\in\mathcal{P}(C^+)\cap C_{\boldsymbol{Q}}^+$. (For the corresponding modular embeddings, and their connection to K_3-surfaces in the case $p=18$, see Satake [5], [9], and Kuga-Satake [1].)

§ 6. The domains of type (*IV*).

Let (V, S) be a (non-degenerate) quadratic space over $\boldsymbol{R}$ where S is of signature $(p, 2)$ $(p\geq 1)$. As in the case of $(I_{p,q})$, the symmetric space associated with $G = Spin(V, S)$ (or $G' = O(V, S)$) can be identified with an open subset of the (real) Grassmannian $\mathcal{G}_2(V)$ formed of 2-dimensional subspaces V_- of V such that $S|V_-\ll 0$. But it will be more convenient to use another realization, in which the complex structure appears more explicitly. To do this, let h_S be the hermitian form on $V_{\boldsymbol{C}}$ extending S, i. e.,

(6. 1) $$h_S(x, y) = S(\bar{x}, y) \qquad (x, y\in V_{\boldsymbol{C}}).$$

Proposition 6. 1. *There exists a one-to-one correspondence between the following two objects:*

(i) V_-: *2-dimensional oriented (real) subspace of* V *such that* $S|V_- \ll 0$;

(ii) W: *1-dimensional (complex) subspace of* V_C *such that*

$$S|W = 0 \quad \text{and} \quad h_S|W \ll 0, \tag{6. 2}$$

by the relation

(6. 3) $V_{-,C} = W \oplus \bar{W}$ *and* $ix \wedge \bar{x}$ $(x \in W,\ x \neq 0)$ *is "positive" with respect to the given orientation of* V_-.

Proof. Let V_- be a subspace of V as described in (i). Then there exist just two complex structures $\pm I$ on V_- belonging to the orthogonal group $O(S|V_-)$. For any non-zero vector $x \in V_-$, Ix and x are always linearly independent, and the orientation of the basis (Ix, x) of V_- is uniquely determined by I. Choose I in such a way that this orientation coincides with the given one, and put $W = V_{-,C}(I; i)$. Then clearly the relation (6. 3) holds. Since ${}^tI = I^{-1} = -I$, $S(x, Iy)$ is alternating. Therefore, for $x, y \in W$, the bilinear form $S(x, Iy) = iS(x, y)$ is at the same time symmetric and alternating, so that $S|W$ is identically equal to zero. That $h_S|W \ll 0$ is clear from the definition. Thus W satisfies (6. 2). Conversely, given a complex subspace W satisfying (6. 2), one has $W \cap \bar{W} = 0$, so that the relation (6. 3) determines uniquely a real 2-dimensional oriented subspace V_- of V, for which it is easy to check that $S|V_- \ll 0$. It is also clear that W can then be recovered from V_- in the way described above, q. e. d.

Now let M be the set of all 1-dimensional complex subspaces W of V_C such that $S|W = 0$. M is a quadratic hypersurface in the projective space $P(V_C)$ attached to V_C and hence a smooth complex manifold of dimension p. By Proposition 6. 1, the open subset $\tilde{\mathcal{D}}$ of M defined by the inequality $h_S|W \ll 0$ consists of two connected components, of which the one containing W^0 corresponding to V_-^0 with the given orientation can be identified with the symmetric domain $\mathcal{D}$ associated with G. (Then the other one can be written as $\bar{\mathcal{D}}$.) Thus, in this case, choosing one out of the two invariant complex structures on $\mathcal{D}$ amounts to choosing one out of the two orientations of V_-^0. The inclusion map $\mathcal{D} \subset M$ may be viewed as the Borel embedding. The symmetric domain $\mathcal{D}$ thus obtained is the *domain of type* (IV_p).

To introduce complex coordinates in $\mathcal{D}$, let us fix the origin $o = W^0$ in $\mathcal{D}$ corresponding to the pair (V_+^0, V_-^0) and the orthogonal basis (e_i). Then the complex structure I on V_-^0 mentioned in the above proof is given by

$$I: \begin{cases} e_{p+1} \longmapsto -e_{p+2}, \\ e_{p+2} \longmapsto e_{p+1}. \end{cases}$$

Hence $W^0 = \{e_{p+1} + ie_{p+2}\}_C$. Let $W = \{\sum z_i e_i\}_C$ be a point in $\mathcal{D}$. Then the condition (6. 2) can be written as

$$\sum_{i=1}^{p} z_i^2 - z_{p+1}^2 - z_{p+2}^2 = 0,$$
$$\sum_{i=1}^{p} |z_i|^2 - |z_{p+1}|^2 - |z_{p+2}|^2 < 0.$$

These conditions imply that z_{p+1} and z_{p+2} are linearly independent over $\boldsymbol{R}$. In fact, if they were linearly dependent over $\boldsymbol{R}$, one would have

$$\sum_{i=1}^{p} |z_i|^2 < |z_{p+1}|^2 + |z_{p+2}|^2 = |z_{p+1}^2 + z_{p+2}^2|$$
$$= \left|\sum_{i=1}^{p} z_i^2\right|,$$

which is absurd. It follows that in $\mathcal{D}$ one has

$$\mathrm{Im}(z_{p+2}/z_{p+1}) > 0,$$

so that we may normalize (z_i) by the condition

(6. 4) $$z_{p+1} - iz_{p+2} = 1.$$

Put $\zeta = z_{p+1} + iz_{p+2}$. Then one has $|\zeta| < 1$ (since ζ is the Cayley transform of z_{p+2}/z_{p+1}), and the above conditions for (z_i) are reduced to

(6. 5) $$\sum_{i=1}^{p} |z_i|^2 < \frac{1}{\sqrt{2}}(1+|\zeta|^2) < 1, \qquad \zeta = \sum_{i=1}^{p} z_i^2.$$

Thus, taking $z = (z_i)_{1 \le i \le p}$ as complex coordinates of W, $\mathcal{D}$ is realized as a (circular) bounded domain in $\boldsymbol{C}^p$, (which may be viewed as the Harish-Chandra embedding of $\mathcal{D}$).

If one puts

$$f = \frac{1}{2}(e_{p+1} + ie_{p+2}) \ (\in W^0),$$

then one has

(6. 6) $$W = \left\{\sum_{i=1}^{p} z_i e_i + f + \zeta \bar{f}\right\}_C.$$

Using the basis $(e_1, \cdots, e_p, f, \bar{f})$ of V_C, we write $g \in GL(V_C)$ in the form

$$g = \begin{pmatrix} A & B \\ C & D \end{pmatrix}$$

with

$$B = (B_1, B_2), \qquad {}^tC = ({}^tC_1, {}^tC_2), \qquad D = \begin{pmatrix} d_{11} & d_{12} \\ d_{21} & d_{22} \end{pmatrix},$$

where $B_1, B_2, {}^tC_1, {}^tC_2 \in \mathcal{M}_{p,1}(\boldsymbol{C})$. Then one has

$$(6.7)\quad \begin{aligned} g\in O(V,S) \Longleftrightarrow g^{-1} &= \begin{pmatrix} 1_p & & \\ & 0 & -1 \\ & -1 & 0 \end{pmatrix} {}^t g \begin{pmatrix} 1_p & & \\ & 0 & -1 \\ & -1 & 0 \end{pmatrix} \\ &= \begin{pmatrix} {}^tA & -{}^tC_2 & -{}^tC_1 \\ -{}^tB_2 & d_{22} & d_{12} \\ -{}^tB_1 & d_{21} & d_{11} \end{pmatrix}, \quad \text{and} \\ A = \bar{A},\ B_2 &= \bar{B}_1,\ C_2 = \bar{C}_1,\ d_{21} = \bar{d}_{12},\ d_{22} = \bar{d}_{11}. \end{aligned}$$

$G'=O(V,S)$ has four connected components defined as follows:

$$(6.8)\quad \begin{cases} G'^0 = \{g\in O(V,S)\,|\det(g)=1,\ \det(D)>0\}, \\ G'^1 = \{g\in O(V,S)\,|\det(g)=1,\ \det(D)<0\}, \\ G'^2 = \{g\in O(V,S)\,|\det(g)=-1,\ \det(D)>0\}, \\ G'^3 = \{g\in O(V,S)\,|\det(g)=-1,\ \det(D)<0\}. \end{cases}$$

$g\in G'$ acts holomorphically on $\tilde{\mathcal{D}}$ by $W\mapsto gW$; it is easy to see that $g\in G'^0\cup G'^2$ leaves each one of the two components $\mathcal{D}$ and $\bar{\mathcal{D}}$ of $\tilde{\mathcal{D}}$ stable, while $g\in G'^1\cup G'^3$ interchanges them. The action of $g\in G'^0\cup G'^2$ on $\mathcal{D}$ is expressed by a quadratic fractional transformation as follows:

$$(6.9)\quad g(z) = (Az+B_1+B_2\zeta)(C_1z+d_{11}+d_{12}\zeta)^{-1}.$$

Note that $\tilde{\mathcal{D}}-\{W\,|\,z_{p+1}-iz_{p+2}=0\}$ can also be expressed by

$$\sum_{i=1}^{p}|z_i|^2<\frac{1}{2}(1+|\zeta|),\qquad |\zeta|\neq 1$$

and the holomorphic action of $g\in G'$ on $\tilde{\mathcal{D}}$ can be given by the same formula as above as long as the denominator in (6.9) does not vanish.

For the symmetric domain $\mathcal{D}$ it is known (II, § 8) that

$$(6.10)\quad \begin{aligned} \mathrm{Hol}(\mathcal{D}) &= (G'^0\cup G'^2)/\{\pm 1\}, \\ \mathrm{Hol}(\mathcal{D})^\circ &= G'^0/\mathrm{Center}. \end{aligned}$$

Thus, if n is even, one has $(\mathrm{Hol}(\mathcal{D}):\mathrm{Hol}(\mathcal{D})^\circ)=2$. On the other hand, one has a canonical anti-holomorphic isomorphism $\varphi_0:\mathcal{D}\to\bar{\mathcal{D}}$ given by $W\mapsto\bar{W}$, or in the above coordinates, $(z_i)\mapsto(\bar{z}_i/\bar{\zeta})$. Therefore we can define an anti-holomorphic action of $g\in G'^1\cup G'^3$ on $\mathcal{D}$ by $g\circ\varphi_0: W\mapsto g\bar{W}$. If we identify $\mathcal{D}$ with the coset space G'/K', K' being a maximal compact subgroup of G', then this anti-holomorphic action of g on $\mathcal{D}$ coincides with the "isometric" action, i. e., the one defined by the left translation on G'/K'.

A tube domain expression of $\mathcal{D}$ is obtained as follows. Let (V_1', V_1'') be a pair of totally isotropic subspaces of V such that $S|V_1'\times V_1''$ is non-degenerate and let V_0 be the orthogonal complement of $V_1'\oplus V_1''$ in V with respect to S. Then one has a direct sum decomposition:

$$(6.11)\quad V = V_1'\oplus V_1''\oplus V_0,$$

where $S|V_0\gg 0$. Let (e_i') be a basis of V such that (e_1', e_{p+1}'), (e_2', e_{p+2}'), and $(e_i')_{3\le i\le p}$ are bases of V_1', V_1'', and V_0, respectively, and

$$S(e_1', e_2') = S(e_{p+1}', e_{p+2}') = -\frac{1}{2}.$$

Let $W = \{z\}_C \in \mathcal{D}$ and write $z = \sum_{i=1}^{p+2} z_i' e_i'$ and $z_0 = \sum_{i=3}^{p} z_i' e_i'$. Then the condition (6. 2) reads

$$\begin{cases} S(z) = -z_1'z_2' - z_{p+1}'z_{p+2}' + S_0(z_0) = 0, \\ h_S(z) = -\mathrm{Re}(\bar{z}_1'z_2') - \mathrm{Re}(\bar{z}_{p+1}'z_{p+2}') + S_0(\bar{z}_0, z_0) < 0. \end{cases}$$

Again it is easy to see that these conditions imply that $z_1', z_2', z_{p+1}', z_{p+2}'$ are not equal to zero, so that one may normalize (z_i') by the condition $z_{p+2}' = 1$. Then one has

$$z_{p+1}' = -z_1'z_2' + S_0(z_0),$$

and the above conditions reduce to

(6. 12) $$\mathrm{Im}\, z_1' \cdot \mathrm{Im}\, z_2' - S_0(\mathrm{Im}\, z_0) > 0.$$

The two connected components $\mathcal{D}$ and $\bar{\mathcal{D}}$ are then defined by the additional conditions

(6. 13) $$\begin{matrix} \mathrm{Im}\, z_1' > 0, & \mathrm{Im}\, z_2' > 0, & \text{and} \\ \mathrm{Im}\, z_1' < 0, & \mathrm{Im}\, z_2' < 0, & \end{matrix}$$

respectively. Thus, taking (z_i') as complex coordinates, we obtain an expression of $\mathcal{D}$ as a tube domain $\boldsymbol{R}^p + i\Omega$, where Ω is a self-dual homogeneous cone of type (IV) defined by

(6. 14) $$\Omega = \{(y_i') \in \boldsymbol{R}^p \mid y_1' \cdot y_2' - S_0(y_3', \cdots, y_p') > 0,\ y_1' > 0\}$$

(I, § 8). In this expression, when n is even, a holomorphic automorphism of $\mathcal{D}$ not belonging to $\mathrm{Hol}(\mathcal{D})^\circ$ is given by $(z_i') \mapsto (z_2', z_1', z_3', \cdots, z_p')$.

Exercise

1. In the case $p=6$, the two spin representations ρ_i $(i=1, 2)$ give rise to the isomorphisms of the associated symmetric domains $(IV_6) \cong (II_4)$ mentioned in IV, § 5. Describe these isomorphisms explicitly in terms of the complex coordinates introduced in the text. Then check if the holomorphic automorphism of $\mathcal{D} = (II_4)$ given by $\rho_2\rho_1^{-1}$ belongs to $\mathrm{Hol}(\mathcal{D})^\circ$.

References

The following list contains mainly papers and books related to symmetric domains, published during the period of 1960–79, along with some classical ones. For convenience of the reader, some (representative or informative) references on closely related topics, such as arithmetic subgroups, automorphic forms, and discrete series representations, are also included.

A. Adler

[1] Antiholomorphic involutions of analytic families of abelian varieties, Trans. Amer. Math. Soc. **254** (1979), 69–94.

A. A. Albert

[1] On the construction of Riemann matrices, I, Ann. of Math. **35** (1934), 1–28; II, ibid. **36** (1935), 376–394.

[2] On a certain algebra of quantum mechanics, Ann. of Math. **35** (1934), 65–73.

[3] A solution of the principal problem in the theory of Riemann matrices, Ann. of Math. **35** (1934), 500–515.

[4] Involutorial simple algebras and real Riemann matrices, Ann. of Math. **36** (1935), 886–964.

[5] "Structure of Algebras", Coll. Publ., Vol. **24**, Amer. Math. Soc., Providence, 1939.

[6] On Jordan algebras of linear transformations, Trans. Amer. Math. Soc. **59** (1946), 524–555.

[7] A structure theory for Jordan algebras, Ann. of Math. **48** (1947), 546–567.

[8] On involutorial algebras, Proc. Nat. Acad. Sci. U. S. A. **41** (1955), 480–482.

A. Andreotti

[1] Théorèmes de dépendence algébrique sur les espaces complexes pseudo-concaves, Bull. Soc. Math. France **91** (1963), 1–38.

A. Andreotti and H. Grauert

[1] Algebraische Körper von automorphen Funktionen, Nachr. Akad. Wiss. Göttingen, 1961, 39–48.

A. Andreotti and E. Vesentini

[1] On deformations of discontinuous groups, Acta Math. **112** (1964), 249–298.

A. N. Andrianov

[1] Dirichlet series with Euler product in the theory of Siegel modular forms of genus 2, Trudy Mat. Inst. Steklov **112** (1971), 70–93.

S. Araki

[1] On root systems and an infinitesimal classification of irreducible symmetric spaces, J. Math. Osaka City Univ. **13** (1962), 1–34.

E. Artin

[1] "Geometric Algebra", Interscience, New York, 1957.

A. Ash, D. Mumford, M. Rapoport, and Y.-S. Tai

[1] "Smooth compactification of locally symmetric varieties", (Lie groups: History, frontiers and applications, **IV**), Math. Sci. Press, Brookline, 1975.

M. Atiyah and W. Schmid

[1] A geometric construction of the discrete series for semisimple Lie groups, Invent. Math. **42** (1977), 1–62; Erratum, ibid. **54** (1979), 189–192.

G. Averous and S. Kobayashi

[1] On automorphisms of spaces of nonpositive curvature with finite volume, "Diff. Geom. and

Relativity", Dordrecht, 1976, 19–26.

W. L. Baily, Jr.

[1] On Satake's compactification of V_n, Amer. J. Math. **80** (1958), 348–364.

[2] On the Hilbert-Siegel modular space, Amer. J. Math. **81** (1959), 846–874.

[3] On the moduli of Jacobian varieties, Ann. of Math. **71** (1960), 303–314.

[4] On the theory of Θ-functions, the moduli of Abelian varieties, and the moduli of curves, Ann. of Math. **75** (1962), 342–381.

[5] On the theory of automorphic functions and the problem of moduli, Bull. Amer. Math. Soc. **69** (1963), 727–732.

[6] Fourier-Jacobi series, Proc. Symp. Pure Math. **9**, "Algebraic Groups and Discontinuous Subgroups", Amer. Math. Soc., Providence, 1966, 296–300.

[7] An exceptional arithmetic group and its Eisenstein series, Bull. Amer. Math. Soc. **75** (1969), 402–406.

[8] Eisenstein series on tube domains, "Problems in Analysis", (A symp. in honor of S. Bochner), Princeton Univ. Press, 1970, 139–156.

[9] An exceptional arithmetic group and its Eisenstein series, Ann. of Math. **91** (1970), 512–549.

[10] "Introductory Lectures on Automorphic Forms", Publ. Math. Soc. Japan **12**, Iwanami, Tokyo and Princeton Univ. Press, 1973.

W. L. Baily, Jr. and A. Borel

[1] On the compactification of arithmetically defined quotients of bounded symmetric domains, Bull. Amer. Math. Soc. **70** (1964), 588–593.

[2] Compactification of arithmetic quotients of bounded symmetric domains, Ann. of Math. **84** (1966), 442–528.

M. Berger

[1] Les espaces symétriques non-compacts, Ann. École Norm. Sup. **74** (1957), 85–177.

S. Bergman

[1] Über die Kernfunktion eines Bereiches und ihr Verhalten am Rande I, J. reine angew. Math. **169** (1933), 1–42.

[2] "The Kernel Function and Conformal Mappings", Math. Surveys, No. **5**, Amer. Math. Soc., New York, 1950.

A. Borel

[1] Les fonctions automorphes de plusieurs variables complexes, Bull. Soc. Math. France **80** (1952), 167–182.

[2] Les espaces hermitiens symétriques, Sém. Bourbaki 1951/52, (Benjamin, New York, 1966), No. 62.

[3] Kählerian coset spaces of semi-simple Lie groups, Proc. Nat. Acad. Sci. U. S. A. **40** (1954), 1147–1151.

[4] Groupes algébriques linéaires, Ann. of Math. **64** (1956), 20–82.

[5] On the curvature tensor of the Hermitian symmetric manifolds, Ann. of Math. **71** (1960), 508–521.

[6] Density properties for certain subgroups of semisimple groups without compact components, Ann. of Math. **72** (1960), 179–188.

[7] Ensembles fondamentaux pour les groupes arithmétiques, "Colloque sur la théorie des groupes algébriques", (Bruxelles 1962), C. B. R. M., 1962, 23–40.

[8] Compact Clifford-Klein forms of symmetric spaces, Topology **2** (1963), 111–122.

[9] Cohomologie et rigidité d'espaces compacts localement symétriques, Sém. Bourbaki 1963/64, (Benjamin, New York, 1966), No. 265.

[10] Density and maximality of arithmetic subgroups, J. reine angew. Math. **224** (1966), 78–89.

[11] Introduction to automorphic forms, Proc. Symp. Pure Math. **9**, "Algebraic Groups and Discontinuous Subgroups", Amer. Math. Soc., Providence, 1966, 199–210.

[12] Sousgroupes discrets de groupes semisimples, Sém. Bourbaki 1968/69, Lect. Notes in Math. **179**, Springer, 1971, No. 358.
[13] "Linear algebraic groups", Benjamin, New York, 1969.
[14] "Introduction aux groupes arithmétiques", Act. Sci. Ind., Hermann, Paris, 1969.
[15] Pseudo-concavité et groupes arithmétiques, "Essays on Topology and Related Topics", (Mémoires dédiés à Georges de Rham), Springer, Berlin-Heidelberg-New York, 1970, 70–84.
[16] Some metric properties of arithmetic quotients of symmetric spaces and an extension theorem, J. Diff. Geom. **6** (1972), 543–560.
[17] Cohomologie de certains groupes discrets et Laplacien P-adique (d'après H. Garland), Sém. Bourbaki 1973/74, Lect. Notes in Math. **431**, Springer, 1975, No. 437.
[18] Stable real cohomology of arithmetic groups, Ann. Sci. École Norm. Sup. **7** (1974), 235–272.

A. Borel and Harish-Chandra

[1] Arithmetic subgroups of algebraic groups, Ann. of Math. **75** (1962), 485–535.

A. Borel and A. Lichnerowitcz

[1] Espaces riemanniens et hermitiens symétriques, C. R. Acad. Sci. Paris **234** (1952), 2332–2334.

A. Borel and J.-P. Serre

[1] Corners and arithmetic groups, Comment. Math. Helv. **48** (1973), 436–491.

A. Borel and J. Tits

[1] Groupes réductifs, Publ. Math., No. **27**, I. H. E. S., 1965, 55–150.

A. Borel and N. Wallach

[1] "Continuous cohomology, discrete subgroups and representations of reductive groups", Ann. of Math. Studies **94**, Princeton Univ. Press, 1980.

R. Bott and H. Samelson

[1] Application of the theory of Morse to symmetric spaces, Amer. J. Math. **80** (1958), 964–1029.

N. Bourbaki

[1] Éléments de Mathématique, Fasc. 24, Livre II, "Algèbre", Ch. 9, Formes sesquilinéaires et formes quadratiques, Hermann, Paris, 1959.
[2] É. M., Fasc. 26, "Groupes et Algèbres de Lie", Ch. 1, Hermann, Paris, 1960.
[3] É. M., Fasc. 37, "Groupes et Algèbres de Lie", Ch. 2–3, Hermann, Paris, 1972.
[4] É. M., Fasc. 34, "Groupes et Algèbres de Lie", Ch. 4–6, Hermann, Paris, 1968.
[5] É. M., Fasc. 38, "Groupes et Algèbres de Lie", Ch. 7–8, Hermann, Paris, 1975.

H. Braun and M. Koecher

[1] "Jordan-Algebren", Springer-Verlag, Berlin-Heidelberg-New York, 1966.

E. Calabi and E. Vesentini

[1] On compact locally symmetric Kähler manifolds, Ann. of Math. **71** (1960), 472–507.

É. Cartan

[1] Les groupes réels simples finis et continus, Ann. École Norm. Sup. **31** (1914), 263–355.
[2] Sur une classe remarquable d'espaces de Riemann, Bull. Soc. Math. France **54** (1926), 214–264 ; ibid. **55** (1927), 114–134.
[3] Sur certaines formes riemanniennes remarquables des géométries à groupe fondamental simple, Ann. École Norm. Sup. **44** (1927), 345–467.
[4] Groupes simples clos et ouverts et géométrie riemannienne, J. Math. pures appl. **8** (1929), 1–33.
[5] Sur les domaines bornés homogènes de l'espace de n variables complexes, Abh. Math. Sem. Hamburg **11** (1935), 116–162.
[6] Notice sur les travaux scientifiques, "Selecta", 1939, 15–112 ; =Oeuvres complètes, Partie I, Groupes de Lie, Vol. 1, Gauthier-Villars, Paris, 1952, 1–98.

H. Cartan

[1] "Sur les groupes de transformations analytiques", Act. Sci. Ind., Hermann, Paris, 1935.
[2] "Théories des fonctions de plusieurs variables : Fonctions automorphes et espaces analyti-

ques", Sém. H. Cartan 1953/54, École Norm. Sup., Paris.
[3] Quotient d'un espace analytique par un groupe d'automorphismes, "Algebraic Geom. and Top.", (A symp. in honor of S. Lefschetz), Princeton Univ. Press, 1957, 90–102.
[4] "Fonctions automorphes" (2 vol.), Sém. H. Cartan 1957/58, École Norm. Sup., Paris.
[5] Prolongement des espaces analytiques normaux, Math. Ann. **136** (1958), 97–110.
[6] Fonctions automorphes et séries de Poincaré, J. Analyse Math. **6** (1958), 169–175.

P. Cartier

[1] Dualité de Tannaka des groupes et des algèbres de Lie, C. R. Acad. Sci. Paris **242** (1956), 322–325.
[2] Remarks on "Lie algebra cohomology and generalized Borel-Weil theorem" by B. Kostant, Ann. of Math. **74** (1961), 388–390.

E. Cattani

[1] On the partial compactification of the arithmetic quotient of a period matrix domain, Bull. Amer. Math. Soc. **80** (1974), 330–333.

E. Cattani and A. Kaplan

[1] Extension of period mappings for Hodge structures of weight two, Duke Math. J. **44** (1977), 1–43.

B.-Y. Chen and T. Nagano

[1] Totally geodesic submanifolds of symmetric spaces, I, Duke Math. J. **44** (1977), 745–755 ; II, ibid. **45** (1978), 405–425.

S. S. Chern

[1] On a generalization of Kähler geometry, "Algebraic Geom. and Top.", (A symp. in honor of S. Lefschetz), Princeton Univ. Press, 1957, 103–121.

S. S. Chern and C. Chevalley

[1] Élie Cartan and his mathematical work, Bull. Amer. Math. Soc. **58** (1952), 217–250.

C. Chevalley

[1] "Theory of Lie Groups", Princeton Univ. Press, 1946.
[2] Algebraic Lie algebras, Ann. of Math. **48** (1947), 91–100.
[3] "The Algebraic Theory of Spinors", Columbia Univ. Press, New York, 1954.
[4] "Théorie des groupes de Lie", II, Groupes algébriques, 1951 ; III, Théorèmes généraux sur les algèbres de Lie, 1955 ; Combined ed., 1968, Act. Sci. Ind., Hermann, Paris.
[5] "Sur la classification des groupes de Lie algébriques" (2 vol.), Sém. Chevalley 1956/58, École Norm. Sup., Paris.

C. Chevalley and R. D. Schafer

[1] The exceptional simple Lie algebras F_4 and E_6, Proc. Nat. Acad. Sci. U. S. A. **36** (1950), 137–141.

U. Christian

[1] Zur Theorie der Hilbert-Siegelschen Modulfunktionen, Math. Ann. **152** (1963), 275–341.
[2] Über die Uniformisierbarkeit nicht-elliptischer Fixpunkte Siegelschen Modulgruppen, J. reine angew. Math. **219** (1965), 97–112.
[3] Über die Uniformisierbarkeit elliptischer Fixpunkte Hilbert-Siegelschen Modulgruppen, J. reine angew. Math. **223** (1966), 113–130.

L. Cohn

[1] The dimension of spaces of automorphic forms on a certain two-dimensional complex domain, Mem. Amer. Math. Soc., No. **158**, 1975.

P. Deligne

[1] Travaux de Griffiths, Sém. Bourbaki 1969/70, Lect. Notes in Math. **180**, Springer, 1971, No. 376.
[2] Travaux de Shimura, Sém. Bourbaki 1970/71, Lect. Notes in Math. **244**, Springer, 1971, No. 389.

[3] Variétés de Shimura : Interprétation modulaire, et techniques de construction de modèles canoniques, Proc. Symp. Pure Math. **33**, "Automorphic Forms, Representations, and *L*-functions", Amer. Math. Soc., Providence, 1979, Part 2, 247–289.

P. Deligne and D. Mumford

[1] The irreducibility of the space of curves of a given genus, Publ. Math., No. **36,** I. H. E. S., 1969, 75–109.

G. de Rham

[1] Sur la réductibilité d'un espace de Riemann, Comment. Math. Helv. **26** (1952), 328–344.

[2] "Variétés différentiables", Act. Sci. Ind., Hermann, Paris, 1955.

M. Deuring

[1] "Algebren" (2-nd ed.), Erg. Math. **41**, Springer-Verlag, Berlin-Heidelberg-New York, 1968.

J. Dieudonné

[1] "La géométrie des groupes classiques" (2-nd ed.), Erg. Math. **5**, Springer-Verlag, Berlin-Göttingen-Heidelberg, 1963.

J. Dorfmeister

[1] Eine Theorie der homogenen, regulären Kegel, Dissertation, Univ. Münster, 1974.

[2] Zur Konstruktion homogener Kegel, Math. Ann. **216** (1975), 79–96.

[3] Infinitesimal automorphisms of homogeneous Siegel domains, Preprint, 1975.

[4] Homogene Siegel-Gebiete, Habilitation, Univ. Münster, 1978.

[5] Theta functions for the special, formally real Jordan algebras (A remark on a paper of H. L. Resnikoff), Invent. Math. **44** (1978), 103–108.

[6] Inductive construction of homogeneous cones, Trans. Amer. Math. Soc. **252** (1979), 321–349.

[7] Algebraic description of homogeneous cones, Trans. Amer. Math. Soc. **255** (1979), 61–89.

[8] Peirce-Zerlegungen und Jordan-Strukturen zu homogenen Kegeln, Math. Z. **169** (1979), 179–194.

[9] Homogeneous Siegel domains, Nagoya Math. J., to appear.

[10] Quasisymmetric Siegel domains and the automorphisms of homogeneous Siegel domains, Amer. J. Math. **102** (1980), 537–563.

J. Dorfmeister and M. Koecher

[1] Relative Invarianten und nicht-assoziative Algebren, Math. Ann. **228** (1977), 147–186.

[2] Reguläre Kegel, Jber. Deutsch. Math.-Verein. **81** (1979), 109–151.

D. Drucker

[1] Exceptional Lie algebras and the structure of hermitian symmetric spaces, Mem. Amer. Math. Soc., No. **208**, 1978.

M. Eichler

[1] "Quadratische Formen und orthogonale Gruppen", Springer-Verlag, Berlin-Heidelberg-New York, 1952.

[2] Modular correspondences and their representations, J. Indian Math. Soc. **20** (1956), 164–206.

[3] "Lectures on modular correspondences", Tata Inst. of Fund. Res., Bombay, 1957.

[4] Eine Verallgemeinerung der Abelschen Integrale, Math. Z. **67** (1957), 267–298.

[5] Quadratische Formen und Modulfunktionen, Acta Arith. **4** (1958), 217–239.

[6] Zur Begründung der Theorie der automorphen Funktionen in mehreren Variablen, Aequationes Math. **3** (1969), 93–111.

[7] Projective varieties and modular forms, Lect. Notes in Math. **210,** Springer, 1970.

E. Freitag

[1] Modulformen zweiten grades zum rationalen und Gaussschen Zahlkörper, Sitzungsber. Heidelberg. Akad. Wiss. Math.-Natur. Kl., 1967, 3–49.

[2] Fortsetzung von automorphen Funktionen, Math. Ann. **177** (1968), 231–247.

[3] Über die Struktur der Funktionenkörper zu hyperabelschen Gruppen, I, J. reine angew. Math. **247** (1971), 97–117 ; II, ibid. **254** (1972), 1–16.

[4] Lokale und globale Invarianten der Hilbertschen Modulgruppe, Invent. Math. **17** (1972), 106–134.

[5] Automorphy factors of Hilbert's modular groups, "Discrete subgroups of Lie groups and applications to moduli", (Bombay Coll. 1973), Oxford Univ. Press, 1975, 9–19.

[6] Holomorphe Differentialformen zu Kongruenzgruppen der Siegelschen Modulgruppe, Invent. Math. **30** (1975), 181–186.

[7] Stabile Modulformen, Math. Ann. **230** (1977), 197–211.

[8] Die Kodairadimension von Körpern automorpher Funktionen, J. reine angew. Math. **296** (1977), 162–170.

[9] Der Körper der Siegelschen Modulfunktionen, Abh. Math. Sem. Hamburg **47** (1978), 25–41.

[10] Ein Verschwindungssatz für automorphe Formen zur Siegelschen Modulgruppe, Math. Z. **165** (1979), 11–18.

E. Freitag and V. Schneider

[1] Bemerkung zu einem Satz von J. Igusa and W. Hammond, Math. Z. **102** (1967), 9–16.

H. Freudenthal

[1] "Oktaven, Ausnahmengruppen und Oktavengeometrie", Lect. Notes, Math. Inst. Rijks-Univ., Utrecht, 1951 (revised in 1962).

H. Freudenthal and H. de Vries

[1] "Linear Lie Groups", Acad. Press, London-New York, 1969.

H. Furstenberg

[1] A Poisson formula for semi-simple Lie groups, Ann. of Math. **77** (1963), 335–386.

[2] Poisson boundaries and envelopes of discrete groups, Bull. Amer. Math. Soc. **73** (1967), 350–356.

[3] Boundaries of Riemannian symmetric spaces, "Symmetric Spaces", (ed. W. Boothby and G. Weiss), Marcel Dekker, New York, 1972.

[4] Boundary theory and stochastic processes, Proc. Symp. Pure Math. **26**, "Harmonic Analysis on Homogeneous Spaces", Amer. Math. Soc., Providence, 1973, 193–229.

H. Garland

[1] A rigidity theorem for discrete subgroups, Trans. Amer. Math. Soc. **129** (1967), 1–25.

[2] *P*-adic curvature and the cohomology of discrete subgroups of *p*-adic groups, Ann. of Math. **97** (1973), 375–423.

H. Garland and M. S. Raghunathan

[1] Fundamental domains for lattices in ($\boldsymbol{R}$-)rank 1 semi-simple Lie groups, Ann. of Math. **92** (1970), 279–326.

I. M. Gel'fand, M. I. Graev, and I. I. Piatetskii-Shapiro

[1] "Representation Theory and Automorphic Functions", (Generalized Functions, Vol. **6**), Nauka, Moscow, 1966; (English transl.) W. B. Saunders, Philadelphia-London-Toronto, 1969.

S. G. Gindikin

[1] Integral formulas for Siegel domains of second kind, Dokl. Akad. Nauk SSSR **141** (1961), 531–534; =Soviet Math. Dokl. **2** (1961), 1480–1483.

[2] Integral formulas for complex homogeneous bounded domains (Russian), Uspehi Mat. Nauk **17** (1962), 209–211.

[3] Analysis in homogeneous domains, Uspehi Mat. Nauk **19** (1964), 3–92; =Russian Math. Surveys **19** (1964), 1–89.

R. Godement

[1] La formule des traces de Selberg considérée comme source de problèmes mathématiques, Sém. Bourbaki 1962/63, (Benjamin, New York, 1966), No. 244.

[2] Domaines fondamentaux des groupes arithmétiques, Sém. Bourbaki 1962/63, (Benjamin, New York, 1966), No. 257.

[3] Analyse spectrale des fonctions modulaires, Sém Bourbaki 1964/65, (Benjamin, New York, 1966), No. 278.

[4] Introduction à la théorie de Langlands, Sém. Bourbaki 1966/67, (Benjamin, New York, 1968), No. 321.

[5] Formes automorphes et produits eulériens, Sém. Bourbaki 1968/69, Lect. Notes in Math. **179**, Springer, 1971, No. 349.

[6] De l'équation de Schrödinger aux fonctions automorphes, Sém. Bourbaki 1974/75, Lect. Notes in Math. **514**, Springer, 1976, No. 467.

M. Goto

[1] Faithful representations of Lie groups, I, Math. Japonicae **1** (1948), 107–119.

E. Gottschling

[1] Explizite Bestimmung der Randflächen des Fundamentalbereiches der Modulgruppe zweiten Grades, Math. Ann. **138** (1959), 103–124.

[2] Über die Fixpunkte der Siegelschen Modulgruppe, Math. Ann. **143** (1961), 111–149.

[3] Über die Fixpunktuntergruppen der Siegelschen Modulgruppe, Math. Ann. **143** (1961), 399–430.

[4] Die Uniformisierbarkeit der Fixpunkte eigentlich diskontinuierlicher Gruppen von biholomorphen Abbildungen, Math. Ann. **169** (1967), 26–54.

[5] Reflections in bounded symmetric domains, Comm. Pure Appl. Math. **22** (1969), 693–714.

P. Griffiths and W. Schmid

[1] Locally homogeneous complex manifolds, Acta Math. **123** (1969), 253–302.

K.-B. Gundlach

[1] Quotientenraum und meromorphe Funktionen zur Hilbertschen Modulgruppe, Nachr. Akad. Wiss. Göttingen, 1960, 77–85.

[2] Some new results in the theory of Hilbert's modular group, "Contributions to function theory", Tata Inst. of Fund. Res., Bombay, 1960, 165–180.

[3] Zusammenhang zwischen Modulformen in einer und in zwei Variablen, Nachr. Akad. Wiss. Göttingen, 1965, 47–88.

[4] Die Bestimmung der Funktionen zu einigen Hilbertschen Modulgruppen, J. reine angew. Math. **220** (1965), 109–153.

R. C. Gunning

[1] The structure of factors of automorphy, Amer. J. Math. **78** (1956), 357–382.

[2] Factors of automorphy and other cohomology groups for Lie groups, Ann. of Math. **69**(1959), 314–326 ; Correction, ibid., 734.

[3] Homogeneous symplectic multipliers, Ill. J. Math. **4** (1960), 575–583.

R. Hall and M. Kuga

[1] Algebraic cycles in a fiber variety, Sci. Papers Coll. Gen. Ed. Univ. Tokyo **25** (1975), 1–6.

W. Hammond

[1] On the graded ring of Siegel modular forms of genus two, Amer. J. Math. **87** (1965), 502–506.

[2] The modular groups of Hilbert and Siegel, Amer. J. Math. **88** (1966), 497–516.

[3] The Hilbert modular surface of a real quadratic field, Math. Ann. **200** (1973), 25–45.

[4] Chern numbers of 2-dimensional Satake compactifications, Preprint.

W. Hammond and F. Hirzebruch

[1] *L*-series, modular imbeddings, and signatures, Math. Ann. **204** (1973), 263–270.

J. Hano

[1] On Kählerian homogeneous spaces of unimodular Lie groups, Amer. J. Math. **79** (1957), 885–900.

[2] Equivariant projective immersion of a complex coset space with non-degenerate canonical hermitian form, Scripta Math. **29** (1971), 125–139.

[3] Homogeneous C^*-bundles and equivariant projective immersions, Amer. J. Math. **95** (1973), 108–144.

J. Hano and Y. Matsushima

[1] Some studies on Kählerian homogeneous spaces, Nagoya Math. J. **11** (1957), 77–92.

G. Harder

[1] A Gauss-Bonnet formula for discrete arithmetically defined groups, Ann. Sci. École Norm. Sup. **4** (1971), 409–455.

[2] On the cohomology of discrete arithmetically defined groups, "Discrete subgroups of Lie groups and applications to moduli", (Bombay Coll. 1973), Oxford Univ. Press, 1975, 129–160.

Harish-Chandra

[1] Lie algebras and the Tannaka duality theorem, Ann. of Math. **51** (1950), 299–330.

[2] Representations of semi-simple Lie groups, IV, Amer. J. Math. **77** (1955), 743–777 ; V, ibid. **78** (1956), 1–41 ; VI, ibid. **78** (1956), 564–628.

[3] Spherical functions on a semi-simple Lie groups, I, Amer. J. Math. **80** (1958), 241–310 ; II, ibid., 553–613.

[4] Discrete series for semi-simple Lie groups, I, Acta Math. **113** (1965), 241–318 ; II, Explicit determination of the characters, ibid. **116** (1966), 1–111.

[5] "Automorphic forms on semi-simple Lie groups", Lect. Notes in Math. **62,** Springer, 1968.

L. A. Harris

[1] Bounded symmetric homogeneous domains in infinite dimensional spaces, Lect. Notes in Math. **364**, Springer, 1973, 13–40.

[2] Operator Siegel domains, Proc. Royal Soc. Edinburgh **79A** (1977), 137–156.

L. A. Harris and W. Kaup

[1] Linear algebraic groups in infinite dimensions, Ill. J. Math. **21** (1977), 666–674.

S. Helgason

[1] "Differential Geometry and Symmetric Spaces", Acad. Press, New York-London, 1962.

[1a] "Differential Geometry, Lie Groups, and Symmetric Spaces", Acad. Press, New York-San Francisco-London, 1978.

[2] A duality for symmetric spaces with applications to group representations, Ad. in Math. **5** (1970), 1–154 ; II. Differential equations and eigenspace representations, ibid. **22** (1976), 187–219 ; III. Tangent space analysis, to appear.

[3] Functions on symmetric spaces, Proc. Symp. Pure Math. **26**, "Harmonic Analysis on Homogeneous Spaces", Amer. Math. Soc., Providence, 1973, 101–146.

[4] The surjectivity of invariant differential operators on symmetric spaces I, Ann. of Math. **98** (1973), 451–479.

K.-H. Helwig

[1] Automorphismengruppen des allgemeinen Kreiskegels und des zugehörigen Halbraumes, Math. Ann. **157** (1964), 1–33.

[2] Zur Koecherschen Reduktionstheorie in Positivitätsbereichen, I, Math. Z. **91** (1966), 152–168 ; II, ibid., 169–178 ; III, ibid., 355–362.

[3] Eine Verallgemeinerung der formal-reellen Jordan-Algebren, Invent. Math. **1** (1966), 18–35.

[4] Über Mutationen von Jordan-Algebren, Math. Z. **103** (1968), 1–7.

[5] Halbeinfache reelle Jordan-Algebren, Math. Z. **109** (1969), 1–28.

[6] Involutionen von Jordan-Algebren, Manusc. Math. **1** (1969), 211–229.

[7] Jordan-Algebren und symmetrische Räume I, Math. Z. **115** (1970), 315–349.

J. C. Hemperly

[1] The parabolic contribution to the number of linearly independent automorphic forms on a certain bounded domain, Amer. J. Math. **94** (1972), 1078–1100.

R. Hermann

[1] Geodesics of bounded symmetric domains, Comment. Math. Helv. **35** (1961), 1–8.
[2] Geometric aspects of potential theory in bounded symmetric domains, I, Math. Ann. **148** (1962), 349–366; II, ibid. **151** (1963), 143–149; III, ibid. **153** (1964), 384–394.

Ch. Hertneck

[1] Positivitätsbereich und Jordan-Strukturen, Math. Ann. **146** (1962), 433–455.

F. Hirzebruch

[1] Characteristic numbers of homogeneous domains, "Sem. on Analytic Functions", Vol. 2, I. A. S., Princeton, 1957, 92–104.
[2] Automorphe Formen und der Satz von Riemann-Roch, "Symp. Int. Top. Alg.", Univ. de Mexico, 1958, 129–144.
[3] The Hilbert modular groups, resolution of the singularities at the cusps and related problems, Sém. Bourbaki 1970/71, Lect. Notes in Math. **244**, Springer, 1971, No. 396.
[4] Hilbert modular surfaces, Enseignement Math. **19** (1973), 183–282; =Monographie No. **21**.
[5] Modulflächen und Modulkurven zur symmetrischen Hilbertschen Modulgruppe, Ann. Sci. École Norm. Sup. **11** (1978), 101–166.

F. Hirzebruch and A. Van de Ven

[1] Hilbert modular surfaces and the classification of algebraic surfaces, Invent. Math. **23** (1974), 1–29.

F. Hirzebruch and D. Zagier

[1] Intersection numbers of curves on Hilbert modular surfaces and modular forms of Nebentypus, Invent. Math. **36** (1976), 57–114.
[2] Classification of Hilbert modular surfaces, "Complex Analysis and Algebraic Geometry", (Papers ded. to K. Kodaira), Iwanami, Tokyo and Cambridge Univ. Press, 1977.

U. Hirzebruch

[1] Halbräume und ihre holomorphen Automorphismen, Math. Ann. **153** (1964), 395–417.
[2] Über Jordan-Algebren und kompakte Riemannsche symmetrische Räume von Rang 1, Math. Z. **90** (1965), 339–354.
[3] Über Jordan-Algebren und beschränkte symmetrische Gebiete, Math. Z. **94** (1966), 387–390.
[4] Über eine Klasse von Lie-Algebren, J. Algebra **11** (1969), 461–467.
[5] Über eine Realisierung der hermiteschen, symmetrischen Räume, Math. Z. **115** (1970), 371–382.

G. Hochschild and G. D. Mostow

[1] Representations and representative functions of Lie groups, Ann. of Math. **66** (1957), 495–542; II, ibid. **68** (1958), 295–313.
[2] Automorphisms of affine algebraic groups, J. Algebra **13** (1969), 535–543.
[3] Analytic and rational automorphisms of complex algebraic groups, J. Algebra **25** (1973), 146–151.

T. Honda

[1] Isogenies, rational points and section points of group varieties, Japanese J. Math. **30** (1960), 84–101.

R. Hotta

[1] On the realization of the discrete series for semisimple Lie groups, J. Math. Soc. Japan **23** (1971), 384–407.

R. Hotta and R. Parthasarathy

[1] Multiplicity formulae for discrete series, Invent. Math. **26** (1974), 133–178.

L.-K. Hua

[1] Harmonic analysis of functions of several complex variables in the classical domains, Science Press, Peking, 1958 (Chinese); =Transl. Math. Mono., Vol. **6**, Amer. Math. Soc., Providence, 1963.

J. Humphreys

[1] Introduction to Lie algebras and representation theory, Grad. Texts in Math. **9**, Springer-Verlag, New York-Heidelberg-Berlin, 1972.
[2] Linear algebraic groups, Grad. Texts in Math. **21**, Springer-Verlag, New York-Heidelberg-Berlin, 1975.

J.-I. Igusa

[1] Arithmetic variety of moduli for genus two, Ann. of Math. **72** (1960), 612–649.
[2] On Siegel modular forms of genus two, Amer. J. Math. **84** (1962), 175–200; II, ibid. **86** (1964), 392–412.
[3] On the graded ring of theta constants, I, Amer. J. Math. **86** (1964), 219–246; II, ibid. **88** (1966), 221–236.
[4] Modular forms and projective invariants, Amer. J. Math. **89** (1967), 817–855.
[5] A desingularization problem in the theory of Siegel modular functions, Math. Ann. **168** (1967), 228–260.
[6] Harmonic analysis and theta-functions, Acta Math. **120** (1968), 187–222.
[7] On certain representations of semi-simple algebraic groups and the arithmetic of the corresponding invariants (1), Invent. Math. **12** (1971), 62–94.
[8] "Theta Functions", Springer-Verlag, Berlin-Heidelberg-New York, 1972.
[9] On the ring of modular forms of degree two over $\boldsymbol{Z}$, Amer. J. Math. **101** (1979), 149–183.

S. Ihara

[1] Holomorphic imbeddings of symmetric domains into a symmetric domain, Proc. Japan Acad. **42** (1966), 193–197.
[2] Holomorphic imbeddings of symmetric domains, J. Math. Soc. Japan **19** (1967), 261–302; Supplement, ibid., 543–544.

M. Ise

[1] Generalized automorphic forms and certain holomorphic vector bundles, Amer. J. Math. **86** (1964), 70–108.
[2] Realization of irreducible bounded symmetric domain of type (V), Proc. Japan Acad. Sci. **45** (1969), 233–237.
[3] Realization of irreducible bounded symmetric domain of type (VI), Proc. Japan Acad. Sci. **45** (1969), 846–849.
[4] On canonical realizations of bounded symmetric domains as matrix-spaces, Nagoya Math. J. **42** (1971), 115–133.
[5] Bounded symmetric domains of exceptional type, J. Fac. Sci. Univ. Tokyo **23** (1976), 75–105.

K. Iyanaga

[1] Arithmetic of special unitary groups and their symplectic representations, J. Fac. Sci. Univ. Tokyo **15** (1968), 35–69.
[2] Cusps of certain symmetric bounded domains, J. Math. Soc. Japan **26** (1974), 735–769.

N. Jacobson

[1] Derivation algebras and multiplication algebras of semi-simple Jordan algebras, Ann. of Math. **50** (1949), 866–874.
[2] Lie and Jordan triple systems, Amer. J. Math. **71** (1949), 149–170.
[3] A theorem on the structure of Jordan algebras, Proc. Nat. Acad. Sci. U. S. A. **42** (1956), 140–147.
[4] Some groups of transformations defined by Jordan algebras, I, J. reine angew. Math. **201** (1959), 178–195; II, ibid. **204** (1960), 74–98; III, ibid. **207** (1961), 61–85.
[5] "Lie Algebras", Interscience, New York, 1962.
[6] Clifford algebras for algebras with involution of type D, J. Algebra **1** (1964), 288–300.
[7] "Structure and Representations of Jordan Algebras", Coll. Publ., Vol. **39**, Amer. Math. Soc., Providence, 1968.
[8] "Lectures on quadratic Jordan algebras", Tata Inst. of Fund. Res., Bombay, 1970.

[9] Abraham Adrian Albert 1905–1972, Bull. Amer. Math. Soc. **80** (1974), 1075–1100.
[10] Structure groups and Lie algebras of Jordan algebras of symmetric elements of associative algebras with involution, Adv. in Math. **20** (1976), 106–150.

F. D. Jacobson and N. Jacobson

[1] Classification and representation of semi-simple Jordan algebras, Trans. Amer. Math. Soc. **65** (1949), 141–169.

H. Jaffee

[1] Real forms of hermitian symmetric spaces, Bull. Amer. Math. Soc. **81** (1975), 456–458.
[2] Anti-holomorphic automorphisms of the exceptional symmetric domains, J. Diff. Geom. **13** (1978), 79–86.

K. D. Johnson

[1] Remarks on a theorem of Korányi and Malliavin on the Siegel upper half plane of rank two, Proc. Amer. Math. Soc. **67** (1977), 351–356.
[2] Differential equations and the Bergman-Šilov boundary on the Siegel upper half plane, Arkiv för Mat. **16** (1978), 95–108.

K. D. Johnson and A. Korányi

[1] The Hua operators on bounded symmetric domains of tube type, Preprint.

P. Jordan, J. von Neumann, and E. Wigner

[1] On an algebraic generalization of the quantum mechanical formalism, Ann. of Math. **35** (1934), 29–64.

D. A. Kajdan

[1] On arithmetic varieties, "Lie Groups and Their Representations", (Budapest 1971, ed. I. M. Gelfand), Bolyai János Math. Soc., John Wiley-Halsted, New York-Toronto, 1975, 151–217.

S. Kaneyuki

[1] On the automorphism groups of homogeneous bounded domains, J. Fac. Sci. Univ. Tokyo **14** (1967), 89–130.
[2] Homogeneous bounded domains and Siegel domains, Lect. Notes in Math. **241**, Springer, 1971.

S. Kaneyuki and T. Nagano

[1] Quadratic forms related to symmetric riemannian spaces, Osaka Math. J. **14** (1962), 241–252.

S. Kaneyuki and M. Sudo

[1] On Šilov boundaries of Siegel domains, J. Fac. Sci. Univ. Tokyo **15** (1968), 131–146.

S. Kaneyuki and T. Tsuji

[1] Classification of homogeneous bounded domains of lower dimension, Nagoya Math. J. **53** (1974), 1–46.

I. L. Kantor

[1] Some generalizations of Jordan algebras (Russian), Trudy Sem. Vektor. Tenzor. Anal. **16** (1972), 407–499.
[2] Models of exceptional Lie algebras, Dokl. Akad. Nauk SSSR **208** (1973), 1276–1279 ; = Soviet Math. Dokl. **14** (1973), 254–258.

M. Karel

[1] Fourier coefficients of certain Eisenstein series, Ann. of Math. **99** (1974), 176–202.
[2] Eisenstein series and fields of definition, Compositio Math. **37** (1978), 121–169.

F. I. Karpelevic

[1] Geodesics and harmonic functions on symmetric spaces (Russian), Dokl. Akad. Nauk SSSR **124** (1959), 1199–1202.
[2] The geometry of geodesics and the eigenfunctions of the Beltrami-Laplace operator on symmetric spaces, Trudy Moskov. Mat. Obšč. **14** (1965), 48–185 ; =Trans. Moscow Math. Soc., 1965, 51–199.

M. Kashiwara, A. Kowata, K. Minemura, K. Okamoto, T. Oshima, and M. Tanaka

[1] Eigenfunctions of invariant differential operators on a symmetric space, Ann. of Math. **107** (1978), 1–39.

K. Katayama

[1] On the Hilbert-Siegel modular group and abelian varieties, J. Fac. Sci. Univ. Tokyo **9** (1962), 261–291 ; II, ibid., 433–467.

W. Kaup

[1] Über das Randverhalten von holomorphen Automorphismen beschränkter Gebiete, Manusc. Math. **3** (1970), 257–270.

[2] Some remarks on the automorphism groups of complex spaces, Rice Univ. Studies **56** (1970), 181–186.

[3] Einige Bemerkungen über polynomiale Vektorfelder, Jordan Algebren und die Automorphismen von Siegelschen Gebieten, Math. Ann. **204** (1973), 131–144.

[4] Über die Automorphismen Grassmannscher Mannigfaltigkeiten unendlicher Dimension, Math. Z. **144** (1975), 75–96.

[5] On the automorphisms of certain symmetric complex manifolds of infinite dimension, Memórias de Matemática **56**, Univ. Fed. Rio de Janeiro, 1975.

[6] Algebraic characterization of symmetric complex Banach manifolds, Math. Ann. **228** (1977), 39–64.

[7] Über die Klassifikation der symmetrischen Hermiteschen Mannigfaltigkeiten unendlichen Dimension, Preprint, 1980.

W. Kaup, Y. Matsushima, and T. Ochiai

[1] On the automorphisms and equivalences of generalized Siegel domains, Amer. J. Math. **92** (1970), 475–497.

W. Kaup and H. Upmeier

[1] Jordan algebras and symmetric Siegel domains in Banach spaces, Math. Z. **157** (1977), 179–200.

P. Kiernan

[1] On the compactifications of arithmetic quotients of symmetric spaces, Bull. Amer. Math. Soc. **80** (1974), 109–110.

P. Kiernan and S. Kobayashi

[1] Satake compactification and extension of holomorphic mappings, Invent. Math. **16** (1972), 237–248.

[2] Comments on Satake compactification and the great Picard theorem, J. Math. Soc. Japan **28** (1976), 577–580.

H. Klingen

[1] Diskontinuierliche Gruppen in symmetrischen Räumen, I, Math. Ann. **129** (1955), 345–369 ; II, ibid. **130** (1955/56), 137–146.

[2] Über analytischen Abbildungen verallgemeinerter Einheitskreis auf sich, Math. Ann. **132** (1956), 134–144.

[3] Zur Theorie der hermitischen Modulfunktionen, Math. Ann. **134** (1958), 355–384.

[4] Analytic automorphisms of bounded symmetric complex domains, Pacific J. Math. **10** (1960), 1327–1332.

[5] Quotientendarstellung Hermitescher Modulfunktionen durch Modulformen, Math. Ann. **143** (1961), 1–18.

[6] Characterisierung der Siegelschen Modulgruppe durch ein endliches System definierender Relationen, Math. Ann. **144** (1961), 64–72.

[7] Volumbestimmung des Fundamentalbereichs der Hilbertschen Modulgruppe n-ten Grades, J. reine angew. Math. **206** (1961), 9–19.

[8] Über die Werte der Dedekindschen Zetafunktion, Math. Ann. **145** (1962), 265–272.

[9] Über einen Zusammenhang zwischen Siegelschen und Hermiteschen Modulfunktionen, Abh. Math. Sem. Hamburg **27** (1964), 1–12.
[10] Zum Darstellungssatz für Siegelsche Modulformen, Math. Z. **102** (1967), 30–43 ; Berichtigung, ibid. **105** (1968), 399–400.
[11] Zur Struktur der Siegelschen Modulgruppe, Math. Z. **136** (1974), 169–178.

A. W. Knapp

[1] Bounded symmetric domains and holomorphic discrete series, "Symmetric Spaces", (ed. W. Boothby and G. Weiss), Marcel Dekker, New York, 1972, 211–254.

S. Kobayashi

[1] Geometry of bounded domains, Trans. Amer. Math. Soc. **92** (1959), 267–290.
[2] On complete Bergman metrics, Proc. Amer. Math. Soc. **13** (1962), 511–518.
[3] Isometric imbeddings of compact symmetric spaces, Tôhoku Math. J. **20** (1968), 21–25.
[4] Volume elements, holomorphic mappings and Schwarz's lemma, Proc. Symp. Pure Math. **11**, "Entire Functions and Related Parts of Analysis", Amer. Math. Soc., Providence, 1968, 253–260.
[5] "Hyperbolic Manifolds and Holomorphic Mappings", Marcel Dekker, New York, 1970.
[6] Intrinsic distance, measures and geometric function theory, Bull. Amer. Math. Soc. **82** (1976), 357–416.
[7] Projectively invariant distances for affine and projective structures, Preprint, 1980.

S. Kobayashi and T. Nagano

[1] On filtered Lie algebras and geometric structures, I, J. Math. Mech. **13** (1964), 875–908 ; II, ibid. **14** (1965), 513–522 ; III, ibid., 679–706.
[2] Riemannian manifolds with abundant isometries, "Differential Geometry" (in honor of K. Yano), Kinokuniya, Tokyo, 1972, 195–219.

S. Kobayashi and K. Nomizu

[1] "Foundation of Differential Geometry", I, *1963* ; II, *1968*, Wiley-Interscience, New York.

S. Kobayashi and T. Ochiai

[1] Satake compactification and the great Picard theorem, J. Math. Soc. Japan **23** (1971), 340–350.
[2] Mappings into compact complex manifolds with negative first Chern class, J. Math. Soc. Japan **23** (1971), 137–148.
[3] Meromorphic mappings onto compact complex spaces of general type, Invent. Math. **31** (1975), 7–16.

S. Kobayashi and H.-H. Wu

[1] On holomorphic sections of certain hermitian vector bundles, Math. Ann. **189** (1970), 1–4.

K. Kodaira

[1] On a differential-geometric method in the theory of analytic stacks, Proc. Nat. Acad. Sci. U. S. A. **39** (1953), 1268–1273.
[2] On Kähler varieties of restricted type (An intrinsic characterization of algebraic varieties), Ann. of Math. **60** (1954), 28–48.

K. Kodaira and D. C. Spencer

[1] On deformations of complex analytic structures, I–II, Ann. of Math. **67** (1958), 328–466 ; III, Stability theorems for complex structures, ibid. **71** (1960), 43–76.

A. Kodama

[1] On generalized Siegel domains, Osaka J. Math. **14** (1977), 227–252.
[2] A remark on bounded Reinhardt domains, Proc. Japan Acad. **54** (1978), 179–182.
[3] On the equivalence problem for a certain class of generalized Siegel domains, Mem. Fac. Ed. Akita Univ., Nat. Sci. **28** (1978), 45–54.

M. Koecher

[1] Positivitätsbereiche im $\boldsymbol{R}^n$, Amer. J. Math. **79** (1957), 575–596.

[2] Analysis in reellen Jordan Algebren, Nachr. Akad. Wiss. Göttingen, 1958, 67–74.
[3] Die Geodätischen von Positivitätsbereichen, Math. Ann. **135** (1958), 192–202.
[4] Beiträge zu einer Reduktionstheorie in Positivitätsbereichen, I, Math. Ann. **141** (1960), 384–432 ; II, ibid. **144** (1961), 175–182.
[5] "Jordan algebras and their applications", Lect. Notes, Univ. of Minnesota, Minneapolis, 1962.
[6] Über eine Gruppe von rationalen Abbildungen, Invent. Math. **3** (1967), 136–171.
[7] Imbedding of Jordan algebras into Lie algebras, I, Amer. J. Math. **89** (1967), 787–816 ; II, ibid. **90** (1968), 476–510.
[8] Gruppen und Lie-Algebren von rationalen Funktionen, Math. Z. **109** (1969), 349–392.
[9] "An elementary approach to bounded symmetric domains", Lect. Notes, Rice Univ., Houston, 1969.
[10] On bounded symmetric domains, Rice Univ. Studies **56** (1970), 63–65.
[11] Jordan algebras and differential geometry, "Actes, Congrès Int. Math. 1970", Gauthier-Villars, Paris, 1971, Vol. 1, 279–283.
[12] Eine Konstruktion von Jordan-Algebren, Manusc. Math. **23** (1978), 387–425.

A. Korányi

[1] The Poisson integral for generalized half-planes and bounded symmetric domains, Ann. of Math. **82** (1965), 332–350.
[2] A generalized Schwarz lemma for bounded symmetric domains, Proc. Amer. Math. Soc. **17** (1966), 210–213.
[3] Analytic invariants of bounded symmetric domains, Proc. Amer. Math. Soc. **19** (1968), 279–284.
[4] Holomorphic and harmonic functions on bounded symmetric domains, C. I. M. E. 1967, "Geometry of homogeneous bounded domains", Cremonese, Roma, 1968, 125–197.
[5] Harmonic functions on Hermitian hyperbolic space, Trans. Amer. Math. Soc. **135** (1969), 507–516.
[6] Boundary behavior of Poisson integrals on symmetric spaces, Trans. Amer. Math. Soc. **140** (1969), 393–409.
[7] Generalizations of Fatou's theorem to symmetric spaces, Rice Univ. Studies **56** (1970), 127–136.
[8] Harmonic functions on symmetric spaces, "Symmetric Spaces", (ed. W. Boothby and G. Weiss), Marcel Dekker, 1972, 379–412.
[9] Poisson integrals and boundary components of symmetric spaces, Invent. Math. **34** (1976), 19–35.
[10] Boundary behavior of holomorphic functions on bounded symmetric domains, Proc. Symp. Pure Math. **30**, "Several Complex Variables", Amer. Math. Soc., Providence, 1977, Part 2, 291–295.

A. Korányi and P. Malliavin

[1] Poisson formula and compound diffusion associated to an overdetermined elliptic system on the Siegel halfplane of rank two, Acta Math. **134** (1975), 185–209.

A. Korányi and E. M. Stein

[1] Fatou's theorem for generalized halfplanes, Ann. Scuola Norm. Sup. Pisa **22** (1968), 107–112.
[2] H^2-spaces of generalized half-planes, Studia Math. **44** (1972), 379–388.

A. Korányi and S. Vági

[1] Isometries of H^p-spaces of bounded symmetric domains, Canad. J. Math. **28** (1976), 330–334.
[2] Rational inner functions on bounded symmetric domains, Trans. Amer. Math. Soc. **254** (1979), 179–193.

A. Korányi and J. A. Wolf

[1] Realization of hermitian symmetric spaces as generalized half-planes, Ann. of Math. **81**

(1965), 265–288.
B. Kostant
[1] A characterization of the classical groups, Duke Math. J. **25** (1958), 107–124.
[2] The principal three-dimensional subgroup and the Betti numbers of a complex simple Lie groups, Amer. J. Math. **81** (1959), 973–1032.
[3] Lie algebra cohomology and the generalized Borel-Weil theorem, Ann. of Math. **74** (1961), 329–387.
[4] On convexity, the Weyl group and the Iwasawa decomposition, Ann. Sci. École Norm. Sup. **6** (1973), 413–455.
B. Kostant and S. Rallis
[1] Orbits and representations associated with symmetric spaces, Amer. J. Math. **93** (1971), 753–809.
J. L. Koszul
[1] Algèbres de Jordan, Sém. Bourbaki 1949/50, No. 31.
[2] Sur la forme hermitienne canonique des espaces homogènes complexes, Canad. J. Math. **7** (1955), 562–576.
[3] "Exposés sur les espaces homogènes symétriques", Soc. Math. São Paulo, 1959.
[4] Domaines bornés homogènes et orbites de groupes de transformations affines, Bull. Soc. Math. France **89** (1961), 515–533.
[5] Ouverts convex homogènes des espaces affines, Math. Z. **79** (1962), 254–259.
T. Kubota
[1] "Elementary Theory of Eisenstein Series", Kodansha, Tokyo and Wiley, New York, 1973.
M. Kuga
[1] "Fiber varieties over a symmetric space whose fibers are abelian varieties", I, II, Lect. Notes, Univ. Chicago, 1963–64.
[2] Fibered variety over symmetric space whose fibers are abelian varieties, "Proc. U. S.-Japan Sem. in Diff. Geom.", (Kyoto 1965), Nippon Hyoronsha, 1966, 72–81.
[3] Hecke's polynomial as a generalized congruence Artin L-function, Proc. Symp. Pure Math. **9**, "Algebraic Groups and Discontinuous Subgroups", Amer. Math. Soc., Providence, 1966, 333–337.
[4] Case of three quaternions (Japanese), Preprint, 1978.
M. Kuga and S. Ihara
[1] Family of families of abelian varieties, "Algebraic Number Theory", (Kyoto 1976), Japan Soc. for Prom. Sci., Tokyo, 1977, 129–142.
M. Kuga and J. V. Leahy
[1] Shimura's abelian varieties as Weil's higher Jacobian varieties, J. Fac. Sci. Univ. Tokyo **16** (1965), 229–253.
M. Kuga and I. Satake
[1] Abelian varieties attached to polarized K_3-surfaces, Math. Ann. **169** (1967), 239–242.
M. Kuga and G. Shimura
[1] On vector differential forms attached to automorphic forms, J. Math. Soc. Japan **12** (1960), 258–270.
[2] On the zeta function of a fibre variety whose fibres are abelian varieties, Ann. of Math. **82** (1965), 478–539.
S. Lang
[1] "$SL_2(\boldsymbol{R})$", Addison Wesley, Reading, 1975.
R. P. Langlands
[1] The dimension of spaces of automorphic forms, Amer. J. Math. **85** (1963), 99–125.
[2] The volume of the fundamental domain for some arithmetical subgroups of Chevalley groups, Proc. Symp. Pure Math. **9**, "Algebraic Groups and Discontinuous Subgroups", Amer. Math.

Soc., Providence, 1966, 143–148.
[3] On the functional equations satisfied by Eisenstein series, Lect. Notes in Math. **544**, Springer, 1976.
[4] Some contemporary problems with origins in the JUGEND TRAUM, Proc. Symp. Pure Math. **28**, "Mathematical Developments Arising from Hilbert Problems", Amer. Math. Soc., Providence, 1976, 401–418.
[5] Automorphic representations, Shimura varieties, and motives, Ein Märchen, Proc. Symp. Pure Math. **33**, "Automorphic Forms, Representations and L-functions", Amer. Math. Soc., Providence, 1979, Part 2, 205–246.

A. Lascoux and M. Berger

[1] Variétés Kählériennes compactes, Lect. Notes in Math. **154**, Springer, 1970.

O. Loos

[1] "Symmetric spaces" (2 vol.), Benjamin, New York-Amsterdam, 1969.
[2] Jordan triple systems, R-spaces, and bounded symmetric domains, Bull. Amer. Math. Soc. **77** (1971), 558–561.
[3] "Jordan triples", Lect. Notes, Univ. of British Columbia, Vancouver, 1971.
[4] Assoziative Tripelsysteme, Manusc. Math. **7** (1972), 103–112.
[5] Alternative Tripelsysteme, Math. Ann. **198** (1972), 205–238.
[6] Representations of Jordan triples, Trans. Amer. Math. Soc. **185** (1973), 199–211.
[7] A structure theory of Jordan pairs, Bull. Amer. Math. Soc. **80** (1974), 67–71.
[8] "Jordan pairs", Lect. Notes in Math. **460**, Springer, 1975.
[9] "Bounded symmetric domains and Jordan pairs", Math. Lect., Univ. of Calif., Irvine, 1977.
[10] On algebraic groups defined by Jordan pairs, Nagoya Math. J. **74** (1979), 23–66.
[11] Homogeneous algebraic varieties defined by Jordan pairs, Preprint.

O. Loos and K. McCrimmon

[1] Speciality of Jordan triple systems, Comm. in Algebra **5** (1977), 1057–1082.

H. Maass

[1] Über die Darstellung der Modulformen n-ten Grades durch Poincarésche Reihen, Math. Ann. **123** (1951), 125–151.
[2] Die Primzahlen in der Theorie der Siegelschen Modulformen, Math. Ann. **124** (1951), 87–122.
[3] Die Differentialgleichungen in der Theorie der Siegelschen Modulformen, Math. Ann. **126** (1953), 44–68.
[4] "Lectures on Siegel's modular functions", Tata Inst. of Fund. Res., Bombay, 1954–55.
[5] Zur Theorie der Kugelfunktionen einer Matrixvariablen, Math. Ann. **135** (1958), 391–416.
[6] Die Multiplikatorsysteme zur Siegelschen Modulgruppe, Nachr. Akad. Wiss. Göttingen, 1964, 125–135.
[7] "Siegel modular forms and Dirichlet series", Lect. Notes in Math. **216**, Springer, 1971.

H. Matsumoto

[1] Quelques remarques sur les groupes algébriques réels, Proc. Japan Acad. **40** (1964), 4–7.
[2] Quelques remarques sur les groupes de Lie algébriques réels, J. Math. Soc. Japan **16** (1964), 419–446.
[3] Sur les sous-groupes arithmétiques des groupes semi-simples déployés, Ann. Sci. École. Norm. Sup. **2** (1969), 1–62.
[4] Quelques remarques sur les espaces riemanniens isotropes, C. R. Acad. Sci. Paris **272** (1971), 316–319.
[5] Sur un théorème de point fixe de É. Cartan, C. R. Acad. Sci. Paris **274** (1972), 955–958.

Y. Matsushima

[1] "Theory of Lie algebras" (Japanese), Kyoritsusha, 1956.
[2] On hermitian symmetric spaces (Japanese), "Summer Sem. in Diff. Geom.", (Akakura), Sugakushinkokai, 1956, 105–128.

[3] Sur les espaces homogènes kaehlériens d'un groupe de Lie réductif, Nagoya Math. J. **11** (1957), 53–60.
[4] Sur certaines variétés homogènes complexes, Nagoya Math. J. **18** (1961), 1–12.
[5] On the first Betti number of compact quotient spaces of higher dimensional symmetric spaces, Ann. of Math. **75** (1962), 312–330.
[6] On Betti numbers of compact, locally symmetric Riemannian manifolds, Osaka Math. J. **14** (1962), 1–20.
[7] A formula for the Betti numbers of compact locally symmetric Riemannian manifolds, J. Diff. Geom. **1** (1967), 99–109.
[8] "Differentiable Manifolds" (Japanese), Shokabo, Tokyo, 1965; (English transl.), Marcel Dekker, New York, 1972.
[9] On tube domains, "Symmetric Spaces", (ed. W. Boothby and G. Weiss), Marcel Dekker, New York, 1972, 255–270.

Y. Matsushima and S. Murakami

[1] On vector bundle valued harmonic forms and automorphic forms on a symmetric Riemannian manifold, Ann. of Math. **78** (1963), 365–416.
[2] On certain cohomology groups attached to hermitian symmetric spaces, Osaka J. Math. **2** (1965), 1–35; II, ibid. **5** (1968), 223–241.

Y. Matsushima and G. Shimura

[1] On the cohomology groups attached to certain vector valued differential forms on the product of the upper half planes, Ann. of Math. **78** (1963), 417–449.

K. McCrimmon

[1] The Freudenthal-Springer-Tits constructions of exceptional Jordan algebras, Trans. Amer. Math. Soc. **139** (1969), 495–510.
[2] The Freudenthal-Springer-Tits constructions revisited, Trans. Amer. Math. Soc. **148** (1970), 293–314.
[3] Jordan algebras and their applications, Bull. Amer. Math. Soc. **84** (1978), 612–627.

K. Meyberg

[1] Jordan-Tripelsysteme und die Koecher-Konstruktion von Lie-Algebren, Math. Z. **115** (1970), 58–78.
[2] Zur Konstruktion von Lie-Algebren aus Jordan-Tripelsystemen, Manusc. Math. **3** (1970), 115–132.
[3] Identitäten und das Radikal in Jordan-Tripelsystemen, Math. Ann. **197** (1972), 203–220.
[4] Von Neumann regularity in Jordan triple systems, Arch. Math. **23** (1972), 589–593.
[5] "Lectures on algebras and triple systems", Lect. Notes, Univ. of Virginia, Charlottesville, 1972.

K. Miyake

[1] On models of certain automorphic function fields, Acta Math. **126** (1971), 245–307.

T. Miyake

[1] On automorphism groups of the fields of automorphic functions, Ann. of Math. **95** (1972), 243–252.

C. C. Moore

[1] Compactification of symmetric spaces, I, Amer. J. Math. **86** (1964), 201–218; II: The Cartan domains, ibid., 358–378.

K. Morita

[1] On the kernel functions of symmetric domains, Sci. Rep. Tokyo Kyoiku Daigaku **5** (1956), 190–212.

Y. Morita

[1] An explicit formula for the dimension of spaces of Siegel modular forms of degree two, J. Fac. Sci. Univ. Tokyo **21** (1974), 167–248.

G. D. Mostow

[1] Some new decomposition theorems for semi-simple groups, Mem. Amer. Math. Soc., No. **14** (1955), 31–54.

[2] Self-adjoint groups, Ann. of Math. **62** (1955), 44–55.

[3] Fully reducible subgroups of algebraic groups, Amer. J. Math. **78** (1956), 200–221.

[4] "Strong rigidity of locally symmetric spaces", Ann. of Math. Studies **78**, Princeton Univ. Press, 1973.

G. D. Mostow and T. Tamagawa

[1] On the compactness of arithmetically defined homogeneous spaces, Ann. of Math. **76** (1962), 440–463.

D. Mumford

[1] Families of abelian varieties, Proc. Symp. Pure Math. **9**, "Algebraic Groups and Discontinuous Subgroups", Amer. Math. Soc., Providence, 1966, 347–351.

[2] On the equations defining abelian varieties, I, Invent. Math. **1** (1966), 287–354 ; II, ibid. **3** (1967), 75–135 ; III, ibid., 215–244.

[3] Abelian quotients of the Teichmüller modular group, J. Analyse Math. **18** (1967), 227–244.

[4] A note of Shimura's paper "Discontinuous groups and abelian varieties", Math. Ann. **181** (1969), 345–351.

[5] "Abelian Varieties", Oxford Univ. Press, 1970.

[6] A new approach to compactifying locally symmetric varieties, "Discrete subgroups of Lie groups and applications to moduli", (Bombay Coll. 1973), Oxford Univ. Press, 1975, 211–224.

[7] Hirzebruch's proportionality theorem in the non-compact case, Invent. Math. **42** (1977), 239–272.

S. Murakami

[1] On the automorphisms of a real semi-simple Lie algebra, J. Math. Soc. Japan **4** (1952), 103–133 ; Supplements and corrections, ibid. **5** (1953), 105–112.

[2] Sur la classification des algèbres de Lie réelles et simples, Osaka J. Math. **2** (1965), 291–307.

[3] Plongements holomorphes de domaines symétriques, C. I. M. E. 1967, "Geometry of homogeneous bounded domains", Cremonese, Roma, 1968, 263–277.

[4] Facteurs d'automorphie associés à un espace hermitien symétrique, C. I. M. E. 1967, "Geometry of homogeneous bounded domains", Cremonese, Roma, 1968, 279–287.

[5] "On automorphisms of Siegel domains", Lect. Notes in Math. **286**, Springer, 1972.

T. Nagano

[1] Transformation groups on compact symmetric spaces, Trans. Amer. Math. Soc. **118** (1965), 428–453.

K. Nakajima

[1] On Tanaka's imbeddings of Siegel domains, J. Math. Kyoto Univ. **14** (1974), 533–548.

[2] Symmetric spaces associated with Siegel domains, Proc. Japan Acad. **50** (1974), 188–191.

[3] Symmetric spaces associated with Siegel domains, J. Math. Kyoto Univ. **15** (1975), 303–349.

[4] Some studies on Siegel domains, J. Math. Soc. Japan **27** (1975), 54–75.

[5] On realization of Siegel domains of the second kind as those of the third kind, J. Math. Kyoto Univ. **16** (1976), 143–166.

[6] On equivariant holomorphic imbeddings of Siegel domains to compact complex homogeneous spaces, J. Math. Kyoto Univ. **19** (1979), 471–480.

Y. Namikawa

[1] On the canonical holomorphic map from the moduli space of stable curves to the Igusa monoidal transform, Nagoya Math. J. **52** (1973), 197–259.

[2] On moduli of stable quasi-abelian varieties, Nagoya Math. J. **58** (1975), 149–214.

[3] A new compactification of the Siegel space and degeneration of abelian varieties, I, Math.

Ann. **221** (1976), 97–141; II, ibid., 201–241.

R. Narasimhan

[1] "Several complex variables", Univ. of Chicago Press, 1971.

[2] Automorphisms of bounded domains in C^n, "Colloque analyse et topologie" (en l'honneur de Henri Cartan), Ast. **32–33**, Soc. Math. France, 1976, 213–225.

M. S. Narasimhan and K. Okamoto

[1] An analogue of the Borel-Weil-Bott theorem for hermitian symmetric pairs of non-compact type, Ann. of Math. **91** (1970), 486–511.

E. Neher

[1] Klassifikation der einfachen reellen Jordan-Tripelsysteme, Dissertation, Univ. Münster, 1978.

[2] Cartan-Involutionen von halbeinfachen Jordan-Tripelsystemen, Math. Z. **169** (1979), 271–292.

K. Nomizu

[1] "Lie groups and differential geometry", Publ. Math. Soc. Japan **2**, Tokyo, 1956.

T. Ochiai

[1] A lemma on open convex cones, J. Fac. Sci. Univ. Tokyo **12** (1966), 231–234.

[2] Transformation groups on Riemannian symmetric spaces, J. Diff. Geom. **3** (1969), 231–236.

[3] Geometry associated with semi-simple flat homogeneous spaces, Trans. Amer. Math. Soc. **152** (1970), 159–193.

R. D. Ogden and S. Vági

[1] Harmonic analysis and H^2-functions on Siegel domains of type II, Proc. Nat. Acad. Sci. U. S. A. **69** (1972), 11–14.

K. Okamoto and H. Ozeki

[1] On square-integrable $\bar{\partial}$-cohomology spaces attached to hermitian symmetric spaces, Osaka J. Math. **4** (1967), 95–110.

T. Ono

[1] Arithmetic of algebraic tori, Ann. of Math. **74** (1961), 101–139.

[2] The Gauss-Bonnet theorem and the Tamagawa number, Bull. Amer. Math. Soc. **71** (1965), 345–348.

R. Parthasarathy

[1] Dirac operator and the discrete series, Ann. of Math. **96** (1972), 1–30.

I. I. Piatetskii-Shapiro

[1] On a problem proposed by É. Cartan, Dokl. Akad. Nauk SSSR **124** (1959), 272–273.

[2] "Geometry of Classical Domains and Theory of Automorphic Functions" (Russian), Fizmatgiz, Moscow, 1961; (French transl.) Dunot, Paris, 1966; (English transl.) Gordon and Breach, New York, 1969.

[3] Classification of bounded homogeneous domains in n-dimensional complex space, Dokl. Akad. Nauk SSSR **141** (1961), 316–319; =Soviet Math. Dokl. **2** (1961), 1460–1463.

[4] On bounded homogeneous domains in n-dimensional complex space, Izv. Akad. Nauk SSSR Ser. Mat. **26** (1962), 107–124; =Amer. Math. Soc. Transl. (2) **43** (1964), 299–320.

[5] Arithmetic groups on complex domains, Uspehi Mat. Nauk **19** (1964), 93–121; =Russian Math. Surveys **19** (1964), 83–109.

[6] The geometry and classification of bounded homogeneous domains, Uspehi Mat. Nauk **20** (1965), 3–51; =Russian Math. Surveys **20** (1965), 1–48.

[7] Automorphic functions and arithmetic groups, "Proc. Int. Congress Math., Moscow 1966", Izdat. Mir, Moscow, 1968, 232–247; =Amer. Math. Soc. Transl. (2) **70** (1968), 185–201.

I. I. Piatetskii-Shapiro and M. E. Novodvorsky

[1] Rankin-Selberg method in the theory of automorphic forms, Proc. Symp. Pure Math. **30**, "Several Complex Variables", Amer. Math. Soc., Providence, 1977, Part 2, 297–301.

M. S. Raghunathan

[1] On the first cohomology of discrete subgroups of semisimple Lie groups, Amer. J. Math. **78** (1965), 103–139.

[2] Cohomology of arithmetic subgroups of algebraic groups, I, Ann. of Math. **86** (1967), 408–424 ; II, ibid. **87** (1968), 279–304.

[3] A note on quotients of real algebraic groups by arithmetic subgroups, Invent. Math. **4** (1968), 318–335.

[4] "Discrete Subgroups of Lie Groups", Erg. Math. **68**, Springer-Verlag, New York-Heidelberg-Berlin, 1972 ; (Russian transl. with an appendix by G. Margulis), Izdat. Mir, Moscow, 1977.

[5] Discrete groups and $\boldsymbol{Q}$-structures on semi-simple Lie groups, "Discrete subgroups of Lie groups and applications to moduli", (Bombay Coll. 1973), Oxford Univ. Press, 1975, 225–321.

[6] On the congruence subgroup problem, Publ. Math., No. **46**, I. H. E. S., 1976, 107–162.

M. Rapoport

[1] Compactifications de l'espace de modules de Hilbert-Blumenthal, Compositio Math., to appear.

H. L. Resnikoff

[1] The maximum modulus principle for tubes over domains of positivity, Math. Ann. **156** (1964), 340–346.

[2] On a class of linear differential equations for automorphic forms in several complex variables, Amer. J. Math. **95** (1973), 321–331.

[3] Automorphic forms of singular weight are singular forms, Math. Ann. **215** (1975), 173–193.

[4] Theta functions for Jordan algebras, Invent. Math. **31** (1975), 87–104.

[5] The structure of spaces of singular automorphic forms, Preprint, 1977.

H. L. Resnikoff and Y.-S. Tai

[1] On the structure of a graded ring of automorphic forms on the 2-dimensional complex ball, Preprint, 1978.

H. Rossi and M. Vergne

[1] Fonctions holomorphes de carré sommable sur un domaine de Siegel et étude de la série discrète holomorphe, C. R. Acad. Sci. Paris **275** (1972), 17–20.

[2] Representations of certain solvable Lie groups on Hilbert spaces of holomorphic functions and the application to the holomorphic discrete series of a semi-simple Lie group, J. Funct. Anal. **13** (1973), 324–389.

[3] Group representations on Hilbert spaces defined in terms of $\bar{\partial}_b$-cohomology on the Šilov boundary of a Siegel domain, Pacific J. Math. **65** (1976), 193–207.

[4] Équations de Cauchy Riemann tangentielles associées à un domaine de Siegel, Ann. Sci. École Norm. Sup. **9** (1976), 31–80.

O. S. Rothaus

[1] Domains of positivity, Abh. Math. Sem. Hamburg **24** (1960), 189–235.

[2] The construction of homogeneous convex cones, Ann. of Math. **83** (1966), 358–376 ; Correction, ibid. **87** (1968), 399.

[3] Automorphisms of Siegel domains, Trans. Amer. Math. Soc. **162** (1971), 351–382.

[4] Ordered Jordan algebras, Amer. J. Math. **100** (1978), 925–939.

[5] Automorphisms of Siegel domains, Amer. J. Math. **101** (1979), 1167–1179.

[6] Siegel domains and representations of Jordan algebras, Preprint, 1979.

I. Satake

[1] On representations and compactifications of symmetric Riemannian spaces, Ann. of Math. **71** (1960), 77–110.

[2] On compactifications of the quotient spaces for arithmetically defined discontinuous groups, Ann. of Math. **72** (1960), 555–580.

[3] On the theory of reductive algebraic groups over a perfect field, J. Math. Soc. Japan **15** (1963), 210–235.

[4] Holomorphic imbeddings of symmetric domains into a Siegel space, Amer. J. Math. **87** (1965), 425–461.

[5] Clifford algebras and families of Abelian varieties, Nagoya Math. J. **27** (1966), 435–446; Corrections, ibid. **31** (1968), 295–296.

[6] Symplectic representations of algebraic groups satisfying a certain analyticity condition, Acta Math. **117** (1967), 215–279.

[7] A note on holomorphic imbeddings and compactification of symmetric domains, Amer. J. Math. **90** (1968), 231–247.

[8] On some properties of holomorphic imbeddings of symmetric domains, Amer. J. Math. **91** (1969), 289–305.

[9] On modular imbeddings of a symmetric domain of type (*IV*), "Global Analysis", (Papers in honor of K. Kodaira; ed. D. C. Spencer and S. Iyanaga), Univ. of Tokyo Press and Princeton Univ. Press, 1970, 341–354.

[10] "Classification theory of semi-simple algebraic groups" (with an appendix by M. Sugiura), Marcel Dekker, New York, 1971.

[11] Fock representations and theta-functions, Ann. of Math. Studies **66**, "Advances in the theory of Riemann surfaces", Princeton Univ. Press, 1971, 393–405.

[12] Factors of automorphy and Fock representations, Adv. in Math. **7** (1971), 83–110.

[13] Unitary representations of a semi-direct product of Lie groups on $\bar{\partial}$-cohomology spaces, Math. Ann. **190** (1971), 177–202.

[14] Mountjoy's abelian varieties attached to a holomorphic immersion, "Number Theory, Algebraic Geometry and Commutative Algebra", (in honor of Y. Akizuki), Kinokuniya, Tokyo, 1973, 45–69.

[15] Linear imbeddings of self-dual homogeneous cones, Nagoya Math. J. **46** (1972), 121–145; Corrections, ibid. **60** (1976), 219.

[16] On the arithmetic of tube domains (Blowing-up of the point at infinity), Bull. Amer. Math. Soc. **79** (1973), 1076–1094.

[17] On classification of quasi-symmetric domains, Nagoya Math. J. **62** (1976), 1–12.

[18] On symmetric and quasi-symmetric Siegel domains, Proc. Symp. Pure Math. **30**, "Several Complex Variables", Amer. Math. Soc., Providence, 1977, Part 2, 309–315.

[19] $\boldsymbol{Q}$-forms of symmetric domains and Jordan triple systems, "Algebraic Number Theory", (Kyoto 1976), Japan Soc. for Prom. Sci., Tokyo, 1977, 163–176.

[20] La déformation des formes hermitiennes et son application aux domaines de Siegel, Ann. Sci. École Norm. Sup. **11** (1978), 445–449.

M. Sato and T. Shintani

[1] On zeta functions associated with prehomogeneous vector spaces, Ann. of Math. **100** (1974), 131–170.

R. D. Schafer

[1] "An Introduction to Non-Associative Algebras", Acad. Press, New York, 1966.

W. Schmid

[1] Die Randwerte holomorpher Funktionen auf Hermitesch symmetrischen Raumen, Invent. Math. **9** (1969/70), 61–80.

[2] On a conjecture of Langlands, Ann. of Math. **93** (1971), 1–42.

[3] On the character of the discrete series (the Hermitian symmetric case), Invent. Math. **30** (1975), 47–144.

[4] Some properties of square-integrable representations of semisimple Lie groups, Ann. of Math. **102** (1975), 535–564.

[5] L^2-cohomology and the discrete series, Ann. of Math. **103** (1976), 375–394.

A. Selberg

[1] Harmonic analysis and discontinuous groups in weakly symmetric Riemannian spaces with applications to Dirichlet series, J. Indian Math. Soc. **20** (1956), 47–87.

[2] On discrete subgroups in higher dimensional symmetric spaces, "Contributions to function theory", Tata Inst. of Fund. Res., Bombay, 1960, 147–164.

[3] Recent developments in the theory of discontinuous groups of motions of symmetric spaces, Lect. Notes in Math. **118**, "Proc. of the 15th Scandinavian Congress Oslo 1968", Springer, 1970, 99–120.

J.-P. Serre

[1] "Cohomologie galoisienne" (3-rd ed.), Lect. Notes in Math. **5**, Springer, 1965.

[2] "Lie algebras and Lie groups", Benjamin, New York, 1965.

[3] "Algèbres de Lie semi-simples complexes", Benjamin, New York, 1966.

[4] Groupes de congruence (d'après H. Bass, H. Matsumoto, J. Mennicke, J. Milnor, C. Moore), Sém. Bourbaki 1966/67 (Benjamin, New York, 1968), No. 330.

[5] Cohomologie de groupes discrets, Ann. of Math. Studies **70**, "Prospects in Mathematics", Princeton Univ. Press, 1971, 77–169.

K.-Y. Shih

[1] Anti-holomorphic automorphisms of arithmetic automorphic function fields, Ann. of Math. **103** (1976), 81–102.

[2] Existence of certain canonical models, Duke Math. J. **45** (1978), 63–66.

[3] Conjugations of arithmetic automorphic function fields, Invent. Math. **44** (1978), 87–102.

[4] Construction of arithmetic automorphic functions for special Clifford groups, Nagoya Math. J. **76** (1979), 153–171.

H. Shimizu

[1] On discontinuous groups operating on the product of the upper half planes, Ann. of Math. **77** (1963), 33–71.

[2] On traces of Hecke operators, J. Fac. Sci. Univ. Tokyo **10** (1963), 1–19.

[3] Theta series and automorphic forms on GL_2, J. Math. Soc. Japan **24** (1972), 638–683; Corrections, ibid. **26** (1974), 374–376.

[4] "Automorphic functions" (Japanese), Kisosugaku, Iwanami, Tokyo, 1978.

G. Shimura

[1] On the theory of automorphic functions, Ann. of Math. **70** (1959), 101–144.

[2] Sur les intégrales attachées aux formes automorphes, J. Math. Soc. Japan **11** (1959), 291–311.

[3] On the zeta-functions of the algebraic curves uniformized by certain automorphic functions, J. Math. Soc. Japan **13** (1961), 275–331.

[4] On Dirichlet series and abelian varieties attached to automorphic forms, Ann. of Math. **76** (1962), 237–294.

[5] Arithmetic of alternating forms and quaternion hermitian forms, J. Math. Soc. Japan **15** (1963), 33–65.

[6] On analytic families of polarized abelian varieties and automorphic functions, Ann. of Math. **78** (1963), 159–192.

[7] On modular correspondences for $Sp(n, Z)$ and their congruence relations, Proc. Nat. Acad. Sci. U. S. A. **49** (1963), 824–828.

[8] Arithmetic of unitary groups, Ann. of Math. **79** (1964), 369–409.

[9] On the field of definition for a field of automorphic functions, I, Ann. of Math. **80** (1964), 160–189; II, ibid. **81** (1965), 124–165; III, ibid. **83** (1966), 377–385.

[10] Class-fields and automorphic functions, Ann. of Math. **80** (1964), 444–463.

[11] Moduli and fibre systems of abelian varieties, Ann. of Math. **83** (1966), 294–338.

[12] Construction of class fields and zeta-functions of algebraic curves, Ann. of Math. **85** (1967),

58–159.

[13] Discontinuous groups and abelian varieties, Math. Ann. **168** (1967), 171–199.

[14] Algebraic number fields and symplectic discontinuous groups, Ann. of Math. **86** (1967), 503–592.

[15] "Automorphic functions and number theory", Lect. Notes in Math. **54**, Springer, 1968.

[16] Local representations of Galois groups, Ann. of Math. **89** (1969), 99–124.

[17] On canonical models of arithmetic quotients of bounded symmetric domains, I, Ann. of Math. **91** (1970), 144–222 ; II, ibid. **92** (1970), 528–549.

[18] On arithmetic automorphic functions, "Actes, Congrès Int. Math. 1970", Gauthier-Villars, Paris, 1971, Vol. 2 , 343–348.

[19] "Introduction to the Arithmetic Theory of Automorphic Functions", Publ. Math. Soc. Japan **11**, Iwanami, Tokyo and Princeton Univ. Press, 1971.

[20] On the zeta-function of an abelian varieties with complex multiplication, Ann. of Math. **94** (1971), 504–533.

[21] On modular forms of half integral weight, Ann. of Math. **97** (1973), 440–481.

[22] On the real points of an arithmetic quotient of a bounded symmetric domain, Math. Ann. **215** (1975), 135–164.

[23] On some arithmetic properties of modular forms of one and several variables, Ann. of Math. **102** (1975), 491–515.

[24] On the Fourier coefficients of modular forms of several variables, Nachr. Akad. Wiss. Göttingen, 1975, 261–268.

[25] On the derivatives of theta functions and modular forms, Duke Math. J. **44** (1977), 365–387.

[26] On certain reciprocity-law for theta functions and modular forms, Acta Math. **141** (1978), 35–71.

[27] The arithmetic of automorphic forms with respect to a unitary group, Ann. of Math. **107** (1978), 569–605.

[28] Automorphic forms and the periods of abelian varieties, J. Math. Soc. Japan **31** (1979), 561–592.

[29] The arithmetic of certain zeta functions and automorphic forms on orthogonal groups, Ann. of Math. **111** (1980), 313–375.

T. Shintani

[1] On zeta-functions associated with the vector space of quadratic forms, J. Fac. Sci. Univ. Tokyo **22** (1975), 25–65.

C. L. Siegel

[1] Einführung in die Theorie der Modulfunktionen *n*-ten Grades, Math. Ann. **116** (1939), 617–657 (=Ges. Abh. II, No. 32).

[2] Einheiten quadratischer Formen, Abh. Math. Sem. Hamburg **13** (1940), 209–239 (=Ges. Abh. II, No. 33).

[3] Note on automorphic functions of several variables, Ann. of Math. **43** (1942), 613-616 (= Ges. Abh. II, No. 40).

[4] "Symplectic Geometry", Amer. J. Math. **65** (1943), 1–86 (=Ges. Abh. II, No. 41) ; Acad. Press, New York-London, 1964.

[5] Discontinuous groups, Ann. of Math. **44** (1943), 674–689 (=Ges. Abh. II, No. 43).

[6] Some remarks on discontinuous groups, Ann. of Math. **46** (1945), 708–718 (=Ges. Abh. III, No. 53).

[7] Indefinite quadratische Formen und Modulfunktionen, Courant Anniv. Vol., 1948, 395–406 (=Ges. Abh. III, No. 55).

[8] "Analytic functions of several complex variables", Lect. Notes, I. A. S., Princeton, 1949.

[9] Indefinite quadratische Formen und Funktionen Theorie, I, Math. Ann. **124** (1951), 17–54 (=Ges. Abh. III, No. 58); II, ibid., 364–387 (=Ges. Abh. III, No. 60).

[10] Die Modulgruppe in einer einfachen involutorischen Algebra, Fest. Akad. Wiss. Göttingen, 1951, 157–167 (=Ges. Abh. III, No. 59).
[11] Meromorphe Funktionen auf kompakten analytischen Mannigfaltigkeiten, Nachr. Akad. Wiss. Göttingen, 1955, 71–77 (=Ges. Abh. III, No. 64).
[12] Zur Theorie der Modulfunktionen *n*-ten Grades, Comm. Pure Appl. Math. **8** (1955), 677–681 (=Ges. Abh. III, No. 65).
[13] Über die algebraische Abhangigkeit von Modulfunktionen *n*-ten Grades, Nachr. Akad. Wiss. Göttingen, 1960, 257–272 (=Ges. Abh. III, No. 75).
[14] "Lectures on Riemann matrices", Tata Inst. of Fund. Res., Bombay, 1963.
[15] Moduln Abelscher Funktionen, Nachr. Akad. Wiss. Göttingen, 1964, 365–427 (=Ges. Abh. III, No. 77).
[16] Über die Fourierschen Koeffizienten der Eisensteinschen Reihen, Mat.-Fys. Medd. Danske Vid. Selsk. **34** (1964), No. 6 (=Ges. Abh. III, No. 79).
[17] "Topics in Complex Function Theory", Vol. I, "Elliptic functions and uniformization theory", 1969 ; Vol. II, "Automorphic functions and Abelian integrals", 1971 ; Vol. III, "Abelian functions and modular functions of several variables", 1973, Wiley-Interscience, New York.

"Sophus Lie"

[1] Séminaire "Sophus Lie", 1954/55, "Théorie des algèbres de Lie, Topologie des groupes de Lie", École Norm. Sup., Paris.

J. Spilker

[1] Algebraische Körper von automorphen Funktionen, Math. Ann. **149** (1963), 341–360.
[2] Kompaktifizierung der Humbertschen Fundamentalbereichs, Math. Ann. **161** (1965), 296–314.
[3] Über den Rand der Siegelschen Halbebene, Math. Z. **90** (1965), 273–285.

T. A. Springer

[1] On a class of Jordan algebras, Indag. Math. **21** (=Nederl. Akad. Wetensch. Proc. Ser. A **62**) (1959), 254–264.
[2] The classification of reduced exceptional simple Jordan algebras, Indag. Math. **22** (1960), 414–422.
[3] Characterization of a class of cubic forms, Indag. Math. **24** (1962), 259–265.
[4] "Oktaven, Jordan-Algebren und Ausnahmegruppen", Lect. Notes, Univ. Göttingen, 1963.
[5] "Jordan Algebras and Algebraic Groups", Erg. Math. **75**, Springer-Verlag, New York-Heidelberg-Berlin, 1973.
[6] Reductive groups, Proc. Symp. Pure Math. **33**, "Automorphic Forms, Representations, and *L*-functions", Amer. Math. Soc., Providence, 1979, Part 1, 3–28.

S. Sternberg and J. A. Wolf

[1] Hermitian Lie algebras and metaplectic representations, I, Trans. Amer. Math. Soc. **238** (1978), 1–43.

M. Sudo

[1] A remark on Siegel domains of second kind, J. Fac. Sci. Univ. Tokyo **22** (1975), 199–203.
[2] On infinitesimal automorphisms of some non-degenerate Siegel domains, J. Fac. Sci. Univ. Tokyo **22** (1975), 429–436.

T. Sunada

[1] On bounded Reinhardt domains, Proc. Japan Acad. **50** (1974), 119–123.
[2] Holomorphic mappings into a compact quotient of symmetric bounded domain, Nagoya Math. J. **64** (1976), 159–175.
[3] Holomorphic equivalence problem for bounded Reinhardt domains, Math. Ann. **235** (1978), 111–128.
[4] Closed geodesics in a locally symmetric space, Tôhoku Math. J. **30** (1978), 59–68.

R. Takagi and M. Takeuchi

[1] Degree of symmetric Kählerian submanifolds of a complex projective space, Osaka J. Math. **14** (1977), 501–518.

M. Takeuchi

[1] On the fundamental group and the group of isometries of a symmetric space, J. Fac. Sci. Univ. Tokyo **10** (1964), 88–123.

[2] Cell decompositions and Morse equalities on certain symmetric spaces, J. Fac. Sci. Univ. Tokyo **12** (1965), 81–192.

[3] On orbits in a compact hermitian symmetric space, Amer. J. Math. **90** (1968), 657–680.

[4] Nice functions on symmetric spaces, Osaka J. Math. **6** (1969), 283–289.

[5] On infinitesimal affine automorphisms of Siegel domains, Proc. Japan Acad. **45** (1969), 590–594.

[6] On the fundamental group of a simple Lie group, Nagoya Math. J. **40** (1970), 147–159.

[7] Polynomial representations associated with symmetric bounded domains, Osaka J. Math. **10** (1973), 441–475.

[8] "Homogeneous Siegel domains", Publ. Study Group Geometry, Vol. **7**, Tokyo, 1973.

[9] On symmetric Siegel domains, Nagoya Math. J. **59** (1975), 9–44.

[10] Homogeneous Kähler submanifolds in complex projective spaces, Japan J. Math. **4** (1978), 171–219.

M. Takeuchi and S. Kobayashi

[1] Minimal embeddings of R-spaces, J. Diff. Geom. **2** (1968), 203–215.

N. Tanaka

[1] On the equivalence problems associated with a certain class of homogeneous spaces, J. Math. Soc. Japan **17** (1965), 103–139.

[2] On infinitesimal automorphisms of Siegel domains, J. Math. Soc. Japan **22** (1970), 180–212.

J. Tits

[1] Espaces homogènes complexes compacts, Comment. Math. Helv. **37** (1962), 111–120.

[2] Une classe d'algèbres de Lie en relation avec les algèbres de Jordan, Indag. Math. **24** (=Nederl. Akad. Wetensch. Proc. Ser. A **65**) (1962), 530–535.

[3] Algèbres alternatives, algèbres de Jordan et algèbres de Lie exceptionnelles I, Construction, Indag. Math. **28** (1966), 223–237.

[4] Classification of algebraic semi-simple groups, Proc. Symp. Pure Math. **9**, "Algebraic Groups and Discontinuous Subgroups", Amer. Math. Soc., Providence, 1966, 33–62.

[5] Formes quadratiques, groupes orthogonaux et algèbres de Clifford, Invent. Math. **5** (1968), 19–41.

[6] Travaux de Margulis sur les sous-groupes discrets de groupe de Lie, Sém. Bourbaki 1975/76, Lect. Notes in Math. **567**, Springer, 1977, No. 482.

L.-C. Tsao

[1] The rationality of the Fourier coefficients of certain Eisenstein series on tube domains (I), Compositio Math. **32** (1976), 225–291.

T. Tsuji

[1] On infinitesimal automorphisms and homogeneous Siegel domains over circular cones, Proc. Japan Acad. **49** (1973), 390–393.

[2] Classification of homogeneous Siegel domains of type II of dimension 9 and 10, Proc. Japan Acad. **49** (1973), 394–396.

[3] Siegel domains over self-dual cones and their automorphisms, Nagoya Math. J. **55** (1974), 33–80.

R. Tsushima

[1] Hyperbolicity of quotients of bounded symmetric domains, "Int. Symp. on Algebraic Geometry", (Kyoto 1977), Kinokuniya, Tokyo, 1978, 687–691.

[2] A formula for the dimension of spaces of cusp forms of degree three, Preprint, 1979.

V. S. Varadarajan

[1] "Lie Groups, Lie Algebras, and Their Representations", Prentice-Hall, Englewood Cliffs, 1974.

M. Vergne and H. Rossi

[1] Analytic continuation of the holomorphic discrete series of a semi-simple Lie group, Acta Math. **136** (1976), 1–59.

J. Vey

[1] Sur la division des domaines de Siegel, Ann. Sci. École Norm. Sup. **3** (1970), 479–506.

[2] Sur les automorphismes affines des ouverts convexes dans les espaces numériques, C. R. Acad. Sci. Paris **270** (1970), 249–251.

J. P. Vigué

[1] Le groupe des automorphismes analytiques d'un domaine borné d'un espace de Banach complexe, Application aux domaines bornés symétriques, Ann. Sci. École Norm. Sup. **9** (1976), 203–282.

È. B. Vinberg

[1] Homogeneous cones, Dokl. Akad. Nauk SSSR **133** (1960), 9–12; =Soviet Math. Dokl. **1** (1961), 787–790.

[2] The Morozov-Borel theorem for real Lie groups, Dokl. Akad. Nauk SSSR **141** (1961), 270–273; =Soviet Math. Dokl. **2** (1961), 1416–1419.

[3] Convex homogeneous domains, Dokl. Akad. Nauk SSSR **141** (1961), 521–524; =Soviet Math. Dokl. **2** (1961), 1470–1473.

[4] Automorphisms of homogeneous convex cones, Dokl. Akad. Nauk SSSR **143** (1962), 265–268; =Soviet Math. Dokl. **3** (1962), 371–374.

[5] The theory of convex homogeneous cones, Trudy Moskov. Mat. Obšč. **12** (1963), 303–358; =Trans. Moscow Math. Soc. 1963, 340–403.

[6] The structure of the group of automorphisms of a homogeneous convex cone, Trudy Moskov. Mat. Obšč. **13** (1965); =Trans. Moscow Math. Soc. 1965, 63–93.

[7] Construction of the exceptional simple Lie algebras (Russian), Trudy Sem. Vektor. Tenzor. Anal. **13** (1966), 7–9.

[8] Discrete groups generated by reflections in Lobačevskiĭ spaces, Mat. Sbornik **72** (1967), 471–488; =Math. USSR-Sb. **1** (1967), 429–444; Correction, ibid. **73** (1967), 303.

[9] Some arithmetical discrete groups in Lobačevskiĭ spaces, "Discrete subgroups of Lie groups and applications to moduli", (Bombay Coll. 1973), Oxford Univ. Press, 1975, 323–348.

È. B. Vinberg, S. G. Gindikin, and I. I. Piatetskii-Shapiro

[1] Classification and canonical realization of complex bounded homogeneous domains, Trudy Moskov. Mat. Obšč. **12** (1963), 359–388; =Trans. Moscow Math. Soc. 1963, 404–437.

A. Weil

[1] Sur les théorèmes de de Rham, Comment. Math. Helv. **26** (1952), 119–145.

[2] "Introduction à l'étude des variétés kählériennes", Act. Sci. Ind., Hermann, Paris, 1958.

[3] Algebras with involutions and the classical groups, J. Indian Math. Soc. **24** (1960), 589–623.

[4] On discrete subgroups of Lie groups, Ann. of Math. **72** (1960), 369–384; II, ibid. **75** (1962), 578–602.

[5] "Adeles and algebraic groups", Lect. Notes, I. A. S., Princeton, 1961.

[6] Un théorème fondamental de Chern en géométrie riemannienne, Sém. Bourbaki 1961/62, No. 239.

[7] Remarks on the cohomology of groups, Ann. of Math. **80** (1964), 149–157.

[8] Sur certains groupes d'opérateurs unitaires, Acta Math. **111** (1964), 143–211.

[9] Sur la formule de Siegel dans la théorie des groupes classiques, Acta Math. **113** (1965), 1–87.

[10] Abelian varieties and the Hodge ring, Coll. Papers, Vol. III, Springer-Verlag, New York-Heidelberg-Berlin, 1979, 421–429.

R. O. Wells, Jr. and J. A. Wolf

[1] Poincaré series and automorphic cohomology on flag domains, Ann. of Math. **105** (1977), 397–448.

H. Weyl

[1] Theorie der Darstellung kontinuierlicher halb-einfacher Gruppen durch lineare Transformationen, I, Math. Z. **23** (1925), 271–309 ; II, ibid. **24** (1926), 238–376 ; III, ibid., 377–395 ; Nachtrag, ibid., 789–791.

[2] "The Classical Groups, Their Invariants and Representaions", Princeton Univ. Press, 1939.

H. Whitney

[1] Elementary structure of real algebraic varieties, Ann. of Math. **66** (1957), 545–556.

E. Witt

[1] Eine Identität zwischen Modulformen zweiten Grades, Abh. Math. Sem. Hamburg **14** (1941), 323–337.

J. A. Wolf

[1] Discrete groups, symmetric spaces, and global holonomy, Amer. J. Math. **84** (1962), 527–542.

[2] On the classification of hermitian symmetric spaces, J. Math. Mech. **31** (1964), 489–496.

[3] "Spaces of Constant Curvature", McGraw-Hill, New York, 1967 ; (4-th. ed.), Publish or Perish, Berkeley, 1977.

[4] The action of a real semi-simple group on a complex flag manifold, I : Orbit structure and holomorphic arc components, Bull. Amer. Math. Soc. **75** (1969), 1121–1237.

[5] Remark on discrete subgroups, Proc. Amer. Math. Soc. **29** (1971), 423–425.

[6] Remark on Siegel domains of type III, Proc. Amer. Math. Soc. **30** (1971), 487–491.

[7] Fine structure of hermitian symmetric spaces, "Symmetric Spaces", (ed. W. Boothby and G. Weiss), Marcel Dekker, New York, 1972, 271–357.

[8] The action of a real semi-simple group on a complex flag manifold, II : Unitary representation on partially holomorphic cohomology spaces, Mem. Amer. Math. Soc., No. **138**, 1974.

J. A. Wolf and A. Korányi

[1] Generalized Cayley transformations of bounded symmetric domains, Amer. J. Math. **87** (1965), 899–939.

H. Yamaguchi

[1] On blow-ups of cusp singularities of 3-dimensional tube domains, Thesis, Univ. of Calif., Berkeley, 1975.

[2] The parabolic contribution to the dimension of the spaces of cusp forms on Siegel space of degree two, Preprint, 1976.

T. Yamazaki

[1] On Siegel modular forms of degree two, Amer. J. Math. **98** (1976), 39–53.

R. Zelow (Lundquist)

[1] On the differential geometry of quasi-symmetric domains, Thesis, Univ. of Calif., Berkeley, 1977.

[2] On the geometry of some Siegel domains, Nagoya Math. J. **73** (1979), 175–195.

Index

Library of Congress Cataloging in Publication Data

Satake, Ichirō, 1927-
Algebraic structures of symmetric domains.

(Kanô memorial lectures; 4) (Publications of the Mathematical Society of Japan; 14)
Includes bibliographical references and index.
1. Symmetric domains. 2. Holomorphic mappings. I. Title. II. Series. III. Series: Nihon Sūgakkai. Publications; 14.
QA331.S346 1980 512'.5 80-7551
ISBN 0-691-08271-5

Ingram Content Group UK Ltd.
Milton Keynes UK
UKHW021842090723
424747UK00003B/175

9 780691 615455